Indian Statistical Institute Series

The *Indian Statistical Institute Series* publishes high-quality content in the domain of mathematical sciences, bio-mathematics, financial mathematics, pure and applied mathematics, operations research, applied statistics and computer science and applications with primary focus on mathematics and statistics. Editorial board comprises of active researchers from major centres of Indian Statistical Institute. Launched at the 125th birth Anniversary of P.C. Mahalanobis, the series will publish textbooks, monographs, lecture notes and contributed volumes. Literature in this series will appeal to a wide audience of students, researchers, educators, and professionals across mathematics, statistics and computer science disciplines.

More information about this series at http://www.springer.com/series/15910

A. A. Ambily · Roozbeh Hazrat ·
B. Sury
Editors

Leavitt Path Algebras and Classical *K*-Theory

Springer

Editors
A. A. Ambily
Department of Mathematics
Cochin University of Science
and Technology
Cochin, Kerala, India

Roozbeh Hazrat
Centre for Research in Mathematics
Western Sydney University
Sydney, NSW, Australia

B. Sury
Statistics and Mathematics Unit
Indian Statistical Institute
Bengaluru, Karnataka, India

ISSN 2523-3114 ISSN 2523-3122 (electronic)
Indian Statistical Institute Series
ISBN 978-981-15-1610-8 ISBN 978-981-15-1611-5 (eBook)
https://doi.org/10.1007/978-981-15-1611-5

Cover photo: Reprography & Photography Unit, Indian Statistical Institute, Kolkata

This Springer imprint is published by the registered company Springer Nature Singapore Pte Ltd.
The registered company address is: 152 Beach Road, #21-01/04 Gateway East, Singapore 189721, Singapore

Preface

This volume is an outcome of the International Workshop on Leavitt Path Algebras and *K*-Theory held at the Department of Mathematics, Cochin University of Science and Technology, Kerala, during July 1–3, 2017. This workshop intended to give an introduction of the newly developing subject Leavitt path algebras (LPAs for short) and the classical *K*-theory. It consists of articles on several aspects of Leavitt path algebras, on the one hand, and related *K*-theory, on the other. The articles on LPAs are mostly of an expository nature. A number of articles dealing with *K*-theory give new proofs of old results and are accessible to and of interest to students and beginners.

The subject of Leavitt path algebras was born about sixty years ago out of a construction by William Leavitt to showcase counterexamples to the invariant basis number problem. Leavitt path algebras were then introduced about fifteen years ago, associating certain algebras to directed graphs. The algebra associated with one vertex and n loops retrieves Leavitt's algebra. The initial impetus came from the theory of C^*-algebras which was already well developed and analogues came to be discovered in the theory of LPAs. A recent book by G. Abrams, P. Ara, and M. Siles Molina titled *Leavitt Path Algebras* has appeared in Springer's *Lecture Notes in Mathematics* series in 2017. It details several aspects of the main thrusts in this subject until 2015, but there has been a mushrooming of questions and ideas in the last five years. At least three conferences have been held recently, and it has been mentioned by several interested mathematicians that it ought to be very useful to have the proceedings of this CUSAT workshop published. In order to introduce the vast possibilities of this subject to graduate students and mathematicians working in somewhat allied areas, we have included surveys on topics that have not been covered in the above text. Development of *K*-theory of LPAs with initial impetus from that of graph C^*-algebras has also begun. The volume also contains articles on *K*-theoretic aspects apart from LPAs.

K. M. Rangaswamy has substantially contributed to this subject for several decades. In his survey, he concentrates on various algebraic aspects and describes some of his very recent results. Especially, since the theory of modules—questions on Morita equivalence, etc.—over these algebras is still in its infancy, the results obtained by Rangaswamy and others are described in clear detail by him. As a sample, we state one result in his article—every one-sided ideal over a Leavitt path algebra $L_K(E)$ is graded, if and only if every simple module over it is graded, and these happen only when $L_K(E)$ is a von Neumann regular ring.

In a lucid, self-contained exposition, Simon Rigby elucidates the groupoid approach to the LPAs. It was noticed a few years back that a new approach using topological groupoids can assist in the study of hitherto difficult questions on LPAs. The key fact here is that the LPA of a graph is graded isomorphic to the Steinberg algebra of the boundary path groupoid. This survey is expected to be useful to all levels of interested mathematicians. The author proves some results in more generality than have appeared in publications so far. One such instance is the uniqueness theorems for LPAs.

Very recently, étale groupoids have shown up in the forefront of several areas of mathematics. Important algebras such as the Cuntz algebra are known to arise as the convolution algebras arising from étale groupoids. The realization that invariants long studied in topological dynamics can be modeled on étale groupoids permits an interaction between analysis and algebra. Lisa Orloff Clark and Roozbeh Hazrat describe in complete detail how the LPAs allow us to treat all these algebras systematically and uniformly.

For a certain finite graph E and for the corresponding finite-dimensional algebra A with a square of the radical equal to zero, Huanhuan Li had constructed a compact generator of the homotopy category of acyclic complexes of injective modules over A—the so-called injective Leavitt complex of E. She gives an overview of the connection between the injective or projective Leavitt complex and the Leavitt path algebra of E.

Müge Kanuni and Suat Sert give an overview of results on the ideal theory of LPAs. In recent times, there has been a large body of work on graded, non-graded, prime, primitive, and maximal ideals of LPAs. Their survey is at an introductory level and narrates the correspondence between the lattice of ideals and the lattice of hereditary and saturated subsets of the graph over which the LPA is constructed.

In their article, Fatemeh Bagherzadeh and Murray Bremner recount the connection with the theory of operads. Gröbner bases for operads had been introduced by Dotsenko, Vallette, and others. The authors consider certain nonsymmetric operads for which they construct Gröbner bases and thereby compute their dimension formulae.

About 50 years ago, Stewart Priddy introduced Koszul algebras and Koszul duality partly in order to construct examples of algebras for which the Peter May spectral sequence is easy to compute and stops early. Steenrod algebra is a classic example. Koszul duality has been generalized to the operad setting also. In an expository article, Neeraj Kumar traces the notions and results on Koszul algebras developed in the last decade or so. These involve connections with combinatorics, geometry of monomial curves, Stanley–Reisner ring, Polya frequency sequence, etc. He also recalls older results and gives modern proofs for some of them like the theorem of Tate on the Poincaré–Betti series of quadratic complete intersection ring.

There are open questions mentioned which will be helpful to young researchers entering this area.

The spectacular work of Quillen and Suslin in solving Serre's conjecture on projective modules led to an explosion of sorts with new approaches designed to study more general problems of a similar nature. The technique of completion of unimodular rows developed by Suslin and Vaserstein led to the theorem on the normality of the elementary subgroup in $SL(n)$ for $n > 2$ over any commutative ring. Led by many experts, including Bak and Bass, more general notions of classical-like groups were defined and analogous questions were posed.

An article written by Ravi A. Rao and Ram Shila establishes an elementary symplectic analogue of Karoubi's linearization process of a polynomial matrix. They prove that one can stably linearize an alternating polynomial matrix by conjugating it with an elementary symplectic matrix.

Bhatoa Joginder Singh and Selby Jose study the action of SL_n on alternating matrices over a commutative ring A and prove (an analogue of the isomorphism of A_3 and D_3 over fields) that there is an injection from $SL_4(A)/E_4(A)$ to $SO_6(A)/EO_6(A)$.

Leonid Vaserstein had shown for two-dimensional rings that the unimodular rows of length three up to elementary transformations have the structure of a Witt group. The Vaserstein symbol mapping these classes of unimodular rows of length three to the elementary symplectic group has been studied in recent times. The non-injectivity of this symbol map for the coordinate ring of the 3-sphere has produced intense interest in the question of injectivity for more general rings. Neena Gupta and Dhvanita Rao had even produced an uncountable family of rings of dimension three over $\mathbb{R}$ for which the symbol is not injective. Neena Gupta, Dhvanita Rao, and Sagar Kolte survey these results as well as related works of Ravi A. Rao, van der Kallen, Richard Swan, and Jean Fasel.

In another paper, Ravi A. Rao and Selby Jose provide two possible approaches to the famous Bass–Suslin conjecture on the completability of unimodular polynomial rows over local rings.

Rabeya Basu, Reema Khanna, and Ravi A. Rao show for a commutative ring that the normality of the relative elementary subgroup is equivalent to the relative Quillen–Suslin local–global principle. They also obtain a relative local–global principle for the transvection subgroups. They use the concept of a Noetherian excision ring.

Reema Khanna, Selby Jose, Sampat Sharma, and Ravi A. Rao study the so-called special unimodular vector group and its elementary unimodular vector subgroup. They use certain ideas of Anthony Bak to deduce that the quotient vector group is nilpotent of class at most the dimension of the ring.

Raja Sridharan and Sunil Yadav reprove some classical theorems in a novel manner. They use the theory of Euler classes to deduce Seshadri's old theorem of the freeness of finitely generated projective modules over $k[X,Y]$. In another article, they deduce Suslin's $n!$ theorem on unimodular rows using Quillen's splitting principle. Along with Sumit Kumar Upadhyay in an article, they describe an algebraic analogue of the Mayer–Vietoris sequence for the part of the sequence that corresponds to the zeroth and the first cohomology.

In yet another article demonstrating the utility of Euler classes in algebra, Anjan Gupta, Raja Sridharan, and Sunil Yadav use it to give a group structure on the equivalence classes of unimodular rows of length three over a two-dimensional ring.

Finally, in an article dealing with Quillen–Suslin's foundational principles, Ravi A. Rao and Sunil Yadav demonstrate that monic inversion is equivalent to the local–global principle as well as to the normality of the elementary subgroup.

Bangalore, India
June 2019

B. Sury
A. A. Ambily

Acknowledgements

This volume is an outcome of the International Workshop on Leavitt Path Algebras and *K*-Theory held at the Department of Mathematics, Cochin University of Science and Technology, Kerala, during July 1–3, 2017. This workshop intended to give an introduction of the newly developing subject Leavitt path algebras and the subject classical *K*-theory.The administrative and logistic support for the workshop was given by the Cochin University of Science and Technology. The organizers gratefully acknowledge this support.

The convener of this workshop Dr. A. A. Ambily gratefully acknowledges the financial support by Kerala State Council for Science, Technology and Environment (KSCSTE)—Scheme for promoting S&T Seminar/Symposia/ Workshop—SSW/377/2016/KSCSTE, Government of Kerala and National Board for Higher Mathematics (NBHM)—Conference Support 02010/16/2017/R&D—II/8117, Department of Atomic Energy, Government of India.

We are also grateful to the faculty, administrative staff, and the research scholars of Cochin University of Science and Technology who had helped in organizing the event as well as the technical assistance during the preparation of the volume for publication. We thank Dr. Neeraj Kumar for helping us in organizing the event. We are also thankful to Mr. Bintu Shyam, Mr. Rahul Rajan, and other research scholars for the initial preparations of the proceedings. We also thank Ms. Aparna Pradeep V. K. for helping us in the editorial process.

Finally, we thank all the authors for contributing to the volume and the reviewers for helping us with the review process.

Cochin, India — A. A. Ambily
Sydney, Australia — Roozbeh Hazrat
Bangalore, India — B. Sury

Contents

Editors and Contributors

About the Editors

A. A. Ambily is Assistant Professor at the Department of Mathematics, Cochin University of Science and Technology, Kerala, India. She received Ph.D. in Mathematics from the Indian Statistical Institute, Bangalore Center, India. Her research interests include algebraic K-theory and noncommutative algebras such as Leavitt path algebras and related topics.

Roozbeh Hazrat is Professor at the School of Computer, Data and Mathematical Sciences, Western Sydney University, Australia. He received Ph.D. in Mathematics from the University of Bielefeld, Germany, in 2002. His research interests include Leavitt path algebras, algebraic K-theory and noncommutative algebras. He has authored three books, including *Mathematica®: A Problem-Centered Approach* published by Springer, and contributed over 50 papers in respected journals. In 2015, he was awarded a one-year fellowship for experienced researchers by Germany's Alexander von Humboldt Foundation.

B. Sury is Professor at the Statistics and Mathematics Unit, Indian Statistical Institute, Bangalore Center, India. He received Ph.D. in Mathematics from the Tata Institute of Fundamental Research, Mumbai, India, in 1991. His research interests include algebraic groups over global and local fields, division algebras, and number theory. He has authored three books and published several research papers in leading international journals. An elected fellow of The National Academy of Sciences, India, Prof. Sury is the national coordinator for the Mathematics Olympiad Program in India.

Contributors

Fatemeh Bagherzadeh Department of Mathematics and Statistics, University of Saskatchewan, Saskatoon, Canada

Rabeya Basu Indian Institute of Science Education and Research, Pune, India

Murray Bremner Department of Mathematics and Statistics, University of Saskatchewan, Saskatoon, Canada

Lisa Orloff Clark School of Mathematics and Statistics, Victoria University of Wellington, Wellington, New Zealand

Anjan Gupta Department of Mathematics, Institute of Science Education and Research Bhopal, Bhopal, India

Neena Gupta Statistics and Mathematics Unit, Indian Statistical Institute, Kolkata, India

Roozbeh Hazrat Centre for Research in Mathematics, Western Sydney University, Sydney, Australia

Selby Jose Department of Mathematics, Institute of Science, Mumbai, India

Müge Kanuni Department of Mathematics, Düzce University, Konuralp, Düzce, Turkey

Reema Khanna Somaiya College, Vidyavihar, Mumbai, India

Sagar Kolte Credit Suisse, Powai, Mumbai, India;
Department of Mathematics, Indian Institute of Technology, Mumbai, India

Neeraj Kumar Department of Mathematics, Indian Institute of Technology Hyderabad, Sangareddy, Kandi, India

Huanhuan Li School of Mathematical Sciences, Nanjing Normal University, Nanjing, China;
Centre for Research in Mathematics, Western Sydney University, Sydney, NSW, Australia

Kulumani M. Rangaswamy Department of Mathematics, University of Colorado, Colorado Springs, CO, USA

Dhvanita R. Rao Bhavan's College, Mumbai, India

Ravi A. Rao School of Mathematics, Tata Institute of Fundamental Research, Mumbai, India

Simon W. Rigby Department of Mathematics and Applied Mathematics, University of Cape Town, Cape Town, South Africa;
Department of Mathematics: Algebra and Geometry, Ghent University, Ghent, Belgium

Suat Sert Department of Mathematics, Düzce University, Konuralp, Düzce, Turkey

Sampat Sharma School of Mathematics, Tata Institute of Fundamental Research, Mumbai, India

Ram Shila School of Physical Sciences, JNU, Delhi, India;
Department of Science, Forbesganj College, Forbesganj, Bihar, India

Bhatoa Joginder Singh Department of Mathematics, Government College, Nani-Daman, India

Raja Sridharan School of Mathematics, Tata Institute of Fundamental Research, Mumbai, India

Sumit Kumar Upadhyay Department of Applied Science, Indian Institute of Information Technology, Allahabad, Uttar Pradesh, India

Sunil K. Yadav Department of Mathematics, Indian Institute of Technology Bombay, Powai, Mumbai, India

Part I
Leavitt Path Algebras

The theory of Leavitt path algebras has created an astonishingly large amount of recent activities in ring theory. Besides a beautiful subject in its own right, it is closely related to several other areas in mathematics, which might explain the burst of activity in the subject. The first part of this volume exclusively deals with Leavitt path algebras and the related areas.

Chapter 1
A Survey of Some of the Recent Developments in Leavitt Path Algebras

Kulumani M. Rangaswamy

1.1 Introduction

Leavitt path algebras are algebraic analogues of graph C^*-algebras and, ever since they were introduced in 2004, have become an active area of research. Many of the initial developments during the 2004–2014 period have been nicely described in the recent book [2] and in the excellent survey article [1]. Our goal in this article is to report on some of the recent developments in the investigation of the algebraic aspects of Leavitt path algebras not included in [1, 2]. Because the Leavitt path algebras grew as algebraic analogues of graph C^*-algebras, their initial investigation involved mostly the ideas and techniques used in the study of graph C^*-algebras such as the graph properties of Conditions (K) and (L), and the ring properties of being simple, purely infinite simple, prime/primitive, etc. An important starting goal in this initial study was to work out the algebraic analogue of the deep and powerful Kirchberg Phillips theorem to classify purely infinite simple Leavitt path algebras $L := L_K(E)$ up to isomorphism or up to Morita equivalence by means of the Grothendieck groups $K_0(L)$ and the sign of the determinant $\det(I - A_E)$ where A_E is the adjacent matrix of the graph E. After such initial progress, there has been an explosion of articles dealing with not only the various different aspects of Leavitt path algebras, but also many natural generalizations such as Leavitt path algebras over commutative rings, of separated graphs, of high-rank graphs, Steinberg algebras and groupoids etc. In the background of many of these investigations is the special feature that every Leavitt path algebra L is endowed with three mutually compatible structures: L is a K-algebra, L is a $\mathbb{Z}$-graded ring and L is a ring with involution *. Our focus in this survey is to describe a selection of recent graded and non-graded ring-theoretic and module-theoretic investigations of Leavitt path algebras. My apologies

---

K. M. Rangaswamy (✉)
Department of Mathematics, University of Colorado,
Colorado Springs, CO 80918, USA
e-mail: kmranga@gmail.com

A. A. Ambily et al. (eds.), *Leavitt Path Algebras and Classical* K*-Theory*,
Indian Statistical Institute Series,
https://doi.org/10.1007/978-981-15-1611-5_1

to authors whose work has not been included due to the constrained focus, limitation of time and length of the paper.

In the first part of this survey, we describe graphical conditions on E under which the corresponding Leavitt path algebra $L_K(E)$ belongs to well-known classes of rings. The interesting fact is that often a single graph property of E seems to imply multiple ring properties of $L_K(E)$ and these properties for general rings are usually independent of each other. The poster child of such a phenomenon is the graph property for a finite graph E that no cycle in E has an exit. In this case, $L_K(E)$ possesses at least nine completely different ring properties! (see Theorem 1.5). Because of such connections between E and $L_K(E)$, Leavitt path algebras can be effective tools in the construction of examples of rings with various desired properties. If we do not impose any graphical conditions on E and just look at $L_K(E)$ as a $\mathbb{Z}$-graded ring, a really interesting result by Hazrat [23] states that $L_K(E)$ is a graded von Neumann regular ring. Because of this, the graded one-sided and two-sided ideals of $L_K(E)$ possess many desirable properties.

The module theory over Leavitt path algebras is still at an infant stage. The second part of this survey gives an account of some of the recent advances in this theory. Naturally, the initial investigations focussed on the simplest of the modules, namely, the simple modules over $L_K(E)$. We begin with outlining a few methods of constructing graded and non-graded simple left/right $L_K(E)$-modules. A special type of simple modules, called Chen simple modules introduced by Chen [19], play an important role. This is followed by characterizing Leavitt path algebras over which all the simple modules possess some special properties, such as, when all the simple modules are flat, or injective, or finitely presented or graded etc. For example, very recently, Ambily, Hazrat and Li [10] have proved that every simple left/right $L_K(E)$-module is flat if and only if $L_K(E)$ is von Neumann regular, thus showing, in the case of Leavitt path algebras, an open question in ring theory has an affirmative answer. Likewise, it was shown in [5] that $Ext^1_{L_K(E)}(S, S) \neq 0$ for a Chen simple module S induced by a cycle. It can then be shown that if all the simple left $L_K(E)$-modules are injective, then $L_K(E)$ is von Neumann regular. The converse easily holds if E is a finite graph, since it that case $L_K(E)$ is semi-simple artinian. In contrast, if R is an arbitrary non-commutative ring, the injectivity of all simple left R-modules need not imply von Neumann regularity of R (see [20]). Our next result in this section describes Leavitt path algebras of finite graphs having only finitely many isomorphism classes of simple modules. Interestingly, this class of algebras turns out to be precisely the class of Leavitt path algebras of finite graphs having finite Gelfand–Kirillov dimension.

The last section deals with one-sided ideals of a Leavitt path algebra L. Four years ago it was shown in [36] that finitely generated two-sided ideals of L are principal ideals. Recently, Abrams, Mantese and Tonolo [6] generalized this by showing that every finitely generated one-sided ideal of L is a principal ideal. Such rings are called Bézout rings. Using a deep theorem of Bergman, Ara and Goodearl [12] showed that one-sided ideals of L are projective. From these two results, it follows that the sum and the intersection of principal one-sided ideals of L are again principal. Thus, the principal one-sided ideals of L form a sublattice of the lattice of all one-sided ideals of L. A well-known theorem, proved originally for graph C^*-algebras and later for Leavitt path algebras $L_K(E)$, states that every two-sided ideal of $L_K(E)$ is a graded ideal if and only if E satisfies Condition (K), equivalently $L_K(E)$ is a weakly regular ring. What happens when every one-sided ideal of $L_K(E)$ is graded? The last theorem of this section answers this question, namely, every one-sided ideal of $L_K(E)$ is graded if and only if every simple $L_K(E)$-module is graded if and only if $L_K(E)$ is a von Neumann regular ring (see [25]).

In summary, this survey is intended to showcase a small sample of some of the recent research on the algebraic aspects of Leavitt path algebras. Hopefully, this provides the reader with some insights into this theory and generates further interest in this exciting and growing field of algebra.

1.2 Preliminaries

For the general notation, terminology and results in Leavitt path algebras, we refer to [1, 2]. We give below an outline of some of the needed basic concepts and results.

A (directed) graph $E = (E^0, E^1, r, s)$ consists of two sets E^0 and E^1 together with maps $r, s : E^1 \to E^0$. The elements of E^0 are called *vertices* and the elements of E^1 *edges*. A vertex v is called a *sink* if it emits no edges and a vertex v is called a *regular vertex* if it emits a non-empty finite set of edges. An *infinite emitter* is a vertex which emits infinitely many edges. For each $e \in E^1$, we call e^* a *ghost* edge. We let $r(e^*)$ denote $s(e)$, and we let $s(e^*)$ denote $r(e)$. A *path* μ of length $n > 0$ is a finite sequence of edges $\mu = e_1 e_2 \cdots e_n$ with $r(e_i) = s(e_{i+1})$ for all $i = 1, \ldots, n-1$. In this case $\mu^* = e_n^* \cdots e_2^* e_1^*$ is the corresponding ghost path. A vertex is considered a path of length 0.

A path $\mu = e_1 \cdots e_n$ in E is *closed* if $r(e_n) = s(e_1)$, in which case μ is said to be *based at the vertex* $s(e_1)$. A closed path μ as above is called *simple* provided it does not pass through its base more than once, i.e., $s(e_i) \neq s(e_1)$ for all $i = 2, ..., n$. The closed path μ is called a *cycle* if it does not pass through any of its vertices twice, that is, if $s(e_i) \neq s(e_j)$ for every $i \neq j$. An *exit* for a path $\mu = e_1 \cdots e_n$ is an edge e such that $s(e) = s(e_i)$ for some i and $e \neq e_i$.

For any vertex v, the *tree* of v is $T_E(v) = \{w \in E^0 : v \geq w\}$. We say there is a *bifurcation* at a vertex v or v is a *bifurcation vertex*, if v emits more than one edge. In a graph E, a vertex v is called a *line point* if there is no bifurcation or a cycle based at any vertex in $T_E(v)$. Thus, if v is a line point, the vertices in $T_E(v)$ arrange themselves on a straight line path μ starting at v (μ could just be v) such as $\bullet_v \to \bullet \cdots \bullet \to \bullet \cdots$ which could be finite or infinite.

If p is an infinite path in E, say, $p = e_1 \cdots e_n e_{n+1} \ldots$, we follow Chen [19] to define, for each $n \geq 1$, $\tau^{\leq n}(p) = e_1 \cdots e_n$ and $\tau^{>n}(p) = e_{n+1} e_{n+2} \cdots$. Two infinite paths p, q are said to be *tail-equivalent* if there are positive integers m, n such that $\tau^{>m}(p) = \tau^{>n}(q)$. This defines an equivalence relation among the infinite paths in E and the equivalence class containing the path p is denoted by $[p]$. An infinite path

p is said to be a *rational path* if it is tail-equivalent to an infinite path $q = ccc\cdots$, where c is a closed path.

Given an arbitrary graph E and a field K, the *Leavitt path algebra* $L_K(E)$ is defined to be the K-algebra generated by a set $\{v : v \in E^0\}$ of pair-wise orthogonal idempotents together with a set of variables $\{e, e^* : e \in E^1\}$ which satisfy the following conditions:

(1) $s(e)e = e = er(e)$ for all $e \in E^1$.
(2) $r(e)e^* = e^* = e^*s(e)$ for all $e \in E^1$.
(3) (The "CK-1 relations") For all $e, f \in E^1$, $e^*e = r(e)$ and $e^*f = 0$ if $e \neq f$.
(4) (The "CK-2 relations") For every regular vertex $v \in E^0$,

$$v = \sum_{e \in E^1, s(e)=v} ee^*.$$

An arbitrary element $a \in L := L_K(E)$ can be written as $a = \sum_{i=1}^{n} k_i \alpha_i \beta_i^*$ where α_i, β_i are paths and $k_i \in K$. Here $r(\alpha_i) = s(\beta_i^*) = r(\beta_i)$.

Every Leavitt path algebra $L_K(E)$ is a $\mathbb{Z}$*-graded algebra*, namely, $L_K(E) = \bigoplus_{n \in \mathbb{Z}} L_n$ induced by defining, for all $v \in E^0$ and $e \in E^1$, $\deg(v) = 0$, $\deg(e) = 1$, $\deg(e^*) = -1$. Here the L_n are abelian subgroups satisfying $L_m L_n \subseteq L_{m+n}$ for all $m, n \in \mathbb{Z}$. Further, for each $n \in \mathbb{Z}$, the homogeneous component L_n is given by $L_n = \{\sum k_i \alpha_i \beta_i^* \in L : \alpha, \beta \in Path(E), |\alpha_i| - |\beta_i| = n\}$. (For details, see Sect. 2.1 in [2]). An ideal I of $L_K(E)$ is said to be a *graded ideal* if $I = \bigoplus_{n \in \mathbb{Z}} (I \cap L_n)$.

Throughout this paper, E will denote an arbitrary graph (with no restriction on the number of vertices or on the number of edges emitted by each vertex) and K will denote an arbitrary field. For convenience in notation, we will denote, most of the times, the Leavitt path algebra $L_K(E)$ by L.

We shall first recall the definition of the Gelfand–Kirillov dimension of associative algebras over a field.

Let A be a finitely generated K-algebra, generated by a finite dimensional subspace $V = Ka_1 \oplus \cdots \oplus Ka_m$. Let $V^0 = K$ and, for each $n \geq 1$, let V^n denote the K-subspace of A spanned by all the monomials of length n in $a_1, \ldots, a_m$. Set $V_n = \sum_{i=0}^{n} V^i$. Then the **Gelfand–Kirillov dimension** of A (for short, the **GK-dimension** of A) is defined by

$$\text{GK-dim}(A) := \limsup_{n \to \infty} \log_n (\dim V_n).$$

It is known that the GK-dim(A) is independent of the choice of the generating subspace V.

If A is an infinitely generated K-algebra, then the GK-dimension of A is defined as

$$\text{GK-dim}(A) := \sup_B \text{GK-dim}(B)$$

where B runs over all the finitely generated K-subalgebras of A.

The GK-dimension of an algebra A measures the growth of the algebra A and might be considered the non-commutative analogue of the classical Krull dimension for commutative algebras. Some useful examples (see [28]) are: The GK-dimension the matrix ring $M_\Lambda(K)$ is 0 and the GK-dimension of the matrix ring $M_{\Lambda'}(K[x, x^{-1}])$ is 1, where Λ, Λ' are arbitrary possibly infinite index sets.

Grading of a matrix ring over a $\mathbb{Z}$-graded ring: We wish to recall the grading of matrices of finite order and then indicate how to extend this to the case of infinite matrices in which at most finitely many entries are non-zero (see [22, 31]). This will be used in Sect. 1.4. In the following, the length of a path p will be denoted by $|p|$.

Let Γ be an additive abelian group, A be a Γ-graded ring and $(\delta_1, \dots, \delta_n)$ an n-tuple where $\delta_i \in \Gamma$. Then $M_n(A)$ is a Γ-graded ring where, for each $\lambda \in \Gamma$, its λ-homogeneous component consists of $n \times n$ matrices

$$M_n(A)(\delta_1, \dots, \delta_n)_\lambda = \begin{pmatrix} A_{\lambda+\delta_1-\delta_1} & A_{\lambda+\delta_2-\delta_1} & \cdots & A_{\lambda+\delta_n-\delta_1} \\ A_{\lambda+\delta_1-\delta_2} & A_{\lambda+\delta_2-\delta_2} & & A_{\lambda+\delta_n-\delta_2} \\ & & & \\ & & & \\ A_{\lambda+\delta_1-\delta_n} & A_{\lambda+\delta_2-\delta_n} & & A_{\lambda+\delta_n-\delta_n} \end{pmatrix}. \tag{1.1}$$

This shows that for each homogeneous element $x \in A$,

$$\deg(e_{ij}(x)) = \deg(x) + \delta_i - \delta_j, \tag{1.2}$$

where $e_{ij}(x)$ is a matrix with x in the ij-position and with every other entry 0.

Now let A be a Γ-graded ring and let I be an arbitrary infinite index set. Denote by $M_I(A)$ the matrix with entries indexed by $I \times I$ having all except finitely many entries non-zero and for each $(i, j) \in I \times I$, the ij-position is denoted by $e_{ij}(a)$ where $a \in A$. Considering a "vector" $\bar{\delta} := (\delta_i)_{i \in I}$ where $\delta_i \in \Gamma$ and following the usual grading on the matrix ring (see (1.1), (1.2)), define, for each homogeneous element a,

$$\deg(e_{ij}(a)) = \deg(a) + \delta_i - \delta_j. \tag{1.3}$$

This makes $M_I(A)$ a Γ-graded ring, which we denote by $M_I(A)(\bar{\delta})$. Clearly, if I is finite with $|I| = n$, then the graded ring coincides (after a suitable permutation) with $M_n(A)(\delta_1, \dots, \delta_n)$.

Suppose E is a finite acyclic graph consisting of exactly one sink v. Let $\{p_i : 1 \leq i \leq n\}$ be the set of all paths ending at v. Then it was shown in (Lemma 3.4, [4])

$$L_K(E) \cong M_n(K) \tag{1.4}$$

under the map $p_i p_j^* \longmapsto e_{ij}$. Now taking into account the grading of $M_n(K)$, it was further shown in (Theorem 4.14, [22]) that the same map induces a graded isomorphism

$$L_K(E) \longrightarrow M_n(K)(|p_1|, \ldots, |p_n|) \tag{1.5}$$

$$p_i p_j^* \longmapsto e_{ij}.$$

In the case of a comet graph E (that is, a finite graph E with a cycle c without exits and a vertex v on c such that every path in E which does not include all the edges in c ends at v), it was shown in (Lemma 2.7.1, [2]) that the map

$$L_K(E) \longrightarrow M_n(K[x, x^{-1}]) \tag{1.6}$$

given by

$$p_i c^k p_j^* \longmapsto e_{ij}(x^k)$$

where the e_{ij} are matrix units, induces an isomorphism. Again taking into account the grading, it was shown in (Theorem 4.20, [22]) that the map

$$L_K(E) \longrightarrow M_n(K[x^{|c|}, x^{-|c|}])(|p_1|, \ldots, |p_n|) \tag{1.7}$$

given by

$$p_i c^k p_j^* \longmapsto e_{ij}(x^{k|c|})$$

induces a graded isomorphism. Later in the paper [3], the isomorphisms (1.4) and (1.6) were extended to infinite acyclic and infinite comet graphs, respectively (see Proposition 3.6 [3]). The same isomorphisms with the grading adjustments will induce graded isomorphisms for Leavitt path algebras of such graphs. We now describe this extension below.

Let E be a graph such that no cycle in E has an exit and such that every infinite path contains a line point or is tail-equivalent to a rational path $ccc\cdots$. where c is a cycle (without exits). Define an equivalence relation in the set of all line points in E by setting $u \sim v$ if $T_E(u) \cap T_E(v) \neq \emptyset$. Let X be the set of representatives of distinct equivalence classes of line points in E, so that for any two line points $u, v \in X$ with $u \neq v$, $T_E(u) \cap T_E(v) = \emptyset$. For each vertex $v_i \in X$, let $\overline{p^{v_i}} := \{p_s^{v_i} : s \in \Lambda_i\}$ be the set of all paths that end at v_i, where Λ_i is an index set which could possibly be infinite. Denote by $|\overline{p^{v_i}}| = \{|p_s^{v_i}| : s \in \Lambda_i\}$.

Let Y be the set of all distinct cycles in E. As before, for each cycle $c_j \in Y$ based at a vertex w_j, let $\overline{q^{w_j}} := \{q_r^{w_j} : r \in \Upsilon_j\}$ be the set of all paths that end at w_j but do not include all the edges of c_j, where Υ_j is an index set which could possibly be infinite. Let $|\overline{q^{w_j}}| := \{|q_r^{w_j}| : r \in \Upsilon_j\}$. Then the isomorphisms (1.5) and (1.7) extend to a $\mathbb{Z}$-graded isomorphism

$$L_K(E) \cong_{gr} \bigoplus_{v_i \in X} M_{\Lambda_i}(K)(|\overline{p^{v_i}}|) \oplus \bigoplus_{w_j \in Y} M_{\Upsilon_j}(K[x^{|c_j|}, x^{-|c_j|}])(|\overline{q^{w_j}}|) \qquad (1.8)$$

where the grading is as in (1.3).

1.3 Leavitt Path Algebras Satisfying a Polynomial Identity

Observe that Leavitt path algebras in general are highly non-commutative. For instance, if the graph E contains an edge e with $u = s(e) \neq r(e) = v$, then $ue = e$, but $eu = 0$. Indeed, it is an easy exercise to conclude that if E is a connected graph, then $L_K(E)$ is a commutative ring if and only if the graph E consists of just a single vertex v or is a loop e, that is a single edge e with $s(e) = r(e) = v$. In this case, $L_K(E)$ is isomorphic to K or $K[x, x^{-1}]$.

Note that to say a ring R is commutative is equivalent to saying that R satisfies the polynomial identity $xy - yx = 0$. An algebra A over a field K is said to **satisfy a polynomial identity** (or simply, a **PI-algebra**), if there is a polynomial $p(x_1, \ldots, x_n)$ in finitely many non-commuting variable $x_1, \ldots, x_n$ with coefficients in K such that $p(a_1, \ldots, a_n) = 0$ for all choices of elements $a_1, \ldots, a_n \in A$. For example, the Amitsur-Levitzky theorem (see [33]) states that the ring $M_n(R)$ of $n \times n$ matrices over a commutative ring R satisfies the so called standard polynomial identity $P_n(x_1, \ldots, x_n) = \sum_{\sigma \in S_n} \epsilon_\sigma x_{\sigma(1)} \cdots x_{\sigma(n)}$ where S_n is the symmetric group of $n!$ permutations of the set $\{1, \ldots, n\}$ and $\epsilon_\sigma = 1$ or -1 according as σ is even or odd. A natural question is to characterize the Leavitt path algebras which satisfy a polynomial identity. This is completely answered in the next theorem.

Theorem 1.1 ([17]) *Let E be an arbitrary graph. Then the following properties are equivalent for $L_K(E)$:*

(a) *$L_K(E)$ satisfies a polynomial identity;*
(b) *No cycle in E has an exit, there is a fixed positive integer d such that the number of distinct paths that end at any vertex v is $\leq d$ and the only infinite paths in E are paths that are eventually of the form $ggg\cdots$, for some cycle g;*
(c) *There is a fixed positive integer d such that $L_K(E)$ is a subdirect product of matrix rings over K or $K[x, x^{-1}]$ of order at most d.*

If the graph E is row-finite, then the Leavitt path algebra $L_K(E)$ in Theorem 1.1 actually decomposes as a ring direct sum of matrix rings over K or $K[x, x^{-1}]$ of order at most a fixed positive integer d. This shows that satisfying a polynomial identity imposes a serious restriction on the structure of Leavitt path algebras.

1.4 Four Important Graphical Conditions

In this section, we shall illustrate how specific graphical conditions on the graph E give rise to various algebraic properties of $L_K(E)$. We illustrate this by choosing four different graph properties of E. Interestingly, a single graph theoretical property of E often implies several different ring properties for $L_K(E)$. It is amazing that a single property that no cycle in a finite graph E has an exit implies that the corresponding Leavitt path algebra $L_K(E)$ possesses several different ring properties such as being directly finite, self-injective, having bounded index of nilpotence, a Baer ring, satisfying a polynomial identity, having GK-dimension ≤ 1, etc. (see Theorem 1.5 below). Consequently, Leavitt path algebras turn out to be useful tools in the construction of various examples of rings. We will also describe the interesting history behind the terms Condition (K) and Condition (L) which play an important role in the investigation of both the graph C^*-algebras and Leavitt path algebras (see [2, 40, 42]).

Recall, a ring R is said to be von Neumann regular if to each element $a \in R$ there is an element $b \in R$ such that $a = aba$. The ring R is said to be π-regular (strongly left or right π-regular) if to each element $a \in R$, there is a $b \in R$ and an integer $n \geq 1$ such that $a^n = a^n b a^n$ ($a^n = a^{n+1}b$ or $a^n = ba^{n+1}$). In general, these ring properties are not equivalent. But as the next theorem shows, they all coincide for Leavitt path algebras.

A graph E is said to be **acyclic** if E contains no cycles. The next theorem characterizes the von Neumann regular Leavitt path algebras.

Theorem 1.2 ([7]) *For an arbitrary graph E, the following conditions are equivalent for* $L := L_K(E)$*:*

(a) *The graph E is acyclic;*
(b) *L is von Neumann regular;*
(c) *L is* π*-regular;*
(d) *L is strongly left/right* π*-regular.*

Another important graph property is Condition (K). In some sense this property is diametrically opposite of being acyclic.

Definition 1.1 A graph E satisfies **Condition (K)** if whenever a vertex v lies on a simple closed path α, v also lies on another simple closed path β distinct from α.

The Condition (K) implies a number of ring properties;

Definition 1.2 (i) A ring R is said to be left/right **weakly regular** if for every left/right ideal I of R, $I^2 = I$;

(ii) A ring R is said to be an **exchange ring** if given any left/right R-module M and two direct decompositions of M as $M = M' \oplus A$ and $M = \bigoplus_{i=1}^{n} A_i$, where $M' \cong R$, there exist submodules $B_i \subseteqq A_i$ such that $M = M' \oplus \bigoplus_{i=1}^{n} B_i$.

Theorem 1.3 ([15, 16, 40]) *Let E be an arbitrary graph. Then the following conditions are equivalent for $L := L_K(E)$:*

(i) *The graph E satisfies Condition (K);*
(ii) *L is an exchange ring;*
(iii) *L is left/right weakly regular;*
(iv) *Every two-sided ideal of L is a graded ideal.*

Definition 1.3 A graph E is said to satisfy **Condition (L)**, if every cycle in E has an exit.

Theorem 1.4 ([35]) *Let E be an arbitrary graph. Then the following are equivalent for $L_K(E)$:*

(i) *E satisfies Condition (L);*
(ii) *L is a* ***Zorn ring****, that is, every (non-nil) right/left ideal I contains a non-zero idempotent.;*
(iii) *Every element $a \in L$ is the von Neumann inverse of another element $b \in L$; that is, to each $a \in L$, there is an element $b \in L$ such that $bab = b$.*

An interesting history of Conditions (K) and (L): One may wonder about the choice of the letters K and L in the terms Condition (K) and Condition (L). There is an interesting narrative about the origins of these terms. I am grateful to Mark Tomforde for outlining this history to me which he will also be including in his forthcoming book on Graph Algebras [42]. Both these two graph conditions were originally introduced by graph C^*-algebraists. It all started when Cuntz and Krieger (whom some consider the founders of graph C^*-algebras), introduced in their original paper [21] a condition on matrices with entries in $\{0, 1\}$ and called it Condition (I). Assuming that the 'I' is the English letter I and not the Roman numeral one, Pask and Raeburn introduced in 1996 a Condition (J) in their paper [32], as J is the letter that follows I in the English alphabet. (They apparently did not recognize that Cüntz and Krieger also introduced a follow-up Condition (II), thus indicating, in their view, I and II stand for Roman numerals.) Conforming to this pattern, when Kumjian, Pask, Raeburn and Renault introduced a new condition in their 1997 paper [29], they chose the letter K to denote this new condition and called it Condition (K). Continuing this pattern yet again, Kumjian, Pask and Raeburn introduced Condition (L) in 1998 [30]. Actually, Astrid an Huef later showed that Condition (L) coincides with Condition (I) for graphs of finite matrices. Moreover, Condition (K) is considered analogous to

Condition (II) for Cuntz-Krieger algebras. In all the investigations that followed in graph C^*-algebras and also in Leavitt path algebras, Conditions (K) and (L) emerged as important graph conditions. Poor Condition (J) remains neglected!

Recall, Condition (L) requires every cycle to have an exit. We next consider a graph property that is diametrically opposite to Condition (L), namely, no cycle in the graph has an exit. This implies several interesting ring/module properties.

First, consider a finite graph E in which no cycle has an exit. In this case, $L_K(E)$ is a ring with identity. We begin recalling a number of ring properties.

A ring R with identity 1 is said to be **directly finite** if for any two elements $x, y, xy = 1$ implies $yx = 1$. This is equivalent to R being not isomorphic to any proper direct summand of R as a left or a right R-module. A ring R with identity is called a **Baer ring** if the left/right annihilator of every subset X of R is generated by an idempotent. A Γ-graded ring R is said to be a **graded Baer ring**, if the left/right annihilator of every subset X of homogeneous elements is generated by a homogeneous idempotent. A ring R is said to have **bounded index of nilpotence** if there is a positive integer n which is such that $a^n = 0$ for every nilpotent element $a \in R$.

Theorem 1.5 ([9, 17, 25, 27, 39, 43]) *For a finite graph E, the following conditions are equivalent for $L := L_K(E)$:*

(i) *No cycle in E has an exit;*
(ii) *L is directly finite;*
(iii) *L is a Baer ring;*
(iv) *L is a graded Baer ring;*
(v) *L is a graded left/right self-injective ring;*
(vi) *L satisfies a polynomial identity;*
(vii) *L has bounded index of nilpotence;*
(viii) *L is graded semi-simple;*
(ix) *L has GK-dimension ≤ 1;*
(x) *L is finite over its center.*

Thus if E is the following graph,

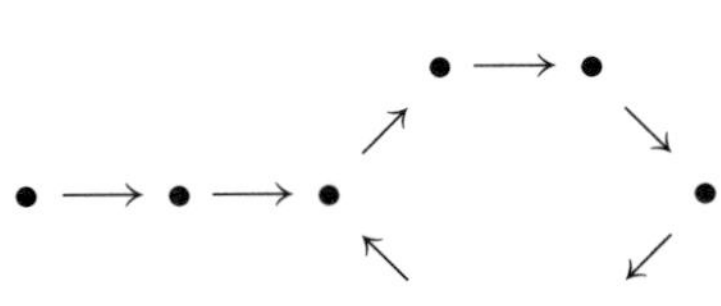

then $L_K(E)$ will possess all the stated nine ring properties.

For a finite graph E, if $L_K(E)$ satisfies any of the equivalent conditions in the preceding theorem, $L_K(E)$ decomposes as a graded direct sum of finitely many matrix rings of finite order over K and/or $K[x, x^{-1}]$ which are given the matrix gradings indicated in Eqs. (1.5) and (1.7) in the Preliminary section.

For a ring R without identity, but with local units, R is said to be directly finite if for every $x, y \in R$ and an idempotent $u \in R$ satisfying $ux = x = xu$, $uy = y = yu$, we have $xy = u$ implies $yx = u$. Every commutative ring is trivially directly finite.

If R is a ring without identity, R is called a **locally Baer ring (locally graded Baer** ring) if for every idempotent (homogeneous idempotent) e, the corner eRe is a Baer (graded Baer) ring.

Theorem 1.6 ([25, 27]) *Let E be an arbitrary graph. Then the following conditions are equivalent for $L := L_K(E)$:*

(i) *No cycle in E has an exit, E is row-finite and every infinite path ends at a sink or a cycle.*

(i) *L is a locally Baer ring;*
(ii) *L is a graded locally Baer ring;*
(iii) *L is a graded left/right self-injective ring;*
(iv) *L is graded isomorphic to a ring direct sum of matrix rings*

$$L_K(E) \cong_{gr} \bigoplus_{v_i \in X} M_{\Lambda_i}(K)(|\overline{p^{v_i}}|) \oplus \bigoplus_{w_j \in Y} M_{\Upsilon_j}(K[x^{t_j}, x^{-t_j}])(|\overline{q^{w_j}}|)$$

where Λ_i, Υ_j are suitable index sets, the t_j are positive integers, X is the set of representatives of distinct equivalence classes of line points in E and Y is the set of all distinct cycles (without exits) in E.

1.5 Simple Modules over Leavitt Path Algebras

In this section, we shall indicate the methods of constructing simple modules over Leavitt path algebras by graphical means.

As noted in [2], every element a of a Leavitt path algebra $L_K(E)$ of a graph E can be written in the form $a = \sum_{i=1}^{n} \alpha_i \beta_i^*$ and that the map $\sum_{i=1}^{n} \alpha_i \beta_i^* \longrightarrow \sum_{i=1}^{n} \beta_i \alpha_i^*$ induces an isomorphism $L_K(E) \longrightarrow (L_K(E))^{op}$. Consequently, $L_K(E)$ is left-right symmetric. So in this and the next section, we shall only be stating results on left ideals and left modules over $L_K(E)$. The corresponding results on right ideals and right modules hold by symmetry.

Definition 1.4 A vertex v is called a **Laurent vertex** if $T_E(v)$ consists of the set of all vertices on a single path $\gamma = \mu c$ where μ is a path without bifurcations starting at v and c is a cycle without exits based on a vertex $u = r(\mu)$.

An easy example of a Laurent vertex is the vertex v in the following graph:

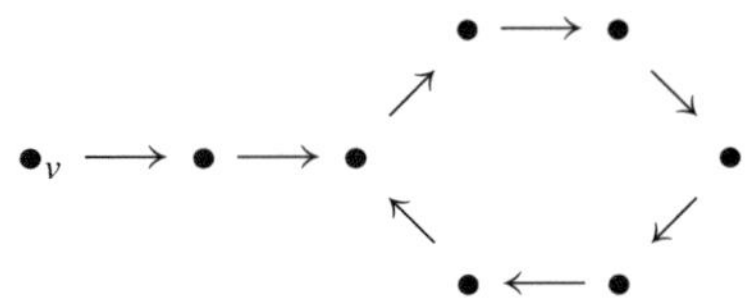

The next theorem gives conditions under which a vertex in the graph E generates a simple left ideal/graded simple left ideal of $L_K(E)$.

Theorem 1.7 ([2, 25]) *Let E be an arbitrary graph and let v be a vertex. Then*

(a) *The left ideal $L_K(E)v$ is a simple/minimal left ideal of $L_K(E)$ if and only if v is a line point;*
(b) *The left ideal $L_K(E)v$ is a graded simple/minimal left ideal of $L_K(E)$ if and only if v is either a line point or a Laurent vertex.*

Next, we shall describe the general methodology used by Chen [19] and extended in [25, 37] to construct left simple and graded simple modules over $L_K(E)$ by using special vertices or cycles in the graph E.

(I) **Definition of the module** A_u: Let u be a vertex in a graph E which is either a sink or an infinite emitter. Let A_u be the K-vector space having as a basis the set $B = \{p : p \text{ is a path in } E \text{ with } r(p) = u\}$. We make A_u a left $L_K(E)$-module as follows: Define, for each vertex v and each edge e in E, linear transformations P_v, S_e and S_{e^*} on A by defining their actions on the basis B are as follows:

For all $p \in B$,

(I) $P_v(p) = \begin{cases} p, \text{ if } v = s(p) \\ 0, \text{ otherwise} \end{cases}$

(II) $S_e(p) = \begin{cases} ep, \text{ if } r(e) = s(p) \\ 0, \text{ otherwise} \end{cases}$

(III) $S_{e^*}(p) = \begin{cases} p', \text{ if} p = ep' \\ 0, \text{ otherwise} \end{cases}$

(IV) $S_{e^*}(u) = 0$.

The endomorphisms $\{P_v, S_e, S_{e^*} : v \in E^0, e \in E^1\}$ satisfy the defining relations (1.1)–(1.4) of the Leavitt path algebra $L_K(E)$. This induces an algebra homomorphism ϕ from $L_K(E)$ to $End_K(A_u)$ mapping v to P_v, e to S_e and e^* to S_{e^*}. Then A_u can be made a left module over $L_K(E)$ via the homomorphism ϕ. We denote this $L_K(E)$-module operation on A_u by $\cdot$.

Theorem 1.8 ([19, 37]) *If the vertex u is either a sink or an infinite emitter, then A_u is a simple left $L_K(E)$-module.*

If the vertex u lies on a cycle without exits, then in the Definition of A_u, define the basis $B = \{pq^* : p, q \text{ path in } E \text{ with } r(q^*) = s(q) = u\}$. We then get the following result.

Theorem 1.9 ([25]) *If a vertex $u \in E$ lies on a cycle without exits, then A_u is a graded simple left $L_K(E)$-module graded isomorphic to the graded minimal left ideal $L_K(E)u$ and A_u is not a simple left $L_K(E)$-module.*

Remark 1.1 In [25], the module A_u is defined by using an algebraic branching system and is denoted as N_{vc}. Here we have defined the module A_u differently, but the proof of the above theorem is just the proof of Theorem 3.5(1) of [25].

With a slight modification of the definition of A_u, Chen [19] shows one more way of constructing simple modules by using the infinite paths that are tail-equivalent to a fixed infinite path in E. Recall, two infinite paths $p = e_1 \cdots e_r \cdots$ and $q = f_1 \cdots f_s \cdots$ are said to be **tail-equivalent** if there exist fixed positive integers m, n such that $e_{n+k} = f_{m+k}$ for all $k \geq 1$. Let $[p]$ denote the tail-equivalence class of all infinite paths equivalent to p. Let $A_{[p]}$ denote the K-vector space having $[p]$ as a basis. As in the definition of A_u, for each vertex v and each edge e in E, define the linear transformations P_v, S_e and S_{e^*} on A by defining their actions on the basis $[p]$ satisfying the conditions (I), (II), (III), but not (IV) above. As before, they satisfy the defining relations of a Leavitt path algebra and thus induce a homomorphism $\varphi : L_K(E) \to A_{[p]}$. The vector space $A_{[p]}$ then becomes a left $L_K(E)$-module via the map φ.

Theorem 1.10 ([19]) *The module $A_{[p]}$ is a simple left $L_K(E)$-module and for two infinite paths p, q, $A_{[p]} \cong A_{[q]}$ if and only if $[p] = [q]$.*

It can be shown (see [37]) that the simple modules A_u, A_v and $A_{[p]}$ corresponding, respectively, to a sink u, an infinite emitter v and an infinite path p, are all non-isomorphic.

A special infinite path is the so-called a **rational infinite path** induced by a simple closed path (and in particular, a cycle) c. This is the infinite path $ccc \cdots$. We denote this path by c^∞. We shall be using the corresponding simple $L_K(E)$-module A_{c^∞} subsequently.

1.6 Leavitt Path Algebras with Simple Modules Having Special Properties

We shall describe when all the simple modules over a Leavitt path algebra are flat or injective or finitely presented or graded etc.

An open problem, raised by Ramamuthi [34] some forty years ago, asks whether a non-commutative ring R with 1 is von Neumann regular if all the simple left R-modules are flat. Using a more general approach of Steinberg algebras, Ambily, Hazrat and Li [10] obtain the following theorem which shows that Ramamurthi's question has an affirmative answer in the case of Leavitt path algebras.

Theorem 1.11 ([10]) *Let E be an arbitrary graph. Then every simple left $L_K(E)$-module is flat if and only if $L_K(E)$ is von Neumann regular.*

Next, we consider the case when $L_K(E)$ is a left V-ring, that is, when every simple left $L_K(E)$-module is injective. Kaplansky showed that if R is a commutative ring, then every simple R-module is injective if and only if R is von Neumann regular. A natural question is under what conditions every simple left $L_K(E)$-module is injective. It was shown in [26] that, in this case, $L_K(E)$ becomes a weakly regular ring. However, recently Abrams, Mantese and Tonolo [5] showed that, if c is a cycle in a graph E, then the corresponding simple left $L_K(E)$-module A_{c^∞} satisfies $Ext^1_{L_K(E)}(A_{c^\infty}, A_{c^\infty}) \neq 0$. This implies that the module A_{c^∞} cannot be an injective module. So, if every simple left $L_K(E)$-module is injective, then necessarily E contains no cycles. Then by [7] $L_K(E)$ must be von Neumann regular. Thus, we obtain the following new result and its corollary.

Theorem 1.12 *Let E be an arbitrary graph. If every simple left $L_K(E)$-module is injective, then $L_K(E)$ is a von Neumann regular ring.*

Conversely, if $L_K(E)$ is a von Neumann regular ring then the graph E contains no cycles [7] and, if E is further a finite graph, then $L_K(E)$ is a direct sum of finitely many matrix rings of finite order over K (Theorem 2.6.17, [2]). In this case, $L_K(E)$ is a direct sum of left/right simple modules and hence every simple left/right $L_K(E)$-module is injective. This leads to the following corollary.

Corollary 1.1 *Let E be a finite graph. Then every simple left/right $L_K(E)$-module is injective if and only if $L_K(E)$ is a von Neumann regular ring.*

When E is an arbitrary graph, it is an open question whether the von Neumann regularity of $L_K(E)$ implies that every simple left/right $L_K(E)$-module is injective.

Next, we consider Leavitt path algebras $L_K(E)$ whose simple modules are all finitely presented. When E is a finite graph, $L_K(E)$ possesses a number of interesting properties as noted in the following theorem.

Theorem 1.13 ([13]) *For any finite graph E, the following properties of the Leavitt path algebra $L := L_K(E)$ are equivalent:*

(i) *Every simple left L-module is finitely presented;*
(ii) *No two cycles in E have a common vertex;*
(iii) *There is a one-to-one correspondence between isomorphism classes of simple L-modules and primitive ideals of L;*
(iv) *The Gelfand–Kirillov dimension of L is finite.*

The preceding theorem has been generalized in [38] to the case when E is an arbitrary graph with several similar equivalent conditions.

It is easy to observe that every simple left module over a Leavitt path algebra $L := L_K(E)$ is of the form Lv/N for some vertex v and a maximal left submodule N of Lv. A natural question is, given a vertex u, can we estimate the number κ_u of distinct maximal left L-submodules M of Lu such that Lu/M is isomorphic to Lv/N? In [38] it is shown that $\kappa_u \leq |uLv\backslash N|$ and consequently the cardinality of the set of all such simple modules corresponding to various vertices is $\leq \sum_{u \in E^0} |uLv \backslash N|$ and thus is $\leq |L|$.

More generally, one may try to estimate the number of non-isomorphic simple modules over a Leavitt path algebra. In this connection, observe the following: Suppose a graph E contains two cycles g, h which share a common vertex v, such as the two cycles in the following graph.

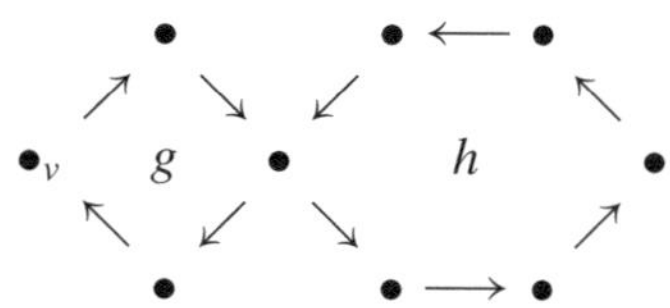

We then wish to show that there are uncountably many non-isomorphic simple modules over the corresponding $L_K(E)$. By Theorem 1.10, we need only to produce uncountably many non-equivalent infinite paths in E. With that in mind, consider the infinite rational path $p = ggg \cdots$ which, for convenience, we write as $p = g_1 g_2 g_3 \cdots$ indexed by the set $\mathbb{P}$ of positive integers, where $g_i = g$ for all i. Now, for every subset S of $\mathbb{P}$, define an infinite path p_S by replacing g_i by h if and only if $i \in S$. Observe that this gives rise to uncountably many distinct infinite paths. From the definition of equivalence of paths, it can be derived that, given an infinite path q there are at most countably many infinite paths that are equivalent to q. From this one can then establish that there are uncountably many non-isomorphic simple left $L_K(E)$-modules.

Leavitt path algebras having only finitely many non-isomorphic simple modules are characterized in the next theorem. Recall, a ring R is called left/right semi-artinian if every non-zero left/right R-module contains a simple submodule.

Theorem 1.14 ([14]) *Let E be an arbitrary graph and K be any field. Then the following are equivalent for the Leavitt path algebra $L = L_K(E)$:*

(i) *L has at most finitely many non-isomorphic simple left/right L-modules;*
(ii) *L is a left and right semi-artinian von Neumann regular ring with finitely many two-sided ideals which form a chain under set inclusion;*
(iii) *The graph E is acyclic and there is a finite ascending chain of hereditary saturated subsets $\{0\} \underset{\neq}{\subseteq} H_1 \underset{\neq}{\subseteq} \cdots \underset{\neq}{\subseteq} H_n = E^0$ such that for each $i < n$, $H_{i+1} \backslash H_i$ is the hereditary satusrated closure of the set of all the line points in $E \backslash H_i$.*

1.7 One-Sided Ideals in a Leavitt Path Algebra

In this section, we shall describe some of the interesting properties of one-sided ideals of a Leavitt path algebra.

Using a deep theorem of Bergman [18], Ara and Goodearl proved the following result that Leavitt path algebras are hereditary.

Theorem 1.15 ([12]) *If E is an arbitrary graph, then every left/right ideal of $L_K(E)$ is a projective left/right $L_K(E)$-module.*

A von Neumann regular ring R has the characterizing property that every finitely generated one-sided ideal of R is a principal ideal generated by an idempotent. Four years ago, it was shown in [36] that every finitely generated two-sided ideal of a Leavitt path algebra is a principal ideal. Recently, Abrams, Mantese and Tonolo have proved that this interesting property holds for one sided ideals too, as indicated in the next theorem.

Theorem 1.16 ([6]) *Let E be an arbitrary graph. Then $L := L_K(E)$ is a Bézout ring, that is, every finitely generated one-sided ideal is a principal ideal.*

Thus, if La and Lb are two principal left ideals of L, then, being finitely generated, $La + Lb = Lc$, a principal left ideal. So the sum of any two principal left ideals of L is again a principal left ideal. What about their intersection? Should the intersection of two principal left/right ideals of L be again a principal left/right ideal? This is answered in the next theorem. Since this result is new, we outline its proof which is straight forward.

Theorem 1.17 *Let E be an arbitrary graph. Then both the sum and the intersection of two principal left/right ideals of $L := L_K(E)$ are again principal left/right ideals. Thus, the principal left/right ideals of L form a sublattice of the lattice of all left/right ideals of L.*

Proof Suppose A, B are two principal left ideals of L. Consider the following exact sequence where the map θ is the additive map $(a, b) \to a + b$

$$0 \longrightarrow K \longrightarrow A \oplus B \xrightarrow{\theta} A + B \longrightarrow 0$$

and where $K = \{(x, -x) : x \in A \cap B\} \cong A \cap B$. Now $A + B$ is a (finitely generated) left ideal of $L_K(E)$. By Theorem 1.15, it is a projective module and hence the above exact sequence splits. Consequently, $A \cap B$ is isomorphic a direct summand of $A \oplus B$. Since $A \oplus B$ is finitely generated, so is $A \cap B$. By Theorem 1.16, $A \cap B$ is then a principal left ideal. The same argument works for principal right ideals of L. □

Remark 1.2 Thus, the above theorem states that, for any two non-zero elements $a, b \in L$, $aL + bL = cL$ and $aL \cap bL = dL$ where c, d are suitable elements of L. Since $c = ax + by$ for some $x, y \in L$, it is easy to see that $c = \textit{right}\ \gcd(a, b)$. (Recall that in a non-commutative ring R, an element c is a **right gcd** of a, b, if $a = xc$, $b = yc$ and if $a = x'c'$, $b = y'c'$, then $c = zc'$.) Similarly, $aL \cap bL = dL$ implies that $d = \textit{right}\ \mathrm{lcm}(a, b)$. Thus, in a Leavitt path algebra L, the right gcd and the right lcm of any two non-zero elements exist. In the same way, one could conclude that the left gcd and the left lcm of any two non-zero elements also exist in L.

As we noted in Theorem 1.3, every two-sided ideal of $L_K(E)$ is graded if and only if the graph E satisfies Condition (K). What happens if every one-sided ideal of $L_K(E)$ is graded? This is answered in the next and the last theorem of this section.

Theorem 1.18 ([25]) *Let E be an arbitrary graph. Then the following properties are equivalent for $L = L_K(E)$:*

(i) *Every left ideal of L is a graded left ideal;*
(ii) *Every simple left L-module is a graded module;*
(iii) *The graph E contains no cycles;*
(iv) *L is a von Neumann regular ring.*

References

1. G. Abrams, Leavitt path algebras: the first decade. Bull. Math. Sci. **5**(1), 59–120 (2015)
2. G. Abrams, P. Ara and M. Siles Molina, *Leavitt Path Algebras*. Lecture Notes in Mathematics, vol. 2191 (Springer, Berlin, 2017)
3. G. Abrams, G. Aranda Pino, F. Perera, M. Siles Molina, Chain conditions for Leavitt path algebras. Forum Math. **22**, 95–114 (2010)
4. G. Abrams, G. Aranda Pino, M. Siles Molina, Finite dimensional Leavitt path algebras. J. Pure Appl. Algebra **209**, 753–762 (2007)
5. G. Abrams, F. Mantese, A. Tonolo, Extensions of simple modules over Leavitt path algebras. J. Algebra **431**, 78–106 (2015)
6. G. Abrams, F. Mantese, A. Tonolo, Leavitt path algebras are Bézout. Israel J. Math. **228**, 53–75 (2018)
7. G. Abrams, K.M. Rangaswamy, Regularity conditions for the Leavitt path algebras of arbitrary graphs. Algebr. Represent. Theory **13**, 319–334 (2010)
8. G. Abrams, K.M. Rangaswamy, M. Siles Molina, Socle series in a Leavitt path algebra. Israel J. Math. **184**, 413–435 (2011)
9. A. Alahmadi, H. Alsulami, S.K. Jain, E. Zelmanov, Leavitt path algebras of finite Gelfand-Kirillov dimension. J. Algebra Appl. **11**, 1250225, 6 pp (2012)
10. A.A. Ambily, R. Hazrat, H. Li, Simple flat Leavitt path algebras are regular (2018), arXiv:1803.01283v1 [math.RA]
11. P. Ara, M. Brustenga, Module theory over Leavitt path algebras and K-theory. J. Pure Appl. Algebr. **214**, 1131–1151 (2010)
12. P. Ara, K. Goodearl, Leavitt path algebras of separated graphs. J. Reine Angew. Math. **669**, 165–224 (2012)
13. P. Ara, K.M. Rangaswamy, Finitely presented simple modules over Leavitt path algebras. J. Algebr. **417**, 333–352 (2014)
14. P. Ara, K.M. Rangaswamy, Leavitt path algebras with at most countably many representations. Rev. Mat. Iberoam **31**, 263–276 (2015)
15. G. Aranda Pino, E. Pardo, M. Siles Molina, Exchange Leavitt path algebras and stable range. J. Algebr. **305**, 912–936 (2006)
16. G. Aranda Pino, K.M. Rangaswamy, M. Siles Molina, Weakly regular and self-injective Leavitt path algebras over arbitrary graphs. Algebr. Represent. Theory **14**, 751–777 (2011)
17. J.P. Bell, T.H. Lenagan, K.M. Rangaswamy, Leavitt path algebras satisfying a polynomial identity. J. Algebr. Appl. **15**(5), 1650084 (13 pages) (2016)
18. G. Bergman, Coproducts and some universal ring constructions. Trans. Am. Math. Soc. **200**, 33–88 (1974)

19. X.W. Chen, Irreducible representations of Leavitt path algebras. Forum Math. **22** (2012)
20. J.H. Cozzens, Homological properties properties of the ring of differential polynomials. Bull. Am. Math. Soc. **76**, 75–79 (1970)
21. J. Cuntz, W. Krieger, A class of C^* -algebras and topological Markov chains. Invent. Math. **56**, 251–268 (1980)
22. R. Hazrat, The graded structure of Leavitt path algebras. Israel J. Math. **195**, 833–895 (2013)
23. R. Hazrat, Leavitt path algebras are graded von Neumann regular rings. J. Algebr. **401**, 220–233 (2014)
24. R. Hazrat, *Graded rings and Graded Grothendieck Groups*, vol. 435, LMS Lecture Notes Series (Cambridge University Press, Cambridge, 2016)
25. R. Hazrat, K.M. Rangaswamy, Graded irreducible representations of Leavitt path algebras. J. Algebr. **450**, 458–496 (2016)
26. R. Hazrat, K.M. Rangaswamy, A. Srivastava, Leavitt path algebras: graded direct finiteness and graded $\sum$-injective. J. Algebr. **503**, 299–328 (2018)
27. R. Hazrat, L. Vás, Baer and Baer *-Ring characterizations of Leavitt path algebras. J. Pure Appl. Algebr. **222**, 39–60 (2018)
28. G.R. Krause, T.H. Lenagan, *Growth of Algebras and Gelfand-Kirillov Dimension*, vol. 22, Graduate Studies in Mathematics (American Mathematical Society, Providence, 2000)
29. A. Kumjian, D. Pask, I. Raeburn, J. Renault, Graphs, groupoids, and Cuntz-Krieger algebras. J. Funct. Anal. **144**, 505–541 (1997)
30. A. Kumjian, D. Pask, I. Raeburn, Cuntz-Krieger algebras of directed graphs. Pac. J. Math. **184**, 161–174 (1998)
31. C. Nastasescu, F. van Oystaeyen, *Graded Ring Theory* (North-Holland, Amsterdam, 1982)
32. D. Pask, I. Raeburn, On the K-theory of Cuntz-Krieger algebras. Publ. Res. Inst. Math. Sci. **32**, 415–443 (1996)
33. C. Procesi, *Rings with Polynomial Identities* (Marcel Dekker, New York, 1973)
34. V.S. Ramamurthi, On the injectivity and flatness of certain cyclic modules. Proc. Am. Math. Soc. **48**, 21–25 (1975)
35. K.M. Rangaswamy, Leavitt path algebras which are Zorn rings. Contemp. Math. **609**, 277–283 (2014)
36. K.M. Rangaswamy, On generators of two-sided ideals of Leavitt path algebras over arbitrary graphs. Commun. Algebr. **42**, 2859–2868 (2014)
37. K.M. Rangaswamy, On simple modules over Leavitt path algebras. J. Algebr. **423**, 239–258 (2015)
38. K.M. Rangaswamy, Leavitt path algebras with finitely presented irreducible representations. J. Algebr. **447**, 624–648 (2016)
39. K.M. Rangaswamy, A. Srivastava, Leavitt path algebras with bounded index of nilpotence. J. Algebr. Appl. **18**, 1950185 (10 pages) (2019)
40. M. Tomforde, Uniqueness theorems and ideal structure of Leavitt path algebras. J. Algebr. **318**, 270–299 (2007)
41. M. Tomforde, Leavitt path algebras with coefficients in a commutative ring. J. Pure Appl. Algebr. **215**, 471–484 (2011)
42. M. Tomforde, *Graph Algebras* (In preparation)
43. L. Vás, Canonical trace and directly-finite Leavitt path algebras. Algebr. Represent. Theory **18**(3), 711–738 (2015)

Chapter 2
The Groupoid Approach to Leavitt Path Algebras

Simon W. Rigby

2.1 Introduction

Leavitt path algebras are $\mathbb{Z}$-graded algebras with involution, whose generators and relations are encoded in a directed graph. Steinberg algebras, on the other hand, are algebras of functions defined on a special kind of topological groupoid, called an ample groupoid. To understand how they are related, it is useful to weave together some historical threads. This historical overview might not be comprehensive, but it is intended to give some idea of the origins of our subject.

2.1.1 Historical Overview: Groupoids, Graphs, and Their Algebras

In the late 1950s and early 1960s, William G. Leavitt [51, 52] showed that there exist simple rings whose finite-rank free modules admit bases of different sizes. In a seemingly unrelated development, in 1977, Joachim Cuntz [33] showed that there exist separable C^*-algebras that are simple and purely infinite. Cuntz's paper was one of the most influential in the history of operator theory. It provoked intense interest (that is still ongoing) in generalising, classifying, and probing the structure of various classes of C^*-algebras. One of the next landmarks was reached in 1980, when Jean Renault [59] defined groupoid C^*-algebras, taking inspiration from the C^*-algebras

S. W. Rigby (✉)
Department of Mathematics and Applied Mathematics,
University of Cape Town, Cape Town, South Africa
e-mail: simon.rigby@ugent.be

Department of Mathematics: Algebra and Geometry,
Ghent University, Ghent, Belgium

A. A. Ambily et al. (eds.), *Leavitt Path Algebras and Classical* K*-Theory*,
Indian Statistical Institute Series,
https://doi.org/10.1007/978-981-15-1611-5_2

that had previously been associated to transformation groups. The Cuntz algebras were interpreted as groupoid C^*-algebras, and from that point onwards there was a new framework and some powerful results with which to pursue new and interesting examples.

In 1997, Kumjian, Pask, Raeburn, and Renault [50] showed how to construct a Hausdorff ample groupoid (and hence a C^*-algebra) from a row- and column-finite directed graph with no sinks. They showed that these C^*-algebras universally satisfy the Cuntz–Krieger relations from [34], which had become significant in the intervening years. Graph C^*-algebras were then studied in depth. Usually, they were conceptualised in terms of the partial isometries that generate them; direct methods, rather than groupoid methods, were used predominantly [16, 58]. Meanwhile, the Cuntz algebras had also been interpreted as inverse semigroup C^*-algebras. Paterson [56, 57], at the turn of the 21st century, organised the situation a bit better. He showed that all graph C^*-algebras (of countable graphs, possibly with sinks, infinite emitters, and infinite receivers) are inverse semigroup C^*-algebras, and that all inverse semigroup C^*-algebras are groupoid C^*-algebras. The key innovation was defining the universal groupoid of an inverse semigroup, which is an ample but not necessarily Hausdorff topological groupoid.

It is unclear when the dots were first connected between Leavitt's algebras and Cuntz's C^*-algebras (probably in [11], a very long time after they first appeared). The Cuntz algebra O_n is the norm completion of the complex Leavitt algebra $L_{n,\mathbb{C}}$. Over any field $\mathbb{K}$, the ring $L_{n,\mathbb{K}}$ and the C^*-algebra O_n are purely infinite simple, and they have the same K_0 group (but these concepts have a different meaning for rings compared to C^*-algebras). This begins a process in which the algebraic community generalises, classifies, and probes the structure of various classes of rings in much the same way as the operator algebra community did with C^*-algebras. Leavitt path algebras were introduced in [2, 12] as universal $\mathbb{K}$-algebras satisfying path algebra relations and Cuntz–Krieger relations. Generalising the relationship between the Leavitt and Cuntz algebras, the graph C^*-algebra of a graph E is the norm completion of the complex Leavitt path algebra of E. The interplay with C^*-algebras is not the only connection between Leavitt path algebras and other, older, areas of mathematics—see for instance [1, Sect. 1] and [55].

Knowing what we know now, the next step was very natural. Is there a way of defining a "groupoid $\mathbb{K}$-algebra" in such a way that:

- When the input is the universal groupoid of an inverse semigroup S, the output is the (discrete) inverse semigroup algebra $\mathbb{K}S$;
- When the input is a graph groupoid $\mathcal{G}_E$, the output is the Leavitt path algebra $L_{\mathbb{K}}(E)$?

This question was asked and answered by Steinberg [63], and Clark, Farthing, Sims, and Tomforde [27]. Consistent with previous experiences of converting operator algebra constructions into $\mathbb{K}$-algebra constructions, they found that the groupoid C^*-algebra is the norm completion of the groupoid $\mathbb{C}$-algebra. It is also worth noting that Steinberg chose a broad scope and defined groupoid R-algebras over any commutative ring R, rather than just fields. We call these groupoid algebras *Steinberg algebras*.

Each of the three parts in this paper can be read separately. However, we work towards Leavitt path algebras as the eventual subject of interest, and this influences the rest of the text. For example, in Sect. 2.2 we try not to impose the Hausdorff assumption on groupoids if it is not necessary, but there are no examples here of non-Hausdorff groupoids. Throughout, we use the graph theory notation and terminology that is conventional in Leavitt path algebras. (In most of the C^*-algebra literature, the orientation of paths is reversed.) And in Sect. 2.3, we ignore some topics like amenability that would be important if we were intending to study the C^*-algebras of boundary path groupoids.

A standing assumption throughout the paper is that R is a commutative ring with 1. We rarely need to draw attention to it or require it to be anything special.

2.1.2 Background: Leavitt Path Algebras

For an arbitrary graph E, there is an R-algebra, $L_R(E)$, called the Leavitt path algebra of E. The role of the graph may seem unclear at the outset, because all it does is serve as a kind of notational device for the generators and relations that define $L_R(E)$. Surprisingly, it turns out that many of the ring-theoretic properties of $L_R(E)$ are controlled by graphical properties of E. For example, the Leavitt path algebra has some special properties if the graph is acyclic, cofinal, downward-directed, has no cycles without exits, etc.

Since 2005, there has been an abundance of research on Leavitt path algebras. One of the main goals has been to characterise their internal properties, ideals, substructures, and modules. As a result, we have a rich supply of algebras with "interesting and extreme properties" [8]. This is useful for generating counterexamples to reasonable-sounding conjectures, e.g. [6, 46], or for supporting other long-standing conjectures by showing they hold within this varied class, e.g. [9, 15].

Another goal has been finding invariants that determine Leavitt path algebras up to isomorphism, or Morita equivalence. This enterprise is known as the classification question for Leavitt path algebras. Of course, something that is easier than classifying *all* Leavitt path algebras is classifying those that have a certain property (like purely infinite simplicity), or classifying the Leavitt path algebras of small graphs. This has led to interesting developments in K-theory (see [7, Sect. 6.3] and [44]) and has motivated the study of substructures of Leavitt path algebras, like the socle [14] and invariant ideals [47].

A third goal is to explain why graph C^*-algebras and Leavitt path algebras have so much in common. One expects *a priori* that these two different structures would have little to do with one another. But in fact, many theorems about Leavitt path algebras resemble theorems about graph C^*-algebras [1, Appendix 1]. For instance, the graphs whose C^*-algebras are C^*-simple are exactly the same graphs whose Leavitt path algebras are simple (over any base field). One conjecture in this general direction is the Isomorphism Conjecture for Graph Algebras [4]: if E and F are two graphs such that $L_{\mathbb{C}}(E) \cong L_{\mathbb{C}}(F)$ as rings, then $C^*(E) \cong C^*(F)$ as $*$-algebras. In the unital case, an affirmative answer has been given in [37, Theorem 14.7].

2.1.3 Background: Steinberg Algebras

An ample groupoid is a special kind of locally compact topological groupoid. The Steinberg algebra of such a groupoid is an R-module of functions defined on it. It becomes an associative R-algebra once it is equipped with a generally noncommutative operation called the convolution (generalising the multiplicative operation on a group algebra). If the groupoid $\mathcal{G}$ is Hausdorff, one can characterise its Steinberg algebra quite succinctly as the convolution algebra of locally constant, compactly supported functions $f : \mathcal{G} \to R$. Steinberg algebras first appeared independently in [27, 63]. The primary motivation was to generalise other classes of algebras, especially inverse semigroup algebras and Leavitt path algebras.

Steinberg algebras do not only unify and generalise some seemingly disparate classes of algebras, but they also provide an entirely new approach to studying them. Many theorems about Leavitt path algebras and inverse semigroup algebras have since been recovered as specialisations of more general theorems about Steinberg algebras. For example, various papers [28, 64, 65, 67] have used groupoid techniques to characterise, in terms of the underlying graph or inverse semigroup, when a Leavitt path algebra or inverse semigroup algebra is (semi)prime, indecomposable, (semi)primitive, noetherian, or artinian.

Simplicity theorems play a very important role in graph algebras and some related classes of algebras. (In contrast, inverse semigroup algebras are never simple.) This theme goes right back to the beginning, when Leavitt proved in [52] that the Leavitt algebras $L_{n,\mathbb{K}}$ ($n \geq 2$) are all simple. Likewise, Cuntz proved in [33] that the Cuntz algebras $\mathcal{O}_n$ ($n \geq 2$) are C^*-simple in the sense that they have no closed two-sided ideals. When Leavitt path algebras were introduced, in the very first paper on the subject, Abrams and Aranda Pino [2] wrote the simplicity theorem for Leavitt path algebras of row-finite graphs. It was extended to Leavitt path algebras of arbitrary graphs, as soon as these were defined in [3].

Once Steinberg algebras appeared on the scene, Brown, Clark, Farthing, and Sims [20] proved a simplicity theorem for Steinberg algebras of Hausdorff ample groupoids over $\mathbb{C}$. That effort led them to unlock a remarkable piece of research in which they derived a simplicity theorem for the C^*-algebras of second-countable, locally compact, Hausdorff étale groupoids. It speaks to the significance of these new ideas, that they were put to use in solving a problem that was open for many decades. The effort has recently been repeated for non-Hausdorff groupoids, in [31], where it is said that "We view Steinberg algebras as a laboratory for finding conditions to characterisze C^*-simplicity for groupoid C^*-algebras."

Besides the ones we have already discussed, there are many interesting classes of algebras that appear as special cases of Steinberg algebras. These include partial skew group rings associated to topological partial dynamical systems [17], and Kumjian–Pask algebras associated to higher -rank graphs [29]. In quite a different application, Nekrashevych [54] has produced Steinberg algebras with prescribed growth properties, including the first examples of simple algebras of arbitrary Gelfand–Kirillov dimension.

2.1.4 Background: Graph Groupoids

There are actually a few ways to associate a groupoid to a graph E; see for example, [50, p. 511], [56, pp. 156–159], and [26, Example 5.4]. The one that we are interested in is called the boundary path groupoid, $\mathcal{G}_E$. Its unit space is the set of all paths that are either infinite or end at a sink or an infinite emitter (i.e., boundary paths). This groupoid was introduced in its earliest form, for row- and column-finite graphs without sinks, by Kumjian, Pask, Raeburn, and Renault [50]. It bears a resemblance to a groupoid studied a few years earlier by Deaconu [35]. The construction was later generalised in a number of different directions, taking a route through inverse semigroup theory [57], and going as far as topological higher-rank graphs [49, 61, 71].

The boundary path groupoid is an intermediate step towards proving that all Leavitt path algebras are Steinberg algebras, and it becomes an important tool for the analysis of Leavitt path algebras. For an arbitrary graph E, there is a $\mathbb{Z}$-graded isomorphism $A_R(\mathcal{G}_E) \cong L_R(E)$, where $A_R(\mathcal{G}_E)$ is the Steinberg algebra of $\mathcal{G}_E$ and $L_R(E)$ is the Leavitt path algebra of E. Consequently, if we understand some property of Steinberg algebras (for example, the centre [30]) then we can understand that property of Leavitt path algebras by translating groupoid terms into graphical terms and applying the isomorphism $A_R(\mathcal{G}_E) \cong L_R(E)$. Similarly, there is an isometric $*$-isomorphism $C^*(\mathcal{G}_E) \cong C^*(E)$, where $C^*(\mathcal{G}_E)$ is the full groupoid C^*-algebra of $\mathcal{G}_E$ and $C^*(E)$ is the graph C^*-algebra of E.

The diagonal subalgebra of a Steinberg algebra (resp., groupoid C^*-algebra) is the commutative subalgebra (resp., C^*-subalgebra) generated by functions supported on the unit space. If two ample groupoids $\mathcal{F}$ and $\mathcal{G}$ are topologically isomorphic, it is immediate that $A_R(\mathcal{F}) \cong A_R(\mathcal{G})$ and the isomorphism sends the diagonal to the diagonal. The converse is a very interesting and current research topic called "groupoid reconstruction". It was shown in [13] that if $\mathcal{F}$ and $\mathcal{G}$ are topologically principal, and R is an integral domain, then $A_R(\mathcal{F}) \cong A_R(\mathcal{G})$ with an isomorphism that preserves diagonals if and only if $\mathcal{F} \cong \mathcal{G}$. This was generalised in [23, 66]. For C^*-algebras, there are results of a similar flavour [60].

For boundary path groupoids, groupoid reconstruction is essentially the question: if E and F are graphs such that $L_R(E) \cong L_R(F)$, does it imply $\mathcal{G}_E \cong \mathcal{G}_F$? Many mathematicians [13, 19, 22, 66] have been working on this and they have given positive answers after imposing various assumptions on the graphs, the ring R, or the types of isomorphism between the Leavitt path algebras. It seems likely that more results will emerge. It is already known from [21, Theorem 5.1] that if there exists a diagonal-preserving isomorphism of graph C^*-algebras $C^*(E) \cong C^*(F)$, then $\mathcal{G}_E \cong \mathcal{G}_F$. It is plausible that the groupoid reconstruction programme for graph groupoids could eventually prove the general Isomorphism Conjecture for Graph Algebras [4].

2.2 The Steinberg Algebra of a Groupoid

Section 2.2 is structured as follows. It begins, in Sect. 2.2.1, by providing some background on groupoids. In Sect. 2.2.2, we develop some facts about topological groupoids and almost immediately specialise to étale and ample groupoids. We give a very brief treatment of inverse semigroups and their role in the subject. In Sect. 2.2.3, we introduce the Steinberg algebra of an ample groupoid, describing it in a few different ways to make the definition more transparent. We develop the basic theory in a self-contained way, paying attention to what can and cannot be said about non-Hausdorff groupoids. In Sect. 2.2.4, we investigate some important properties, showing that these algebras are locally unital and enjoy a kind of symmetry that comes from an involution (in other words, they are $*$-algebras). In Sect. 2.2.5, we investigate the effects of groupoid-combining operations like products, disjoint unions, and directed unions, and find applications with finite-dimensional Steinberg algebras and the Steinberg algebras of approximately finite groupoids. In Sect. 2.2.6, we discuss graded groupoids and graded Steinberg algebras.

2.2.1 Groupoids

This classical definition of a groupoid is modified from [59]. We have chosen to paint a complete picture; indeed, some parts of the definition can be derived from other parts.

Definition 2.1 A **groupoid** is a system $(\mathcal{G}, \mathcal{G}^{(0)}, \boldsymbol{d}, \boldsymbol{c}, \boldsymbol{m}, \boldsymbol{i})$ such that:

(G1) $\mathcal{G}$ and $\mathcal{G}^{(0)}$ are nonempty sets, called the *underlying set* and *unit space*, respectively;

(G2) $\boldsymbol{d}, \boldsymbol{c}$ are maps $\mathcal{G} \to \mathcal{G}^{(0)}$, called *domain* and *codomain*;

(G3) $\boldsymbol{m}$ is a partially defined binary operation on $\mathcal{G}$ called *composition*: specifically, it is a map from the set of *composable pairs*

$$\mathcal{G}^{(2)} = \left\{(g, h) \in \mathcal{G} \times \mathcal{G} \mid \boldsymbol{d}(g) = \boldsymbol{c}(h)\right\}$$

onto $\mathcal{G}$, written as $\boldsymbol{m}(g,h) = gh$, with the properties:

- $\boldsymbol{d}(gh) = \boldsymbol{d}(h)$ and $\boldsymbol{c}(gh) = \boldsymbol{c}(g)$ whenever the composition gh is defined;
- $(gh)k = g(hk)$ whenever either side is defined;

(G4) For every $x \in \mathcal{G}^{(0)}$ there is a unique identity $1_x \in \mathcal{G}$ such that $1_x g = g$ whenever $\boldsymbol{c}(g) = x$, and $h1_x = h$ whenever $\boldsymbol{d}(h) = x$;

(G5) $\boldsymbol{i} : \mathcal{G} \to \mathcal{G}$ is a map called *inversion*, written as $\boldsymbol{i}(g) = g^{-1}$, such that $g^{-1}g = 1_{\boldsymbol{c}(g)}$, $gg^{-1} = 1_{\boldsymbol{d}(g)}$, and $(g^{-1})^{-1} = g$.

The definition can be summarised by saying: a *groupoid* is a small category in which every morphism is invertible. Having said this, the elements of $\mathcal{G}$ will usually be called morphisms.

Remark 2.1 We always identify $x \in \mathcal{G}^{(0)}$ with $1_x \in \mathcal{G}$, so $\mathcal{G}^{(0)}$ is considered a subset of $\mathcal{G}$. The elements of $\mathcal{G}^{(0)}$ are called *units*.

Many authors write $\boldsymbol{s}$ (source) and $\boldsymbol{r}$ (range) instead of $\boldsymbol{d}$ and $\boldsymbol{c}$ in the definition of a groupoid. Our notation is chosen to avoid confusion in the context of graphs, where s and r refer to the source and range, respectively, of edges and directed paths.

A *homomorphism* between groupoids $\mathcal{G}$ and $\mathcal{H}$ is a functor $F : \mathcal{G} \to \mathcal{H}$; that is, a map sending units of $\mathcal{G}$ to units of $\mathcal{H}$ and mapping all the morphisms in $\mathcal{G}$ to morphisms in $\mathcal{H}$ in a way that respects the structure. A *subgroupoid* is a subset $\mathcal{S} \subseteq \mathcal{G}$ that is a groupoid with the structure that it inherits from $\mathcal{G}$. For $x \in \mathcal{G}^{(0)}$, we use the notation $x\mathcal{G} = \boldsymbol{c}^{-1}(x)$, $\mathcal{G}x = \boldsymbol{d}^{-1}(x)$, and $x\mathcal{G}y = \boldsymbol{c}^{-1}(x) \cap \boldsymbol{d}^{-1}(y)$. The set $x\mathcal{G}x$ is a group, called the *isotropy group* based at x, and the set $\mathrm{Iso}(\mathcal{G}) = \bigcup_{x \in \mathcal{G}^{(0)}} x\mathcal{G}x$ is a subgroupoid, called the *isotropy subgroupoid* of $\mathcal{G}$. If $\mathrm{Iso}(\mathcal{G}) = \mathcal{G}^{(0)}$ then $\mathcal{G}$ is called *principal*. We say that $\mathcal{G}$ is *transitive* if for every pair of units $x, y \in \mathcal{G}^{(0)}$ there is at least one morphism in $x\mathcal{G}y$.

The *conjugacy class* of $g \in \mathrm{Iso}(\mathcal{G})$ is the set $\mathrm{Cl}_{\mathcal{G}}(g) = \{hgh^{-1} \mid h \in \mathcal{G}\boldsymbol{c}(g)\}$. The set of conjugacy classes partitions $\mathrm{Iso}(\mathcal{G})$. The conjugacy class of a unit is called an *orbit*, and the set of orbits partitions $\mathcal{G}^{(0)}$. Equivalently, the orbit of $x \in \mathcal{G}^{(0)}$ is $\mathrm{Cl}_{\mathcal{G}}(x) = \boldsymbol{c}(\boldsymbol{d}^{-1}(x)) = \boldsymbol{d}(\boldsymbol{c}^{-1}(x))$, or the unit space of the maximal transitive subgroupoid containing x. A subset $U \subseteq \mathcal{G}^{(0)}$ is *invariant* if for all $g \in \mathcal{G}$, $\boldsymbol{d}(g) \in U$ implies $\boldsymbol{c}(g) \in U$, which is to say that U is a union of orbits. If $x, y \in \mathcal{G}^{(0)}$ belong to the same orbit, then the isotropy groups $x\mathcal{G}x$ and $y\mathcal{G}y$ are isomorphic. In fact, there can be many isomorphisms $x\mathcal{G}x \to y\mathcal{G}y$. For every $g \in y\mathcal{G}x$ there is an "inner" isomorphism $x\mathcal{G}x \to y\mathcal{G}y$ given by $x \mapsto gxg^{-1}$. This allows us to speak of the isotropy group of an orbit.

Example 2.1 Many familiar mathematical objects are essentially groupoids:

(a) Any **group** G with identity ε can be viewed as a groupoid with unit space $\{\varepsilon\}$. Conjugacy classes are conjugacy classes in the usual sense.

(b) If $\{G_i \mid i \in I\}$ is a family of groups with identities $\{\varepsilon_i \mid i \in I\}$, then the disjoint union $\bigsqcup_{i \in I} G_i$ has a groupoid structure with $\boldsymbol{d}(g) = \boldsymbol{c}(g) = \varepsilon_i$ for every $g \in G_i$.

The composition, defined only for pairs $(g, h) \in \bigsqcup_{i \in I} G_i \times G_i$, is just the relevant group law. This is known as a **bundle of groups**. The isotropy subgroupoid of any groupoid is a bundle of groups.

(c) Let X be a set with an equivalence relation $\sim$. We define the **groupoid of pairs** $\mathcal{G}_X = \{(x, y) \in X \times X \mid x \sim y\}$ with unit space X, and view (x, y) as a morphism with domain y, codomain x, and inverse $(x, y)^{-1} = (y, x)$. A pair of morphisms (x, y), (w, z) is composable if and only if $y = w$, and composition is defined as $(x, y)(y, z) = (x, z)$. Every principal groupoid is isomorphic to a groupoid of pairs. If $\sim$ is the indiscrete equivalence relation (where $x \sim y$ for all $x, y \in X$) then $\mathcal{G}_X$ is called the *transitive principal groupoid on* X.

(d) Let G be a group with a left action on a set X. There is a groupoid structure on $G \times X$, where the unit space is $\{\varepsilon\} \times X$, or simply just X. We understand that the morphism (g, x) has domain $g^{-1}x$ and codomain x. Composition is defined as $(g, x)(h, g^{-1}x) = (gh, x)$, and inversion as $(g, x)^{-1} = (g^{-1}, g^{-1}x)$. The isotropy group at x is isomorphic to the stabiliser subgroup associated to x. Orbits are orbits in the usual sense, and the groupoid is transitive if and only if the action is transitive. This is called the **transformation groupoid** associated to the action of G on X.

(e) The **fundamental groupoid** of a topological space X is the set of homotopy path classes on X. The unit space of this groupoid is X itself, and the isotropy group at $x \in X$ is the fundamental group $\pi_1(X, x)$. The groupoid is transitive if and only if X is path-connected, and it is principal if and only if every path component is simply connected.

2.2.2 *Topological Groupoids*

Briefly, here are some of our topological conventions. We use the word *base* to mean a collection of open sets, called *basic open sets*, that generates a topology by taking unions. A neighbourhood base is a filter for the set of neighbourhoods of a point. In this paper, the word *basis* is reserved for linear algebra. A *compact* topological space is one in which every open cover has a finite subcover, and a *locally compact* topological space is one in which every point has a neighbourhood base of compact sets. If X and Y are topological spaces, a *local homeomorphism* is a map $f : X \to Y$ with the property: every point in X has an open neighbourhood U such that $f|_U$ is a homeomorphism onto an open subset of Y. Every local homeomorphism is open and continuous.

The definition of a topological groupoid is straightforward, but there is some inconsistency in the literature on what it means for a groupoid to be étale or locally compact. While some papers require germane conditions, our definitions are chosen to be classical and minimally restrictive. We are mainly concerned with étale and ample groupoids. Roughly speaking, étale groupoids are topological groupoids whose topology is locally determined by the unit space.

Definition 2.2 A groupoid $\mathcal{G}$ is

(a) a **topological groupoid** if its underlying set has a topology, and the maps $\boldsymbol{m}$ and $\boldsymbol{i}$ are continuous, with the understanding that $\mathcal{G}^{(2)}$ inherits its topology from $\mathcal{G} \times \mathcal{G}$;
(b) an **étale groupoid** if it is a topological groupoid and $\boldsymbol{d}$ is a local homeomorphism.

Some pleasant consequences follow from these two definitions. In any topological groupoid, $\boldsymbol{i}$ is a homeomorphism because it is a continuous involution, and $\boldsymbol{d}$ and $\boldsymbol{c}$ are both continuous because $\boldsymbol{d}(g) = \boldsymbol{m}(\boldsymbol{i}(g), g)$ and $\boldsymbol{c} = \boldsymbol{di}$. If $\mathcal{G}$ is étale, then $\boldsymbol{d}$, $\boldsymbol{c}$, and $\boldsymbol{m}$ are local homeomorphisms, and $\mathcal{G}^{(0)}$ is open in $\mathcal{G}$ (the openness of $\mathcal{G}^{(0)}$ is proved from first principles in [38, Proposition 3.2]). If $\mathcal{G}$ is a Hausdorff topological groupoid, then $\mathcal{G}^{(0)}$ is closed. Indeed (and this neat proof is from [62]) if $(x_i)_{i \in I}$ is a net in $\mathcal{G}^{(0)}$ with $x_i \to g \in \mathcal{G}$, then $x_i = \boldsymbol{c}(x_i) \to \boldsymbol{c}(g)$ because $\boldsymbol{c}$ is continuous, so $g = \boldsymbol{c}(g) \in \mathcal{G}^{(0)}$ by uniqueness of limits. If $\mathcal{G}$ is any topological groupoid, the maps $\boldsymbol{d} \times \boldsymbol{c} : \mathcal{G} \times \mathcal{G} \to \mathcal{G}^{(0)} \times \mathcal{G}^{(0)}$ and $(\boldsymbol{d}, \boldsymbol{c}) : \mathcal{G} \to \mathcal{G}^{(0)} \times \mathcal{G}^{(0)}$ are both continuous. If $\mathcal{G}^{(0)}$ is Hausdorff, the diagonal $\Delta = \{(x, x) \mid x \in \mathcal{G}^{(0)}\}$ is closed in $\mathcal{G}^{(0)} \times \mathcal{G}^{(0)}$; consequently, $\mathcal{G}^{(2)} = (\boldsymbol{d} \times \boldsymbol{c})^{-1}(\Delta)$ is closed in $\mathcal{G} \times \mathcal{G}$ and $\mathrm{Iso}(\mathcal{G}) = (\boldsymbol{d}, \boldsymbol{c})^{-1}(\Delta)$ is closed in $\mathcal{G}$.

Let $\mathcal{G}$ be a topological groupoid. If $U \subseteq \mathcal{G}$ is an open set such that $\boldsymbol{c}|_U$ and $\boldsymbol{d}|_U$ are homeomorphisms onto open subsets of $\mathcal{G}^{(0)}$, then U is called an *open bisection*. If $\mathcal{G}$ is étale and $U \subseteq \mathcal{G}$ is open, the restrictions $\boldsymbol{c}|_U$ and $\boldsymbol{d}|_U$ are continuous open maps, so they need only be injective for U to be an open bisection. An equivalent definition of an *étale groupoid* is a topological groupoid that has a base of open bisections. If $\mathcal{G}$ is étale and $\mathcal{G}^{(0)}$ is Hausdorff, then $\mathcal{G}$ is locally Hausdorff, because all the open bisections are homeomorphic to subspaces of $\mathcal{G}^{(0)}$. Another property of étale groupoids is that for any $x \in \mathcal{G}^{(0)}$, the fibres $x\mathcal{G}$ and $\mathcal{G}x$ are discrete spaces. Consequently, a groupoid with only one unit (i.e., a group) is étale if and only if it has the discrete topology.

Definition 2.3 An **ample groupoid** is a topological groupoid with Hausdorff unit space and a base of compact open bisections.

If $\mathcal{G}$ is an ample groupoid, the notation $B^{\mathrm{co}}(\mathcal{G})$ stands for the set of all nonempty compact open bisections in $\mathcal{G}$, and $\mathcal{B}(\mathcal{G}^{(0)})$ stands for the set of nonempty compact open subsets of $\mathcal{G}^{(0)}$.

Recall that a topological space is said to be *totally disconnected* if the only nonempty connected subsets are singletons, and 0-*dimensional* if every point has a neighbourhood base of clopen (i.e., closed and open) sets. These two notions are equivalent if the space is locally compact and Hausdorff [70, Theorems 29.5 and 29.7]. The following proposition is similar to [39, Proposition 4.1]. It is useful for reconciling slightly different definitions in the literature (e.g., [27]) and for checking when an étale groupoid is ample.

Proposition 2.1 *Let $\mathcal{G}$ be an étale groupoid such that $\mathcal{G}^{(0)}$ is Hausdorff. Then the following are equivalent:*

(1) *$\mathcal{G}$ is an ample groupoid;*
(2) *$\mathcal{G}^{(0)}$ is locally compact and totally disconnected;*
(3) *Every open bisection is locally compact and totally disconnected.*

Proof (1) $\Rightarrow$ (2) Let $U \subseteq \mathcal{G}^{(0)}$ be open. Since $\mathcal{G}$ is ample and $\mathcal{G}^{(0)}$ is open, for every $x \in U$ there is a compact open bisection B such that $x \in B \subseteq U \subseteq \mathcal{G}^{(0)}$. Moreover, $\mathcal{G}^{(0)}$ is Hausdorff, so B is closed. This shows that $\mathcal{G}^{(0)}$ is locally compact and 0-dimensional (hence totally disconnected).

(2) $\Rightarrow$ (3) Every open bisection is homeomorphic to an open subspace of $\mathcal{G}^{(0)}$, so it is totally disconnected and locally compact.

(3) $\Rightarrow$ (1) Let U be open in $\mathcal{G}$, and $x \in U$. Since $\mathcal{G}$ is étale, it has a base of open bisections, so there is an open bisection B with $x \in B \subseteq U$. Moreover, B is Hausdorff, locally compact, and totally disconnected, so x has a compact neighbourhood $W \subseteq B$ and a clopen neighbourhood $V \subseteq W$. Since B is Hausdorff and V is closed in W, it follows that V is compact. Moreover, V is an open bisection because B is an open bisection. So, V is a compact open bisection. This shows that $\mathcal{G}$ has a base of compact open bisections, so $\mathcal{G}$ is ample. □

Remark 2.2 If $\mathcal{G}$ is a topological groupoid and $\mathcal{E}$ is a subgroupoid of $\mathcal{G}$, then $\mathcal{E}$ is automatically a topological groupoid with the topology it inherits from $\mathcal{G}$. If $\mathcal{G}$ is étale, then so is $\mathcal{E}$. However, if $\mathcal{G}$ is ample, then it is not guaranteed that $\mathcal{E}$ is ample. Indeed, by Proposition 2.1 (2), a subgroupoid $\mathcal{E}$ of an ample groupoid $\mathcal{G}$ is ample if and only if $\mathcal{E}^{(0)}$ is locally compact. In particular, $\mathcal{E}$ is ample if $\mathcal{G}$ is ample and $\mathcal{E}^{(0)}$ is either open or closed in $\mathcal{G}^{(0)}$.

The following lemma is similar to [56, Proposition 2.2.4], but with slightly different assumptions.

Lemma 2.1 *Let $\mathcal{G}$ be an étale groupoid where $\mathcal{G}^{(0)}$ is Hausdorff. If $A, B, C \subseteq \mathcal{G}$ are compact open bisections, then*

(1) *$A^{-1} = \{a^{-1} \mid a \in A\}$ and $AB = \{ab \mid (a, b) \in (A \times B) \cap \mathcal{G}^{(2)}\}$ are compact open bisections.*
(2) *If $\mathcal{G}$ is Hausdorff, then $A \cap B$ is a compact open bisection.*

Proof (1) Firstly, $A^{-1} = \boldsymbol{i}(A)$ is compact and open because $\boldsymbol{i}$ is a homeomorphism. Clearly, A^{-1} is an open bisection. Secondly, note that AB might be empty, in which case it is trivially a compact open bisection. Otherwise, $(A \times B) \cap \mathcal{G}^{(2)}$ is compact because $\mathcal{G}^{(2)}$ is closed in $\mathcal{G} \times \mathcal{G}$, and $AB = \boldsymbol{m}\big((A \times B) \cap \mathcal{G}^{(2)}\big)$ is compact because $\boldsymbol{m}$ is continuous. Since $\boldsymbol{m}$ is a local homeomorphism, it is an open map, and $AB = \boldsymbol{m}\big((A \times B) \cap \mathcal{G}^{(2)}\big)$ is open. To prove that it is a bisection, suppose (a, b) is a composable pair in $A \times B$ and $\boldsymbol{d}(ab) = x$. Since A and B are bisections, b is the unique element in B having $\boldsymbol{d}(b) = x$, and a is the unique element of A having $\boldsymbol{d}(a) = \boldsymbol{c}(b)$. So, $\boldsymbol{d}|_{AB}$ is injective. Similarly, $\boldsymbol{c}|_{AB}$ is injective.

(2) It is trivial that $A \cap B$ is an open bisection. The Hausdorff property on $\mathcal{G}$ implies A and B are closed, so $A \cap B$ is closed, hence compact. □

Lemma 2.1 remains true if the words "compact" or "open", or both, are removed throughout the statement. Using Lemma 2.1 (2) with mathematical induction shows that when an ample groupoid is Hausdorff, its set of compact open bisections is closed under finite intersections. The converse to this statement is also true: an ample groupoid is Hausdorff if the set of compact open bisections is closed under finite intersections (see [63, Proposition 3.7]).

The main takeaway from Lemma 2.1 (1) is that the compact open bisections in an ample groupoid are important for two reasons: they generate the topology, and they can be multiplied and inverted in a way that is consistent with an algebraic structure called an inverse semigroup. An *inverse semigroup* is a semigroup S such that every $s \in S$ has a unique *inverse* $s^* \in S$ with the property $ss^*s = s$ and $s^*ss^* = s^*$.

Example 2.2 If X is a set, a *partial symmetry* of X is a bijection $s : \mathrm{dom}(s) \to \mathrm{cod}(s)$ where $\mathrm{dom}(s)$ and $\mathrm{cod}(s)$ are (possibly empty) subsets of X. Two partial symmetries s and t are composed in the way that binary relations are composed, so that $st : \mathrm{dom}(st) \to \mathrm{cod}(st)$ is the map $st(x) = s(t(x))$ for all $x \in X$ such that $s(t(x))$ makes sense. It is *not* necessary to have $\mathrm{dom}(s) = \mathrm{cod}(t)$ in order to compose s and t. The semigroup $\mathcal{I}_X$ of partial symmetries on X is called the **symmetric inverse semigroup** on X. The Wagner–Preston Theorem is an analogue of Cayley's Theorem for groups: every inverse semigroup S has an embedding into $\mathcal{I}_S$.

The following result is an adaptation of [56, Proposition 2.2.3].

Proposition 2.2 *If $\mathcal{G}$ is an ample groupoid, $B^{\mathrm{co}}(\mathcal{G})$ is an inverse semigroup with the inversion and composition rules displayed in Lemma 2.1* (1).

Proof Lemma 2.1 (1) proves that $B^{\mathrm{co}}(\mathcal{G})$ is a semigroup and that $A \in B^{\mathrm{co}}(\mathcal{G})$ implies $A^{-1} \in B^{\mathrm{co}}(\mathcal{G})$. If $A \in B^{\mathrm{co}}(\mathcal{G})$ then $AA^{-1} = \boldsymbol{c}(A)$ because all composable pairs in $A \times A^{-1}$ are of the form (a, a^{-1}) for some $a \in A$. Therefore $AA^{-1}A = \boldsymbol{c}(A)A = A$ and $A^{-1}AA^{-1} = A^{-1}\boldsymbol{c}(A) = A^{-1}\boldsymbol{d}(A^{-1}) = A^{-1}$. To show that the inverses are unique, suppose $B \in B^{\mathrm{co}}(\mathcal{G})$ satisfies $ABA = A$ and $BAB = B$. Then for all $a \in A$ there exists $b \in B$ such that $aba = a$. But then $b = a^{-1}aa^{-1} = a^{-1}$. This shows $A^{-1} \subseteq B$. Similarly, $BAB = B$ implies $B^{-1} \subseteq A$ and consequently $B \subseteq A^{-1}$. Therefore $B = A^{-1}$. □

The proposition above has shown how to associate an inverse semigroup to an ample groupoid. The connections between ample groupoids and inverse semigroups run much deeper than this. There are at least two ways to associate an ample groupoid $\mathcal{G}$ to an inverse semigroup S. The first is the *underlying groupoid* $\mathcal{G}_S$, where the underlying set is S, the topology is discrete, the unit space is the set of idempotents in S, and $\boldsymbol{d}(s) = s^*s$ while $\boldsymbol{c}(s) = ss^*$, for every $s \in S$. Composition in $\mathcal{G}_S$ is the binary operation from S, just restricted to composable pairs. The second way to associate an ample groupoid to an inverse semigroup S is more complicated. It is called the *universal groupoid* of S, and it only differs from the underlying groupoid when S is large (i.e., fails to have some finiteness conditions). The universal groupoid has a topology that makes it ample but not necessarily Hausdorff. The universal groupoid of S is quite powerful (as shown in [63]) because its Steinberg algebra $A_R(\mathcal{G}(S))$ is isomorphic to the inverse semigroup algebra RS. This takes us beyond our scope and, after all, we still need to define Steinberg algebras.

2.2.3 *Introducing Steinberg Algebras*

The purpose of this section is to define and characterise the Steinberg algebra of an ample groupoid over a unital commutative ring R. Throughout this section, assume $\mathcal{G}$ is an ample groupoid. In order to make sense of continuity for R-valued functions, assume R has the discrete topology. The *support* of a function $f : X \to R$ is defined as the set $\operatorname{supp} f = \{x \in X \mid f(x) \neq 0\}$. When X has a topology, we say that f is *compactly supported* if $\operatorname{supp} f$ is compact. If every point $x \in X$ has an open neighbourhood N such that $f|_N$ is constant, then f is called *locally constant*. It is easy to prove that $f : X \to R$ is locally constant if and only if it is continuous. We use the following notation for the *characteristic function* of a subset U of $\mathcal{G}$:

$$\mathbf{1}_U : \mathcal{G} \to R, \qquad \mathbf{1}_U(g) = \begin{cases} 1 & \text{if } g \in U \\ 0 & \text{if } g \notin U. \end{cases}$$

Let $R^{\mathcal{G}}$ be the set of all functions $f : \mathcal{G} \to R$. Canonically, $R^{\mathcal{G}}$ has the structure of an R-module with operations defined pointwise.

Definition 2.4 *(The Steinberg algebra)* Let $A_R(\mathcal{G})$ be the R-submodule of $R^{\mathcal{G}}$ generated by the set:

$$\{\mathbf{1}_U \mid U \text{ is a Hausdorff compact open subset of } \mathcal{G}\}.$$

The *convolution* of $f, g \in A_R(\mathcal{G})$ is defined as

$$f * g(x) = \sum_{\substack{y \in \mathcal{G} \\ \boldsymbol{d}(y) = \boldsymbol{d}(x)}} f(xy^{-1})g(y) = \sum_{\substack{(z,y) \in \mathcal{G}^{(2)} \\ zy = x}} f(z)g(y) \quad \text{for all } x \in \mathcal{G}. \tag{2.1}$$

The R-module $A_R(\mathcal{G})$, with the convolution, is called the **Steinberg algebra** of $\mathcal{G}$ over R.

Example 2.3 If Γ is a discrete group, then $A_R(\Gamma)$ is isomorphic to $R\Gamma$, the usual **group algebra** of Γ with coefficients in R.

We have yet to justify the definition of the convolution in Eq. 2.1. The two sums in the formula are equal, by substituting $z = xy^{-1}$. But it should not be taken for granted that the sum is finite, that $*$ is associative, or even that $A_R(\mathcal{G})$ is closed under $*$. These facts will be proved later. First, we prove the following result (inspired by [63]) that leads to some alternative descriptions of $A_R(\mathcal{G})$ as an R-module.

Proposition 2.3 *Let $\mathcal{B}$ be a base for $\mathcal{G}$ consisting of Hausdorff compact open sets, with the property:*

$$\left\{\bigcap_{i=1}^{n} B_i \mid B_i \in \mathcal{B}, \bigcup_{i=1}^{n} B_i \text{ is Hausdorff}\right\} \subseteq \mathcal{B} \cup \{\emptyset\}.$$

Then $A_R(\mathcal{G}) = \operatorname{span}_R\{\mathbf{1}_B \mid B \in \mathcal{B}\}$.

Proof Let $A = \operatorname{span}_R\{\mathbf{1}_B \mid B \in \mathcal{B}\}$. From the definition of $A_R(\mathcal{G})$, we have $A \subseteq A_R(\mathcal{G})$. To prove the other containment, suppose U is a Hausdorff compact open subset of $\mathcal{G}$. It is sufficient to prove that $\mathbf{1}_U$ is an R-linear combination of finitely many $\mathbf{1}_{B_i}$, where each $B_i \in \mathcal{B}$. Since $\mathcal{B}$ is a base for the topology on $\mathcal{G}$, we can write U as a union of sets in $\mathcal{B}$, and use the compactness of U to reduce it to a finite union $U = B_1 \cup \cdots \cup B_n$, where $B_1, \ldots, B_n \in \mathcal{B}$. By the principle of inclusion–exclusion:

$$\mathbf{1}_U = \sum_{k=1}^{n} (-1)^{k-1} \sum_{\substack{I \subseteq \{1,\ldots,n\} \\ |I|=k}} \mathbf{1}_{\bigcap_{i \in I} B_i}.$$

The main assumption ensures that the sets $\bigcap_{i \in I} B_i$ on the right hand side are either empty or in $\mathcal{B}$. Therefore $A_R(\mathcal{G}) \subseteq A$. □

Corollary 2.1 *If $\mathcal{G}$ is Hausdorff and $\mathcal{B}$ is a base of compact open sets that is closed under finite intersections, then* $A_R(\mathcal{G}) = \operatorname{span}_R\{\mathbf{1}_B \mid B \in \mathcal{B}\}$.

We remarked after Lemma 2.1 that if $\mathcal{G}$ is non-Hausdorff, $B^{\mathrm{co}}(\mathcal{G})$ is *not* closed under finite intersections. Strange things can happen in non-Hausdorff spaces and the problem lies in the fact that compact sets are not always closed, and the intersection of two compact sets is not always compact. However, $B^{\mathrm{co}}(\mathcal{G})$ does satisfy the hypothesis of Proposition 2.3.

Corollary 2.2 ([63, Proposition 4.3]) *The Steinberg algebra is generated as an R-module by characteristic functions of compact open bisections. That is,*

$$A_R(\mathcal{G}) = \operatorname{span}_R\{\mathbf{1}_B \mid B \in B^{\mathrm{co}}(\mathcal{G})\}.$$

Proof If $B_1, \ldots, B_n \in B^{\mathrm{co}}(\mathcal{G})$, and $U = \bigcup_i B_i$ is Hausdorff, then each B_i is closed in U because U is compact, so $\bigcap_i B_i$ is closed in U. And, B_1 is a compact set containing the closed set $\bigcap_i B_i$, so $\bigcap_i B_i$ is compact. Clearly $\bigcap_i B_i$ is an open bisection, so $\bigcap_i B_i \in B^{\mathrm{co}}(\mathcal{G})$. □

Remark 2.3 If $\mathcal{G}$ is an ample groupoid and $\mathcal{E}$ is an open subgroupoid, then $\mathcal{E}$ is also ample (see Remark 2.2). Let $\iota : \mathcal{E} \hookrightarrow \mathcal{G}$ be the inclusion homomorphism. There is a canonical monomorphism $m : A_R(\mathcal{E}) \hookrightarrow A_R(\mathcal{G})$, linearly extended from $\mathbf{1}_U \mapsto \mathbf{1}_{\iota(U)}$ for every Hausdorff compact open set $U \subseteq \mathcal{E}$. If $\mathcal{E}$ is closed, m has a left inverse $e : A_R(\mathcal{G}) \twoheadrightarrow A_R(\mathcal{E})$, linearly extended from $\mathbf{1}_U \mapsto \mathbf{1}_{U \cap \mathcal{E}}$ for every Hausdorff compact open set $U \subseteq \mathcal{G}$.

We still owe a proof that the convolution, from Eq. 2.1, is well-defined and gives an R-algebra structure to $A_R(\mathcal{G})$. The next two results are similar to [63, Propositions 4.5 and 4.6].

Lemma 2.2 *Let $A, B, C \in B^{\mathrm{co}}(\mathcal{G})$ and $r, s \in R$. Then:*

(1) $\mathbf{1}_{A^{-1}}(x) = \mathbf{1}_A(x^{-1})$ *for all* $x \in \mathcal{G}$;
(2) $\mathbf{1}_A * \mathbf{1}_B = \mathbf{1}_{AB}$;

Proof (1) We have $x \in A^{-1}$ if and only if $x^{-1} \in A$.
(2) Let $x \in \mathcal{G}$. By definition:

$$\mathbf{1}_A * \mathbf{1}_B(x) = \sum_{\substack{y \in \mathcal{G} \\ \boldsymbol{d}(y) = \boldsymbol{d}(x)}} \mathbf{1}_A(xy^{-1})\mathbf{1}_B(y) = \sum_{\substack{y \in B \\ \boldsymbol{d}(y) = \boldsymbol{d}(x)}} \mathbf{1}_A(xy^{-1}). \tag{2.2}$$

Assume x is of the form $x = ab$ where $a \in A$ and $b \in B$. Since B is a bisection, b is the only element of B having $\boldsymbol{d}(b) = \boldsymbol{d}(x)$, and it follows that

$$\mathbf{1}_A * \mathbf{1}_B(x) = \mathbf{1}_A(xb^{-1}) = \mathbf{1}_A(a) = 1.$$

On the other hand, assume $x \notin AB$. If there is $y \in B$ such that $\boldsymbol{d}(y) = \boldsymbol{d}(x)$, then $xy^{-1} \notin A$, for if it were, then $xy^{-1}y = x$ would be in AB. Therefore Eq. 2.2 yields $\mathbf{1}_A * \mathbf{1}_B(x) = 0$. □

Lemma 2.2 (2) implies that characteristic functions of compact open subsets of the unit space can be multiplied pointwise. That is, if $V, W \in \mathcal{B}(\mathcal{G}^{(0)})$ then $VW = V \cap W = WV$ and $\mathbf{1}_V * \mathbf{1}_W(x) = \mathbf{1}_V(x)\mathbf{1}_W(x)$ for all $x \in \mathcal{G}$. As $\mathcal{G}^{(0)}$ is open in any ample groupoid $\mathcal{G}$, by Remark 2.3, there is a commutative subalgebra $A_R(\mathcal{G}^{(0)}) \hookrightarrow A_R(\mathcal{G})$.

The ingredients of an *R-algebra* are an R-module A and a binary operation $A \times A \to A$. The binary operation should be R-linear in the first and second arguments (that is, bilinear), and it should be associative. There does not need to be a multiplicative identity. It is tedious to prove that $*$ is associative from its definition in Eq. 2.1, so a proof was omitted in [63].

Proposition 2.4 *The R-module $A_R(\mathcal{G})$, equipped with the convolution, is an R-algebra.*

Proof We need to show that the image of $* : A_R(\mathcal{G}) \times A_R(\mathcal{G}) \to R^{\mathcal{G}}$ is contained in $A_R(\mathcal{G})$, and that $*$ is associative and bilinear. Bilinearity can be proved quite easily from Eq. 2.1. Recall from Corollary 2.2 that the elements of $A_R(\mathcal{G})$ are R-linear combinations of characteristic functions of compact open bisections. If $f = \sum_i a_i \mathbf{1}_{A_i}$, $g = \sum_j b_j \mathbf{1}_{B_j}$, and $h = \sum_k c_k \mathbf{1}_{C_k}$, where the sums are finite, and $A_i, B_j, C_k \in B^{\mathrm{co}}(\mathcal{G})$ while $a_i, b_j, c_k \in R$ for all i, j, k, then

$$(f * g) * h = \sum_i \sum_j \sum_k a_i b_j c_k \mathbf{1}_{(A_i B_j) C_k} = \sum_i \sum_j \sum_k a_i b_j c_k \mathbf{1}_{A_i (B_j C_k)} = f * (g * h),$$

using Lemma 2.2 (2) and the bilinearity of $*$. This proves $*$ is associative. Evidently, $f * g = \sum_{i,j} a_i b_j \mathbf{1}_{A_i B_j} \in A_R(\mathcal{G})$, so $A_R(\mathcal{G})$ is closed under $*$. □

It is often useful to think of $*$ simply as the extension of the rule $\mathbf{1}_A * \mathbf{1}_B = \mathbf{1}_{AB}$ for all pairs $A, B \in B^{co}(\mathcal{G})$, rather than the more complicated-looking Eq. 2.1 that we first defined it with. Moreover, one can infer from it that $A_R(\mathcal{G})$ is a homomorphic image of the semigroup algebra of $B^{co}(\mathcal{G})$ with coefficients in R.

Proposition 2.5 *If $\mathcal{G}$ is Hausdorff and ample, then*

$$A_R(\mathcal{G}) = \{f : \mathcal{G} \to R \mid f \text{ is locally constant, compactly supported}\}. \tag{2.3}$$

Moreover, if $\mathcal{B}$ is a base for $\mathcal{G}$ consisting of compact open sets, such that $\mathcal{B}$ is closed under finite intersections and relative complements, then every nonzero $f \in A_R(\mathcal{G})$ is of the form $f = \sum_{i=1}^{m} r_i \mathbf{1}_{B_i}$, where $r_1, \ldots, r_n \in R \setminus \{0\}$ and $B_1, \ldots, B_n \in \mathcal{B}$ are mutually disjoint.

Proof Let A be the set of locally constant, compactly supported R-valued functions on $\mathcal{G}$. Let $\mathcal{B}$ be a base of compact open sets for $\mathcal{G}$, such that $\mathcal{B}$ is closed under finite intersections and relative complements. (A worthy candidate for $\mathcal{B}$ is $B^{co}(\mathcal{G})$.) If $0 \neq f \in A_R(\mathcal{G})$ then according to Corollary 2.1, $f = \sum_{i=1}^{n} s_i \mathbf{1}_{D_i}$ for some basic open sets $D_i \in \mathcal{B}$ and non-zero scalars $s_i \in R$. We aim to rewrite it as a linear combination of characteristic functions of *disjoint* open sets. If $s \in \operatorname{im} f \setminus \{0\}$, then we have the expression:

$$f^{-1}(s) = \bigcup_{\substack{I \subseteq \{1,\ldots,n\} \\ s = \sum_{i \in I} s_i}} B_I, \qquad \text{where} \qquad B_I = \bigcap_{\substack{i \in I \\ j \notin I}} D_i \setminus D_j. \tag{2.4}$$

By assumption, each nonempty B_I in the expression is an element of $\mathcal{B}$; in particular, each B_I is compact and open. Finite unions preserve openness and compactness, so $f^{-1}(s)$ is open and compact for every nonzero $s \in \operatorname{im} f$. It follows that $f^{-1}(0) = \mathcal{G} \setminus \left(\bigcup_{s \in \operatorname{im} f \setminus \{0\}} f^{-1}(s)\right)$ is open. Therefore f is locally constant. As f is a linear combination of n characteristic functions, it is clear that $|\operatorname{im} f \setminus \{0\}| \leq 2^n$. Being a finite union of compact sets, $\operatorname{supp} f = \bigcup_{s \in \operatorname{im} f \setminus \{0\}} f^{-1}(s)$ is compact. Thus $f \in A$, and this shows $A_R(\mathcal{G}) \subseteq A$. To prove the other containment, that $A \subseteq A_R(\mathcal{G})$, suppose $f \in A$. As f is continuous and $\operatorname{supp} f$ is compact, $f(\operatorname{supp} f) = \operatorname{im} f \setminus \{0\}$ is compact in R, so it must be finite. Let $\operatorname{im} f \setminus \{0\} = \{r_1, \ldots, r_n\}$. Then each set $U_i = f^{-1}(r_i)$ is clopen because f is continuous, and compact because $U_i \subseteq \operatorname{supp} f$. Hence $f = \sum_{i=1}^{n} r_i \mathbf{1}_{U_i} \in A_R(\mathcal{G})$, and this shows $A \subseteq A_R(\mathcal{G})$.

To prove the "moreover" part, we look again at Eq. 2.4. If $I, J \subseteq \{1, \ldots, n\}$ and $I \neq J$ then $B_I \cap B_J = \emptyset$. Therefore, $f \in A_R(\mathcal{G})$ can be written as an R-linear combination of characteristic functions of disjoint basic open sets in $\mathcal{B}$:

$$f = \sum_{s \in \operatorname{im} f \setminus \{0\}} s \mathbf{1}_{f^{-1}(s)} = \sum_{s \in \operatorname{im} f \setminus \{0\}} \sum_{\substack{I \subseteq \{1,\ldots,n\} \\ s = \sum_{i \in I} s_i}} s \mathbf{1}_{B_I}.$$ □

2.2.4 Properties of Steinberg Algebras

It is useful to know when $A_R(\mathcal{G})$ is unital or has some property that is nearly as good. The answer is quite easy, and we show it below. We use the definition that a ring (or R-algebra) A is *locally unital* if there is a set of commuting idempotents $E \subseteq A$, called *local units*, with the property: for every finite subset $\{a_1, \ldots, a_n\} \subseteq A$, there is a local unit $e \in E$ with $ea_i = a_i = a_i e$ for every $1 \le i \le n$. Equivalently, A is the direct limit of unital subrings: $A = \varinjlim_{e \in E} eAe$. The directed system is facilitated by the partial order, $e \le e'$ if $ee' = e = e'e$, and the connecting homomorphisms (which need not be unit-preserving) are the inclusions $eAe \hookrightarrow e'Ae'$ for $e \le e'$.

In many respects, working with locally unital rings is like working with unital rings. Every locally unital ring A is idempotent (i.e., $A^2 = A$) and if $I \subseteq A$ is an ideal, then $AI = I = IA$. If A is an R-algebra with local units, then the ring ideals of A are always R-algebra ideals (which, by definition, should be R-submodules of A). These facts are not true in general for arbitrary non-unital rings. Locally unital rings and algebras are always *homologically unital*, in the sense of [53, Definition 1.4.6], which essentially means that they have well-behaved homology. The classical Morita Theorems, with slight adjustments, are valid for rings with local units (see [10]).

Proposition 2.6 ([63, Proposition 4.11], [32, Lemma 2.6]) *Let $\mathcal{G}$ be an ample groupoid. Then $A_R(\mathcal{G})$ is locally unital. Moreover, $A_R(\mathcal{G})$ is unital if and only if $\mathcal{G}^{(0)}$ is compact.*

Proof We prove the "moreover" part first. If $\mathcal{G}^{(0)}$ is compact, then it is a compact open bisection, and $\mathbf{1}_{\mathcal{G}^{(0)}} \in A_R(\mathcal{G})$. Following Lemma 2.2 (2), $\mathbf{1}_{\mathcal{G}^{(0)}} * \mathbf{1}_B = \mathbf{1}_{\mathcal{G}^{(0)}B} = \mathbf{1}_B = \mathbf{1}_{B\mathcal{G}^{(0)}} = \mathbf{1}_B * \mathbf{1}_{\mathcal{G}^{(0)}}$, for every $B \in B^{\mathrm{co}}(\mathcal{G})$. Since $\{\mathbf{1}_B \mid B \in B^{\mathrm{co}}(\mathcal{G})\}$ spans $A_R(\mathcal{G})$, it follows by linearity that $\mathbf{1}_{\mathcal{G}^{(0)}} * f = f = f * \mathbf{1}_{\mathcal{G}^{(0)}}$ for every $f \in A_R(\mathcal{G})$. This proves that $\mathbf{1}_{\mathcal{G}^{(0)}}$ is the multiplicative identity in $A_R(\mathcal{G})$.

Conversely, suppose $A_R(\mathcal{G})$ has a multiplicative identity called ξ. The first step is to show that $\xi = \mathbf{1}_{\mathcal{G}^{(0)}}$. Let $x \in \mathcal{G}$ and let $V \subseteq \mathcal{G}^{(0)}$ be a compact open set containing $\boldsymbol{d}(x)$. Then V must be Hausdorff because $\mathcal{G}^{(0)}$ is, so $\mathbf{1}_V \in A_R(\mathcal{G})$. If $x \notin \mathcal{G}^{(0)}$, then

$$0 = \mathbf{1}_V(x) = \xi * \mathbf{1}_V(x) = \sum_{y \in \mathcal{G}\boldsymbol{d}(x)} \xi(xy^{-1})\mathbf{1}_V(y) = \sum_{y \in V \cap \mathcal{G}\boldsymbol{d}(x)} \xi(xy^{-1}) = \xi(x)$$

because $V \cap \mathcal{G}\boldsymbol{d}(x) = \{\boldsymbol{d}(x)\}$. Similarly, if $x \in \mathcal{G}^{(0)}$ then $x = \boldsymbol{d}(x) \in V$ and

$$1 = \mathbf{1}_V(x) = \xi * \mathbf{1}_V(x) = \xi(x).$$

This shows that $\xi = \mathbf{1}_{\mathcal{G}^{(0)}}$. The second step is to show that $\mathbf{1}_{\mathcal{G}^{(0)}} \in A_R(\mathcal{G})$ implies $\mathcal{G}^{(0)}$ is compact. By the definition of $A_R(\mathcal{G})$, there exist scalars $r_1, \ldots, r_n \in R \setminus \{0\}$ and compact open sets $U_1, \ldots, U_n \subseteq \mathcal{G}$ such that $\mathbf{1}_{\mathcal{G}^{(0)}} = r_1\mathbf{1}_{U_1} + \cdots + r_n\mathbf{1}_{U_n}$. Then $\mathcal{G}^{(0)} \subseteq U_1 \cup \cdots \cup U_n$ and consequently $\mathcal{G}^{(0)} = \boldsymbol{d}(U_1) \cup \cdots \cup \boldsymbol{d}(U_n)$. Each of the sets $\boldsymbol{d}(U_1), \ldots, \boldsymbol{d}(U_n)$ is compact (because $\boldsymbol{d}$ is continuous), so $\mathcal{G}^{(0)}$ is compact.

To show that $A_R(\mathcal{G})$ is locally unital for all ample groupoids $\mathcal{G}$, suppose $F = \{f_1, \ldots, f_m\}$ is a finite subset of $A_R(\mathcal{G})$. Since $A_R(\mathcal{G})$ is spanned by $\{\mathbf{1}_B \mid B \in B^{\mathrm{co}}(\mathcal{G})\}$, there exist finite subsets $\{B_1, \ldots, B_n\} \subseteq B^{\mathrm{co}}(\mathcal{G})$ and $\{r_{i,j} \mid 1 \leq i \leq n, 1 \leq j \leq m\} \subseteq R$ such that $f_j = r_{1,j}\mathbf{1}_{B_1} + \cdots + r_{n,j}\mathbf{1}_{B_n}$ for all $1 \leq j \leq m$. Let $X = \boldsymbol{d}(B_1) \cup \cdots \cup \boldsymbol{d}(B_n) \cup \boldsymbol{c}(B_1) \cup \cdots \cup \boldsymbol{c}(B_n)$. Then $X \subseteq G^{(0)}$ is compact and open because it is a finite union of compact open sets, and X is Hausdorff because it is a subset of $\mathcal{G}^{(0)}$, so $\mathbf{1}_X \in A_R(\mathcal{G})$. Clearly, $XB_i = B_i = B_iX$, so $\mathbf{1}_X * \mathbf{1}_{B_i} = \mathbf{1}_{B_i} = \mathbf{1}_{B_i} * \mathbf{1}_X$, for all $1 \leq i \leq n$. By linearity, $\mathbf{1}_X * f_j = f_j = f_j * \mathbf{1}_X$ for all $1 \leq j \leq m$. The conclusion is that $E = \{\mathbf{1}_X \mid X \in \mathcal{B}(\mathcal{G}^{(0)})\}$ is a set of local units for $A_R(\mathcal{G})$. □

The *characteristic* of a ring A, written $\operatorname{char} A$, is defined as the least positive integer n such that $n \cdot a = 0$ for all $a \in A$, or 0 if no such n exists. If A has a set of local units E, the characteristic of A can be defined as the least n such that $n \cdot e = 0$ for all $e \in E$, or 0 if no such n exists.

Proposition 2.7 *For any ample groupoid $\mathcal{G}$,* $\operatorname{char} A_R(\mathcal{G}) = \operatorname{char} R$.

Proof If n is a positive integer, $n \cdot \mathbf{1}_U = 0$ for all $U \in \mathcal{B}(\mathcal{G}^{(0)})$ if and only if $n \cdot 1 = 0$. □

Given a topological groupoid $(\mathcal{G}, \mathcal{G}^{(0)}, \boldsymbol{d}, \boldsymbol{c}, \boldsymbol{m}, \boldsymbol{i})$, the *opposite groupoid* is:

$$\mathcal{G}^{\mathrm{op}} = (\mathcal{G}, \mathcal{G}^{(0)}, \boldsymbol{d}^{\mathrm{op}}, \boldsymbol{c}^{\mathrm{op}}, \boldsymbol{m}^{\mathrm{op}}, \boldsymbol{i})$$

where $\boldsymbol{d}^{\mathrm{op}} = \boldsymbol{c}$, $\boldsymbol{c}^{\mathrm{op}} = \boldsymbol{d}$, and $\boldsymbol{m}^{\mathrm{op}}(x, y) = \boldsymbol{m}(y, x)$ for any x, y with $\boldsymbol{c}(x) = \boldsymbol{d}(y)$. We call the opposite groupoid $\mathcal{G}^{\mathrm{op}}$ to distinguish it from $\mathcal{G}$, even though they have the same underlying sets. We assume $\mathcal{G}^{\mathrm{op}}$ has the same topology as $\mathcal{G}$. Naturally, the inversion map $\boldsymbol{i} : \mathcal{G} \to \mathcal{G}^{\mathrm{op}}$ is an isomorphism of topological groupoids.

If A is a ring, an *involution* on A is an additive, anti-multiplicative map $\tau : A \to A$ such that $\tau^2 = \mathrm{id}_A$. If A has an involution, it is called an *involutive ring* or $*$*-ring*. If $\mathcal{G}$ is an ample groupoid, $f \mapsto f \circ \boldsymbol{i}$ is a canonical involution on $A_R(\mathcal{G})$ that makes it a $*$-algebra. More generally, if there is an involution $\overline{} : R \to R$, written as $r \mapsto \overline{r}$, then $f \mapsto \overline{f \circ \boldsymbol{i}}$ is an involution on $A_R(\mathcal{G})$. To summarise:

Proposition 2.8 *Let $\mathcal{G}$ be an ample groupoid. There are canonical isomorphisms $\mathcal{G} \cong \mathcal{G}^{\mathrm{op}}$ and $A_R(\mathcal{G}) \cong A_R(\mathcal{G}^{\mathrm{op}}) \cong A_R(\mathcal{G})^{\mathrm{op}}$. Moreover, to each involution $\overline{} : R \to R$ is associated a canonical involution on $A_R(\mathcal{G})$, namely, $f \mapsto \overline{f \circ \boldsymbol{i}}$ for all $f \in A_R(\mathcal{G})$.*

This kind of symmetry is very nice to work with. It implies, for example, that the category of left $A_R(\mathcal{G})$-modules is isomorphic to the category of right $A_R(\mathcal{G})$-modules, and the lattice of left ideals in $A_R(\mathcal{G})$ is isomorphic to the lattice of right ideals. Many important notions, like left and right primitivity, are equivalent for involutive algebras (or more generally, self-opposite algebras).

2.2.5 First Examples

One or two of the results in this section will be useful later on, but mostly they are just interesting in their own right. Presumably, most of this content is already known, but we do not adhere closely to any references.

Given two groupoids $(\mathcal{G}_1, \boldsymbol{d}_1, \boldsymbol{c}_1, \boldsymbol{m}_1, \boldsymbol{i}_1)$ and $(\mathcal{G}_2, \boldsymbol{d}_2, \boldsymbol{c}_2, \boldsymbol{m}_2, \boldsymbol{i}_2)$, their *disjoint union* $\mathcal{G}_1 \sqcup \mathcal{G}_2$ has the structure of a groupoid with unit space $\mathcal{G}_1^{(0)} \sqcup \mathcal{G}_2^{(0)}$, set of composable pairs $\mathcal{G}_1^{(2)} \sqcup \mathcal{G}_2^{(2)}$, and the following structure maps: for all $x_1, y_1 \in \mathcal{G}_1$ and $x_2, y_2 \in \mathcal{G}_2$,

$$\boldsymbol{d}(x_i) = \boldsymbol{d}_i(x_i), \quad \boldsymbol{c}(x_i) = \boldsymbol{c}_i(x_i), \quad \boldsymbol{i}(x_i) = \boldsymbol{i}_i(x_i), \quad \boldsymbol{m}(x_i, y_i) = \boldsymbol{m}_i(x_i, y_i).$$

The *product* $\mathcal{G}_1 \times \mathcal{G}_2$ also has the structure of a groupoid with unit space $\mathcal{G}_1^{(0)} \times \mathcal{G}_2^{(0)}$, and the following structure maps: for all $x_1, y_1 \in \mathcal{G}_1$ and $x_2, y_2 \in \mathcal{G}_2$,

$$\begin{aligned} \boldsymbol{d}(x_1, x_2) &= (\boldsymbol{d}_1(x_1), \boldsymbol{d}_2(x_2)), & \boldsymbol{c}(x_1, x_2) &= (\boldsymbol{c}_1(x_1), \boldsymbol{c}_2(x_2)), \\ \boldsymbol{i}(x_1, x_2) &= (\boldsymbol{i}_1(x_1), \boldsymbol{i}_2(x_2)), & \boldsymbol{m}((x_1, x_2), (y_1, y_2)) &= (\boldsymbol{m}_1(x_1, y_1), \boldsymbol{m}_2(x_2, y_2)). \end{aligned}$$

These constructions work just as well for the disjoint union or product of arbitrarily many (even infinitely many) groupoids. If $\mathcal{G}_1$ and $\mathcal{G}_2$ are topological groupoids, then $\mathcal{G}_1 \sqcup \mathcal{G}_2$ (with the coproduct topology) and $\mathcal{G}_1 \times \mathcal{G}_2$ (with the product topology) are again topological groupoids. The properties of being étale or ample are preserved by arbitrary disjoint unions and finite products.

Proposition 2.9 *Let $\mathcal{G}_1$ and $\mathcal{G}_2$ be ample groupoids. The Steinberg algebra of $\mathcal{G}_1 \sqcup \mathcal{G}_2$ is a direct sum of two ideals: $A_R(\mathcal{G}_1 \sqcup \mathcal{G}_2) \cong A_R(\mathcal{G}_1) \oplus A_R(\mathcal{G}_2)$.*

Proof Let $I_1 = \{f_1 \in A_R(\mathcal{G}_1 \sqcup \mathcal{G}_2) \mid \operatorname{supp} f_1 \subseteq \mathcal{G}_1\}$ and $I_2 = \{f_2 \in A_R(\mathcal{G}_1 \sqcup \mathcal{G}_2) \mid \operatorname{supp} f_2 \subseteq \mathcal{G}_2\}$. Recall from Remark 2.3 that $I_1 \cong A_R(\mathcal{G}_1)$ and $I_2 \cong A_R(\mathcal{G}_2)$. Every $f \in A_R(\mathcal{G}_1 \sqcup \mathcal{G}_2)$ decomposes as $f = f_1 + f_2$ where $f_i \in I_i$ are defined as:

$$f_i(x) = \begin{cases} f(x) & \text{if } x \in \mathcal{G}_i \\ 0 & \text{if } x \notin \mathcal{G}_i \end{cases}$$

for $i = 1, 2$. We claim I_1 and I_2 are orthogonal ideals (that is, $I_1 * I_2 = 0$). For all $f_1 \in I_1$, $f_2 \in I_2$, and $x \in \mathcal{G}_1 \sqcup \mathcal{G}_2$, $f_1 * f_2(x) = \sum_{ab=x} f_1(a) f_2(b)$. So, $\operatorname{supp}(f_1 * f_2) \subseteq \operatorname{supp}(f_1)\operatorname{supp}(f_2) \subseteq \mathcal{G}_1\mathcal{G}_2 = \emptyset$. This implies I_1 and I_2 are ideals, and $A_R(\mathcal{G}_1 \sqcup \mathcal{G}_2) = I_1 \oplus I_2 \cong A_R(\mathcal{G}_1) \oplus A_R(\mathcal{G}_2)$. □

By mathematical induction, the Steinberg algebra of a finite disjoint union of ample groupoids is isomorphic to the direct sum of their respective Steinberg algebras.

Like in [7, Notation 2.6.3], we have reasons to consider matrix rings of a slightly more general nature than usual.

Definition 2.5 *(Matrix rings)* Let A be a ring (not necessarily commutative or unital). If n is a positive integer, we write $M_n(A)$ for the ring of $n \times n$ matrices with entries in A. If Λ is a set (not necessarily finite) we define $M_\Lambda(A)$ to be the ring of square matrices, with rows and columns indexed by Λ, having entries in A and only finitely many non-zero entries.

Note that $M_\Lambda(A)$ is the direct limit of the finite-sized matrix rings associated to finite subsets of Λ. Also, $M_\Lambda(A)$ is unital if and only if A is unital and Λ is finite. The notation $[a_{ij}]$ stands for the matrix in $M_n(A)$, or $M_\Lambda(A)$, with a_{ij} in its (i, j)-entry. Let $\mathcal{N} = \{1, \ldots, n\}^2$ be the transitive principal groupoid on n elements, with the discrete topology, as seen in Example 2.1 (c).

Proposition 2.10 *If $\mathcal{G}$ is a Hausdorff ample groupoid, then $A_R(\mathcal{N} \times \mathcal{G}) \cong M_n(A_R(\mathcal{G}))$.*

Proof Define the map $F : A_R(\mathcal{N} \times \mathcal{G}) \to M_n(A_R(\mathcal{G}))$:

$$F(f) = [f_{ij}],$$

where $f_{ij}(x) = f\big((i, j), x\big)$ for all $f \in A_R(\mathcal{N} \times \mathcal{G})$, $(i, j) \in \mathcal{N}$, and $x \in \mathcal{G}$. If $f \in A_R(\mathcal{N} \times \mathcal{G})$, then f is compactly supported and locally constant. The restriction of f to a clopen subset, such as $\{(i, j)\} \times \mathcal{G}$ for some $(i, j) \in \mathcal{N}$, is also compactly supported and locally constant. Therefore $f_{i,j} \in A_R(\mathcal{G})$ for all $(i, j) \in \mathcal{N}$. Clearly, F is bijective. Now, let $f, g \in A_R(\mathcal{N} \times \mathcal{G})$. For all $(i, j) \in \mathcal{N}$ and $x \in \mathcal{G}$, the convolution formula yields

$$\begin{aligned}(f * g)_{ij}(x) = f * g\big((i, j), x\big) &= \sum_{\substack{(k,\ell,y) \in \mathcal{N} \times \mathcal{G} \\ (\ell, \boldsymbol{d}(y)) = (j, \boldsymbol{d}(x))}} f\big[((i, j), x)((k, \ell), y)^{-1}\big] g((k, \ell), y) \\ &= \sum_{1 \le k \le N} \sum_{\substack{y \in \mathcal{G} \\ \boldsymbol{d}(y) = \boldsymbol{d}(x)}} f((i, k), xy^{-1}) g((k, j), y) \\ &= \sum_{1 \le k \le n} f_{ik} * g_{kj}(x)\end{aligned}$$

This shows $F(f * g) = F(f)F(g)$, so F is an isomorphism. □

Remark 2.4 As a specialisation of Proposition 2.10, we obtain $A_R(\mathcal{N}) \cong M_n(R)$. It is well—known that when A is an R-algebra, $M_n(A) \cong M_n(R) \otimes_R A$ (see [18, Example 4.22]). It is also well-known (see [18, Example 4.20]) that if G and H are groups, then $R(G \times H) \cong RG \otimes_R RH$. One can show using the standard techniques that when $\mathcal{G}_1$ and $\mathcal{G}_2$ are arbitrary ample groupoids, there is a surjective homomorphism $A_R(\mathcal{G}_1) \otimes_R A_R(\mathcal{G}_2) \to A_R(\mathcal{G}_1 \times \mathcal{G}_2)$. An interesting question is: under what circumstances is it an isomorphism?

Suppose $\mathcal{G}$ is a topological groupoid and $\{\mathcal{G}_i\}_{i \in I}$ is a family of open subgroupoids indexed by a directed set $(I, \le)$, such that $\mathcal{G} = \bigcup_{i \in I} \mathcal{G}_i$ and $\mathcal{G}_i \subseteq \mathcal{G}_j$ whenever $i \le j$ in I. If this happens, we say that $\mathcal{G}$ is the *directed union* of the subgroupoids $\{\mathcal{G}_i\}_{i \in I}$.

Proposition 2.11 *If a Hausdorff ample groupoid $\mathcal{G}$ is the directed union of a family of open subgroupoids $\{\mathcal{G}_i\}_{i\in I}$, then $A_R(\mathcal{G})$ is the direct limit of subalgebras $\{A_R(\mathcal{G}_i)\}_{i\in I}$.*

Proof For all $i \le j$ in I, let $\varphi_{ij} : A_R(\mathcal{G}_i) \hookrightarrow A_R(\mathcal{G}_j)$ and $m_i : A_R(\mathcal{G}_i) \hookrightarrow A_R(\mathcal{G})$ be the canonical embeddings (see Remark 2.3). We claim that for every $f \in A_R(\mathcal{G})$, there exists $j \in I$ such that $f \in m_j(A_R(\mathcal{G}_j))$. If $f \in A_R(\mathcal{G})$ then $\operatorname{supp} f$ is compact and open. Thus, there is a finite subcover of $\{\mathcal{G}_i\}_{i\in I}$ that covers $\operatorname{supp} f$. If $\operatorname{supp} f \subseteq \mathcal{G}_{i_1} \cup \cdots \cup \mathcal{G}_{i_n}$, then there exists $j \in I$ with $i_1, \ldots, i_n \le j$, using the fact that $(I, \le)$ is directed. Thus, $\operatorname{supp} f \subseteq \mathcal{G}_j$, and $f|_{\mathcal{G}_j}$ is compactly supported and locally constant, whereby $f|_{\mathcal{G}_j} \in A_R(\mathcal{G}_j)$. Finally, this shows $f = m_j(f|_{\mathcal{G}_j}) \in m_j(A_R(\mathcal{G}_j))$.

Now assume B is an R-algebra and $\{\beta_i\}_{i\in I}$ is a family of R-homomorphisms $\beta_i : A_R(\mathcal{G}_i) \to B$, such that $\beta_i = \beta_j \varphi_{ij}$ for all $i \le j$. Then, since every $\varphi_{ij} : A_R(\mathcal{G}_i) \to A_R(\mathcal{G}_j)$ is injective, β_j is an extension of β_i whenever $i \le j$. Since $A_R(\mathcal{G}) = \bigcup_{i\in I} m_i(A_R(\mathcal{G}_i))$, it follows that there is a unique homomorphism $\beta : A_R(\mathcal{G}) \to B$ such that $\beta_i = \beta m_i$ for all $i \in I$. As such, $A_R(\mathcal{G})$ has the universal property for the directed system $\{A_R(\mathcal{G}_i)\}_{i\in I}$, so we can conclude it is the direct limit of that system. □

We can now extend Propositions 2.9 and 2.10 to allow infinite index sets. This could have been proved directly, mentioning that the functions in $A_R(\mathcal{G})$ have compact supports, but it is nice to demonstrate direct limits.

Proposition 2.12 *Let $\mathcal{G}$ be a Hausdorff ample groupoid, and let Λ be an infinite set.*

(1) *If $\mathcal{D} = \Lambda^2$ is the transitive principal groupoid on Λ, equipped with the discrete topology, then $A_R(\mathcal{D} \times \mathcal{G}) \cong M_\Lambda(A_R(\mathcal{G}))$.*
(2) *If $\mathcal{G} = \bigsqcup_{\lambda\in\Lambda} \mathcal{G}_\lambda$ is the disjoint union of an infinite family of clopen subgroupoids $\{\mathcal{G}_\lambda\}_{\lambda\in\Lambda}$, then $A_R(\mathcal{G}) \cong \bigoplus_{\lambda\in\Lambda} A_R(\mathcal{G}_\lambda)$.*

Proof (1) Note that $\mathcal{D} \times \mathcal{G}$ is the directed union of the subgroupoids $\mathcal{D}_F \times \mathcal{G}$, where $\mathcal{D}_F = \{(d_1, d_2) \in \mathcal{D} \mid d_1, d_2 \in F\}$, as F ranges over all the finite subsets of Λ ordered by inclusion. By Propositions 2.10 and 2.11, $A_R(\mathcal{D} \times \mathcal{G})$ is the direct limit of matrix algebras $A_R(\mathcal{D}_F \times \mathcal{G}) \cong M_F(A_R(\mathcal{G}))$, and this direct limit is isomorphic to $M_\Lambda(A_R(\mathcal{G}))$.

(2) Note that $\mathcal{G}$ is the directed union of the subgroupoids $\mathcal{G}_F = \bigsqcup_{\lambda\in F} \mathcal{G}_\lambda$, as F ranges over finite subsets of Λ ordered by inclusion. By Propositions 2.9 and 2.11, $A_R(\mathcal{G})$ is the direct limit of the subalgebras $A_R(\mathcal{G}_F) \cong \bigoplus_{\lambda\in F} A_R(\mathcal{G}_\lambda)$, and this direct limit is isomorphic to $\bigoplus_{\lambda\in\Lambda} A_R(\mathcal{G}_\lambda)$. □

Here we describe a class of principal groupoids, called *approximately finite* groupoids, that was defined by Renault in his influential monograph [59].

Example 2.4 Let X be a locally compact, totally disconnected Hausdorff space. Consider it as a groupoid with unit space X and no morphisms outside the unit space. Then $A_R(X)$ is the commutative R-algebra of locally constant, compactly supported functions $f : X \to R$, with pointwise addition and multiplication. We adopt the notation $A_R(X) = C_R(X)$ and drop the $*$ notation for products, because this serves as a reminder that $C_R(X)$ is commutative. An ample groupoid is called **elementary** if it is of the form $(\mathcal{N}_1 \times X_1) \sqcup \cdots \sqcup (\mathcal{N}_t \times X_t)$, where $\mathcal{N}_1, \ldots \mathcal{N}_t$ are discrete, finite, transitive principal groupoids on $n_1, \ldots, n_t$ elements, respectively, and $X_1, \ldots, X_n$ are locally compact, totally disconnected, Hausdorff topological spaces. Using the results of this section:

$$A_R\left(\bigsqcup_{i=1}^{n}(\mathcal{N}_i \times X_i)\right) \cong \bigoplus_{i=1}^{t} M_{n_i}\big(C_R(X_i)\big). \tag{2.5}$$

A groupoid is called **approximately finite** if it is the directed union of an increasing sequence of elementary groupoids. The Steinberg algebra of an approximately finite groupoid is a direct limit of matricial algebras, each resembling Eq. 2.5.

Definition 2.6 A ring A is called **von Neumann regular** if for every $x \in A$ there exists $y \in A$ such that $x = xyx$.

If $y \in A$ satisfies $x = xyx$ then y is called a *von Neumann inverse* of x. If R is a commutative von Neumann regular ring, then for every $r \in R$ there exists a unique element $s \in R$ such that $r = r^2 s$ and $s = s^2 r$ (see [42, Proposition 3.6]).

Proposition 2.13 *If $\mathcal{F}$ is an approximately finite groupoid and R is a von Neumann regular unital commutative ring, then $A_R(\mathcal{F})$ is von Neumann regular.*

Proof Let X be a locally compact, totally disconnected, Hausdorff topological space, and suppose R is von Neumann regular. To verify that $C_R(X)$ is von Neumann regular, take $f \in C_R(X)$ and for every $x \in X$ define $g(x)$ to be the unique element of R such that $f(x) = f(x)^2 g(x)$ and $g(x) = g(x)^2 f(x)$. Note that $g \in C_R(X)$ and $fgf = f$. Now, $C_R(X)$ being regular implies $M_n(C_R(X))$ is regular (this could be argued carefully with Morita equivalence, but one finds in [48, Theorem 24] a clever direct proof by induction). A direct sum of regular rings is regular, so any ring of the form Eq. 2.5 is regular, provided R is regular. A direct limit of regular rings is regular: each element in the direct limit must belong to a regular subring, and the von Neumann inverse can be chosen from that same subring. Therefore $A_R(\mathcal{F})$ is von Neumann regular. □

Note that we did not use the assumption that $\mathcal{F}$ is a countable directed union of elementary groupoids; any directed union will do. It is an open problem to characterise von Neumann regularity for Steinberg algebras in groupoid terms; partial progress is achieved in [9].

This next result is a "baby version" of [65, Proposition 3.1], with a new proof. In preparation for it, we briefly remark that every transitive groupoid $\mathcal{G}$ is (algebraically, but not necessarily topologically) isomorphic to the product of a transitive principal groupoid and a group. To construct such an isomorphism, fix a unit $b \in \mathcal{G}^{(0)}$. Let $\Gamma = b\mathcal{G}b$ be the isotropy group based at b, and let $\mathcal{P} = [\mathcal{G}^{(0)}]^2$ be the transitive principal groupoid on $\mathcal{G}^{(0)}$. Fix a morphism $h_y \in b\mathcal{G}y$ for every $y \in \mathcal{G}^{(0)}$, and define the groupoid isomorphisms:

$$F : \mathcal{G} \to \mathcal{P} \times \Gamma, \quad F(g) = \Big(\big(\boldsymbol{c}(g), \boldsymbol{d}(g) \big), h_{\boldsymbol{c}(g)} g h_{\boldsymbol{d}(g)}^{-1} \Big) \quad \text{for all } g \in \mathcal{G};$$
$$F^{-1} : \mathcal{P} \times \Gamma \to \mathcal{G}, \quad F^{-1}\big((x, y), \gamma \big) = h_x^{-1} \gamma h_y \quad \text{for all } x, y \in \mathcal{G}^{(0)}, \gamma \in \Gamma.$$

Proposition 2.14 *Let $\mathbb{K}$ be a field and $\mathcal{G}$ an ample groupoid. Then $A_{\mathbb{K}}(\mathcal{G})$ is finite-dimensional if and only if $\mathcal{G}$ is finite and has the discrete topology. If $O_1, \dots, O_t$ are the orbits of $\mathcal{G}$, and $\Gamma_1, \dots, \Gamma_t$ are the corresponding isotropy groups, then*

$$A_{\mathbb{K}}(\mathcal{G}) \cong \bigoplus_{i=1}^{t} M_{O_i}(R\Gamma_i).$$

Proof First of all, if $\mathcal{G}$ is discrete, then $\dim_{\mathbb{K}} A_{\mathbb{K}}(\mathcal{G}) = |\mathcal{G}|$, because $\{\mathbf{1}_{\{g\}} \mid g \in \mathcal{G}\}$ is a basis for $A_{\mathbb{K}}(\mathcal{G})$, by Corollary 2.1. Thus, $A_{\mathbb{K}}(\mathcal{G})$ is finite-dimensional if $\mathcal{G}$ is finite and discrete. Conversely, suppose $A_{\mathbb{K}}(\mathcal{G})$ is finite-dimensional, and let $\{f_1, \dots, f_n\}$ be a basis. The image of each f_i is finite, so $|\operatorname{im} f_1 \cup \dots \cup \operatorname{im} f_n|$ is bounded by some $M < \infty$. If $|\mathcal{G}^{(0)}| > M^n$ then, by the pigeonhole principle, there exists $u \neq v$ in $\mathcal{G}^{(0)}$ such that $f_i(u) = f_i(v)$ for all $1 \le i \le n$, and thus $f(u) = f(v)$ for all $f \in A_{\mathbb{K}}(\mathcal{G})$. But $\mathcal{G}^{(0)}$ is Hausdorff, locally compact, and totally disconnected, so there is a compact open subset $U \subseteq \mathcal{G}^{(0)}$ with $u \in U$ and $v \notin U$. Since $\mathcal{G}^{(0)}$ is open in $\mathcal{G}$, it follows that U is a compact open bisection in $\mathcal{G}$, so $\mathbf{1}_U \in A_{\mathbb{K}}(\mathcal{G})$. We arrive at a contradiction, because $\mathbf{1}_U(u) \neq \mathbf{1}_U(v)$. Therefore $|\mathcal{G}^{(0)}| \le M^n < \infty$. A finite Hausdorff space is discrete, so $\mathcal{G}^{(0)}$ is discrete. As $\mathcal{G}$ is étale, it must also be discrete. Thus $\dim_{\mathbb{K}} A_{\mathbb{K}}(\mathcal{G}) = n = |\mathcal{G}|$. Given that $\mathcal{G}$ is finite and discrete, it is isomorphic to a disjoint union of transitive groupoids (one for each orbit), each of which is isomorphic to the product of a transitive principal groupoid (with as many elements as the corresponding orbit), and a finite group (the isotropy group of that orbit). The expression giving the structure of $A_{\mathbb{K}}(\mathcal{G})$ follows from Propositions 2.9 and 2.10. □

2.2.6 Graded Groupoids and Graded Steinberg Algebras

Just as the Steinberg algebra of a groupoid inherits an involution from the groupoid, so it can inherit a graded structure. Many well-studied examples of Steinberg algebras receive a canonical group-grading that comes from a grading on the groupoid itself. We first introduce the concepts and terminology of graded groupoids and graded algebras.

A standing assumption is that Γ is a group with identity ε. A ring A is called a Γ-*graded ring* if it decomposes as a direct sum of additive subgroups $A = \bigoplus_{\gamma \in \Gamma} A_\gamma$ such that $A_\gamma A_\delta \subseteq A_{\gamma\delta}$ for every $\gamma, \delta \in \Gamma$. The meaning of $A_\gamma A_\delta$ is the additive subgroup generated by all products ab where $a \in A_\gamma, b \in A_\delta$. The additive group A_γ is called the *γ-component* of A. The elements of $\bigcup_{\gamma\in\Gamma} A_\gamma$ in a graded ring A are called *homogeneous elements*. The non-zero elements of A_γ are called *γ-homogeneous*, and we write $\deg(a) = \gamma$ for $a \in A_\gamma \setminus \{0\}$. When it is clear from context that a ring A is graded by the group Γ, we simply say that A is a *graded ring*. If A is an R-algebra, then A is called a *graded algebra* if it is a graded ring and each A_γ is an R-submodule.

An ideal $I \subseteq A$ is a *graded ideal* if $I \subseteq \sum_{\gamma\in\Gamma} I \cap A_\gamma$. Graded left ideals, graded right ideals, graded subrings, and graded subalgebras are defined in a similar manner. If H is a set of homogeneous elements in A, the ideal generated by H is a graded ideal. Likewise, the left and right ideals generated by H are graded. A *graded homomorphism* of Γ-graded rings is a homomorphism $f : A \to B$ such that $f(A_\gamma) \subseteq B_\gamma$ for every $\gamma \in \Gamma$. Finally, we say that a Γ-graded ring A has *homogeneous local units* (or *graded local units*) if A is locally unital, and the set of local units can be chosen to be a subset of A_ε.

A topological groupoid $\mathcal{G}$ is called *Γ-graded* if it can be partitioned by clopen subsets $\mathcal{G} = \bigsqcup_{\gamma\in\Gamma} \mathcal{G}_\gamma$, such that $\mathcal{G}_\gamma\mathcal{G}_\delta \subseteq \mathcal{G}_{\gamma\delta}$ for every $\gamma, \delta \in \Gamma$. Equivalently $\mathcal{G}$ is Γ-graded if there is a continuous homomorphism $\kappa : \mathcal{G} \to \Gamma$. We can show the definitions are equivalent by setting $\mathcal{G}_\gamma = \kappa^{-1}(\{\gamma\})$. If $\kappa : \mathcal{G} \to \Gamma$ defines the grading on $\mathcal{G}$, we call it the *degree map*. We use the notation $\mathcal{G}_\gamma x = \mathcal{G}_\gamma \cap \mathcal{G}x$ and $x\mathcal{G}_\gamma = x\mathcal{G} \cap \mathcal{G}_\gamma$ for $x \in \mathcal{G}^{(0)}$ and $\gamma \in \Gamma$.

We say a subset $X \subseteq \mathcal{G}$ is γ-homogeneous if $X \subseteq \mathcal{G}_\gamma$. Obviously, the unit space is ε-homogeneous and if X is γ-homogeneous then X^{-1} is γ^{-1}-homogeneous. Moreover, ${\mathcal{G}_\gamma}^{-1} = \mathcal{G}_{\gamma^{-1}}$ for all $\gamma \in \Gamma$. For a Γ-graded ample groupoid, we write $B^{\mathrm{co}}_\gamma(\mathcal{G})$ for the set of all γ-homogeneous compact open bisections of $\mathcal{G}$. For the set of all homogeneous compact open bisections, we use the notation:

$$B^{\mathrm{co}}_*(\mathcal{G}) = \bigcup_{\gamma\in\Gamma} B^{\mathrm{co}}_\gamma(\mathcal{G}) \subseteq B^{\mathrm{co}}(\mathcal{G}).$$

In Proposition 2.2, we proved that $B^{\mathrm{co}}(\mathcal{G})$ is an inverse semigroup, and it is readily apparent that $B^{\mathrm{co}}_*(\mathcal{G})$ is an inverse subsemigroup of $B^{\mathrm{co}}(\mathcal{G})$. In addition, $B^{\mathrm{co}}_*(\mathcal{G})$ is a base of compact open bisections for $\mathcal{G}$. Indeed, since $B^{\mathrm{co}}(\mathcal{G})$ is a base for $\mathcal{G}$, it suffices to show that every $B \in B^{\mathrm{co}}(\mathcal{G})$ is a union of sets in $B^{\mathrm{co}}_*(\mathcal{G})$. This is almost trivial, for if $B \in B^{\mathrm{co}}(\mathcal{G})$ then $B = \bigcup_{\gamma\in\Gamma} B \cap \mathcal{G}_\gamma$ and $B \cap \mathcal{G}_\gamma \in B^{\mathrm{co}}_\gamma(\mathcal{G})$. The next two results are from [24, Lemma 3.1].

Proposition 2.15 *If $\mathcal{G} = \bigsqcup_{\gamma\in\Gamma} \mathcal{G}_\gamma$ is a Γ-graded ample groupoid, then $A_R(\mathcal{G}) = \bigoplus_{\gamma\in\Gamma} A_R(\mathcal{G})_\gamma$ is a Γ-graded algebra with homogeneous local units, where:*

$$A_R(\mathcal{G})_\gamma = \left\{f \in A_R(\mathcal{G}) \mid \operatorname{supp} f \subseteq \mathcal{G}_\gamma\right\} \qquad \text{for all } \gamma \in \Gamma.$$

Proof From Proposition 2.3, it follows that

$$A_R(\mathcal{G}) = \text{span}_R\{\mathbf{1}_B \mid B \in B_*^{\text{co}}(\mathcal{G})\} = \sum_{\gamma\in\Gamma} \text{span}_R\{\mathbf{1}_B \mid B \in B_\gamma^{\text{co}}(\mathcal{G})\} = \sum_{\gamma\in\Gamma} A_R(\mathcal{G})_\gamma.$$

It is clear that $A_R(\mathcal{G})_\gamma \cap \left(\sum_{\delta\neq\gamma} A_R(\mathcal{G})_\delta\right) = \{0\}$ for all $\gamma \in \Gamma$, so we have $A_R(\mathcal{G}) = \bigoplus_{\gamma\in\Gamma} A_R(\mathcal{G})_\gamma$. Now for all $f \in A_R(\mathcal{G})_\gamma$ and $g \in A_R(\mathcal{G})_\delta$, we have $\text{supp}(f * g) \subseteq \text{supp}(f)\text{supp}(g) \subseteq \mathcal{G}_\gamma\mathcal{G}_\delta \subseteq \mathcal{G}_{\gamma\delta}$, and thus $f * g \in A_R(\mathcal{G})_{\gamma\delta}$. Therefore $A_R(\mathcal{G})_\gamma * A_R(\mathcal{G})_\delta \subseteq A_R(\mathcal{G})_{\gamma\delta}$. It follows from Proposition 2.6, and the fact that $\mathcal{G}^{(0)} \subseteq \mathcal{G}_\varepsilon$, that $A_R(\mathcal{G})$ has homogeneous local units. □

Lemma 2.3 *If $\mathcal{G}$ is a Γ-graded Hausdorff ample groupoid, every $f \in A_R(\mathcal{G})$ can be expressed as a finite sum $f = \sum_{i=1}^n r_i\mathbf{1}_{B_i}$, where $r_1, \ldots, r_n \in R$, and $B_1, \ldots, B_n \in B_*^{\text{co}}(\mathcal{G})$ are mutually disjoint.*

Proof Since $\mathcal{G}$ is Hausdorff, every homogeneous compact open bisection is closed, so $B_*^{\text{co}}(\mathcal{G})$ is closed under finite intersections and relative complements. The statement now follows from Proposition 2.5. □

Example 2.5 Recall, from Example 2.1 (d), the definition of the transformation groupoid $G \times X$, associated to a group G and a G-set X. Now assume that X is a locally compact, totally disconnected, Hausdorff topological space, and for each $g \in G$ the map $\rho_g : X \to X$, $\rho_g(x) = g \cdot x$, is continuous. If we assign the discrete topology to G and the product topology to $G \times X$, then $G \times X$ is an ample groupoid. It is easy to verify that this is a G-graded groupoid with homogeneous components $(G \times X)_g = \{g\} \times X$ for all $g \in G$. The Steinberg algebra of $G \times X$ turns out (see [17]) to be the **skew group ring** $C_R(X) \star G$, associated to a certain action of G on $C_R(X)$, canonically induced by the action of G on X.

One can generalise this example quite profitably, by replacing the group action with something more general called a *partial group action* (see [40, Definition 2.1]). In doing so, one obtains a class of algebras so general that it includes all Leavitt path algebras (see [41, Theorem 3.3]) and other interesting things, like the *partial group algebras* that were studied in [36, 45].

2.3 The Path Space and Boundary Path Groupoid of a Graph

Section 2.3 is structured as follows. In Sect. 2.3.1, we define directed graphs and introduce some terminology. In Sect. 2.3.2, we introduce a topological space called the *path space* of a graph. The path space of a graph is the set of all finite and infinite paths, with a topology described explicitly by a base of open sets. Generalising [69, Theorem 2.1], we prove in Theorem 2.1 that for graphs of any cardinality, the path space is locally compact and Hausdorff. We also determine which graphs have a second-countable, first-countable, or σ-compact path space. In Sect. 2.3.3, we use the path space (or more precisely, a closed subspace called the boundary path space) to define the *boundary path groupoid* associated to a graph. We prove it is ample and study its local structure from a topological and an algebraic point of view.

Remark 2.5 Perhaps as an artefact of its history, many fundamental properties of the boundary path groupoid were absorbed into folklore. Some proofs were never written, and others were written at a higher level of generality, and not all in one place, making them difficult to relate back to our present needs. For instance, we could not find a complete proof that the boundary path groupoid is an ample groupoid, even though this fact was used in all the early papers that pioneered the use of groupoid methods for Leavitt path algebras [24, 28, 30]. The groupoid approach to Leavitt path algebras is particularly well-suited, compared to traditional, purely algebraic methods, for dealing with graphs of large cardinalities. Therefore, it is important to make sure that the theorems used to justify these methods can be proved without assuming graphs are countable. This is something that we achieve here, in Theorems 2.1 and 2.4.

2.3.1 Graphs

In this section, we introduce the necessary terminology and conventions pertaining to graphs. We always use the word graph to mean a *directed* graph, defined as follows.

Definition 2.7 A **graph** is a system $E = (E^0, E^1, r, s)$, where E^0 is a set whose elements are called *vertices*, E^1 is a set whose elements are called *edges*, $r : E^1 \to E^0$ is a map that associates a *range* to every edge, and $s : E^1 \to E^0$ is a map that associates a *source* to every edge.

A *countable graph* is one where E^0 and E^1 are countable sets. A *row-finite* (resp., *row-countable*) graph is one in which $s^{-1}(v)$ is finite (resp., countable) for every $v \in E^0$. If e is an edge with $s(e) = v$ and $r(e) = w$ then we say that v *emits* e and w *receives* e. A *sink* is a vertex that emits no edges and an *infinite emitter* is a vertex that emits infinitely many edges. If $v \in E^0$ is either a sink or an infinite emitter (that is, $s^{-1}(v)$ is either empty or infinite) then v is called *singular*, and if v is not singular then it is called *regular*. A vertex that neither receives nor emits any edges is called an *isolated vertex*.

A *finite path* is a finite sequence of edges $\alpha = \alpha_1\alpha_2 \dots \alpha_n$ such that $r(\alpha_i) = s(\alpha_{i+1})$ for all $i = 1, \dots, n-1$. The *length* of the path α is $|\alpha| = n$. Reusing notation and terminology, we shall say that $s(\alpha) = s(\alpha_1)$ is the *source* of the path, and $r(\alpha) = r(\alpha_n)$ is the *range* of the path. By convention, vertices $v \in E^0$ are regarded as finite paths of zero length, with $r(v) = s(v) = v$. If $v, w \in E^0$, we write $v \geq w$ if there exists a finite path α with $s(\alpha) = v$ and $r(\alpha) = w$. If a finite path α of positive length satisfies $r(\alpha) = s(\alpha) = v$, then α is called a *closed path* based at v. A closed path α with the property that none of the vertices $s(\alpha_1), \dots, s(\alpha_{|\alpha|})$ are repeated is called a *cycle*, and a graph that has no cycles is called *acyclic*. An *exit* for a finite path α is an edge $f \in E^1$ with $s(f) = s(\alpha_i)$ for some $1 \leq i \leq |\alpha|$, but $f \neq \alpha_i$.

An *infinite path* is, predictably, an infinite sequence of edges $p = p_1 p_2 p_3 \dots$ such that $r(p_i) = s(p_{i+1})$ for $i = 1, 2, \dots$. Again, $s(p) = s(p_1)$ is called the source of the infinite path p. We let $|p| = \infty$ if p is an infinite path. We use the notation $E^\star$ for the set of finite paths (including vertices), and E^∞ for the set of infinite paths.

Paths can be concatenated if their range and source agree. If $\alpha, \beta \in E^\star$ have positive length and $r(\alpha) = s(\beta)$, then $\alpha\beta = \alpha_1 \dots \alpha_{|\alpha|}\beta_1 \dots \beta_{|\beta|} \in E^\star$. If $p \in E^\infty$ has $r(\alpha) = s(p)$, then $\alpha p = \alpha_1 \dots \alpha_{|\alpha|} p_1 p_2 \dots \in E^\infty$. If $v \in E^0$ and $x \in E^\star \cup E^\infty$ has $s(x) = v$, then $vx = x$ by convention. Likewise, if $\alpha \in E^\star$ has $r(\alpha) = v$ then $\alpha v = \alpha$. If $\alpha \in E^\star$, $x \in E^\star \cup E^\infty$, and $x = \alpha x'$ for some $x' \in E^\star \cup E^\infty$, then we say that α is an *initial subpath* of x. In particular, $s(\alpha)$ is considered an initial subpath of α.

Let $E^0_{\text{sing}} = \{v \in E^0 \mid v \text{ is singular}\}$ and $E^0_{\text{reg}} = \{v \in E^0 \mid v \text{ is regular}\}$. Using the terminology of [69], we define the set of *boundary paths* as

$$\partial E = E^\infty \cup \left\{\alpha \in E^\star \mid r(\alpha) \in E^0_{\text{sing}}\right\}.$$

We employ the following notation from now on:

$$vE^1 = \{e \in E^1 \mid s(e) = v\}, \qquad vE^\star = \{\alpha \in E^\star \mid s(\alpha) = v\},$$
$$vE^\infty = \{p \in E^\infty \mid s(p) = v\}, \qquad v\partial E = \{x \in \partial E \mid s(x) = v\},$$
$$E^\star \times_r E^\star = \left\{(\alpha, \beta) \in E^\star \times E^\star \mid r(\alpha) = r(\beta)\right\}.$$

2.3.2 The Path Space of a Graph

Throughout this section, assume $E = (E^0, E^1, r, s)$ is an arbitrary graph. The *path space* of E is $E^\star \cup E^\infty$, the set of all finite and infinite paths, and the *boundary path space* is ∂E, the set of paths that are either infinite or end at a singular vertex. We now set out to define a suitable topology on the path space. For a finite path $\alpha \in E^\star$, we define the *cylinder* set

$$C(\alpha) = \left\{\alpha x \mid x \in E^\star \cup E^\infty, r(\alpha) = s(x)\right\} \subseteq E^\star \cup E^\infty. \tag{2.6}$$

It is easy to see that the intersection of two cylinders is either empty or a cylinder. Indeed, if $x \in C(\alpha) \cap C(\beta)$ then $x = \alpha y = \beta z$ for some $y, z \in E^\star \cup E^\infty$. If $|\alpha| \leq |\beta|$ then α is an initial subpath of β, implying $C(\beta) \subseteq C(\alpha)$. In symbols:

$$C(\alpha) \cap C(\beta) = \begin{cases} C(\beta) & \text{if } \alpha \text{ is an initial subpath of } \beta \\ C(\alpha) & \text{if } \beta \text{ is an initial subpath of } \alpha \\ \emptyset & \text{otherwise.} \end{cases}$$

This is all we need to conclude that the collection of cylinder sets is a base for a topology on $E^\star \cup E^\infty$. As the authors of [50] have stated, the subspace $E^\infty \subseteq E^\star \cup E^\infty$ with the cylinder set topology is homeomorphic (in the canonical way) to a subspace of $\prod_{n=1}^{\infty} E^1$, where E^1 is discrete and the product has the product topology. In particular, the cylinder sets generate a Hausdorff topology on E^∞, and if E is row-finite, that topology is locally compact. However, the cylinder set topology generated by the sets Eq. 2.6 is not Hausdorff (or even T_1) on the whole set $E^\star \cup E^\infty$, because a finite path cannot be separated from a proper initial subpath. In order to have enough open sets in hand for a Hausdorff topology, we define a base of open sets called *generalised cylinder* sets:

$$C(\alpha, F) = C(\alpha) \setminus \bigcup_{e \in F} C(\alpha e); \qquad \alpha \in E^\star,\ F \subseteq r(\alpha)E^1 \text{ is finite.} \tag{2.7}$$

We shall write $F \subseteq_{\text{finite}} vE^1$ to mean that F is a finite subset of vE^1. The next lemma (a generalisation of [50, Lemma 2.1]) shows that the collection of generalised cylinders is closed under intersections, so it is a base for a topology on $E^\star \cup E^\infty$. With the generalised cylinder set topology on $E^\star \cup E^\infty$, every finite path is an isolated point unless its range is an infinite emitter.

Lemma 2.4 *If* $\alpha, \beta \in E^\star$, $|\alpha| \le |\beta|$, $F \subseteq_{\text{finite}} r(\alpha)E^1$, *and* $H \subseteq_{\text{finite}} r(\beta)E^1$, *then*

$$C(\alpha, F) \cap C(\beta, H) = \begin{cases} C(\beta, F \cup H) & \textit{if } \beta = \alpha \\ C(\beta, H) & \textit{if } \exists\, \delta \in E^\star,\ |\delta| \ge 1,\ \beta = \alpha\delta, \textit{ and } \delta_1 \notin F \\ \emptyset & \textit{otherwise.} \end{cases}$$

Proof By definition of $C(\alpha, F)$ and $C(\beta, H)$, we have

$$C(\alpha, F) \cap C(\beta, H) = C(\alpha) \cap C(\beta) \setminus \left(\bigcup_{e \in F} C(\alpha e) \cup \bigcup_{e \in H} C(\beta e) \right). \tag{2.8}$$

If $\beta = \alpha$, the right hand side of Eq. 2.8 is $C(\beta, F \cup H)$. If $\beta = \alpha\delta$ ($|\delta| \ge 1$) and $\delta_1 \notin F$ then $C(\beta) \cap C(\alpha) = C(\beta)$ does not meet $\bigcup_{e \in F} C(\alpha e)$, so the right hand side of Eq. 2.8 is $C(\beta, H)$. If $\beta = \alpha\delta$ and $\delta_1 \in F$, then $C(\beta) \cap C(\alpha) = C(\beta) = C(\alpha\delta_1 \ldots \delta_{|\delta|}) \subseteq C(\alpha\delta_1) \subseteq \bigcup_{e \in F} C(\alpha e)$, so the right hand side of Eq. 2.8 is empty. If α is not an initial subpath of β then $C(\alpha) \cap C(\beta) = \emptyset$. □

To apply Steinberg's theory from Sect. 2.2, it is critical that the induced topology on the boundary path space $\partial E \subseteq E^\star \cup E^\infty$ is locally compact and Hausdorff. We proceed by proving that the topology on the path space $E^\star \cup E^\infty$, generated by the base in Eq. 2.7, is locally compact and Hausdorff, and that ∂E is closed in $E^\star \cup E^\infty$. As it were, this base is well -chosen: the basic open sets themselves are compact in the Hausdorff topology that they generate.

The proof of the theorem below is essentially the same as [69, Theorem 2.1], just written slightly differently so that it does not use any assumptions of countability.

The main idea is to equip $\mathbb{P}(E^\star)$, i.e., the power set of $E^\star$, with a compact Hausdorff topology, and show that $E^\star \cup E^\infty$ is homeomorphic to a locally compact subspace $\mathbb{S} \subset \mathbb{P}(E^\star)$.

Theorem 2.1 *The collection Eq. 2.7 of generalised cylinder sets is a base of compact open sets for a locally compact Hausdorff topology on* $E^\star \cup E^\infty$.

Proof Let $\{0, 1\}$ have the discrete topology. The product space $\{0, 1\}^{E^\star}$ is compact by Tychonoff's Theorem, and Hausdorff because products preserve the Hausdorff property. There is a canonical bijection from $\mathbb{P}(E^\star)$ to $\{0, 1\}^{E^\star}$, which transfers a compact Hausdorff topology to $\mathbb{P}(E^\star)$. For the first part of the proof, we work entirely in the space $\mathbb{P}(E^\star)$. The topology on $\mathbb{P}(E^\star)$, by definition, is generated by the base of open sets:

$$[P, N] = \{A \in \mathbb{P}(E^\star) \mid P \subseteq A,\ N \subseteq E^\star \setminus A\}; \qquad P, N \subseteq_{\text{finite}} E^\star.$$

Note that $[P, N] = \emptyset$ if $P \cap N \neq \emptyset$. Define the subspace $\mathbb{S} \subset \mathbb{P}(E^\star)$ to be the set of subsets $A \subseteq E^\star$ such that:

- $A \neq \emptyset$ and for all $\alpha \in A$, every initial subpath of α is in A;
- For every $0 \leq n < \infty$, there is at most one path of length n in X.

We claim that $\mathbb{S} \cup \{\emptyset\}$ is closed in $\mathbb{P}(E^\star)$. Suppose $A \in \mathbb{P}(E^\star) \setminus (\mathbb{S} \cup \{\emptyset\})$. If A contains two distinct paths α and β of the same length, then $[\{\alpha, \beta\}, \emptyset]$ is open, contains A, and does not meet $\mathbb{S} \cup \{\emptyset\}$. If there is some $\alpha \in A$ and a proper initial subpath β of α such that $\beta \notin A$, then $[\{\alpha\}, \{\beta\}]$ is open, contains A, and does not meet $\mathbb{S} \cup \{\emptyset\}$. Failing this, $A \in \mathbb{S} \cup \{\emptyset\}$, which we assumed is false. Therefore $\mathbb{S} \cup \{\emptyset\}$ is closed in $\mathbb{P}(E^\star)$, which implies it is compact.

We now work out what the subspace topology is on $\mathbb{S}$. Let $P, N \subseteq_{\text{finite}} E^\star$. If $[P, N] \cap \mathbb{S} \neq \emptyset$ then P contains a unique path ρ of maximal length (because of the way $\mathbb{S}$ is defined) and $[P, N] \cap \mathbb{S} = [\{\rho\}, N'] \cap \mathbb{S}$ where

$$N' = \{\eta \in N \mid \rho \text{ is an initial subpath of } \eta\}.$$

Therefore, the topology on $\mathbb{S}$ is generated by basic open sets of the form $[\{\rho\}, N'] \cap \mathbb{S}$ where $\rho \in E^\star$ and $N' \subseteq E^\star$ is a finite set of paths that are proper extensions of ρ.

Note that $\mathbb{S} = \bigsqcup_{v \in E^0} [\{v\}, \emptyset] \cap \mathbb{S}$. For each $v \in E^0$, the set $[\{v\}, \emptyset]$ is closed in $\mathbb{P}(E^\star)$ because $\mathbb{P}(E^\star) \setminus [\{v\}, \emptyset] = [\emptyset, \{v\}]$ is open. Since $[\{v\}, \emptyset] \cap \mathbb{S} = [\{v\}, \emptyset] \cap (\mathbb{S} \cup \{\emptyset\})$ and $\mathbb{S} \cup \{\emptyset\}$ is closed in $\mathbb{P}(E^\star)$, we have that $[\{v\}, \emptyset] \cap \mathbb{S}$ is closed in $\mathbb{P}(E^\star)$, and therefore compact. This proves that $\mathbb{S}$ is locally compact, because it is Hausdorff and every point has a compact neighbourhood.

Now we show that $E^\star \cup E^\infty$ is homeomorphic to $\mathbb{S}$. Define the map

$$\Psi : E^\star \cup E^\infty \to \mathbb{S},$$
$$\Psi(x) = \{\nu \in E^\star \mid \nu \text{ is an initial subpath of } x\}.$$

It is clear that Ψ is a bijection. Let $\rho \in E^\star$ and let $N' \subseteq E^\star$ be a finite set of paths that properly extend ρ. Then

$$\Psi^{-1}\big([\{\rho\}, N'] \cap \mathbb{S}\big) = C(\rho) \setminus \bigcup_{\rho\beta \in N'} C(\rho\beta) = \bigcap_{\rho\beta \in N'} C(\rho) \setminus C(\rho\beta).$$

It is not difficult to see that for each $\rho\beta \in N'$, the set

$$C(\rho) \setminus C(\rho\beta) = C\big(\rho, \{\beta_1\}\big) \cup C\big(\rho\beta_1, \{\beta_2\}\big) \cup \cdots \cup C\big(\rho\beta_{|\beta|-1}, \{\beta_{|\beta|}\}\big)$$

is open. Therefore $\Psi^{-1}\big([\{\rho\}, N'] \cap \mathbb{S}\big)$ is open in $E^\star \cup E^\infty$. Consequently, Ψ is continuous. If $\alpha \in E^\star$ and $F \subseteq_{\text{finite}} r(\alpha)E^1$, then $C(\alpha, F)$ is mapped to an open set in $\mathbb{S}$:

$$\Psi\big(C(\alpha, F)\big) = \big[\{\alpha\}, N'\big] \cap \mathbb{S}$$

where $N' = \{\alpha e \mid e \in F\}$. It follows that Ψ is a homeomorphism and $E^\star \cup E^\infty$ is Hausdorff.

Since we showed that $\big[\{v\}, \emptyset\big] \cap \mathbb{S}$ is compact, it follows that

$$C(v) = \Psi^{-1}\big([\{v\}, \emptyset] \cap \mathbb{S}\big)$$

is compact, for all $v \in E^0$. To show that $C(\alpha)$ is compact for all $\alpha \in E^\star$, we proceed by induction on the length of α. If $e \in E^1$, then $C(s(e)) \setminus C(e) = C(s(e), \{e\})$ is a basic open set, so $C(e)$ is closed in $C(s(e))$, hence compact. Assume $C(\alpha)$ is compact for any $\alpha \in E^\star$ with $|\alpha| = n$. If $\mu \in E^\star$ has $|\mu| = n + 1$ then let $\mu' = \mu_1\mu_2 \dots \mu_n$. We have that $C(\mu') \setminus C(\mu) = C(\mu', \{\mu_{n+1}\})$ is a basic open set, so $C(\mu)$ is closed in $C(\mu')$, hence compact. By induction, $C(\alpha)$ is compact for arbitrary $\alpha \in E^\star$. Finally, if $F \subseteq_{\text{finite}} r(\alpha)E^1$ then $C(\alpha) \setminus C(\alpha, F) = \bigcup_{e \in F} C(\alpha e)$ is open, so $C(\alpha, F)$ is compact. □

Recall that a topological space is called *second-countable* if it has a countable base, *first-countable* if every point has a countable neighbourhood base, and σ*-compact* if it is a countable union of compact subsets.

Theorem 2.2 *The path space $E^\star \cup E^\infty$ is:*

(1) *second-countable if and only if E is a countable graph;*
(2) *first-countable if and only if E is a row-countable graph;*
(3) σ*-compact if and only if E^0 is countable.*

Proof (1) If E is a countable graph (i.e., $E^0 \cup E^1$ is countable) then $E^\star$ is countable. The base of open sets Eq. 2.7 is countable too, because there are only countably many pairs (α, F) where $\alpha \in E^\star$ and $F \subseteq_{\text{finite}} r(\alpha)E^1$. This proves the topology is second-countable. Conversely, if one of E^0 or E^1 is uncountable, then one of $\{C(v) \mid v \in E^0\}$ or $\{C(e) \mid e \in E^1\}$ is an uncountable set of pairwise disjoint open sets, so $E^\star \cup E^\infty$ is not second-countable.

(2) Notice that the following sets are neighbourhood bases at $\alpha \in E^\star$ and $p \in E^\infty$ respectively,:

$$\mathcal{N}_\alpha = \{C(\alpha, F) \mid F \subseteq_{\text{finite}} r(\alpha)E^1\}, \qquad \mathcal{N}_p = \{C(p_1 \dots p_m) \mid m \geq 1\}.$$

Regardless of the graph, $\mathcal{N}_p$ is countable for every $p \in E^\infty$. If a finite path $\alpha \in E^\star$ has the property that $r(\alpha)E^1$ is countable, then $\mathcal{N}_\alpha$ is countable, because there are only countably many finite subsets F of $r(\alpha)E^1$. So, for every row-countable graph E, the path space $E^\star \cup E^\infty$ is first-countable. Conversely, suppose there exists $v \in E^0$ such that vE^1 is uncountable. Towards a contradiction, assume v has a countable neighbourhood base $\mathcal{B}_v = \{B_1, B_2, \dots, \}$. By replacing B_n, for all $n \geq 1$, with a set of the form $C(v, F_n) \subseteq B_n$, where $F_n \subseteq_{\text{finite}} vE^1$, we have a countable neighbourhood base for v of the form $\mathcal{C}_v = \{C(v, F_1), C(v, F_2), \dots\}$. Since $\bigcup_{n=1}^\infty F_n$ is countable, one can choose $e \in vE^1 \setminus \bigcup_{n=1}^\infty F_n$. Then every neighbourhood of v contains e, which is absurd, because the space is Hausdorff. Therefore $E^\star \cup E^\infty$ is first-countable if and only if E is row-countable.

(3) If E^0 is countable then the path space is σ-compact, because $E^\star \cup E^\infty = \bigcup_{v \in E^0} C(v)$ and $C(v)$ is compact for every $v \in E^0$, by Theorem 2.1. For the converse, suppose $E^\star \cup E^\infty$ is σ-compact. Then there is a sequence of compact subsets $(K_n)_1^\infty$ such that $E^\star \cup E^\infty = \bigcup_{n=1}^\infty K_n$. Each K_n is compact, so it can be covered by a finite subcover of $\{C(v) \mid v \in E^0\}$, implying that there is a countable set $S \subseteq E^0$ such that $E^\star \cup E^\infty = \bigcup_{v \in S} C(v)$. But this implies $S = E^0$ because $C(v)$ and $C(w)$ are disjoint unless $v = w$. □

We now prove an easy fact that forms a bridge to the next section, where we shall construct a groupoid with unit space $\partial E = E^\infty \cup \{\alpha \in E^\star \mid r(\alpha) \in E^0_{\text{sing}}\}$.

Proposition 2.16 *The boundary path space ∂E is closed in $E^\star \cup E^\infty$.*

Proof The complement of ∂E consists of isolated points. Indeed, if $\mu \in (E^\star \cup E^\infty) \setminus \partial E$, then $r(\mu)$ is a regular vertex, and $C(\mu, r(\mu)E^1) = \{\mu\}$ is open in $E^\star \cup E^\infty$. □

An immediate consequence of Theorem 2.1 and Proposition 2.16 is that ∂E is a locally compact Hausdorff space with the base of compact open sets:

$$Z(\alpha, F) = C(\alpha, F) \cap \partial E; \qquad \alpha \in E^\star,\ F \subseteq_{\text{finite}} r(\alpha)E^1.$$

For $\alpha \in E^\star$, we define $Z(\alpha) = Z(\alpha, \emptyset)$, which is the same as $Z(\alpha) = C(\alpha) \cap \partial E$. As it were, the sets $Z(\alpha, F)$ are very rarely empty. In particular, $Z(\alpha) \neq \emptyset$ for all $\alpha \in E^\star$; in other words, every finite path can be extended to a boundary path.

Lemma 2.5 *Let $\alpha \in E^\star$ and let $F \subseteq_{\text{finite}} r(\alpha)E^1$. Then $Z(\alpha, F) = \emptyset$ if and only if $r(\alpha)$ is a regular vertex and $F = r(\alpha)E^1$.*

Proof ($\Rightarrow$) Assume $Z(\alpha, F) = \emptyset$. If $r(\alpha)$ were a singular vertex then it would imply $\alpha \in Z(\alpha, F)$. Therefore $r(\alpha)$ is regular, so $r(\alpha)E^1 \neq \emptyset$. Towards a contradiction, assume F is a proper subset of $r(\alpha)E^1$. Then there exists some $x_1 \in r(\alpha)E^1 \setminus F$.

Assume that we have a path $x_1 x_2 \dots x_n \in r(\alpha)E^\star$. If $r(x_n)$ is a sink, let $x = x_1 \dots x_n$. Otherwise, let $x_{n+1} \in r(x_n)E^1$. Inductively, this constructs $x \in r(\alpha)\partial E$ such that $\alpha x \in Z(\alpha, F)$. Since this is a contradiction, it proves $F = r(\alpha)E^1$.

($\Leftarrow$) If $r(\alpha)$ is regular, then $Z(\alpha) = \bigcup_{e \in r(\alpha)E^1} Z(\alpha e)$, so $Z(\alpha, r(\alpha)E^1) = \emptyset$. □

Theorem 2.3 *The boundary path space ∂E is:*

(1) *second-countable if and only if E is a countable graph;*
(2) *first-countable if and only if E is a row-countable graph;*
(3) *σ-compact if and only if E^0 is countable.*

Proof Together with Lemma 2.5, the proof is almost identical to Theorem 2.2. □

2.3.3 The Boundary Path Groupoid

In this section, we define the boundary path groupoid of a graph (see [24, Example 2.1]) and investigate some of its algebraic and topological properties. Throughout, let $E = (E^0, E^1, r, s)$ be an arbitrary graph.

Define the *one-sided shift map* $\sigma : \partial E \setminus E^0 \to \partial E$ as follows:

$$\sigma(x) = \begin{cases} r(x) & \text{if } x \in E^\star \cap \partial E \text{ and } |x| = 1 \\ x_2 \dots x_{|x|} & \text{if } x \in E^\star \cap \partial E \text{ and } |x| \geq 2 \\ x_2 x_3 \dots & \text{if } x \in E^\infty \end{cases}$$

The n-fold composition σ^n is defined on paths of length $\geq n$ and we understand that $\sigma^0 : \partial E \to \partial E$ is the identity map.

Definition 2.8 Let k be an integer and let $x, y \in \partial E$. We say that x and y are **tail equivalent with lag** k, written $x \sim_k y$, if there exists some $n \geq \max\{0, k\}$ such that

$$\sigma^n(x) = \sigma^{n-k}(y).$$

If an integer k exists such that $x \sim_k y$, we say that x and y are *tail equivalent*, and write $x \sim y$.

An equivalent definition is that $x \sim_k y$ if there exists $(\alpha, \beta) \in E^\star \times_r E^\star$ and $z \in r(\alpha)\partial E$, such that $x = \alpha z$, $y = \beta z$, and $|\alpha| - |\beta| = k$. Something that is potentially counter-intuitive about these relations is that the lag is not necessarily unique: it is possible to have $x \sim_k y$ and $x \sim_\ell y$ even when $k \neq \ell$. It is straightforward to prove from the definition that for all $x, y, z \in \partial E$:

$$\begin{aligned} x &\sim_0 x, \\ x \sim_k y &\implies y \sim_{-k} x, \\ x \sim_k y \text{ and } y \sim_\ell z &\implies x \sim_{k+\ell} z, \\ x \sim_k y &\implies x, y \in E^\star \text{ or } x, y \in E^\infty. \end{aligned}$$

This shows that $\sim$ is an equivalence relation on ∂E that respects the partition between finite and infinite paths.

Definition 2.9 The **boundary path groupoid** of a graph E is

$$\begin{aligned}\mathcal{G}_E &= \big\{(x, k, y) \mid x, y \in \partial E,\ x \sim_k y\big\} \\ &= \big\{(\alpha x, |\alpha| - |\beta|, \beta x) \mid (\alpha, \beta) \in E^\star \times_r E^\star, x \in r(\alpha)\partial E\big\}\end{aligned}$$

where a morphism $(x, k, y) \in \mathcal{G}_E$ has domain y and codomain x. The composition of morphisms and their inverses are defined by the formulae:

$$(x, k, y)(y, l, z) = (x, k + l, z), \qquad (x, k, y)^{-1} = (y, -k, x).$$

The unit space is $\mathcal{G}_E^{(0)} = \{(x, 0, x) \mid x \in \partial E\}$, which we silently identify with ∂E (see Remark 2.1). The orbits in ∂E are tail equivalence classes.

Example 2.6 Consider this graph, called the *rose with two petals*:

$R_2 \quad =$

A standard diagonal argument proves that ∂R_2 is an uncountable set. There are uncountably many orbits in ∂R_2, but the topology on ∂R_2 is second-countable and even metrisable. In fact, it can be shown that ∂R_2 is homeomorphic to the Cantor set $\{0, 1\}^{\mathbb{N}}$.

A boundary path $p \in \partial E$ is called *eventually periodic* if it is of the form $p = \mu\epsilon\epsilon \ldots \in E^\infty$ where $\mu, \epsilon \in E^\star$ and ϵ is a closed path of positive length (note that ϵ is not necessarily a cycle). The following result is [65, Proposition 4.2], but we prove it a bit more formally here.

Proposition 2.17 *If E is a graph and $p \in \partial E$, then the isotropy group at p is:*

(1) *infinite cyclic if p is eventually periodic;*
(2) *trivial if p is not eventually periodic.*

Proof (1) Assume $p = \mu\epsilon\epsilon \ldots \in E^\infty$ where $\mu, \epsilon \in E^\star$, $r(\mu) = s(\epsilon) = r(\epsilon)$, and assume ϵ is minimal in the sense that it has no initial subpath δ such that $\epsilon = \delta^n$ for some $n > 1$. Let $(p, k, p) \in p(\mathcal{G}_E)p$ and suppose $k \geq 0$. Then $p \sim_k p$ implies that for all sufficiently large $n \geq 0$, we have $\sigma^{|\mu|+n|\epsilon|+k}(p) = \sigma^{|\mu|+n|\epsilon|}(p)$. This yields:

$$\sigma^{|\mu|+n|\epsilon|+k}(p) = \sigma^k(\epsilon\epsilon \ldots) = \sigma^{|\mu|+n|\epsilon|}(p) = \epsilon\epsilon \ldots.$$

Let $m = k \mod |\epsilon|$. Then $0 \leq m < |\epsilon|$ and

$$\sigma^k(\epsilon\epsilon \ldots) = \sigma^m(\epsilon\epsilon \ldots) = \epsilon_{m+1} \ldots \epsilon_{|\epsilon|}\epsilon\epsilon \ldots = \epsilon_1 \ldots \epsilon_m\epsilon\epsilon \ldots.$$

Since ϵ is minimal, this implies $m = 0$, so $k \mid |\epsilon|$. On the other hand, if $k < 0$ then $(p, -k, p) = (p, k, p)^{-1} \in p(\mathcal{G}_E)p$ and the same argument establishes $k \mid |\epsilon|$. The conclusion is that $p(\mathcal{G}_E)p$ is the infinite cyclic group generated by $(p, |\epsilon|, p)$.

(2) Let $(p, k, p) \in p(\mathcal{G}_E)p$. Then $p \sim_k p$ implies $p = \alpha x = \beta x$ for some $(\alpha, \beta) \in E^\star \times_r E^\star$ and $x \in r(\alpha)\partial E$, with $|\alpha| - |\beta| = k$. If p is finite, this implies $\alpha = \beta$, so $k = 0$. That is, the isotropy group at p is trivial. On the other hand, suppose p is infinite and not eventually periodic. If $|\alpha| < |\beta|$, then $\beta = \alpha\beta'$ for some $\beta' \in E^\star$. But then $p = \alpha x = \beta x = \alpha\beta' x$, so $x = \beta' x = \beta'\beta' x = \beta'\beta'\beta' \ldots$, and this proves p is eventually periodic, a contradiction. Similarly, assuming $|\beta| < |\alpha|$ reaches the same contradiction. Therefore, $|\alpha| = |\beta|$ and $k = 0$, implying that the isotropy group at p is trivial. □

The next step is to define a topology on $\mathcal{G}_E$. Let $(\alpha, \beta) \in E^\star \times_r E^\star$, and let $F \subseteq_{\text{finite}} r(\alpha)E^1$. Define the sets[1]:

$$\mathcal{Z}(\alpha, \beta) = \big\{(\alpha x, |\alpha| - |\beta|, \beta x) \mid x \in r(\alpha)\partial E\big\};$$
$$\mathcal{Z}(\alpha, \beta, F) = \mathcal{Z}(\alpha, \beta) \setminus \bigcup_{e \in F} \mathcal{Z}(\alpha e, \beta e).$$

Obviously, $\mathcal{Z}(\alpha, \beta) = \mathcal{Z}(\alpha, \beta, \emptyset)$. Next we present a pair of technical lemmas (generalising [50, Lemma 2.5]) which prove that the collection of sets of the form $\mathcal{Z}(\alpha, \beta, F)$ is closed under pairwise intersections, so it can serve as a base for a topology on $\mathcal{G}_E$.

Lemma 2.6 *Let $(\alpha, \beta), (\gamma, \delta) \in E^\star \times_r E^\star$. Then*

$$\mathcal{Z}(\alpha, \beta) \cap \mathcal{Z}(\gamma, \delta) = \begin{cases} \mathcal{Z}(\alpha, \beta) & \text{if } \exists\, \kappa \in E^\star,\ \alpha = \gamma\kappa,\ \beta = \delta\kappa \\ \mathcal{Z}(\gamma, \delta) & \text{if } \exists\, \kappa \in E^\star,\ \gamma = \alpha\kappa,\ \delta = \beta\kappa \\ \emptyset & \text{otherwise.} \end{cases}$$

Proof We prove that when the intersection of the two sets is nonempty, then it must be one of the first two cases in the piecewise expression. To this end, let $(\alpha x, |\alpha| - |\beta|, \beta x) = (\gamma x', |\gamma| - |\delta|, \delta x') \in \mathcal{Z}(\alpha, \beta) \cap \mathcal{Z}(\gamma, \delta)$, where $x \in r(\alpha)\partial E$ and $x' \in r(\gamma)\partial E$. Assume $|\gamma| \le |\alpha|$, which implies $|\delta| \le |\beta|$; if not, rearrange. Since $\alpha x = \gamma x'$, it must be that $\alpha = \gamma\kappa$ where κ is the initial subpath of x' of length $|\alpha| - |\gamma|$. Similarly, $\beta = \delta\kappa$. So we are in the first case (or the second case, if a rearrangement took place). In the first two cases in the piecewise expression, it is clear from the definitions what the intersection of $\mathcal{Z}(\alpha, \beta)$ and $\mathcal{Z}(\gamma, \delta)$ must be. □

Lemma 2.7 *Suppose $(\alpha, \beta), (\gamma, \delta) \in E^\star \times_r E^\star$, $F \subseteq_{\text{finite}} r(\alpha)E^1$, and $H \subseteq_{\text{finite}} r(\gamma)E^1$. Then*

[1] Note the subtle difference in notation: we were using Z for basic open sets in ∂E and now we are using $\mathcal{Z}$ for basic open sets in $\mathcal{G}_E$.

$$\mathcal{Z}(\alpha,\beta,F)\cap\mathcal{Z}(\gamma,\delta,H)=\begin{cases}\mathcal{Z}(\alpha,\beta,F\cup H) & \textit{if } \alpha=\gamma,\ \beta=\delta\\ \mathcal{Z}(\alpha,\beta,F) & \textit{if } \exists\,\kappa\in E^\star,\ |\kappa|\geq 1,\ \alpha=\gamma\kappa,\ \beta=\delta\kappa,\ \kappa_1\notin H\\ \mathcal{Z}(\gamma,\delta,H) & \textit{if } \exists\,\kappa\in E^\star,\ |\kappa|\geq 1,\ \gamma=\alpha\kappa,\ \delta=\beta\kappa,\ \kappa_1\notin F\\ \emptyset & \textit{otherwise.}\end{cases}$$

Proof We make a calculation and then proceed by cases:

$$\begin{aligned}\mathcal{Z}(\alpha,\beta,F)\cap\mathcal{Z}(\gamma,\delta,H)&=\left[\mathcal{Z}(\alpha,\beta)\setminus\bigcup_{e\in F}\mathcal{Z}(\alpha e,\beta e)\right]\bigcap\left[\mathcal{Z}(\gamma,\delta)\setminus\bigcup_{e\in H}\mathcal{Z}(\gamma e,\delta e)\right] \quad (2.9)\\ &=\left[\mathcal{Z}(\alpha,\beta)\cap\mathcal{Z}(\gamma,\delta)\right]\setminus\left[\bigcup_{e\in F}\mathcal{Z}(\alpha e,\beta e)\cup\bigcup_{e\in H}\mathcal{Z}(\gamma e,\delta e)\right].\end{aligned}$$

Case 1: If $\alpha=\gamma$ and $\beta=\delta$, Eq. 2.9 yields $\mathcal{Z}(\alpha,\beta,F)\cap\mathcal{Z}(\gamma,\delta,H)=\mathcal{Z}(\alpha,\beta,F\cup H)$.

Case 2: If there exists $\kappa\in E^\star\setminus E^0$ such that $\alpha=\gamma\kappa$ and $\beta=\delta\kappa$ then after applying Lemma 2.6, the right hand side of Eq. 2.9 becomes

$$\mathcal{Z}(\alpha,\beta)\setminus\left[\bigcup_{e\in F}\mathcal{Z}(\alpha e,\beta e)\cup\bigcup_{e\in H}\mathcal{Z}(\gamma e,\delta e)\right].$$

Moreover, $\mathcal{Z}(\alpha,\beta)\cap\mathcal{Z}(\gamma e,\delta e)=\emptyset$ for all $e\in H$, provided $e\neq\kappa_1$. If $e=\kappa_1$ then $\mathcal{Z}(\alpha,\beta)\cap\mathcal{Z}(\gamma e,\delta e)=\mathcal{Z}(\alpha,\beta)$. Therefore Eq. 2.9 becomes $\mathcal{Z}(\alpha,\beta,F)$ if $\kappa_1\notin H$ and $\emptyset$ if $\kappa_1\in H$.

Case 3: If there exists $\kappa\in E^\star\setminus E^0$ such that $\gamma=\alpha\kappa$ and $\delta=\beta\kappa$ then the situation is symmetric to the second case.

Case 4: Otherwise, $\mathcal{Z}(\alpha,\beta)\cap\mathcal{Z}(\gamma,\delta)=\emptyset$, by Lemma 2.6. □

From now on, we assume $\mathcal{G}_E$ has the topology generated by all the sets:

$$\mathcal{Z}(\alpha,\beta,F);\qquad (\alpha,\beta)\in E^\star\times_r E^\star,\ F\subseteq_{\text{finite}} r(\alpha)E^1. \tag{2.10}$$

Some of our references give a different base for the topology on $\mathcal{G}_E$, but all the different bases that we know of contain the sets $\mathcal{Z}(\alpha,\beta,F)$. There are advantages to working with a base that is not too large, which is why we have chosen to focus on this one.

Let E be a graph and consider $\mathbb{Z}$ with the discrete topology. The map

$$\theta:\mathcal{G}_E\to\mathbb{Z},\qquad (x,k,y)\mapsto k,$$

is a continuous groupoid homomorphism. In fact, it is a degree map giving $\mathcal{G}_E$ the structure of a $\mathbb{Z}$-graded groupoid. Some parts of this lemma are reminiscent of [50, Proposition 2.6].

Lemma 2.8 *Let E be a graph.*

(1) *The topology on $\mathcal{G}_E$ is Hausdorff.*
(2) *$\boldsymbol{d} : \mathcal{G}_E \to \partial E$ is a local homeomorphism.*
(3) *If $(\alpha, \beta) \in E^\star \times_r E^\star$ and $F \subseteq_{\text{finite}} r(\alpha)E^1$, then $\mathcal{Z}(\alpha, \beta, F)$ is compact.*

Proof (1) Take $(x, k, y) \neq (w, \ell, z)$ in $\mathcal{G}_E$. If $k \neq \ell$ then $\theta^{-1}(k)$ and $\theta^{-1}(\ell)$ are disjoint open sets separating the two points. Otherwise, either $x \neq w$ or $y \neq z$. If $w \neq x$ then either: w and x must differ on some initial segment, or one must be an initial subpath of the other. Using Lemma 2.7, it is not difficult to separate the two points by disjoint open sets. If $y \neq z$, the same reasoning applies.

(2) For $(\alpha, \beta) \in E^\star \times_r E^\star$, define

$$h_{\alpha,\beta} : Z(\beta) \to \mathcal{Z}(\alpha, \beta), \qquad \beta x \mapsto (\alpha x, |\alpha| - |\beta|, \beta x).$$

Clearly, $h_{\alpha,\beta}$ is a bijection. By Lemma 2.7, the basic open sets contained in $\mathcal{Z}(\alpha, \beta)$ are all of the form $\mathcal{Z}(\alpha\kappa, \beta\kappa, F')$ where $\kappa \in r(\alpha)E^\star$ and $F' \subseteq_{\text{finite}} r(\kappa)E^1$. Clearly

$$h_{\alpha,\beta}^{-1}\big(\mathcal{Z}(\alpha\kappa, \beta\kappa, F')\big) = Z(\beta\kappa, F')$$

is open in $Z(\beta)$, so $h_{\alpha,\beta}$ is continuous. A continuous map from a compact space to a Hausdorff space is a closed map, so $h_{\alpha,\beta}$ is a closed map. Therefore $h_{\alpha,\beta}$ is a homeomorphism. This proves that $\boldsymbol{d}|_{\mathcal{Z}(\alpha,\beta)}$ is a homeomorphism onto its image (because $\boldsymbol{d}|_{\mathcal{Z}(\alpha,\beta)}^{-1} = h_{\alpha,\beta}$).

(3) According to item (2), $\boldsymbol{d}$ restricts to a homeomorphism $\mathcal{Z}(\alpha, \beta, F) \approx Z(\beta, F)$, and $Z(\beta, F)$ is compact by Theorem 2.1. □

Since $\mathcal{Z}(\alpha, \beta, F) \approx Z(\beta, F)$, Lemma 2.5 implies that $\mathcal{Z}(\alpha, \beta, F) = \emptyset$ if and only if $r(\alpha) = r(\beta)$ is a regular vertex and $F = r(\alpha)E^1$.

Remark 2.6 The groupoid $\mathcal{G}_E$ admits continuous maps

$$\boldsymbol{c} : (x, k, y) \mapsto x, \qquad \theta : (x, k, y) \mapsto k, \qquad \boldsymbol{d} : (x, k, y) \mapsto y,$$

so it is tempting to think that the topology on $\mathcal{G}_E$ coincides with the relative topology that it gets from being a subset of the product space $\partial E \times \mathbb{Z} \times \partial E$. However, this is not the case: the topology on $\mathcal{G}_E$ is much finer than the relative topology from $\partial E \times \mathbb{Z} \times \partial E$.

The main theorem that follows is not new, and it has been in use for some time. Indeed, it is implied by [61, Lemma 2.1], although not in a trivial way (see also [57, Theorem 3.5] and [71, Theorem 3.16]). However, this is the first self-contained proof that we know of that applies to ordinary directed graphs, and does not require the graph to be countable.

Theorem 2.4 *Let E be a graph. The groupoid $\mathcal{G}_E$ is a Hausdorff ample groupoid with the base of compact open bisections given in Eq. 2.10.*

Proof The most technical part that remains is showing that the composition map $\boldsymbol{m}$ is continuous. If $x, z \in E^\star \cap \partial E$ are tail equivalent finite paths, then $(x, |x| - |z|, z)$ has a neighbourhood base of open sets, $\mathcal{N}_{(x,|x|-|z|,z)} = \{\mathcal{Z}(x, z, F) \mid F \subseteq_{\text{finite}} r(x)E^1\}$. If $x, z \in E^\infty$ are tail equivalent infinite paths, with lag t, then there exists $N \geq 0$ such that $\sigma^{N+t}(x) = \sigma^N(z)$. Consequently (x, t, z) has a neighbourhood base of open sets, $\mathcal{N}_{(x,t,z)} = \{\mathcal{Z}(x_1 \dots x_{n+t}, z_1 \dots z_n) \mid n > N\}$.

Now suppose U is an open set in $\mathcal{G}_E$ containing a product of two morphisms $(x, k + \ell, z) = (x, k, y)(y, \ell, z)$. It must be that x, y, z are all finite paths or they are all infinite paths. If x, y, z are finite paths, then they must have $r(x) = r(y) = r(z)$ and U must contain some $\mathcal{Z}(x, z, F) \in \mathcal{N}_{(x,|x|-|z|,z)}$. Then $\big((x, k, y), (y, \ell, z)\big)$ is contained in the open set $\big(\mathcal{Z}(x, y, F) \times \mathcal{Z}(y, z, F)\big) \cap \mathcal{G}_E^{(2)}$ which is mapped bijectively by $\boldsymbol{m}$ into $Z(x, z, F) \subseteq U$. Otherwise x, y, z are all infinite paths, and there must exist n large enough that $\sigma^{n+k+\ell}(x) = \sigma^{n+\ell}(y) = \sigma^n(z)$. Making n even larger if necessary, we can assume U contains some $\mathcal{Z}(x_1 \dots x_{n+k+\ell}, z_1 \dots z_n) \in \mathcal{N}_{(x,k+\ell,z)}$. Define:

$$x' = x_1 \dots x_{n+k+\ell}, \qquad y' = y_1 \dots y_{n+\ell}, \qquad z' = z_1 \dots z_n.$$

Then $\big((x, k, y), (y, \ell, z)\big)$ is contained in the open set $\big(\mathcal{Z}(x', y') \times \mathcal{Z}(y', z')\big) \cap \mathcal{G}_E^{(2)}$, which is mapped bijectively by $\boldsymbol{m}$ into $Z(x', z') \subseteq U$. Since $(x, k + \ell, z) = (x, k, y)(y, \ell, z)$ was an arbitrary product in U, this shows that $\boldsymbol{m}^{-1}(U)$ is open in $\mathcal{G}_E^{(2)}$, so $\boldsymbol{m}$ is continuous. It is much easier to show that the inversion map $\boldsymbol{i}$ is continuous, because $\boldsymbol{i}$ puts $Z(\alpha, \beta, F)$ in bijection with $Z(\beta, \alpha, F)$. We have proved $\mathcal{G}_E$ is a topological groupoid. In Lemma 2.8 (2), it is shown that $\boldsymbol{d}$ is a local homeomorphism. Therefore, $\mathcal{G}_E$ is an étale groupoid. The remaining facts from Lemma 2.8 establish that $\mathcal{G}_E$ is a Hausdorff ample groupoid and that the base described in Eq. 2.10 consists of compact open bisections. □

2.4 The Leavitt Path Algebra of a Graph

In Sect. 2.4.1, we define the Leavitt path algebra of a graph. We define it in terms of its universal property, and then describe how it can be realised as the quotient of a path algebra. Path algebras are, in some sense, the definitive examples of $\mathbb{Z}$-graded algebras, and the $\mathbb{Z}$-grading survives in their Leavitt path algebra quotients. In Sect. 2.4.2, we prove the Graded Uniqueness Theorem for Leavitt path algebras. In Sect. 2.4.3, we prove the cornerstone result: the Leavitt path algebra of a graph is isomorphic to the Steinberg algebra of its boundary path groupoid. Through this lens, we rederive some fundamentals of Leavitt path algebras, and classify finite-dimensional Leavitt path algebras. In Sect. 2.4.4, we prove the Graded and Cuntz–Krieger Uniqueness Theorems for Steinberg algebras and use them to prove the Cuntz–Krieger Uniqueness Theorem for Leavitt path algebras.

Remark 2.7 Historically, the theory of Leavitt path algebras was developed for the case when R is a field, and E is a row-finite countable graph. Later, the methods were improved and R could be any unital commutative ring if E is a countable graph [3, 68]. Alternatively, E could be an arbitrary graph if R is a field [7, 43]. The proofs of some key results, including the fact that the relations on $L_R(E)$ do not collapse the algebra to zero ([43, Lemma 1.5] and [68, Proposition 3.4]) and the Graded Uniqueness Theorem ([43, Proposition 3.6] and [68, Theorem 5.3]), have not yet been recorded for the case where simultaneously E is uncountable and R is not a field. Here, we fix this and complete the picture.

2.4.1 Introducing Leavitt Path Algebras

Let $E = (E^0, E^1, r, s)$ be a graph. We introduce the set of formal symbols $(E^1)^* = \{e^* \mid e \in E^1\}$ and call the elements of $(E^1)^*$ *ghost edges*. For clarity, we will sometimes refer to the elements of E^1 as *real edges*. If $\alpha = \alpha_1 \dots \alpha_{|\alpha|} \in E^\star$ is a finite path of positive length, we define α^* to be the sequence $\alpha^*_{|\alpha|} \dots \alpha^*_1$, and call it a *ghost path*. We also define $v^* = v$ for every $v \in E^0$.

Definition 2.10 [68] Let E be a graph and let A be a ring. Assume

$$\{v, e, e^* \mid v \in E^0, e \in E^1\}$$

is a subset of A; in other words, there is a function $E^0 \sqcup E^1 \sqcup (E^1)^* \to A$ whose image inherits the notation of its domain. Then $\{v, e, e^* \mid v \in E^0, e \in E^1\} \subset A$ is called a **Leavitt E-family** if the following conditions are satisfied:

(V) $v^2 = v$ and $vw = 0$ for all $v, w \in E^0, v \neq w$;
(E1) $s(e)e = er(e) = e$ for all $e \in E^1$;
(E2) $e^*s(e) = r(e)e^* = e^*$ for all $e \in E^1$;
(CK1) $e^*e = r(e)$ and $e^*f = 0$ for all $e, f \in E^1, e \neq f$;
(CK2) $v = \sum_{e \in vE^1} ee^*$ for all $v \in E^0_{\text{reg}}$.

The interpretation of (V) is that $\{v \in A \mid v \in E^0\}$ is a set of pairwise orthogonal idempotents. The relations (CK1) and (CK2) are called the Cuntz–Krieger relations, and they originate from operator theory. The relevant interpretation, at least in that setting, is that vertices are represented by projections, and edges are represented by partial isometries with mutually orthogonal ranges.

In any algebra A containing a Leavitt E-family $\{v, e, e^* \mid v \in E^0, e \in E^1\}$, one can consider paths $\mu = \mu_1 \dots \mu_{|\mu|}$ and ghost paths $\mu^* = \mu^*_{|\mu|} \dots \mu^*_1$ as elements of A in the obvious way: products of their constituent real edges and ghost edges respectively. The following lemma is straightforward to prove using the relations (E1), (E2), and (CK1). It is so fundamental that we will usually use the result without referring to it.

Lemma 2.9 *If A is an R-algebra generated by a Leavitt E-family*

$$\{v, e, e^* \mid v \in E^0, e \in E^1\},$$

the elements of A obey the rule:

$$(r\mu\nu^*)(r'\gamma\lambda^*) = \begin{cases} (rr')\mu\kappa^*\lambda^* & \text{if } \gamma \text{ is an initial subpath of } \nu, \text{ with } \nu = \gamma\kappa \\ (rr')\mu\kappa\lambda^* & \text{if } \nu \text{ is an initial subpath of } \gamma, \text{ with } \gamma = \nu\kappa \\ 0 & \text{otherwise} \end{cases}$$

for all $r, r' \in R$ and all $\mu, \nu, \gamma, \lambda \in E^\star$, with $r(\mu) = r(\nu)$ and $r(\gamma) = r(\lambda)$.

Corollary 2.3 *Every R-algebra generated by a Leavitt E-family is generated, as an abelian group, by the set $\{r\alpha\beta^* \mid r \in R, (\alpha, \beta) \in E^\star \times_r E^\star\}$.*

Proof By Lemma 2.9, every word in the generators $\{v, e, e^* \mid v \in E^0, e \in E^1\}$ reduces to an expression of the form $\alpha\beta^*$ where $\alpha, \beta \in E^\star$. Moreover, $\alpha\beta^* = 0$ unless $r(\alpha) = r(\beta)$, by (V), (E1), and (E2). □

Let B be an R-algebra generated by a Leavitt E-family $\{v, e, e^* \mid v \in E^0, e \in E^1\}$. We say that B is *universal* (for Leavitt E-families) if every R-algebra A containing a Leavitt E-family $\{a_v, b_e, c_{e^*} \mid v \in E^0, e \in E^1\}$ admits a unique R-algebra homomorphism $\pi : B \to A$ such that $\pi(v) = a_v$, $\pi(e) = b_e$, and $\pi(e^*) = c_{e^*}$ for every $v \in E^0$ and $e \in E^1$. The universal property determines B up to isomorphism.

Definition 2.11 Let E be a graph. The **Leavitt path algebra** of E with coefficients in R, denoted by $L_R(E)$, is the universal R-algebra generated by a Leavitt E-family.

Technically, $L_R(E)$ is an isomorphism class in the category of R-algebras. If B is a specific R-algebra having the universal property for Leavitt E-families, then B is a *model* of $L_R(E)$. However, it is customary and natural to refer to $L_R(E)$ as if it were a specific model with the standard generators $\{v, e, e^* \mid v \in E^0, e \in E^1\}$. Every element $x \in L_R(E)$, so to speak, is a finite sum of the form $x = \sum r_i\alpha_i\beta_i^*$ where $r_i \in R$ and $(\alpha, \beta) \in E^\star \times_r E^\star$ for all i. Such an expression for x is not necessarily unique, owing to the (CK2) relation. If we have reason to consider a different model of $L_R(E)$, say another R-algebra B, then we would write $L_R(E) \cong B$.

Example 2.7 Examples 7 [7, Sect. 1.3]. Sometimes $L_R(E)$ can be recognised as a more familiar algebra. Four fundamental examples of Leavitt path algebras are:

(a) The *finite line graph* with n vertices is the graph pictured below:

$A_n =$

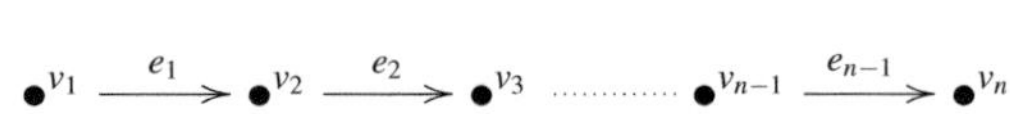

It turns out that $L_R(A_n) \cong M_n(R)$, the **matrix algebra** of $n \times n$ matrices over R. Explicitly, the set of standard matrix units $\{E_{i,j} \mid 1 \le i, j \le n\} \subset M_n(R)$ contains a Leavitt E-family $\{a_v, b_e, c_{e^*} \mid v \in A_n^0, e \in A_n^1\}$, where:

$$a_{v_i} = E_{i,i}, \qquad b_{e_j} = E_{j,j+1}, \qquad c_{e_j^*} = E_{j+1,j}, \qquad 1 \le i \le n,\ 1 \le j \le n-1.$$

(b) The *rose with n petals* is the graph pictured below (see also Example 6):

$R_n =$ 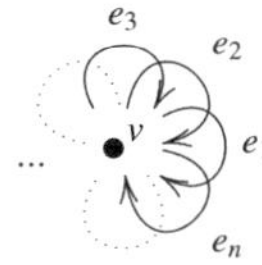

The Leavitt path algebra $L_R(R_n)$, for $n \ge 2$, is isomorphic to the **Leavitt algebra** $L_{n,R}$, discovered by W. G. Leavitt in [51, Sect. 3]. It is from this example that the Leavitt path algebras get their name.

(c) The *rose with 1 petal*,

$R_1 =$ 

gives rise to the algebra of **Laurent polynomials** $R[x, x^{-1}]$.

(d) The *Toeplitz graph*,

$T =$ 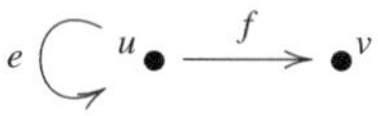

gives rise to the **Toeplitz R-algebra**, which has the presentation $R\langle x, y \mid xy = 1\rangle$. The isomorphism $R\langle x, y \mid xy = 1\rangle \to L_K(T)$ maps $x \mapsto e^* + f^*$ and $y \mapsto e + f$.

As an alternative to Definition 2.11, it is popular to define the Leavitt path algebra of a graph as a certain quotient of a path algebra. The path algebra of a graph (also called the quiver algebra of a quiver) is an older concept, familiar to a wider audience of algebraists and representation theorists. We have defined $L_R(E)$ by its universal property, so we look towards path algebras to provide a model of $L_R(E)$, thereby proving that $L_R(E)$ exists.

Let $E = (E^0, E^1, r, s)$ be a graph. The *path algebra* of E with coefficients in R is the free R-algebra generated by $E^0 \sqcup E^1$, modulo the ideal generated by the relations (V) and (E1). The *extended graph* of E is defined as $\widehat{E} = (E^0, E^1 \sqcup (E^1)^*, r', s')$, where r' and s' are extensions of r and s, respectively:

$$r'(e) = r(e) \text{ for all } e \in E^1, \qquad r'(e^*) = s(e) \text{ for all } e^* \in (E^1)^*,$$
$$s'(e) = s(e) \text{ for all } e \in E^1, \qquad s'(e^*) = r(e) \text{ for all } e^* \in (E^1)^*.$$

In other words, $\widehat{E}$ is formed from E by adding a new edge e^* for each edge e, such that e^* has the opposite direction to e. The path algebra $R\widehat{E}$ can be characterised as the free R-algebra generated by $E^0 \sqcup E^1 \sqcup (E^1)^*$, subject to the relations (V), (E1), and (E2). Let $\mathcal{A}$ be the quotient of $R\widehat{E}$ by the ideal generated by the relations (CK1) and (CK2). By virtue of its construction, $\mathcal{A}$ has the universal property for

Leavitt E-families, and consequently $\mathcal{A} \cong L_R(E)$. The path algebra model is useful for proving the following fact.

Proposition 2.18 *The Leavitt path algebra $L_R(E) = \bigoplus_{n\in\mathbb{Z}} L_R(E)_n$ is a $\mathbb{Z}$-graded algebra, where the homogeneous components are:*

$$L_R(E)_n = \operatorname{span}_R \left\{\mu\nu^* \mid (\mu, \nu) \in E^\star \times_r E^\star, |\mu| - |\nu| = n\right\}.$$

Proof Naturally, the free R-algebra $R\langle E^0 \cup E^1 \cup (E^1)^*\rangle$ is $\mathbb{Z}$-graded by setting $\deg(v) = 0$ for all $v \in E^0$, and $\deg(e) = 1$, $\deg(e^*) = -1$ for all $e \in E^1$. Extending the degree map (in the only possible way) yields $\deg(a_1 \dots a_n) = \sum_{i=1}^n \deg(a_n)$ for any word $a_1 \dots a_n \in R\langle E^0 \cup E^1 \cup (E^1)^*\rangle$. The relations (V), (E1), and (E2) are all homogeneous with respect to the grading on $R\langle E^0 \cup E^1 \cup (E^1)^*\rangle$, so they generate a graded ideal, and the quotient $R\widehat{E}$ is $\mathbb{Z}$-graded. Similarly, relations (CK1) and (CK2) are homogeneous with respect to the grading on $R\widehat{E}$, so they generate a graded ideal, and the quotient $L_R(E)$ is $\mathbb{Z}$-graded. The word $\mu\nu^*$ has degree $|\mu| - |\nu|$ in $R\langle E^0 \cup E^1 \cup (E^1)^*\rangle$, which gives the expression for the homogeneous components of $L_R(E)$. □

2.4.2 Uniqueness Theorems for Leavitt Path Algebras

Research on graph algebras has made extensive use of two main kinds of uniqueness theorems: the Cuntz–Krieger uniqueness theorems, and the graded uniqueness theorems. (In the analytic setting, graded uniqueness theorems are replaced by gauge invariant uniqueness theorems.) These theorems give sufficient conditions for a homomorphism to be injective, so they are very useful for establishing isomorphisms between a graph algebra and another algebra that comes from somewhere else. They are also very useful for studying structural properties like primeness and simplicity. Appropriate versions of these theorems have been proved not just for Leavitt path algebras but also (and we refer to [26, 29, 58, 61]) for graph C^*-algebras, as well as Cohn path algebras, higher—rank graph algebras, and even algebras of topological higher—rank graphs.

This section provides a brief account of the uniqueness theorems for Leavitt path algebras. For the Graded Uniqueness Theorem, we adhere to Tomforde's proof from [68].

Lemma 2.10 ([68, Lemma 5.1]) *Let I be a graded ideal of $L_R(E)$, where E is a graph. Then I is generated as an ideal by its 0-component $I_0 = I \cap L_R(E)_0$.*

Proof Since I is a graded ideal, $I = \sum_{k\in\mathbb{Z}} I_k$, where $I_k = I \cap L_R(E)_k$. Let $k > 0$ and $x \in I_k$. By Corollary 2.3, we can write $x = \sum_{i=1}^n \alpha_i x_i$ where each $x_i \in L_R(E)_0$, and each $\alpha_i \in E^\star$ is distinct with $|\alpha_i| = k$. Then for $1 \leq j \leq n$, we have $x_j = \alpha_j^* \left(\sum_{i=1}^n \alpha_i x_i\right) = \alpha_j^* x \in I_0$. So, I_k is spanned by elements of the form $\alpha_j x_j$ where $\alpha_j \in L_R(E)_k$ and $x_j \in I_0$. That is, $I_k = L_R(E)_k I_0$. Similarly, if $y \in I_{-k}$ then we can write $y = \sum_{i=1}^{m} y_i\beta_i^*$ where each $y_i \in L_R(E)_0$, and each $\beta_i \in E^\star$ is distinct with $|\beta_i| = k$. Then for $1 \leq j \leq n$, we have $y_j = \left(\sum_{i=1}^{n} y_i\beta_i^*\right)\beta_j = y\beta_j \in I_0$. Therefore I_{-k}, is spanned by elements of the form $y_j\beta_j$ where $\beta_j \in L_R(E)_{-k}$ and $y_j \in I_0$. That is, $I_{-k} = I_0 L_R(E)_{-k}$. Since $I = \sum_{n\in\mathbb{Z}} I_k$, this shows I is the ideal generated by I_0. □

The next lemma is a slight variation of the Reduction Theorem [7, Theorem 2.2.11]. The lemma needs the assumption that $rv \in L_R(E)$ is nonzero for every $r \in R \setminus \{0\}$ and $v \in E^0$. In fact, this is always true, but we shall only prove it later.

Lemma 2.11 ([68, Lemma 5.2]) *Let E be an arbitrary graph. Assume $rv \in L_R(E)$ is nonzero for every $r \in R \setminus \{0\}$ and $v \in E^0$. If $x \in L_R(E)_0$ is nonzero, then there exists $(\alpha, \beta) \in E^\star \times_r E^\star$ and $s \in R \setminus \{0\}$, such that $\alpha^* x\beta = sr(\alpha)$.*

Proof The set $\mathcal{M}_n = \operatorname{span}_R\{\alpha\beta^* \mid 1 \leq |\alpha| = |\beta| \leq n\}$ is an R-submodule of $L_R(E)_0$, and indeed $L_R(E)_0 = \bigcup_{n=0}^{\infty} \mathcal{M}_n$. The strategy is to prove inductively that for all $n \geq 0$ the claim holds: for all $0 \neq x \in \mathcal{M}_n$ there exists $(\alpha, \beta) \in E^\star \times_r E^\star$ and $s \in R \setminus \{0\}$ such that $\alpha^* x\beta = sr(\alpha)$. The base case is $n = 0$. If $x \in \mathcal{M}_0$ then x is a linear combination of vertices. Say $x = \sum_i r_i v_i$ with the v_i being distinct vertices and the $r_i \in R \setminus \{0\}$. Then $v_1 x v_1 = r_1 v_1$ proves the claim. Now assume the claim holds for $n - 1$. Let $0 \neq x \in \mathcal{M}_n$. We can write

$$x = \sum_{i=1}^{p} r_i\alpha_i\beta_i^* + \sum_{j=1}^{q} s_j v_j \tag{2.11}$$

where for all $1 \leq i \leq p$ and all $1 \leq j \leq q$: $r_i, s_j \in R \setminus \{0\}$, $(\alpha_i, \beta_i) \in E^\star \times_r E^\star$ with $1 \leq |\alpha_i| = |\beta_i| \leq n$, and $v_j \in E^0$. Further assume that all the (α_i, β_i) are distinct and all the v_j are distinct. In the first case, if v_j is a sink for some $1 \leq j \leq q$, then $v_j x v_j = s_j v_j$ proves the claim. In the second case, if v_j is an infinite emitter for some $1 \leq j \leq q$, then there is an edge $e \in v_j E^1 \setminus \{(\alpha_1)_1, \ldots, (\alpha_p)_1\}$ and $e^* x e = s_j r(e)$ proves the claim. Otherwise, in the third case, every v_j is a regular vertex. Applying (CK2), it is possible to expand $v_j = \sum_{e\in v_j E^1} ee^*$ for all $1 \leq j \leq q$. Then Eq. 2.11 can be rewritten as

$$x = \sum_{i=1}^{p} t_i e_i \mu_i \nu_i^* f_i^* \tag{2.12}$$

where $t_i \in R \setminus \{0\}$, $e_i, f_i \in E^1$, and $(e_i\mu_i, f_i\nu_i) \in E^\star \times_r E^\star$ for all $i \leq 1 \leq p$. It is safe to assume that

$$\sum_{\substack{1\leq j\leq p\\ e_j=e_1, f_j=f_1}} t_j e_j \mu_j \nu_j^* f_j^* = e_1\Bigg(\sum_{\substack{1\leq j\leq p\\ e_j=e_1, f_j=f_1}} t_j \mu_j \nu_j^*\Bigg) f_1^* \neq 0,$$

otherwise it could just be removed from the sum in Eq. 2.12. Then, define

$$x' = \sum_{\substack{1 \le j \le p \\ e_j = e_1, f_j = f_1}} t_j \mu_j \nu_j^*,$$

noting that $0 \neq x' \in \mathcal{M}_{n-1}$. By the inductive assumption there exists $(\alpha, \beta) \in E^\star \times_r E^\star$ and $s \in R \setminus \{0\}$ such that $\alpha^* x' \beta = sr(\alpha)$. Clearly $x' = e_1^* x f_1$, so $\alpha^* x' \beta = \alpha^* e_1^* x f_1 \beta = sr(\alpha)$. By assumption, $sr(\alpha) \neq 0$; this implies $e_1 \alpha$ and $f_1 \beta$ are legitimate paths with the same range. The claim is now proved for n, and by mathematical induction it holds for all $n \geq 0$. □

Combining these lemmas proves the Graded Uniqueness Theorem for Leavitt path algebras. This generalises both [68, Theorem 5.3] and [43, Theorem 3.2] by removing any restrictions on the cardinality of E, and by not requiring R to be a field. However, we emphasise that this is essentially Tomforde's proof with the insight that countability is not required.

Theorem 2.5 (Graded Uniqueness Theorem for Leavitt path algebras) *Let E be a graph, and R a unital commutative ring. If A is a $\mathbb{Z}$-graded ring and $\pi : L_R(E) \to A$ is a graded homomorphism with the property that $\pi(rv) \neq 0$ for every $v \in E^0$ and every $r \in R \setminus \{0\}$, then π is injective.*

Proof The first observation is that $rv \neq 0$ (because $\pi(rv) \neq 0$) for every $v \in E^0$ and $r \in R \setminus \{0\}$. The second observation is that $\ker \pi$ is a graded ideal, because π is a graded homomorphism. Suppose $x \in (\ker \pi)_0 = \ker \pi \cap L_R(E)_0$. If $x \neq 0$, then by Lemma 2.11, there exists $(\alpha, \beta) \in E^\star \times_r E^\star$ and $s \in R \setminus \{0\}$, such that $\alpha^* x \beta = sr(\alpha)$. Then $\pi(sr(\alpha)) = \pi(\alpha^*)\pi(x)\pi(\beta) = 0$, which is a contradiction. Therefore $x = 0$, so $(\ker \pi)_0 = 0$. Lemma 2.10 proves that $\ker \pi$ is generated as an ideal by $(\ker \pi)_0 = 0$; consequently, $\ker \pi = 0$, so π is injective. □

Corollary 2.4 *For every non-zero graded ideal I of $L_R(E)$, there exists $r \in R \setminus \{0\}$ and $v \in E^0$ such that $rv \in I$.*

In fact, all of the uniqueness theorems have a corollary of this sort. We will not always write it so explicitly. The Cuntz–Krieger Uniqueness Theorem is similar in spirit to the Graded Uniqueness Theorem. We do not require the homomorphism to be graded, this time, but pay the price of an extra condition on the graph, called Condition (L).

Definition 2.12 A graph E satisfies **Condition (L)** if every cycle has an exit.

Note that E satisfies Condition (L) if and only if every closed path has an exit; this is fairly intuitive and it is proved in [2, Lemma 2.5]. Combining [68, Theorem 6.5] and [43, Theorem 3.6] (see also [7, Theorem 2.2.16]) produces a version of the Cuntz–Krieger Uniqueness Theorem for Leavitt path algebras.

Theorem 2.6 *Let E be a graph satisfying Condition (L) and let R be a unital commutative ring, such that either E is countable or R is a field. If A is a ring and $\psi : L_R(E) \to A$ is a homomorphism with the property that $\psi(rv) \neq 0$ for every $v \in E^0$ and every $r \in R \setminus \{0\}$, then ψ is injective.*

This theorem can be proved for a field $R = \mathbb{K}$, using the Reduction Theorem [7, Theorem 2.2.11]. However, we shall prove it later using groupoid methods instead. In doing so, we remove the awkward restrictions on E and R.

2.4.3 The Steinberg Algebra Model

Here, we prove the existence of a Steinberg algebra model for Leavitt path algebras, and use it to prove some fundamental facts.

Theorem 2.7 ([24]) *Let E be a graph and R a unital commutative ring. Then $L_R(E)$ and $A_R(\mathcal{G}_E)$ are isomorphic as $\mathbb{Z}$-graded R-algebras.*

Proof For $v \in E^0$ and $e \in E^1$, define

$$a_v = \mathbf{1}_{Z(v)}, \qquad b_e = \mathbf{1}_{Z(e,r(e))}, \qquad b_{e^*} = \mathbf{1}_{Z(r(e),e)}.$$

We can routinely validate that $\{a_v, b_e, b_{e^*} \mid v \in E^0, e \in E^1\}$ is a Leavitt E-family. For all $e, f \in E^1$, $v, w \in E^0$, and $u \in E^0_{\text{reg}}$:

$$a_v a_w = \mathbf{1}_{Z(v)}\mathbf{1}_{Z(w)} = \mathbf{1}_{Z(v)\cap Z(w)} = \delta_{v,w}\mathbf{1}_{Z(v)}, \tag{V}$$

$$a_{s(e)} b_e a_{r(e)} = \mathbf{1}_{Z(s(e))Z(e,r(e))Z(r(e))} = \mathbf{1}_{Z(e,r(e))} = b_e, \tag{E1}$$

$$a_{r(e)} b_{e^*} a_{s(e)} = \mathbf{1}_{Z(r(e))Z(r(e),e)Z(s(e))} = \mathbf{1}_{Z(r(e),e)} = b_{e^*}, \tag{E2}$$

$$b_{e^*} b_f = \mathbf{1}_{Z(r(e),e)Z(f,r(f))} = \delta_{e,f}\mathbf{1}_{Z(r(e))} = \delta_{e,f} a_{r(e)}, \tag{CK1}$$

$$\mathbf{1}_{Z(u)} = \mathbf{1}_{\bigsqcup_{e\in uE^1} Z(e)} = \sum_{e\in uE^1} \mathbf{1}_{Z(e)} = \sum_{e\in uE^1} b_e b_{e^*}. \tag{CK2}$$

By the universal property of Leavitt path algebras, there is a unique homomorphism of R-algebras $\pi : L_R(E) \to A_R(\mathcal{G}_E)$ such that

$$\pi(v) = a_v, \qquad \pi(e) = b_e, \qquad \pi(e^*) = b_{e^*},$$

for all $v \in E^0$ and $e \in E^1$. Evidently, π is a graded homomorphism. The Graded Uniqueness Theorem for Leavitt path algebras implies π is injective. For a path $\mu \in E^\star$, if we define $b_\mu = b_{\mu_1} \dots b_{\mu_{|\mu|}}$ and $b_{\mu^*} = b_{\mu^*_{|\mu|}} \dots b_{\mu^*_1}$ then it turns out that $b_\mu = \mathbf{1}_{Z(\mu,r(\mu))}$ and $b_{\mu^*} = \mathbf{1}_{Z(r(\mu),\mu)}$. Moreover, if $\nu \in E^\star$ is another path with $r(\mu) = r(\nu)$, then $b_\mu b_{\nu^*} = \mathbf{1}_{Z(\mu,\nu)}$. If $F \subseteq_{\text{finite}} r(\mu)E^1$, this yields

$$\mathbf{1}_{Z(\mu,\nu,F)} = \mathbf{1}_{Z(\mu,\nu)} - \sum_{e\in F}\mathbf{1}_{Z(\mu e,\nu e)} = b_\mu b_{\nu^*} - \sum_{e\in F} b_{\mu e} b_{e^*\nu^*} = \pi\Big(\mu\nu^* - \sum_{e\in F}\mu e e^*\nu^*\Big). \tag{2.13}$$

Therefore, $\mathbf{1}_{\mathcal{Z}(\mu,\nu,F)}$ is in the image of π. Corollary 2.1 implies that $A_R(\mathcal{G})$ is generated by functions of the form Eq. 2.13. We conclude that π is surjective. Therefore, π is an isomorphism. □

In the following, we generalise [68, Propositions 3.4 and 4.9] and [43, Lemmas 1.5 and 1.6] by removing restrictions on the graph and the base ring.

Corollary 2.5 *Let E be a graph and R a unital commutative ring.*

(1) *$L_R(E)$ has homogeneous local units, and it has a unit if and only if E^0 is finite.*
(2) *The set $\{\mu, \mu^* \in L_R(E) \mid \mu \in E^\star\}$ is R-linearly independent in $L_R(E)$.*
(3) *For every $v \in E^0$ and $r \in R \setminus \{0\}$, $rv \neq 0$.*
(4) *If $r \mapsto \overline{r}$ is an involution on R, then there exists a unique involution $L_R(E) \to L_R(E)$ such that $r\mu\nu^* \mapsto \overline{r}\nu\mu^*$ for every $r \in R$ and $(\mu, \nu) \in E^\star \times_r E^\star$.*

Proof (1) From Proposition 2.6, $L_R(E)$ has homogeneous local units, and it has a unit if and only if ∂E is compact. Since $\partial E = \bigsqcup_{v\in E^0} Z(v)$, and each $Z(v)$ is compact and open, it is clear that ∂E is compact if and only if E^0 is finite.

(2) Since $L_R(E) = \bigoplus_{n\in\mathbb{Z}} L_R(E)_n$, it suffices to show that $\{\mu \mid \mu \in E^\star, |\mu| = n\}$ and $\{\mu^* \mid \mu \in E^\star, |\mu| = n\}$ are linearly independent in $L_R(E)$, for every $n\in\mathbb{Z}$. Equivalently, $\{\mathbf{1}_{\mathcal{Z}(\mu,r(\mu))} \mid \mu \in E^\star, |\mu| = n\}$ and $\{\mathbf{1}_{\mathcal{Z}(r(\mu),\mu)} \mid \mu \in E^\star, |\mu| = n\}$ are linearly independent in $A_R(\mathcal{G}_E)$, for every $n \in \mathbb{Z}$. This is clearly true, since $\mathcal{Z}(\mu, r(\mu)), \mathcal{Z}(\nu, r(\nu)) \neq \emptyset$ and $\mathcal{Z}(\mu, r(\mu)) \cap \mathcal{Z}(\nu, r(\nu)) = \emptyset$ for every $\mu, \nu \in E^\star$ such that $\mu \neq \nu$ and $|\mu| = |\nu|$.

(3) This follows directly from (2), or just the fact that $Z(v) \neq \emptyset$ for all $v \in E^0$.

(4) The existence follows from Proposition 2.8. The uniqueness follows from the universal property of $L_R(E)$. □

Item (3) in Corollary 2.5 is entirely disarmed by the Steinberg algebra model. It was noticed in the early years of Leavitt path algebras that a nontrivial proof was needed for Corollary 2.5 (3). The first proofs were written, separately, by Goodearl [43] and Tomforde [68] and they involved a representation of $L_R(E)$ on a free R-module of infinite rank $\aleph \geq \mathrm{card}(E^0 \sqcup E^1)$. Here is another result from the early years of Leavitt path algebras.

Proposition 2.19 ([5, Proposition 3.5]) *If E is a graph and $\mathbb{K}$ a field, then $L_{\mathbb{K}}(E)$ is finite-dimensional if and only if E is acyclic and $E^0 \cup E^1$ is finite. In this case, if $v_1, \ldots, v_t$ are the sinks and $n(v_i) = |\{\alpha \in E^\star \mid r(\alpha) = v_i\}|$, then*

$$L_{\mathbb{K}}(E) \cong \bigoplus_{i=1}^{t} M_{n(v_i)}(\mathbb{K}).$$

Proof From Proposition 2.14 we have that $L_{\mathbb{K}}(E)$ is finite-dimensional if and only if $\mathcal{G}_E$ is finite and discrete. If E had a cycle c, then the isotropy group based at $ccc\ldots \in \partial E$ would be infinite. If either E^0 or E^1 were infinite, then ∂E would be infinite, because $\partial E = \bigsqcup_{v\in E^0} Z(v) = E^0_{\mathrm{sing}} \sqcup \left(\bigsqcup_{e\in E^1} Z(e)\right)$. Thus, $\mathcal{G}_E$ is finite only if E is acyclic and $E^0 \cup E^1$ is finite. Conversely, if E is acyclic and $E^0 \cup E^1$ is finite, then there are no infinite paths, and only finitely many finite paths, so $\mathcal{G}_E$ is finite and discrete. To prove the final sentence, note that there are t orbits of sizes $n(v_1), \ldots, n(v_t)$, all with trivial isotropy groups. The structure of $L_{\mathbb{K}}(E)$ is now apparent from Proposition 2.14. □

2.4.4 Uniqueness Theorems for Steinberg Algebras

Steinberg algebras also support a Cuntz–Krieger Uniqueness Theorem and a Graded Uniqueness Theorem. These were first investigated in [20] and later improved in [25, 64]. One can think of the Cuntz–Krieger Uniqueness Theorems as saying that a certain property of a graph, namely, Condition (L), or a certain property of an ample groupoid, namely, *effectiveness*, forces a homomorphism to be injective—provided it does not annihilate any scalar multiples of a local unit. This is interesting as a first example of how a Leavitt path algebra theorem translates into the more general setting of Steinberg algebras.

Briefly, this is the order of events in this section. First, we prove the Graded Uniqueness Theorem for Steinberg algebras of graded ample groupoids. Any groupoid can be graded by the trivial group, and this simple trick obtains the Cuntz–Krieger Uniqueness Theorem for Steinberg algebras. We then use the Cuntz–Krieger Uniqueness Theorem for Steinberg algebras to prove the Cuntz–Krieger Uniqueness Theorem for Leavitt path algebras.

Definition 2.13 An étale groupoid is

(1) **effective** if $\mathrm{Iso}(\mathcal{G})^\circ = \mathcal{G}^{(0)}$, where $^\circ$ denotes the interior in $\mathcal{G}$;
(2) **topologically principal** if $\{x \in \mathcal{G}^{(0)} \mid x\mathcal{G}x = \{x\}\}$ is dense in $\mathcal{G}^{(0)}$.

Recall that a groupoid is called principal if the isotropy group at every unit is trivial. Being topologically principal amounts to having a dense set of units with trivial isotropy groups. Obviously, principal implies topologically principal. Effective does not imply topologically principal, with counterexamples in [20, Examples 6.3 and 6.4], and topologically principal does not imply effective, with counterexamples in [31, Sect. 5.1]. For a deeper understanding of effective groupoids, the upcoming lemma is essential. We state and prove the lemma for more general groupoids than just ample groupoids, mainly because there was an error in its original proof and this is an opportunity to correct it.

First, some topological comments are needed. Sets with compact closure are called *precompact*. A locally compact, Hausdorff étale groupoid $\mathcal{G}$ need not have a base of compact open bisections, but it does have a base of precompact open bisections [20]. Indeed, $\mathcal{G}$ has a base of open bisections. Since it is locally compact and Hausdorff, $\mathcal{G}$ has a base of open bisections, each of which is contained in a (necessarily closed) compact set, and thus has compact closure.

Lemma 2.12 ([20, Lemma 3.1]) *Let $\mathcal{G}$ be a locally compact Hausdorff étale groupoid. Then the following are equivalent:*

(1) $\mathrm{Iso}(\mathcal{G}) \setminus \mathcal{G}^{(0)}$ *has empty interior in* $\mathcal{G}$*;*
(2) $\mathcal{G}$ *is effective;*
(3) *Every nonempty open bisection* $B \subseteq \mathcal{G} \setminus \mathcal{G}^{(0)}$ *contains a morphism* $g \notin \mathrm{Iso}(\mathcal{G})$*;*
(4) *For every compact set* $K \subseteq \mathcal{G} \setminus \mathcal{G}^{(0)}$ *and every nonempty open* $U \subseteq \mathcal{G}^{(0)}$*, there exists an open subset* $V \subseteq U$ *such that* $VKV = \emptyset$.

Proof (1) $\Rightarrow$ (2) Since $\mathcal{G}$ is étale and Hausdorff, $\mathcal{G}^{(0)}$ is clopen in $\mathcal{G}$, so $\mathcal{G}^{(0)} \subseteq \mathrm{Iso}(\mathcal{G})^\circ$. Now assume $(\mathrm{Iso}(\mathcal{G}) \setminus \mathcal{G}^{(0)})^\circ = \emptyset$. If $S \subseteq \mathrm{Iso}(\mathcal{G})$ is open, then S is a disjoint union of two open sets: $S \cap \mathcal{G}^{(0)}$ and $S \cap (\mathcal{G} \setminus \mathcal{G}^{(0)})$. But $S \cap (\mathcal{G} \setminus \mathcal{G}^{(0)}) \subseteq \big(\mathrm{Iso}(\mathcal{G}) \setminus \mathcal{G}^{(0)}\big)^\circ = \emptyset$, so $S \subseteq \mathcal{G}^{(0)}$. This shows $\mathrm{Iso}(\mathcal{G}) = \mathcal{G}^{(0)}$, which means $\mathcal{G}$ is effective.

(2) $\Rightarrow$ (3) Suppose $\mathcal{G}$ is effective. If $B \subseteq \mathcal{G} \setminus \mathcal{G}^{(0)}$ is an open bisection, then $B \subseteq \mathrm{Iso}(\mathcal{G})$ implies $B \subseteq \mathrm{Iso}(\mathcal{G})^\circ = \mathcal{G}^{(0)}$ and therefore $B = \emptyset$.

(3) $\Rightarrow$ (1) If there are no nonempty open bisections contained in $\mathrm{Iso}(\mathcal{G}) \setminus \mathcal{G}^{(0)}$, then there are no nonempty open subsets of $\mathrm{Iso}(\mathcal{G}) \setminus \mathcal{G}^{(0)}$, and therefore $\mathrm{Iso}(\mathcal{G}) \setminus \mathcal{G}^{(0)}$ has empty interior.

(3) $\Rightarrow$ (4) We begin by proving a claim: if $B \subseteq \mathcal{G} \setminus \mathcal{G}^{(0)}$ is an open bisection and $U \subseteq \mathcal{G}^{(0)}$ is open and nonempty, then there exists a nonempty open subset $V \subseteq U$ such that $VBV = \emptyset$. If $UBU = \emptyset$, then set $U = V$ and we are done. Otherwise, $UBU \subseteq B \subseteq \mathcal{G} \setminus \mathcal{G}^{(0)}$ is a nonempty open bisection. Applying (3), there exists some $g \in UBU$ with $\boldsymbol{d}(g) \neq \boldsymbol{c}(g)$. Naturally, $\boldsymbol{d}(g), \boldsymbol{c}(g) \in U$. By the Hausdorff property, there exist disjoint open sets $W, W' \subseteq U$ with $\boldsymbol{c}(g) \in W$ and $\boldsymbol{d}(g) \in W'$. Set $V = W \cap \boldsymbol{c}(BW')$. Then $\boldsymbol{c}(g) \in V$, so V is nonempty, and

$$VB = \big(W \cap \boldsymbol{c}(BW')\big)B = WB \cap \boldsymbol{c}(BW')B = WB \cap BW'.$$

The last equality uses the fact that B is a bisection, so $\boldsymbol{c}(BW')B = BW'$. Therefore,

$$VBV = (WB \cap BW')V \subseteq (BW')V \subseteq (BW')W = \emptyset,$$

because $W'W = W' \cap W = \emptyset$. This proves the claim.

Now, let $K \subseteq \mathcal{G} \setminus \mathcal{G}^{(0)}$ be a compact set, and let $U \subseteq \mathcal{G}^{(0)}$ be open and nonempty. We set out to construct a nonempty open subset $V \subseteq U$ such that $VKV = \emptyset$. The set K, being compact, can be covered by finitely many open bisections: $K \subseteq B_1 \cup \cdots \cup B_n$. The claim in the previous paragraph proves the existence of a nonempty open set $V_1 \subseteq U$, such that $V_1 B_1 V_1 = \emptyset$. Similarly, there is a nonempty open $V_2 \subseteq V_1$ such that $V_2 B_2 V_2 = \emptyset$. Inductively, this produces a chain of open sets $\emptyset \neq V_n \subseteq V_{n-1} \subseteq \cdots \subseteq V_1 \subseteq U$ such that $V_i B_i V_i = \emptyset$ for $1 \leq i \leq n$. Setting $V = V_n$, we have

$$VKV \subseteq V(B_1 \cup \cdots \cup B_n)V \subseteq V_1 B_1 V_1 \cup \cdots \cup V_n B_n V_n = \emptyset.$$

(4) $\Rightarrow$ (3) Suppose (3) does not hold, so there is a nonempty open bisection $B_0 \subseteq \mathcal{G} \setminus \mathcal{G}^{(0)}$ with $B_0 \subseteq \mathrm{Iso}(\mathcal{G})$. By shrinking it if necessary, we can assume B_0

is precompact. Let $K_0 = \overline{B_0}$, the closure of B_0. As $\mathrm{Iso}(\mathcal{G})$ is closed in $\mathcal{G}$, we have that $K_0 \subseteq \mathrm{Iso}(\mathcal{G})$. Let $U_0 = \boldsymbol{c}(B_0)$ and take any $\emptyset \neq V \subseteq U_0$. Since $K_0 \subseteq \mathrm{Iso}(\mathcal{G})$, it follows that $VK_0 = K_0V \neq \emptyset$, so $VK_0V \neq \emptyset$. Therefore (4) does not hold, because there is no $V \subseteq U_0$ such that $VK_0V = \emptyset$. □

Remark 2.8 The original proof of the "(3) ⇒ (4)" part of [20, Lemma 3.1], does not appear to be correct. There are examples for which the set V defined in the proof is empty. Fortunately, this problem is resolved by defining V inductively, as we have done in the proof of Lemma 2.12.

Lemma 2.13 ([60, Proposition 3.6 (i)]) *If a Hausdorff étale groupoid $\mathcal{G}$ is topologically principal, then it is effective.*

Proof Suppose $\mathcal{G}$ is topologically principal: the set $D = \{x \in \mathcal{G}^{(0)} \mid {}^x\mathcal{G}^x = \{x\}\}$ is dense in $\mathcal{G}^{(0)}$. If $U \subseteq \mathrm{Iso}(\mathcal{G}) \setminus \mathcal{G}^{(0)}$ is an open bisection (i.e., open in $\mathcal{G}$) then $\boldsymbol{d}(U)$ is an open subset of $\mathcal{G}^{(0)} \setminus D$, but D is dense in $\mathcal{G}^{(0)}$, so $\boldsymbol{d}(U) = \emptyset$, which implies $U = \emptyset$. This proves $\mathrm{Iso}(\mathcal{G}) \setminus \mathcal{G}^{(0)}$ has empty interior, which implies $\mathcal{G}$ is effective (noting that the proof of (1) ⇒ (2) in Lemma 2.12 only requires $\mathcal{G}$ to be Hausdorff and étale). □

The following result is an analogue of [7, Corollary 2.2.13], and it is just an alternative way of presenting some content from [25, 64].

Proposition 2.20 *Let $\mathcal{G}$ be a Γ-graded Hausdorff ample groupoid such that $\mathcal{G}_\varepsilon$ is effective. Given a non-zero homogeneous element $h \in A_R(\mathcal{G})_\gamma$, there exists $C \in B^{\mathrm{co}}_{\gamma^{-1}}(\mathcal{G})$, nonempty $V \in \mathcal{B}(\mathcal{G}^{(0)})$, and nonzero $r \in R$ such that $\mathbf{1}_C * h * \mathbf{1}_V = r\mathbf{1}_V$.*

Proof *Step 1* [25, Lemma 3.1]: We show that there exists $B \in B^{\mathrm{co}}_{\gamma^{-1}}(\mathcal{G})$ such that the function $f = \mathbf{1}_B * h$ is ε-homogeneous and its support has nonempty intersection with $\mathcal{G}^{(0)}$. Applying Lemma 2.3, we can write $h = \sum_{i=1}^n r_i \mathbf{1}_{D_i}$, where $r_1, \ldots, r_n \in R \setminus \{0\}$ and $D_1, \ldots, D_n \in B^{\mathrm{co}}_*(\mathcal{G})$ are mutually disjoint. Since the D_i are disjoint and the r_i are nonzero, we can assume each $D_i \subseteq \mathcal{G}_\gamma$. Let $B = D_1^{-1}$ and define $f = \mathbf{1}_B * h$. Then

$$f = \mathbf{1}_B * h = \sum_{i=1}^n r_i \mathbf{1}_B * \mathbf{1}_{D_i} = \sum_{i=1}^n r_i \mathbf{1}_{BD_i} = r_1 \mathbf{1}_{BB^{-1}} + \sum_{i=2}^n r_i \mathbf{1}_{BD_i} \in A_R(\mathcal{G})_\varepsilon.$$

Note that $BD_1, \ldots, BD_n \in B^{\mathrm{co}}_\varepsilon(\mathcal{G})$ are mutually disjoint. Indeed, if $x \in B$ and $y \in D_i$ are composable, then $xy \in BD_j$ implies $y = x^{-1}xy \in B^{-1}BD_j = \boldsymbol{d}(B)D_j \subseteq D_j$. But $y \in D_i \cap D_j$ implies $i = j$ because $D_1, \ldots, D_n$ are disjoint. To show that $(\operatorname{supp} f) \cap \mathcal{G}^{(0)} \neq \emptyset$, let $x \in B$. Then $xx^{-1} \in BD_i$ if and only if $i = 1$. Consequently, $f(xx^{-1}) = r_1 \neq 0$, so $xx^{-1} \in (\operatorname{supp} f) \cap \mathcal{G}^{(0)}$.

Step 2 [25, 64]: We show that there exists $V \in \mathcal{B}(\mathcal{G}^{(0)})$ such that $\mathbf{1}_V * f * \mathbf{1}_V = r_1 \mathbf{1}_V$, where f is from Step 1. The set $K = (\operatorname{supp} f) \setminus BB^{-1} = BD_2 \cup \cdots \cup BD_n$ is a compact subset of $\mathcal{G}_\varepsilon \setminus \mathcal{G}^{(0)}$. Since $\mathcal{G}_\varepsilon$ is effective, Lemma 2.12 (4) proves that a nonempty open set $V \subseteq BB^{-1} = \boldsymbol{c}(B)$ exists such that $VKV = \emptyset$. By shrinking if necessary, we can assume V is compact. This yields

$$\mathbf{1}_V * f * \mathbf{1}_V = r_1 \mathbf{1}_{V(BB^{-1})V} + \sum_{i=2}^{n} r_i \mathbf{1}_{V(BD_i)V} = r_1 \mathbf{1}_V.$$

For completion: set $C = VB$ and $r = r_1$. Then $C \in B^{\mathrm{co}}_{\gamma^{-1}}(\mathcal{G})$, $V \in \mathcal{B}(\mathcal{G}^{(0)})$ is nonempty, $r \in R$ is nonzero, and $\mathbf{1}_C * h * \mathbf{1}_V = \mathbf{1}_V * \mathbf{1}_B * h * \mathbf{1}_V = \mathbf{1}_V * f * \mathbf{1}_V = r\mathbf{1}_V$. □

We are now in a position to prove the Graded Uniqueness Theorem for Steinberg algebras, from [25, Theorem 3.4].

Theorem 2.8 (Graded Uniqueness Theorem for Steinberg algebras) *Let $\mathcal{G}$ be a Γ-graded Hausdorff ample groupoid such that $\mathcal{G}_\varepsilon$ is effective. If A is a Γ-graded ring and $\phi : A_R(\mathcal{G}) \to A$ is a graded homomorphism with the property that $\phi(r\mathbf{1}_V) \neq 0$ for every nonempty $V \in \mathcal{B}(\mathcal{G}^{(0)})$ and every $r \in R \setminus \{0\}$, then ϕ is injective.*

Proof The kernel of ϕ is a graded ideal. Let $h \in (\ker \phi)_\gamma$. If $h \neq 0$ then, according to Proposition 2.20, there exists a compact open bisection $C \subseteq \mathcal{G}_{\gamma^{-1}}$ and a nonempty compact open set $V \subseteq \mathcal{G}^{(0)}$ such that $\mathbf{1}_C * h * \mathbf{1}_V = r\mathbf{1}_V$ for some $r \neq 0$. Then $\phi(r\mathbf{1}_V) = \phi(\mathbf{1}_C)\phi(h)\phi(\mathbf{1}_V) = 0$, which contradicts the assumption about ϕ. Therefore $h = 0$, so $(\ker \phi)_\gamma = 0$. Since this is true for every $\gamma \in \Gamma$, $\ker \phi = \bigoplus_{\gamma\in\Gamma}(\ker \phi)_\gamma = 0$. □

Remark 2.9 If $\mathcal{G} = \mathcal{G}_E$ is the groupoid of a graph E, then

$$\mathcal{G}_0 = \bigcup \big\{ \mathcal{Z}(\alpha, \beta) \mid (\alpha, \beta) \in E^\star \times_r E^\star, |\alpha| = |\beta| \big\}$$

so $\mathrm{Iso}(\mathcal{G}_0) = \mathrm{Iso}(\mathcal{G}_0)^\circ = \mathcal{G}^{(0)}$, which shows that $\mathcal{G}$ satisfies the hypotheses of Theorem 2.8. The Graded Uniqueness Theorem for Steinberg algebras is a generalisation of the Graded Uniqueness Theorem for Leavitt path algebras, notwithstanding the fact that the latter theorem is usually called upon to prove that all Leavitt path algebras are Steinberg algebras.

Any groupoid can be graded by the trivial group $\{\varepsilon\}$. With this observation, we immediately obtain the Cuntz–Krieger Uniqueness Theorem for Steinberg algebras [25, Theorem 3.2].

Corollary 2.6 (Cuntz–Krieger Uniqueness Theorem for Steinberg algebras) *Let $\mathcal{G}$ be an effective Hausdorff ample groupoid. If A is a ring and $\phi : A_R(\mathcal{G}) \to A$ is a homomorphism with the property that $\phi(r\mathbf{1}_V) \neq 0$ for every nonempty $V \in \mathcal{B}(\mathcal{G}^{(0)})$ and every $r \in R \setminus \{0\}$, then ϕ is injective.*

We now show how Condition (L) translates to the groupoid setting.

Proposition 2.21 *If E is a graph, then $\mathcal{G}_E$ is effective if and only if $\mathcal{G}_E$ is topologically principal, if and only if E satisfies Condition (L).*

Proof [67] Assume that E satisfies Condition (L), so that every closed path has an exit. Then every basic open set in ∂E contains a path that is not eventually periodic. Such paths have trivial isotropy groups in $\mathcal{G}$, by Proposition 2.17, so $\mathcal{G}^{(0)}$ has a dense subset with trivial isotropy. This implies $\mathcal{G}$ is topologically principal, hence effective, by Lemma 2.13. On the other hand, if E does not satisfy Condition (L), then there exists a cycle c without an exit, and $\mathcal{G}_E$ is not effective because there is an open set: $\mathcal{Z}(cc, |c|, c) = \{(ccc\ldots, |c|, ccc\ldots)\} \subseteq \text{Iso}(\mathcal{G}) \setminus \mathcal{G}^{(0)}$. □

Having proved the Cuntz–Krieger Uniqueness Theorem for Steinberg algebras, we can prove the Cuntz–Krieger Uniqueness Theorem for Leavitt path algebras (see Theorem 2.6), once and for all, in its full generality.

Theorem 2.9 (Cuntz–Krieger Uniqueness Theorem for Leavitt path algebras) *Let E be a graph satisfying Condition (L) and let R be a unital commutative ring. If A is a ring and $\psi : L_R(E) \to A$ is a homomorphism with the property that $\psi(rv) \neq 0$ for every $v \in E^0$ and every $r \in R \setminus \{0\}$, then ψ is injective.*

Proof First of all, suppose $r \in R \setminus \{0\}$, $\mu \in E^\star$, and F is a finite proper subset of $r(\mu)E^1$. Let $x = r\mu\mu^* - r\sum_{e\in F}\mu ee^*\mu^*$. Then $0 \neq x \in L_R(E)_0$, so Lemma 2.11 yields $(\alpha, \beta) \in E^\star \times_r E^\star, v \in E^0$, and $s \in R \setminus \{0\}$ such that $\alpha^* x\beta = sv$. This implies that $\psi(\alpha^*)\psi(x)\psi(\beta) = \psi(sv) \neq 0$, so $\psi(x) \neq 0$.

By Proposition 2.21, the groupoid $\mathcal{G}_E$ is effective. Let $\phi : A_R(\mathcal{G}_E) \to A$ be the map $\phi = \psi \circ \pi^{-1}$, where $\pi : L_R(E) \to A_R(\mathcal{G}_E)$ is the isomorphism from Theorem 2.7. Suppose $V \subseteq \partial E$ is compact and open, and $r \in R \setminus \{0\}$. We can find $\mu \in E^\star$ and $F \subseteq_{\text{finite}} r(\mu)E^1$ such that $Z(\mu, F)$ is a nonempty open subset of V. Then $Z(\mu, F)V = Z(\mu, F) \cap V = Z(\mu, F)$, so $r\mathbf{1}_{Z(\mu,F)} = \mathbf{1}_{Z(\mu,F)} * r\mathbf{1}_V$. Noting that $\pi^{-1}\left(r\mathbf{1}_{Z(\mu,F)}\right) = r\mu\mu^* - r\sum_{e\in F}\mu ee^*\mu^*$, the first paragraph proves that $0 \neq \psi \circ \pi^{-1}(r\mathbf{1}_{Z(\mu,F)}) = \phi\left(r\mathbf{1}_{Z(\mu,F)}\right) = \phi\left(\mathbf{1}_{Z(\mu,F)}\right)\phi(r\mathbf{1}_V)$; consequently $\phi(r\mathbf{1}_V) \neq 0$. Applying Corollary 2.6, the map ϕ is injective. Conclude that $\psi = \phi \circ \pi$ is injective. □

Acknowledgements I thank Juana Sánchez Ortega, for her valuable advice and guidance throughout my Masters degree. I also thank Giang Nam Tran for finding an important error in an earlier version of the paper and suggesting a way to fix it. Finally, I thank the two examiners of my Masters thesis, Pere Ara and Aidan Sims, who wrote very insightful comments that led to an improvement of this work.

I acknowledge the support of the National Research Foundation of South Africa.

References

1. G. Abrams, Leavitt path algebras: the first decade. Bull. Math. Sci. **5**(1), 59–120 (2015)
2. G. Abrams, G. Aranda Pino, The Leavitt path algebra of a graph. J. Algebra **293**(2), 319–334 (2005)
3. G. Abrams, G. Aranda Pino, The Leavitt path algebras of arbitrary graphs. Houst. J. Math. **34**(2), 423–442 (2008)

4. G. Abrams, M. Tomforde, Isomorphism and Morita equivalence of graph algebras. Trans. Am. Math. Soc. **363**(7), 3733–3767 (2011)
5. G. Abrams, G. Aranda Pino, M. Siles Molina, Finite-dimensional Leavitt path algebras. J. Pure Appl. Algebra **209**(3), 753–762 (2007)
6. G. Abrams, J. Bell, K.M. Rangaswamy, On prime nonprimitive von Neumann regular algebras. Trans. Am. Math. Soc. **366**(5), 2375–2392 (2014)
7. G. Abrams, P. Ara, M. Siles Molina, *Leavitt Path Algebras*. Lecture Notes in Mathematics, vol. 2191 (Springer, Berlin, 2017)
8. A. Alahmedi, H. Alsulami, S. Jain, E.I. Zelmanov, Structure of Leavitt path algebras of polynomial growth. Proc. Natl. Acad. Sci. USA **110**(38), 15222–15224 (2013)
9. A.A. Ambily, R. Hazrat, H. Li, Simple flat Leavitt path algebras are von Neumann regular. Commun. Algebra, 1–13 (2019). https://doi.org/10.1080/00927872.2018.1513015
10. P. Ánh, L. Márki, Morita equivalence for rings without identity. Tsukuba J. Math. **11**(1), 1–16 (1987)
11. P. Ara, K.R. Goodearl, E. Pardo, K_0 of purely infinite simple regular rings. K-Theory **26**(1), 69–100 (2002)
12. P. Ara, M.A. Moreno, E. Pardo, Nonstable K-theory for graph algebras. Algebras Represent. Theory **10**(2), 157–178 (2007)
13. P. Ara, J. Bosa, R. Hazrat, A. Sims, Reconstruction of graded groupoids from graded Steinberg algebras. Forum Math. **29**(5), 1023–1037 (2017)
14. G. Aranda Pino, D. Martín Barquero, C. Martín González, M. Siles Molina, Socle theory for Leavitt path algebras of arbitrary graphs. Rev. Mat. Iberoam. **26**(2), 611–638 (2010)
15. G. Aranda Pino, K.M. Rangaswamy, L. Vaš, $*$-Regular Leavitt path algebras of arbitrary graphs. Acta Math. Sin. Engl. Ser. **28**(5), 957–968 (2012)
16. T. Bates, D. Pask, I. Raeburn, W. Szymański, The C^*-algebras of row-finite graphs. N. Y. J. Math. **6**, 307–324 (2000)
17. V. Beuter, D. Gonçalves, The interplay between Steinberg algebras and skew group rings. J. Algebra **487**, 337–362 (2018)
18. M. Brešar, *Introduction to Noncommutative Algebra*, Universitext (Springer, Berlin, 2014)
19. J.H. Brown, L.O. Clark, A. an Huef, Diagonal-preserving ring $*$-isomorphisms of Leavitt path algebras. J. Pure Appl. Algebra **221**(10), 2458–2481 (2017)
20. J. Brown, L.O. Clark, C. Farthing, A. Sims, Simplicity of algebras associated to étale groupoids. Semigroup Forum **88**(2), 433–452 (2014)
21. N. Brownlowe, T.M. Carlsen, M.F. Whittaker, Graph algebras and orbit equivalence. Ergod. Theory Dyn. Syst. **37**(2), 389–417 (2017)
22. T.M. Carlsen, $*$-isomorphism of Leavitt path algebras over $\mathbb{Z}$. Adv. Math. **324**, 326–335 (2018)
23. T.M. Carlsen, J. Rout, Diagonal-preserving graded isomorphisms of Steinberg algebras. Commun. Contemp. Math. 1750064 (2017)
24. L.O. Clark, A. Sims, Equivalent groupoids have Morita equivalent Steinberg algebras. J. Pure Appl. Algebra **219**(6), 2062–2075 (2015)
25. L.O. Clark, C. Edie-Michell, Uniqueness theorems for Steinberg algebras. Algebra Represent. Theory **18**(4), 907–916 (2015)
26. L.O. Clark, Y.E. Pangalela, Cohn path algebras of higher-rank graphs. Algebra Represent. Theory **20**(1), 47–70 (2017)
27. L.O. Clark, C. Farthing, A. Sims, M. Tomforde, A groupoid generalisation of Leavitt path algebras. Semigroup Forum **89**(3), 501–517 (2014)
28. L.O. Clark, D. Martín Barquero, C. Martín González, M. Siles Molina, Using Steinberg algebras to study decomposability of Leavitt path algebras. Forum Math. **29**(6), 1311–1324 (2016)
29. L.O. Clark, Y.E. Pangalela, Kumjian-Pask algebras of finitely aligned higher-rank graphs. J. Algebra **482**, 364–397 (2017)
30. L.O. Clark, D. Martín Barquero, C. Martín González, M. Siles Molina, Using the Steinberg algebra model to determine the center of any Leavitt path algebra. Isr. J. Math. 1–22 (2018). https://doi.org/10.1007/s11856-018-1816-8

31. L.O. Clark, R. Exel, E. Pardo, A. Sims, C. Starling, Simplicity of algebras associated to non-Hausdorff groupoids (2018). https://doi.org/10.1090/tran/7840
32. L.O. Clark, R. Exel, E. Pardo, A generalized uniqueness theorem and the graded ideal structure of Steinberg algebras. Forum Math. **30**(3), 533–552 (2018)
33. J. Cuntz, Simple C^*-algebras generated by isometries. Commun. Math. Phys. **57**(2), 173–185 (1977)
34. J. Cuntz, W. Krieger, A class of C^*-algebras and topological Markov chains. Inven. Math. **56**(3), 251–268 (1980)
35. V. Deaconu, Groupoids associated with endomorphisms. Trans. Am. Math. Soc. **347**(5), 1779–1786 (1995)
36. M. Dokuchaev, R. Exel, P. Piccione, Partial representations and partial group algebras. J. Algebra **226**(1), 505–532 (2000)
37. S. Eilers, G. Restorff, E. Ruiz, A.P. Sørensen, The complete classification of unital graph C^*-algebras: Geometric and strong (2016), arXiv:1611.07120v1
38. R. Exel, Inverse semigroups and combinatorial C^*-algebras. Bull. Braz. Math. Soc. New Ser. **39**(2), 191–313 (2008)
39. R. Exel, Reconstructing a totally disconnected groupoid from its ample semigroup. Proc. Am. Math. Soc. **138**(8), 2991–3001 (2010)
40. R. Exel, *Partial Dynamical Systems, Fell Bundles and Applications*. Mathematical Surveys and Monographs, vol. 224 (American Mathematical Society, Providence, 2017)
41. D. Gonçalves, D. Royer, Leavitt path algebras as partial skew group rings. Commun. Algebra **42**(8), 3578–3592 (2014)
42. K.R. Goodearl, *Von Neumann Regular Rings* (Pitman, London, 1979)
43. K.R. Goodearl, Leavitt path algebras and direct limits. Contemp. Math. **480**(200), 165–187 (2009)
44. R. Hazrat, The graded Grothendieck group and the classification of Leavitt path algebras. Math. Ann. **355**(1), 273–325 (2013)
45. R. Hazrat, H. Li, Graded Steinberg algebras and partial actions. J. Pure Appl. Algebra **222**(12), 3946–3967 (2018)
46. R. Hazrat, L. Vaš, Baer and Baer *-ring characterizations of Leavitt path algebras. J. Pure Appl. Algebra **222**(1), 39–60 (2018)
47. M. Kanuni, D. Martín Barquero, C. Martín González, M. Siles Molina, Classification of Leavitt path algebras with two vertices (2017). http://www.mathjournals.org/mmj/2019-019-003/2019-019-003-004.html
48. I. Kaplansky, *Fields and Rings*, 2nd ed. Chicago Lectures in Mathematics (University of Chicago Press, Chicago, 1972)
49. A. Kumjian, D. Pask, Higher rank graph C^*-algebras. N. Y. J. Math. **6**(1), 1–20 (2000)
50. A. Kumjian, D. Pask, I. Raeburn, J. Renault, Graphs, groupoids, and Cuntz-Krieger algebras. J. Funct. Anal. **144**(2), 505–541 (1997)
51. W.G. Leavitt, The module type of a ring. Trans. Am. Math. Soc. **103**(1), 113–130 (1962)
52. W.G. Leavitt, The module type of homomorphic images. Duke Math. J. **32**(2), 305–311 (1965)
53. J.-L. Loday, *Cyclic Homology*. Grundlehren der mathematischen Wissenschaften, vol. 301, 2nd ed. (Springer, Berlin, 1998)
54. V. Nekrashevych, Growth of étale groupoids and simple algebras. Int. J. Algebra Comput. **26**(2), 375–397 (2016)
55. E. Pardo, The isomorphism problem for Higman-Thompson groups. J. Algebra **344**, 172–183 (2011)
56. A.L.T. Paterson, *Groupoids, Inverse Semigroups, and Their Operator Algebras*. Progress in Mathematics, vol. 170 (Springer, Berlin, 1999)
57. A.L.T. Paterson, Graph inverse semigroups, groupoids and their C^*-algebras. J. Oper. Theory **48**(3), 645–662 (2002)
58. I. Raeburn, *Graph Algebras*. CBMS Regional Conference Series in Mathematics, vol. 103 (American Mathematical Society, Providence 2005)

59. J. Renault, *A Groupoid Approach to C^*-Algebras*. Lecture Notes in Mathematics, vol. 793 (Springer, Berlin, 1980)
60. J. Renault, Cartan subalgebras in C^*-algebras. Bull. Ir. Math. Soc. **61**, 29–63 (2008)
61. J. Renault, A. Sims, D. Williams, T. Yeend, Uniqueness theorems for topological higher-rank graph C^*-algebras. Proc. Am. Math. Soc. **146**(2), 669–684 (2018)
62. A. Sims, Étale groupoids and their C^*-algebras (2017), arXiv:1710.10897v1
63. B. Steinberg, A groupoid approach to discrete inverse semigroup algebras. Adv. Math. **223**(2), 689–727 (2010)
64. B. Steinberg, Simplicity, primitivity and semiprimitivity of étale groupoid algebras with applications to inverse semigroup algebras. J. Pure Appl. Algebra **220**(3), 1035–1054 (2016)
65. B. Steinberg, Chain conditions on étale groupoid algebras with applications to Leavitt path algebras and inverse semigroup algebras. J. Aust. Math. Soc. **104**(3), 403–411 (2018)
66. B. Steinberg, Diagonal-preserving isomorphisms of étale groupoid algebras. J. Algebra **518**, 412–439 (2019)
67. B. Steinberg, Prime étale groupoid algebras with applications to inverse semigroup and Leavitt path algebras. J. Pure Appl. Algebra **223**(6), 2474–2488 (2019)
68. M. Tomforde, Leavitt path algebras with coefficients in a commutative ring. J. Pure Appl. Algebra **215**(4), 471–484 (2011)
69. S. Webster, The path space of a directed graph. Proc. Am. Math. Soc. **142**(1), 213–225 (2014)
70. S. Willard, *General Topology*. Addison-Wesley Series in Mathematics (Addison-Wesley, Boston, 1970)
71. T. Yeend, Groupoid models for the C^*-algebras of topological higher-rank graphs. J. Oper. Theory **57**(1), 95–120 (2007)

Chapter 3
Étale Groupoids and Steinberg Algebras a Concise Introduction

Lisa Orloff Clark and Roozbeh Hazrat

Keep fibbing and you'll end up with the truth!
No truth's ever been discovered without fourteen fibs along the way, if not one hundred and fourteen, and there's honour in that.

Dostoyevsky, Crime and Punishment.

3.1 Introduction

In the last couple of years, étale groupoids have become a focal point in several areas of mathematics. The convolution algebras arising from étale groupoids, considered both in analytical setting [50] and algebraic setting [23, 54], include many deep and important examples such as Cuntz algebras [27] and Leavitt algebras [40] and allow systematic treatment of them. Partial actions and partial symmetries can also be realised as étale groupoids (via inverse semigroups), allowing us to relate convolution algebras to partial crossed products [28, 30].

L. O. Clark
School of Mathematics and Statistics, Victoria University of Wellington, Wellington, New Zealand
e-mail: lisa.clark@vuw.ac.nz

R. Hazrat (✉)
Centre for Research in Mathematics, Western Sydney University, Sydney, Australia
e-mail: r.hazrat@westernsydney.edu.au

A. A. Ambily et al. (eds.), *Leavitt Path Algebras and Classical* K*-Theory*, Indian Statistical Institute Series,
https://doi.org/10.1007/978-981-15-1611-5_3

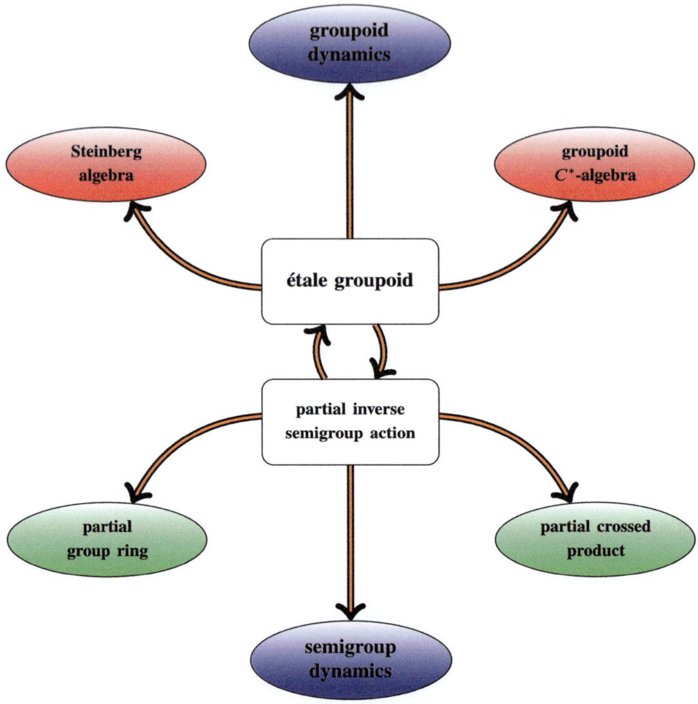

Realising that the invariants long studied in topological dynamics can be modelled on étale groupoids (such as homology, full groups and orbit equivalence [41]) and that these are directly related to invariants long studied in analysis and algebra (such as K-theory) allows interaction between areas; we can use techniques developed in algebra in analysis and vice versa. The étale groupoid is the Rosetta stone.

The study of representations of étale groupoids on Hilbert space and the associated C^*-algebras was pioneered by Renault in [50]. In this seminal work, he showed that Cuntz algebras can be realised using groupoid machinery. In [38] the authors associated an étale groupoid to a directed graph and the subject of graph C^*-algebras was born. The universal construction of these graph C^*-algebras via generators and relations was then established in [6]. The analytic activities then exploded in several directions; to describe the properties of the graph C^*-algebras directly from the geometry of the graph, to classify these algebras and to extend the definition to other types of graphs (such a higher rank graphs [37]).

There has long been a trend of 'algebraisation' of concepts from operator theory into a purely algebraic context. This seems to have started with von Neumann and Kaplansky who devised ways of seeing operator algebraic properties in underlying discrete structures [35]. As Berberian puts it in [8], 'if all the functional analysis is stripped away...what remains should stand firmly as a substantial piece of algebra, completely accessible through algebraic avenues'.

This translation did happen in the setting of graph algebras in the reverse order (and with about 30 year lag): in [2, 3] the algebraic analogue of graph C^*-algebras were defined using generators and relations (called Leavitt path algebras) and then the algebraic analogue of groupoid C^*-algebras was developed in [23, 54] (now called Steinberg algebras).

Another strand is the work of Exel on partial actions of groups on spaces and their corresponding C^*-algebras. Again, the algebraic version of this theory is developed [28] and the close connections with groupoids are established [7, 30].

This survey exclusively concentrates on étale groupoids with totally disconnected unit spaces (aka ample groupoids) and their convolution algebras (aka, Steinberg algebras). One reason over such groupoids our R-algebras are just R-valued continuous functions with compact support over the groupoid and there is a known universal description for such algebras, at least when the groupoid is Hausdorff. We will briefly describe the situation when the groupoid is not Hausdorff as well. We describe their connections with groupoid C^*-algebras and Exel's partial constructions. The concepts of inverse semigroup and groupoid are tightly related (as the diagram in the first page shows) and are models for partial symmetries. In Sects. 3.2 and 3.3 we study these concepts with a view towards the algebras that arise from them which we describe in Sect. 3.4.

The use of groupoids extends to many areas of mathematics, from ergodic theory and functional analysis (such as work of Connes in noncommutative geometry [25]) to homotopy theory [12], algebraic geometry, differential geometry and group theory. The reader is encouraged to consult [13, 34, 60] for more details on the history and the development of groupoids.

3.2 Inverse Semigroups

There is a tight relation between the notion of groupoids and its 'dual' inverse semigroups. We start the survey with a description of inverse semigroups.

3.2.1 Inverse Semigroups

Recall that a semigroup is a non-empty set with an associative binary operation. For a semigroup S, the element $x \in S$ is called *regular* if $xyx = x$. In this case, we can arrange that $xyx = x$ and $yxy = y$ and we say x has an *inner inverse*. We say a semigroup is *regular* if each element has an inner inverse.

An inverse semigroup is a semigroup that each element has a unique inner inverse. Namely, an *inverse semigroup* is a semigroup S such that, for each $s \in S$, there exists a unique element $s^* \in S$ such that

$$ss^*s = s \text{ and } s^*ss^* = s^*.$$

The uniqueness guarantees that the map $s \to s^*$ induces an involution on S. One can check that

$$E(S) := \{ss^* \mid s \in S\},$$

is the set of idempotents of S and is an abelian subsemigroup. One way to prove that a semigroup is an inverse semigroup is to show that it is a regular semigroup and the set of idempotents are abelian. In fact $E(S)$ is a meet semilattice with respect to the partial ordering $e \leq f$ if $ef = e$; the meet is the product. The partial order extends to the entire inverse semigroup by putting $s \leq t$ if $s = te$ for some idempotent $e \in E(S)$ (or, equivalently, $s = ft$ for some $f \in E(S)$). This partial order is preserved under multiplication and inversion.

Most of the inverse semigroups we encounter have a zero element. An inverse semigroup S has a zero element 0 if $0x = 0 = x0$ for all $x \in S$. The zero element is unique when it exists and often corresponds to the empty set in our concrete examples. Any semigroup homomorphism $p\colon S \to T$ of inverse semigroups automatically preserves the involution, i.e., $p(s^*) = p(s)^*$.

Parallel to the group of symmetries and the theorem of Cayley, we next define the inverse semigroup of partial symmetries and recall the theorem of Wagner–Preston. Let X be a set and $A, B \subseteq X$. A bijective map $f : A \to B$ is called a *partial symmetry* of X. Denote by $\mathcal{I}(X)$ the collection of all partial symmetries of X. The set $\mathcal{I}(X)$ is an inverse semigroup with zero under the operation given by the composition of functions in the largest domain in which the composition may be defined. The zero element corresponds to the map assigned to an empty set. The Wagner–Preston theorem guarantees that any inverse semigroup is a subsemigroup of $\mathcal{I}(X)$ for some set X.

A majority of inverse semigroups we encounter here are naturally 'graded'. If S is an inverse semigroup with possibly 0 and Γ is a discrete group, then S is called a Γ-*graded inverse semigroup* if there is a map $c : S \setminus \{0\} \to \Gamma$ such that $c(st) = c(s)c(t)$, whenever $st \neq 0$. For $\gamma \in \Gamma$, if we set $S_\gamma := c^{-1}(\gamma)$, then S decomposes as a disjoint union

$$S \setminus \{0\} = \bigsqcup_{\gamma \in \Gamma} S_\gamma,$$

and we have $S_\beta S_\gamma \subseteq S_{\beta\gamma}$, if the product is not zero. We say that S is *strongly graded* if $S_\beta S_\gamma = S_{\beta\gamma}$, for all β, γ, understanding that we exclude the zero if a product is zero. The reader is referred to Mark Lawson's book [39] for the theory of inverse semigroups.

3.2.2 *Examples of Inverse Semigroups*

Clearly, any group is an inverse semigroup without zero unless it is a trivial group. The Theorem of Wagner–Preston shows that the partial symmetries are the 'mothers' of all inverse semigroups.

Example 3.1 (*Graph inverse semigroups*) Directed graphs provide concrete examples for constructing a variety of combinatorial structures, such as semigroups, groupoids and algebras. We briefly recall the definition of a directed graph and the construct the first combinatorial structure out of them, namely, graph inverse semigroups.

A *directed graph* E is a tuple (E^0, E^1, r, s), where E^0 and E^1 are sets and r, s are maps from E^1 to E^0. We think of each $e \in E^1$ as an arrow pointing from $s(e)$ to $r(e)$. We use the convention that a (finite) path p in E is a sequence $p = \alpha_1\alpha_2 \cdots \alpha_n$ of edges α_i in E such that $r(\alpha_i) = s(\alpha_{i+1})$ for $1 \leq i \leq n-1$. We define $s(p) = s(\alpha_1)$, and $r(p) = r(\alpha_n)$. If $s(p) = r(p)$, then p is said to be closed. If p is closed and $s(\alpha_i) \neq s(\alpha_j)$ for $i \neq j$, then p is called a *cycle*. An edge α is an exit of a path $p = \alpha_1 \cdots \alpha_n$ if there exists i such that $s(\alpha) = s(\alpha_i)$ and $\alpha \neq \alpha_i$. A graph E is called *acyclic* if there are no closed path in E. For a path p, we denote by $|p|$ the length of p, with the convention that $|v| = 0$.

A directed graph E is said to be *row-finite* if for each vertex $u \in E^0$, there are at most finitely many edges in $s^{-1}(u)$. A vertex u for which $s^{-1}(u)$ is empty is called a *sink*, whereas $u \in E^0$ is called an *infinite emitter* if $s^{-1}(u)$ is infinite. If $u \in E^0$ is neither a sink nor an infinite emitter, then it is called a *regular vertex*.

Definition 3.1 Let $E = (E^0, E^1, r, s)$ be a directed graph. The *graph inverse semigroup* S_E is the semigroup with zero generated by the sets E^0 and E^1, together with a set $E^* = \{e^* \mid e \in E^1\}$, satisfying the following relations:

(0) $uv = \delta_{u,v}v$ for every $u, v \in E^0$;
(1) $s(e)e = er(e) = e$ for all $e \in E^1$;
(2) $r(e)e^* = e^* = e^*s(e)$ for all $e \in E^1$;
(3) $e^*f = \delta_{e,f}r(e)$ for all $e, f \in E^1$.

For a path $p = e_1e_2 \cdots e_n$, denoting $p^* = e_n^* \cdots e_2^*e_1^*$, one can show that elements of S_E are of the form pq^* for some paths p and q and the unique inner inverse of pq^* is qp^*.

This definition was first given in [5] and then in [48] in relation with groupoids and groupoids C^*-algebras. The fact that Definition 3.1 gives an inverse semigroup was checked in details in [48, Propositions 3.1, 3.2]. The graph inverse semigroup associated with a graph with one vertex and n loops is called *Cuntz inverse semigroup* and it was defined in [50, p. 141]. We remark that the universal groupoid of S_E (see [54]) is the graph groupoid $\mathcal{G}_E$ which will be studied in Sect. 3.3.5.

For a graph E, the inverse semigroup S_E has a natural $\mathbb{Z}$-grading where $c(pq^*) = |p| - |q|$. We also refer the reader to [44] for further study on these inverse semigroups.

Example 3.2 (*Exel's inverse semigroup associated to a group*) Any group is an inverse semigroup. In [29], Exel defined a semigroup $S(G)$ associated to the partial actions of the group G on sets (Example 3.5) and proved that this semigroup is, in fact, an inverse semigroup. He then established that the partial actions of G on a set X are in one-to-one correspondence with the action of $S(G)$ on X. As the construction of $S(G)$ is very natural we give it here.

Let G be a group with unit ε. We define $S(G)$ to be the semigroup generated by $\{[g] \mid g \in G\}$ subject to the following relations: for $g, h \in G$:

(i) $[g^{-1}][g][h] = [g^{-1}][gh]$;
(ii) $[g][h][h^{-1}] = [gh][h^{-1}]$; and
(iii) $[g][\varepsilon] = [g]$.

Observe that $[\varepsilon][g] = [gg^{-1}][g] = [g][g^{-1}][g] = [g][g^{-1}g] = [g][\varepsilon] = [g]$. Then $S(G)$ is a semigroup with unit $[\varepsilon]$. It was proved in [29, Theorem 3.4] that $S(G)$ is an inverse semigroup and each element of $x \in S(G)$ can be written uniquely as $x = [t_1][t_1^{-1}] \cdots [t_r][t_r^{-1}][g]$. This gives that $S(G)$ is also a G-graded inverse semigroup.

Further in [16] Buss and Exel showed that starting from an inverse semigroup G, a similar construction as above (replacing g^{-1} by g^*) is also an inverse semigroup.

3.3 Groupoids

3.3.1 Groupoids

The use of groupoids to study structures whose operations are partially defined is firmly recognised [12, 34, 39, 60]. We start by recalling the definition of a groupoid with a suitable topology, i.e., an ample groupoid. We will eventually describe a ring of R-valued continuous functions on an ample groupoid, where R is a (commutative, unital) ring. These are the main objects of this survey, namely Steinberg algebras.

A *groupoid* is a small category in which every morphism is invertible. It can also be viewed as a generalisation of a group which has a partially defined binary operation. Let $\mathcal{G}$ be a groupoid. If $x \in \mathcal{G}$, $\boldsymbol{d}(x) := x^{-1}x$ is the *domain* of x and $\boldsymbol{r}(x) := xx^{-1}$ is its *range*. Thus, the pair (x, y) in the category $\mathcal{G}$ is composable if and only if $\boldsymbol{r}(y) = \boldsymbol{d}(x)$ and in this case $xy \in \mathcal{G}$. Denote $\mathcal{G}^{(2)} := \{(x, y) \in \mathcal{G} \times \mathcal{G} : \boldsymbol{d}(x) = \boldsymbol{r}(y)\}$. The set $\mathcal{G}^{(0)} := \boldsymbol{d}(\mathcal{G}) = \boldsymbol{r}(\mathcal{G})$ is called the *unit space* of $\mathcal{G}$. Note that we identify the objects of the category $\mathcal{G}$ with $\mathcal{G}^{(0)}$. which are the identity morphisms of the category $\mathcal{G}$ in the sense that $x\boldsymbol{d}(x) = x$ and $\boldsymbol{r}(x)x = x$ for all $x \in \mathcal{G}$.

The collection of morphisms whose domain and range are a fixed unit $u \in \mathcal{G}^{(0)}$ is a group and the collection of all of these groups is called the isotropy bundle $\mathrm{Iso}(\mathcal{G})$, that is,

$$\mathrm{Iso}(\mathcal{G}) := \{\gamma \in \mathcal{G} : \boldsymbol{d}(\gamma) = \boldsymbol{r}(\gamma)\}.$$

For subsets $U, V \subseteq \mathcal{G}$, we define

$$UV = \big\{xy \mid x \in U, y \in V \text{ and } \boldsymbol{d}(x) = \boldsymbol{r}(y)\big\}, \tag{3.1}$$

and

$$U^{-1} = \big\{x^{-1} \mid x \in U\big\}. \tag{3.2}$$

If $\mathcal{G}$ is a groupoid and Γ is a group, then $\mathcal{G}$ is called a Γ-*graded groupoid* if there is functor $c : \mathcal{G} \to \Gamma$, i.e., there is a function $c : \mathcal{G} \to \Gamma$ such that $c(x)c(y) = c(xy)$ for all $(x, y) \in \mathcal{G}^{(2)}$. For $\gamma \in \Gamma$, if we set $\mathcal{G}_\gamma := c^{-1}(\gamma)$, then $\mathcal{G}$ decomposes as a disjoint union

$$\mathcal{G} = \bigsqcup_{\gamma \in \Gamma} \mathcal{G}_\gamma,$$

and we have $\mathcal{G}_\beta \mathcal{G}_\gamma \subseteq \mathcal{G}_{\beta\gamma}$. We say that $\mathcal{G}$ is *strongly graded* if $\mathcal{G}_\beta \mathcal{G}_\gamma = \mathcal{G}_{\beta\gamma}$, for all β, γ. For $\gamma \in \Gamma$, we say that $X \subseteq \mathcal{G}$ is γ-graded if $X \subseteq \mathcal{G}_\gamma$. We have $\mathcal{G}^{(0)} \subseteq \mathcal{G}_\varepsilon$, so $\mathcal{G}^{(0)}$ is ε-graded, where ε is the identity of the group Γ. Graded groupoids were studied in [20].

3.3.2 Topological Groupoids

A topological groupoid is a groupoid endowed with a topology under which the inverse map is continuous, and composition is continuous with respect to the relative product topology on $\mathcal{G}^{(2)}$. An *étale* groupoid is a topological groupoid $\mathcal{G}$ such that the domain map $\boldsymbol{d}$ is a local homeomorphism. In this case, the range map $\boldsymbol{r}$ is also a local homeomorphism. Further, for a fixed $u \in \mathcal{G}^{(0)}$, $\boldsymbol{d}^{-1}(u)$ and $\boldsymbol{r}^{-1}(u)$ are both discrete with respect to the subspace topology.[1] An *open bisection* of $\mathcal{G}$ is an open subset $U \subseteq \mathcal{G}$ such that $\boldsymbol{d}|_U$ and $\boldsymbol{r}|_U$ are homeomorphisms onto an open subset of $\mathcal{G}^{(0)}$. Notice that a groupoid is étale if and only if it has a basis of open bisections.

We say that a topological groupoid $\mathcal{G}$ is *ample* if there is a basis of compact open bisections. An ample groupoid is automatically étale, locally compact and $\mathcal{G}^{(0)}$ is an open subset of $\mathcal{G}$. The terminology in the literature is inconsistent: sometimes the term ‘étale’ also includes the assumptions of local compactness and $\mathcal{G}^{(0)}$ Hausdorff. We will focus on the situation where $\mathcal{G}$ is Hausdorff ample so these two assumptions are automatically true.

In an ample Hausdorff groupoid, compact open bisections are also closed so that any finite collection of such sets can be ‘disjointified’ to form a disjoint finite collection whose union is equal to the original collection. This is very powerful. We discuss non-Hausdorff groupoids briefly in Sect. 3.7.

In the topological setting, we call a groupoid $\mathcal{G}$, a Γ-graded groupoid, if the functor $c : \mathcal{G} \to \Gamma$ is continuous with respect to the discrete topology on Γ; such a function c is called a *cocycle* on $\mathcal{G}$.

Lemma 3.1 *Let $\mathcal{G}$ be an étale groupoid. If $\mathcal{G}^{(0)}$ is finite, then $\mathcal{G}$ is a discrete topological space.*

---

[1]Historically, the term *r-discrete* was used in place of étale and there are some inconsistencies in the literature surrounding these terms.

Proof Since $\mathcal{G}^{(0)}$ is finite and Hausdorff, it is discrete. Fix $\gamma \in \mathcal{G}$. We show that $\{\gamma\}$ is open. Since $\mathcal{G}$ is étale, γ is contained in an open bisection U. Also, since $\boldsymbol{r}$ is continuous, $\boldsymbol{r}^{-1}(\{\boldsymbol{r}(\gamma)\})$ is open. But $\boldsymbol{r}$ is injective on U so

$$\boldsymbol{r}^{-1}(\{\boldsymbol{r}(\gamma)\}) \cap U = \{\gamma\}.$$

□

We will determine the Steinberg algebra associated to finite $\mathcal{G}^{(0)}$ in Proposition 3.1.

3.3.3 Examples of Groupoids

Example 3.3 (*Transitive groupoids*) A groupoid is called *connected* or *transitive* if for any $u, v \in \mathcal{G}^{(0)}$, there is a $x \in \mathcal{G}$ such that $u = \boldsymbol{d}(x)$ and $\boldsymbol{r}(x) = y$.

Let G be a group and I a non-empty set. The set $I \times G \times I$, considered as morphisms, forms a groupoid where the composition defined by $(i, g, j)(j, h, k) = (i, gh, k)$. One can show that this is a transitive groupoid and any transitive groupoid is of this form [39, Chap. 3.3, Proposition 6]. If $I = \{1, \dots, n\}$, we denote $I \times G \times I$ by $n \times G \times n$. Note that this groupoid is naturally strongly G-graded. This seems to be the first appearance of groupoids after they were introduced by Brandt in 1926 [11] (see in [13] for a nice history of groupoids).

In the next three examples, we explore how a group action on a (combinatorial) structure can be naturally captured by a groupoid. The first example is the action of a group on a set, and we then continue with a partial action of a group and inverse semigroup on a set. Although the (partial) action of an inverse semigroup on a set would be the most general case covering the previous two examples, for a pedagogical reason, we introduce these step by step.

Example 3.4 (*Transformation groupoid arising from a group action*) Let G be a group acting on a set X, i.e., there is a group homomorphism $G \to \operatorname{Iso}(X)$, where $\operatorname{Iso}(X)$ consists of bijective maps from X to X which is a group with respect to composition. Let

$$\mathcal{G} = G \times X \tag{3.3}$$

and define the groupoid structure: $(g, hy) \cdot (h, y) = (gh, y)$, and $(g, x)^{-1} = (g^{-1}, gx)$. Then $\mathcal{G}$ is a groupoid, called the *transformation groupoid* arising from the action of G on X (for short, $G \curvearrowright X$). The unit space $\mathcal{G}^{(0)}$ is canonically identified with X via the map $(\varepsilon, x) \mapsto x$. The natural cocycle $\mathcal{G} \to G$, $(g, x) \mapsto g$, makes $\mathcal{G}$ a strongly G-graded groupoid. Note that the range and source map would distinguish an element of this groupoid up to the stabiliser. Namely, $\boldsymbol{d}(g, x) = x = \boldsymbol{d}(h, x)$ and $\boldsymbol{r}(g, x) = gx = hx = \boldsymbol{r}(h, x)$. But when we consider the grading then we can distinguish these elements as well. When X is a Hausdorff topological space and G is a discrete group, then $\mathcal{G}$ is an étale topological groupoid with respect to the product topology. If, in addition, X has a basis of compact open sets, then $\mathcal{G}$ is ample.

Example 3.5 (*Transformation groupoid arising from a partial group action*) A *partial action* of a group G on a set X is a data $\phi = (\phi_g, X_g, X)_{g \in G}$, where for each $g \in G$, X_g is a subset of X and $\phi_g : X_{g^{-1}} \to X_g$ is a bijection such that

(i) $X_\varepsilon = X$ and ϕ_ε is the identity on X, where ε is the identity of the group G;
(ii) $\phi_g(X_{g^{-1}} \cap X_h) = X_g \cap X_{gh}$ for all $g, h \in G$;
(iii) $\phi_g(\phi_h(x)) = \phi_{gh}(x)$ for all $g, h \in G$ and $x \in X_{h^{-1}} \cap X_{h^{-1}g^{-1}}$.

Although the above construction gives a well-define map $\pi : G \to \mathcal{I}(X), g \mapsto \phi_g$, this map is not a homomorphism, i.e. $\phi_{gh} \neq \phi_g\phi_h$. However there is a one-to-one correspondence between these partial actions and the actions of inverse semigroup $S(G)$ on X (see Example 3.2)

Let $\phi = (\phi_g, X_g, X)_{g \in G}$ be a partial action of G on X. Consider the G-graded groupoid

$$\mathcal{G}_\phi = \bigcup_{g \in G} g \times X_g, \tag{3.4}$$

whose composition and inverse maps are given by $(g, x)(h, y) = (gh, x)$ if $y = \phi_{g^{-1}}(x)$ and $(g, x)^{-1} = (g^{-1}, \phi_{g^{-1}}(x))$. Here the range and source maps are given by $\boldsymbol{r}(g, x) = (\varepsilon, x)$, $\boldsymbol{d}(g, x) = (\varepsilon, \phi_{g^{-1}}(x))$ with ε the identity of G. The unit space of $\mathcal{G}_\phi$ is identified with X.

In case that X is a topological space, we assume $X_g \subseteq X$ is an open subset and each $\phi_g : X_{g^{-1}} \to X_g$ is a homeomorphism, for $g \in G$. In order to obtain an ample groupoid, we further assume that X is a Hausdorff topological space that has a basis of compact open sets, each X_g is a clopen subset of X, and G is a discrete group. The topology of $\mathcal{G}_\phi$ which inherited from the product topology $G \times X$ gives us an Hausdorff ample groupoid.

In fact, one can further generalise this to the setting of partial action of an inverse semigroup on sets, topological spaces and rings. In Sect. 3.6 we will relate partial inverse semigroup rings coming out of this partial actions to Steinberg algebras.

Example 3.6 (*Transformation groupoid arising from an inverse semigroup action; groupoid of germs*) We start with a more concrete example of the groupoid of germs and then move to a more abstract construction of the groupoid of germs of an inverse semigroup acting on a space. In the topological setting, these are one of the main sources of étale groupoids.

Let X be a non-empty set and let $S = \mathcal{I}(X)$ be the inverse semigroup of partial symmetries. The S-germ is a pair $(s, x) \in S \times X$, where $x \in \mathrm{dom}(s)$, modulo the equivalence relation of germs $(s, x) \sim (t, y)$ if $x = y$ and the restriction of s and t coincides on a subset containing x. The groupoid operations defined by

$$(s, ty)(t, y) = (st, y), \ \ (s, x)^{-1} = (s^{-1}, sx).$$

In a more abstract setting, let S be an inverse semigroup acting on a set X, i.e., there is a semigroup homomorphism $S \to \mathcal{I}(X)$. Let

$$\mathcal{G} = \bigcup_{s \in S} s \times X_{s^*s}. \tag{3.5}$$

and define the groupoid structure: $(s, ty) \cdot (t, y) = (st, y)$, and $(s, x)^{-1} = (s^*, sx)$. One can check that these operations are well-defined and $\mathcal{G}$ is a groupoid. However, this groupoid is too large for us and we need to invoke the equivalence of germs.

The *groupoid of germs* $\mathcal{G} = S \ltimes X$ is defined (with an abuse of notation) as $\mathcal{G}$ modulo the equivalence relation $(s, x) \sim (t, y)$ if $x = y$ and there exists an idempotent e such that $x \in X_e$ and $se = te$. We denote the equivalence class of (s, x) by $[s, x]$ and call it the germ of s at x. It is a routine exercise to show that $\mathcal{G}$ with $[s, ty][t, y] = [st, y]$ and $[s, x]^{-1} = [s^*, sx]$ is in fact a groupoid. Note that if S is a group, then there are no identifications and we retrieve the transformation groupoid of Example 3.4.

When X is a Hausdorff topological space, one can show that $[s \times U] := \{[s, x] \mid x \in U\}$, where $U \subseteq X_{s^*s}$ is open, is a basis for a topology on $\mathcal{G}$. With this topology, $\mathcal{G}$ is étale and $[s \times U]$ is an open bisection. If X has a basis of compact open sets, then $\mathcal{G}$ is Hausdorff ample. Further, by [30, Proposition 6.2] if S is a semilattice and X_e are clopen for $e \in E(S)$, then $\mathcal{G}$ is Hausdorff.

Example 3.7 (*Underlying groupoid of an inverse semigroup*) Let S be an inverse semigroup. The maps

$$\begin{aligned} \boldsymbol{d} : S &\longrightarrow E(S) \\ s &\longmapsto s^*s \end{aligned} \qquad\qquad \begin{aligned} \boldsymbol{r} : S &\longrightarrow E(S) \\ s &\longmapsto ss^* \end{aligned}$$

considered as the source and range maps make S into a groupoid with the product of the semigroup as the composition of the groupoid. The unit space is $E(S)$. Note that if S is graded inverse semigroup so is the underlying groupoid of S and the strongly graded property passes from one structure to another.

3.3.4 Inverse Semigroup of Bisections of a Groupoid

Given an ample Hausdorff groupoid, the inverse semigroup made up of all the compact open bisections plays an important role. In fact, the Steinberg algebra associated to a Hausdorff ample groupoid is the inverse semigroup ring of compact open bisections modulo their unions (see Definition 3.2). In the following, we describe this inverse semigroup for a graded topological groupoid. Both the grading and the topology can be stripped away.

Let $\mathcal{G}$ be a Γ-graded Hausdorff ample groupoid. Set

$$\mathcal{G}^h = \{U \mid U \text{ is a graded compact open bisection of } \mathcal{G}\}. \tag{3.6}$$

Then $\mathcal{G}^h$ is an inverse semigroup under the multiplication $U.V = UV$ and inner inverse $U^* = U^{-1}$ as in (3.1) and (3.2) (see [47, Proposition 2.2.4]). Furthermore, the map $c : \mathcal{G}^h \backslash \emptyset \to \Gamma$, $U \mapsto \gamma$, if $U \subseteq \mathcal{G}_\gamma$, makes $\mathcal{G}^h$ a graded inverse semigroup with $\mathcal{G}^h_\gamma = c^{-1}(\gamma)$, $\gamma \in \Gamma$, as the graded components. Observe that in the inverse semigroup $\mathcal{G}^h$, $B \leq C$ if and only if $B \subseteq C$ for $B, C \in \mathcal{G}^h$. If from the outset we consider $\mathcal{G}$ as a trivially graded groupoid, then we have an inverse semigroup consisting of all compact open bisections. In this case we denote the inverse semigroup by $\mathcal{G}^a$. There are other notations for this semigroup in literature, such as $\mathcal{S}(\mathcal{G})$ in [30] or $\mathcal{G}^{co}$ in [47].

There is a natural action of inverse semigroup $\mathcal{G}^h$ on the $\mathcal{G}^{(0)}$. In fact the groupoid of germs (as in Example 3.6) of this action is $\mathcal{G}$ itself and this allows us to relate the partial crossed product construction to the concept of Steinberg algebras (see Theorem 3.9). We describe this action next. In fact, in what follows we will construct a homomorphism of semigroups $\pi : \mathcal{G}^h \to \mathcal{I}(\mathcal{G}^{(0)})$.

For each $B \in \mathcal{G}^h$, BB^{-1} and $B^{-1}B$ are compact open subsets of $\mathcal{G}^{(0)}$. Define

$$\begin{aligned} \pi_B : B^{-1}B &\longrightarrow BB^{-1} \\ u &\longmapsto \boldsymbol{r}(Bu) \end{aligned} \tag{3.7}$$

Since B is a bisection, Bu consists of only one element of $\mathcal{G}$ and thus the map π_B is well-defined. Observe that π_B is a bijection with inverse $\pi_{B^{-1}}$. We claim that π_B is a homeomorphism for each $B \in \mathcal{G}^h$. Take any open subset $O \subseteq U_B$. Observe that $\pi_B^{-1}(O) = \boldsymbol{d}(\boldsymbol{r}^{-1}(O) \cap B)$ is an open subset of $U_{B^{-1}}$. Thus, π_B is continuous. Similarly, π_B^{-1} is continuous. One can check that for compact open bisections B and C, $\pi_B \pi_C = \pi_{BC}$, and thus $\pi : \mathcal{G}^h \to \mathcal{I}(\mathcal{G}^{(0)})$ is a homomorphism of inverse semigroups. If the grade group Γ is considered to be trivial, then we have a homomorphism $\pi : \mathcal{G}^a \to \mathcal{I}(\mathcal{G}^{(0)})$. This homomorphism is injective if, in some sense, there isn't too much *isotropy* which we show in Lemma 3.2 after introducing some more terminologies.

We say a topological groupoid $\mathcal{G}$ is *effective* if the interior of the isotropy bundle is just the unit space, that is

$$\mathrm{Iso}(\mathcal{G})^\circ = \mathcal{G}^{(0)}.$$

Thus, in an effective ample groupoid, if we have a compact open bisection B such that every element $\gamma \in B$ has the property $s(\gamma) = \boldsymbol{r}(\gamma)$, then $B \subseteq \mathcal{G}^{(0)}$.

We say a subset U of the unit space $\mathcal{G}^{(0)}$ of $\mathcal{G}$ is *invariant* if $\boldsymbol{d}(\gamma) \in U$ implies $\boldsymbol{r}(\gamma) \in U$; equivalently,

$$\boldsymbol{r}(\boldsymbol{d}^{-1}(U)) = U = \boldsymbol{d}(\boldsymbol{r}^{-1}(U)).$$

For an invariant $U \subseteq \mathcal{G}^{(0)}$, we write $\mathcal{G}_U := \boldsymbol{d}^{-1}(U)$ which coincides with the restriction

$$\mathcal{G}|_U = \{x \in \mathcal{G} \mid \boldsymbol{d}(x) \in U, \boldsymbol{r}(x) \in U\}.$$

Notice that $\mathcal{G}_U$ is a groupoid with unit space U.

We say $\mathcal{G}$ is *strongly effective* if for every non-empty closed invariant subset D of $\mathcal{G}^{(0)}$, the groupoid $\mathcal{G}_D$ is effective. These assumptions play important roles when classifying ideals of Steinberg algebras (see Sect. 3.4.5).

Lemma 3.2 *Let $\mathcal{G}$ be an ample groupoid. Then the morphism $\pi : \mathcal{G}^h \to \mathcal{I}(\mathcal{G}^{(0)})$ is injective if and only if $\mathcal{G}$ is effective.*

For more equivalences of effective groupoids, see [14, Lemma 3.1].

3.3.5 Graph Groupoids

Our next goal is to describe groupoids associated to directed graphs. There is a general construction of a groupoid from a topological space X and a local homeomorphism $\sigma : X \to X$, called a *Deaconu–Renault groupoid* (see [51]). The graph groupoids are a special case. We briefly recall this general construction.

Let $\sigma : X \to X$ be a local homeomorphism. Consider

$$\mathcal{G}(X, \sigma) = \{(x, m - n, y) \mid m, n \in \mathbb{N}, \sigma^m(x) = \sigma^n(y)\}, \tag{3.8}$$

with the groupoid structure inherited from the transitive groupoid $X \times \mathbb{Z} \times X$. Note that $\mathcal{G}(X, \sigma)$ is not transitive in general.

When X is a Hausdorff space, sets of the form

$$Z(U, m, n, V) = \{(x, m - n, y) \mid (x, y) \in U \times V, \sigma^m(x) = \sigma^n(y)\},$$

where U and V are open subsets of X are a basis for a topology on $\mathcal{G}(X, \sigma)$ making it a Hausdorff étale groupoid. When X also has a basis of compact open sets, the groupoid is Hausdorff ample.

To any graph, E one can associate a groupoid $\mathcal{G}_E$, called the *boundary path groupoid*, which we will just call the *graph groupoid* of E. This is the groupoid that relates the Steinberg algebras to the subject of Leavitt path algebras, as it's foundation is to relate graph C^*-algebras and groupoid C^*-algebras. To be precise, one can show there is a $\mathbb{Z}$-graded $*$-isomorphism $A_R(\mathcal{G}_E) \cong L_R(E)$ (see Example 3.9).

Let $E = (E^0, E^1, r, s)$ be a directed graph (see Example 3.1). We denote by E^∞ the set of infinite paths in E and by E^* the set of finite paths in E. Set

$$X := E^\infty \cup \{\mu \in E^* \mid r(\mu) \text{ is not a regular vertex}\}.$$

Let

$$\mathcal{G}_E := \{(\alpha x, |\alpha| - |\beta|, \beta x) \mid \alpha, \beta \in E^*, x \in X, r(\alpha) = r(\beta) = s(x)\}.$$

We view each $(x, k, y) \in \mathcal{G}_E$ as a morphism with range x and source y. The formulas $(x, k, y)(y, l, z) = (x, k + l, z)$ and $(x, k, y)^{-1} = (y, -k, x)$ define composition and inverse maps on $\mathcal{G}_E$ making it a groupoid with

$$\mathcal{G}_E^{(0)} = \{(x, 0, x) \mid x \in X\},$$

which we will identify with the set X.

Next, we describe a topology on $\mathcal{G}_E$ which is ample and Hausdorff. For $\mu \in E^*$ define

$$Z(\mu) = \{\mu x \mid x \in X, r(\mu) = s(x)\} \subseteq X.$$

For $\mu \in E^*$ and a finite $F \subseteq s^{-1}(r(\mu))$, define

$$Z(\mu \setminus F) = Z(\mu) \setminus \bigcup_{\alpha \in F} Z(\mu\alpha).$$

The sets $Z(\mu \setminus F)$ constitute a basis of compact open sets for a locally compact Hausdorff topology on $X = \mathcal{G}_E^{(0)}$ (see [59, Theorem 2.1]).

For $\mu, \nu \in E^*$ with $r(\mu) = r(\nu)$, and for a finite $F \subseteq E^*$ such that $r(\mu) = s(\alpha)$ for $\alpha \in F$, we define

$$Z(\mu, \nu) = \{(\mu x, |\mu| - |\nu|, \nu x) \mid x \in X, r(\mu) = s(x)\},$$

and then

$$Z((\mu, \nu) \setminus F) = Z(\mu, \nu) \setminus \bigcup_{\alpha \in F} Z(\mu\alpha, \nu\alpha).$$

The sets $Z((\mu, \nu) \setminus F)$ constitute a basis of compact open bisections for a topology under which $\mathcal{G}_E$ is a Hausdorff ample groupoid.

In the case of the graph groupoid $\mathcal{G}_E$, the topological assumptions on $\mathcal{G}_E$ can be described in terms of the geometry of the graph E. We collect them here.

Theorem 3.1 *Let E be a directed graph and $\mathcal{G}_E$ the graph groupoid associated to E. We have the following:*

1. *The unit space $\mathcal{G}_E^{(0)}$ is finite if and only if E is a no exit finite graph [56].*
2. *The unit space $\mathcal{G}_E^{(0)}$ is compact if and only if E has finite number of vertices.*
3. *The unit space $\mathcal{G}_E^{(0)}$ is topologically transitive if and only if E is downward directed [57].*
4. *The unit space $\mathcal{G}_E^{(0)}$ is effective if and only if E satisfies condition (L) [57].*

The following table summarises the properties of the graph E and the corresponding properties of the graph groupoid $\mathcal{G}_E$.

Graph E property	**Groupoid $\mathcal{G}_E$ property**
no cycles	principal (istropy trivial)
condition (L)	effective
condition (K)	strongly effective
cofinal	minimal
E^0 finite	$\mathcal{G}^{(0)}$ compact
E finite and no cycles	discrete

3.4 Steinberg Algebras

3.4.1 Steinberg Algebras

Steinberg algebras (for Hausdorff ample groupoids) are universal algebras that can be defined in terms of inverse semigroup algebras. We present the details below and then provide a concrete realisation as a convolution algebra consisting of certain continuous functions with compact support. In the last section, we describe the algebra when the groupoid is not Hausdorff.

Recall that if R is a commutative ring with unit, then the *semigroup algebra RS* of an inverse semigroup S is defined as the R-algebra with basis S and multiplication extending that of S via the distributive law. If S is an inverse semigroup with zero element z, then the *contracted* semigroup algebra is $R_0S = RS/Rz$. For a Γ-graded groupoid $\mathcal{G}$, recall the graded inverse semigroup $\mathcal{G}^h$ from Sect. 3.3.4.

Definition 3.2 Let $\mathcal{G}$ be a Γ-graded Hausdorff ample groupoid with the inverse semigroup $\mathcal{G}^h$. Given a commutative ring R with identity, the *Steinberg R-algebra* associated to $\mathcal{G}$, and denoted $A_R(\mathcal{G})$, is the contracted semigroup algebra $R_0\mathcal{G}^h$, modulo the ideal generated by $B + D - B \cup D$, where $B, D, B \cup D \in \mathcal{G}^h_\gamma$, $\gamma \in \Gamma$ and $B \cap D = \emptyset$.

So the Steinberg algebra $A_R(\mathcal{G})$, is the algebra generated by the set $\{t_B \mid B \in \mathcal{G}^h\}$ with coefficients in R, subject to the relations

(R1) $t_\emptyset = 0$;
(R2) $t_B t_D = t_{BD}$, for all $B, D \in \mathcal{G}^h$; and
(R3) $t_B + t_D = t_{B\cup D}$, whenever B and D are disjoint elements of $\mathcal{G}^h_\gamma$, $\gamma \in \Gamma$, such that $B \cup D$ is a bisection.

Thus, the Steinberg algebra is universal with respect to the relations (R1), (R2) and (R3) in that if A is any algebra containing $\{T_B : B \in \mathcal{G}^h\}$ satisfying (R1), (R2) and (R3), then there is a homomorphism from $A_R(\mathcal{G})$ to A that sends t_B to T_B. The uniqueness theorems would tell us when this natural homomorphism is injective (Sect. 3.4.4).

Example 3.8 (*Classical groupoid algebras*) If $\mathcal{G}$ is a groupoid and A is a ring then A is said to be a $\mathcal{G}$-graded ring if $A = \bigoplus_{\gamma\in\mathcal{G}} A_\gamma$, where A_γ is an additive subgroup of A and $A_\beta A_\gamma \subseteq A_{\beta\gamma}$, if $(\beta, \gamma) \in \mathcal{G}^{(2)}$ and $A_\beta A_\gamma = 0$, otherwise. A prototype example of $\mathcal{G}$-graded rings are classical groupoid algebras which we describe next. Let R be a ring. Let $R\mathcal{G}$ be a free left R-module with basis $\mathcal{G}$, i.e., $R\mathcal{G} = \bigoplus_{\gamma \in \mathcal{G}} R\gamma$, where $R\gamma = R$. We define a multiplication as follows:

$$\sum_{\sigma \in \mathcal{G}} r_\sigma \sigma \,.\, \sum_{\tau \in \mathcal{G}} s_\tau \tau = \sum_{\sigma, \tau \in \mathcal{G}} r_\sigma s_\tau \sigma \tau,$$

when $(\sigma, \tau) \in \mathcal{G}^{(2)}$, and 0 otherwise. This makes $R\mathcal{G}$ an associative ring and setting $(R\mathcal{G})_\gamma = R\gamma$, clearly gives this ring a $\mathcal{G}$-graded structure.

It is not difficult to see that if $\mathcal{G}^{(0)}$ is finite, then $R\mathcal{G}$ is a direct sum of matrix rings over corresponding isotropy group rings as follows: Let $O_1, \ldots, O_k$ be the orbits of $\mathcal{G}^{(0)}$. Note that for $x, y \in O_i$, there is an isomorphism between the isotropy groups $\mathcal{G}_x^x \cong \mathcal{G}_y^y$. Choosing $x_i \in O_i$, $1 \leq i \leq k$, we then have

$$R\mathcal{G} \cong \bigoplus_{i=1}^{k} \mathbb{M}_{n_i}(RG_i), \tag{3.9}$$

where $G_i = \mathcal{G}_{x_i}^{x_i}$ and $n_i = |O_i|$.

Now if $\mathcal{G}$ has a discrete topology, one can easily establish that $A_R(\mathcal{G}) \cong R\mathcal{G}$. On the other hand, for the case of étale groupoid, finiteness of $\mathcal{G}^{(0)}$ implies $\mathcal{G}$ is discrete (Lemma 3.1). Putting these together we have the following.

Proposition 3.1 [56, Proposition 3.1] *Let $\mathcal{G}$ be an ample groupoid with $\mathcal{G}^{(0)}$ finite. Let $O_1, \ldots, O_k$ be the orbits of $\mathcal{G}^{(0)}$ and let G_i be isotropy group of O_i and $n_i = |O_i|$. Then*

$$A_R(\mathcal{G}) \cong \bigoplus_{i=1}^{k} \mathbb{M}_{n_i}(RG_i),$$

Consider the set $I = \{1, \ldots, n\}$, $n \in \mathbb{N}$ and $G = \{e\}$ a trivial group. Then the transitive groupoid $I \times I$ (see Example 3.3) with discrete topology is ample. Proposition 3.1 now immediately gives

$$A_R(I \times I) \cong \mathbb{M}_n(R).$$

Example 3.9 (*Leavitt path algebras*) Let E be a graph. The Leavitt path algebra associated to the graph E was introduced as a purely algebraic version of the graph C^*-algebras. We refer the reader to the book [1] for a general introduction to the theory and [53] for an excellent survey on the connection of these algebras with Steinberg algebras. We briefly give an account of how to model Leavitt path algebras as Steinberg algebras.

For a graph E, let $\mathcal{G}_E$ be the associated graph groupoid (see Sect. 3.3.5). By [22, Example 3.2] the map

$$\begin{aligned} \pi_E : L_R(E) &\longrightarrow A_R(\mathcal{G}_E), \\ \mu\nu^* - \sum_{\alpha \in F} \mu\alpha\alpha^*\nu^* &\longrightarrow 1_{Z((\mu,\nu)\setminus F)} \end{aligned} \tag{3.10}$$

extends to a $\mathbb{Z}$-graded algebra isomorphism. Observe that the isomorphism of algebras in (3.10) satisfies

$$\pi_E(v) = 1_{Z(v)}, \quad \pi_E(e) = 1_{Z(e,r(e))}, \quad \pi_E(e^*) = 1_{Z(r(e),e)}, \tag{3.11}$$

for each $v \in E^0$ and $e \in E^1$.

If $w : E^1 \to \Gamma$ is a function, we extend w to E^* by defining $w(v) = 0$ for $v \in E^0$, and $w(\alpha_1 \cdots \alpha_n) = w(\alpha_1) \cdots w(\alpha_n)$. Thus $L_R(E)$ is a Γ-graded ring. On the other hand, defining $\widetilde{w} : \mathcal{G}_E \to \Gamma$ by

$$\widetilde{w}(\alpha x, |\alpha| - |\beta|, \beta x) = w(\alpha)w(\beta)^{-1}, \tag{3.12}$$

gives a cocycle [36, Lemma 2.3] and thus $A_R(\mathcal{G})$ is a Γ-graded ring as well. A quick inspection of isomorphism (3.10) shows that π_E respects the Γ-grading.

3.4.2 Convolution Algebra of Continuous Functions From $\mathcal{G}$ to R

In this subsection we give an alternative definition for Steinberg algebras. Let $C_c(\mathcal{G}, R)$ be the algebra of continuous functions from $\mathcal{G}$ to R (where R is a topological discrete space) that vanish outside a compact set. For $f \in C_c(\mathcal{G}, R)$ we have that the support of f denoted $\operatorname{supp}(f) := \{\gamma \in \mathcal{G} \mid f(\gamma) \neq 0\}$ is compact. Since R is discrete,

$$C_c(\mathcal{G}, R) = \{f : \mathcal{G} \to R \mid f \text{ is locally constant and has compact support}\}.$$

Note that when $R = \mathbb{C}$, $C_c(\mathcal{G}, \mathbb{C})$ is not the same as the usual $C_c(\mathcal{G})$, which is the set of continuous functions from $\mathcal{G}$ to $\mathbb{C}$ with the standard topology that vanish outside a compact set.

Addition and scalar multiplication are defined pointwise in $C_c(\mathcal{G}, R)$ and multiplication is given by *convolution* where

$$f * g(\gamma) = \sum_{\alpha\beta=\gamma} f(\alpha)g(\beta).$$

That convolution is well-defined, in that the sum is always finite, uses that $\mathcal{G}$ is étale: for a fixed γ, we are summing over the set

$$\{\alpha\beta \mid \alpha\beta = \gamma,\ f(\alpha) \neq 0 \text{ and } g(\beta) \neq 0\}. \tag{3.13}$$

Since γ is fixed, so is $\boldsymbol{r}(\gamma)$ and $\boldsymbol{d}(\gamma)$. We claim that

$$\{\alpha \mid \boldsymbol{r}(\alpha) = \boldsymbol{r}(\gamma) \text{ and } f(\alpha) \neq 0\} = \boldsymbol{r}^{-1}(\boldsymbol{r}(\gamma)) \cap \operatorname{supp}(f) \tag{3.14}$$

is finite. Since $\mathcal{G}$ is étale, $\boldsymbol{r}^{-1}(\boldsymbol{r}(\gamma))$ is closed and discrete. Thus, (3.14) is the intersection of a discrete closed subspace and a compact set so is finite as claimed. Similarly, the set $\{\beta : \boldsymbol{d}(\beta) = \boldsymbol{d}(\gamma) \text{ and } g(\beta) \neq 0\}$ is finite and hence (3.13) is finite as well.

Since $\mathcal{G}$ is Hausdorff, for each $B \in \mathcal{G}^h$, B is clopen. So the characteristic function 1_B (where $1_B(\gamma) = 1$ for $\gamma \in B$ and 0 otherwise) is in $C_c(\mathcal{G}, R)$. For $B, D \in \mathcal{G}^h$, one can check that $1_B * 1_D = 1_{BD}$ and the set of all characteristic functions $\{1_B : B \in \mathcal{G}\}$ satisfies the relations (R1), (R2) and (R3). So the universal property gives us a homomorphism from $A_R(\mathcal{G})$ to $C_c(\mathcal{G}, R)$ that takes t_B to 1_B for $B \in \mathcal{G}^h$. The range of this homomorphism is the subalgebra

$$\operatorname{span}\{1_B : B \in \mathcal{G}^h\}.$$

Further, this homomorphism is bijective by [54, Theorem 6.3] and hence an isomorphism. Thus, one can write

$$\begin{aligned} A_R(\mathcal{G}) &= \operatorname{span}\big\{1_B \mid B \text{ is a homogeneous compact open bisection}\big\}, \\ &= \Big\{\sum_{B \in F} r_B 1_B \mid F: \text{ mutually disjoint finite collection of} \\ &\qquad\qquad \text{homogeneous compact open bisections}\Big\}, \end{aligned}$$

- addition and scalar multiplication of functions are pointwise,
- multiplications on the generators are $1_B 1_D = 1_{BD}$.

Remark 3.1 One reason we restrict our attention to ample groupoids is to maintain a connection with groupoid *C^*-algebras*, which is the completion of $C_c(\mathcal{G})$ with respect to a particular norm. If we only require our groupoids to be étale, there might not be any locally constant functions so $C_c(\mathcal{G}, \mathbb{C})$ might be empty. On the other hand, when $\mathcal{G}$ is ample, $C_c(\mathcal{G}, \mathbb{C})$ is a dense subset of $C_c(\mathcal{G})$.

3.4.3 Centre of a Steinberg Algebra

The centre of Steinberg algebras were determined in [54] and it has a very pleasant description. We describe it here. We view $A_R(\mathcal{G})$ as $C_c(\mathcal{G}, R)$ where the elements are functions. We say $f \in A_R(\mathcal{G})$ is a *class function* if f satisfies the following conditions:

1. if $f(x) \neq 0$ then $\boldsymbol{d}(x) = \boldsymbol{r}(x)$;
2. if $\boldsymbol{d}(x) = \boldsymbol{r}(x) = \boldsymbol{d}(z)$ then $f(zxz^{-1}) = f(x)$.

Proposition 3.2 [54, Proposition 4.13] *The centre of $A_R(\mathcal{G})$ is the set of class functions.*

Note that if f is a class function, then $\operatorname{supp}(f) \subseteq \operatorname{Iso}(\mathcal{G})^\circ$. Thus if $\mathcal{G}$ is effective, then the centre is contained in the diagonal subalgebra $A_R(\mathcal{G}^{(0)})$. The diagonal preserving isomorphisms play an important role in realising groupoids from the algebra isomorphisms (see Theorem 3.7).

3.4.4 Uniqueness Theorems

A *uniqueness theorem* gives criteria under which a homomorphism from the Steinberg algebra to another R-algebra is injective. Uniqueness theorems are useful when studying other concrete realisations of Steinberg algebras. The most general uniqueness theorem is the following which is [24, Theorem 3.1]:

Theorem 3.2 *Let $\mathcal{G}$ be a second countable, ample, Hausdorff groupoid and let R be a unital commutative ring. Suppose that A is an R-algebra and that $\pi : A_R(\mathcal{G}) \to A$ is a ring homomorphism. Then π is injective if and only if π is injective on $A_R(\operatorname{Iso}(\mathcal{G})^o)$, the subalgebra generated by elements of $\mathcal{G}^h$ that are also contained in the interior of the isotropy bundle.*

Theorem 3.2 has the assumption of second countability because the proof requires the unit space to be a 'Baire space'. The graded uniqueness theorem, below (which is [21, Theorem 3.4]) does not have this assumption. Instead it requires a particular graded structure.

Theorem 3.3 *Let $\mathcal{G}$ be a Hausdorff, ample groupoid, R a commutative ring with identity, Γ a discrete group, and $c : \mathcal{G} \to \Gamma$ a continuous functor such that $\mathcal{G}_e$ is effective. Suppose $\pi : A_R(\mathcal{G}) \to A$ is a graded ring homomorphism. Then π is injective if and only if $\pi(rt_K) \neq 0$ for every nonzero $r \in R$ and compact open $K \subseteq \mathcal{G}^{(0)}$.*

3.4.5 Ideal Structures of Steinberg Algebras

There is a satisfactory description of ideals of a Steinberg algebra $A_k(\mathcal{G})$ based on the geometry of the groupoid $\mathcal{G}$, where the algebra is over a field k (so the ideals of the coefficient ring does not interfere) and the groupoid is Hausdorff ample and (strongly) effective.

The first result is the simplicity of these algebras.

Theorem 3.4 [14] *Let $\mathcal{G}$ be an Hausdorff, ample groupoid, and k a field. Then $A_k(\mathcal{G})$ is simple if and only if $\mathcal{G}$ is effective and $\mathcal{G}^{(0)}$ has no open invariant subsets.*

A glance at the table of properties of the graph versus the graph groupoids (Sect. 3.3.5) shows that Theorem 3.4 is parallel to the first theorem proved in the theory of Leavitt path algebras, namely, for an arbitrary graph E, the Leavitt path algebra $L_k(E)$ is simple if and only if E satisfies condition (L) and E^0 has no non-trivial saturated hereditary subsets [2], [1, Sect. 2.9].

For an invariant $U \subseteq \mathcal{G}^{(0)}$, one can easily see that the set

$$I(U) := \operatorname{span}\{t_B \mid s(B) \subseteq U\},$$

is an ideal of $A_k(\mathcal{G})$. In fact, if the groupoid is strongly effective, this is the only way one can construct ideals in these algebras.

Theorem 3.5 [18] *Suppose $\mathcal{G}$ is a strongly effective ample groupoid. Then the correspondence*

$$U \longmapsto I(U),$$

is a lattice isomorphism from the lattice of open invariant subsets of $\mathcal{G}^{(0)}$ onto the lattice of ideals of $A_k(\mathcal{G})$.

Theorem 3.6 [24] *Let $\mathcal{G}$ be a Γ-graded ample groupoid such that $\mathcal{G}_\varepsilon$ is strongly effective. Then the correspondence*

$$U \longmapsto I(U),$$

is an isomorphism from the lattice of open invariant subsets of $\mathcal{G}^{(0)}$ onto to the lattice of graded ideals in $A_k(G)$.

We refer the reader to [18, 24] for further results on the ideal theory of Steinberg algebras.

3.5 Combinatorial and Dynamical Invariants of étale Groupoids

There are several 'combinatorial' invariants one can associate to a groupoid such as the full group and homology groups. For certain groupoids these combinatorial invariants are related to very interesting Higman–Thompson groups or K-groups as we describe next in this section.

3.5.1 Full Groups

For an étale groupoid $\mathcal{G}$ with the compact unit space $\mathcal{G}^{(0)}$, the full group $[[\mathcal{G}]]$ was defined by Matui [41]. The full group $[[\mathbb{Z} \times X]]$ of the transformation groupoid of the action of $\mathbb{Z}$ on a Cantor set X via a minimal homeomorphism (see Example 3.3) coincides with the full group defined and studied in [32]. We define the full group of a groupoid here and collect results related to this group.

Recall that $\mathcal{G}^a$ is the inverse semigroup of compact open bisections and $\pi : \mathcal{G}^a \to \mathcal{I}(\mathcal{G}^{(0)})$ the action of $\mathcal{G}^a$ on $\mathcal{G}^{(0)}$ (Sect. 3.3.4). Let $G := \{U \in \mathcal{G}^a \mid \boldsymbol{d}(U) = \boldsymbol{r}(U) = \mathcal{G}^{(0)}\}$ be a subgroup of $\mathcal{G}^a$ (here $\mathcal{G}^{(0)}$ considered to be compact). Then the *full group of* $\mathcal{G}$, denoted by $[[\mathcal{G}]]$ is $\pi(G)$. In fact, for a noncompact $\mathcal{G}^{(0)}$, one can give a generalised version of this notion and define the *full inverse semigroup* by $[[\mathcal{G}]] = \pi(\mathcal{G}^a)$.

If the Hausdorff ample groupoids $\mathcal{G}$ and $\mathcal{H}$ are isomorphism, then clearly $A_R(\mathcal{G}) \cong A_R(\mathcal{H})$. Further, since from the outset, $\mathcal{G} \cong \mathcal{H}$ induces $\mathcal{G}^{(0)} \cong \mathcal{H}^{(0)}$, we have the *diagonal isomorphism* $A_R(\mathcal{G}^{(0)}) \cong A_R(\mathcal{H}^{(0)})$ as well. Renault in [52] established the converse of this statement for certain groupoid C^*-algebras. Several recent papers progressively improved this result in the algebraic setting (see [4, 17, 58]). Combining with the full group invariant, we have the following theorem, relating groupoids, inverse semigroups and algebras.

Theorem 3.7 *Let R be a unital commutative ring without nontrivial idempotents and let $\mathcal{G}$ and $\mathcal{H}$ be Hausdorff effective ample groupoids. Then the following are equivalent.*

(1) *$\mathcal{G}$ and $\mathcal{H}$ are isomorphic;*
(2) *the inverse semigroups $\mathcal{G}^a$ and $\mathcal{H}^a$ are isomorphic;*
(3) *the inverse semigroups $[[G]]$ and $[[H]]$ are isomorphic;*
(4) *there exists a diagonal-preserving isomorphism between the Steinberg algebras $A_R(\mathcal{G})$ and $A_R(\mathcal{H})$.*

Consider a graph with one vertex and two loops

$E :$ 

and its graph groupoid $\mathcal{G}_E$ as described in Sect. 3.3.5. The unit space $\mathcal{G}_E^{(0)}$ is compact (Theorem 3.1) and we can consider the full group $[[\mathcal{G}_E]]$. On the other hand, let $L_k(E)$ be the Leavitt path algebra associated to E over a field k. An element $a \in L_k(E)$ is called unitary if $aa^* = a^*a = 1$. Consider the set

$$P_{2,1} = \Big\{a \in L_k(E) \text{ is unitary} \mid a = \sum_{i=1}^{l} \alpha_i \beta_i^* \Big\},$$

where α_i, β_i are distinct paths in E and the coefficients in the sum are all 1. One can prove that

$$P_{2,1} \cong [[\mathcal{G}_E]],$$

and they are isomorphic to the Thompson group $T_{2,1}$, which was constructed in 1965 and was the first infinite finitely presented simple group. In fact, considering a graph with one vertex and n loops, we retrieve Higman–Thompson groups $G_{n,1}$ constructed by Higman in 1974. We refer the reader to [15, 46] for this line of research.

3.5.2 Homology and K-Theory

The homology theory for étale groupoids was introduced by Crainic and Moerdijk [26] who showed these groups are invariant under Morita equivalences of étale groupoids and established some spectral sequences which used for the computation of these homologies. Matui [41–43] considered this homology theory in relation to the dynamical properties of groupoids and their full groups. In [41] Matui proved, using Lindon–Hochschild–Serre spectral sequence established by Crainic and Moerdijk that for an étale groupoid $\mathcal{G}$ arising from subshifts of finite type, the homology groups $H_0(\mathcal{G})$ and $H_1(\mathcal{G})$ coincide with K-groups $K_0(C^*(\mathcal{G}))$ and $K_1(C^*(\mathcal{G}))$, respectively. Here $C^*(\mathcal{G})$ is the groupoid C^*-algebra associated to $\mathcal{G}$ which were first systematically studied by Renault in his seminal work [50]. In the language of graphs, Matui proved that for a finite graph E with no sinks

$$K_0(C^*(E)) \cong H_0(\mathcal{G}_E) \text{ and, } K_1(C^*(E)) \cong H_1(\mathcal{G}_E). \tag{3.15}$$

In this section, we recall the construction of the homology of an ample groupoids and recount Matui's conjecture relating K-groups of groupoid C^*-algebras to the homology of its groupoid (Conjecture 3.1). It would be very desirable to establish a relation between the homology groups and K-groups of Steinberg algebras, as here we have higher K-theories available and strong K-theory machinery which works on them [49].

Let X be a locally compact Hausdorff space and R a topological abelian group. Denote by $C_c(X, R)$ the set of R-valued continuous functions with compact support. With pointwise addition, $C_c(X, R)$ is an abelian group. Let $\pi : X \to Y$ be a local homeomorphism between locally compact Hausdorff spaces. For $f \in C_c(X, R)$, define the map $\pi_*(f) : Y \to R$ by

$$\pi_*(f)(y) = \sum_{\pi(x)=y} f(x).$$

Thus, π_* is a homomorphism from $C_c(X, R)$ to $C_c(Y, R)$ which makes $C_c(-, R)$ a functor from the category of locally compact Hausdorff spaces with local homeomorphisms to the category of abelian groups.

Let $\mathcal{G}$ be an étale groupoid. For $n \in \mathbb{N}$, we write $\mathcal{G}^{(n)}$ for the space of composable strings of n elements in $\mathcal{G}$, that is,

$$\mathcal{G}^{(n)} = \{(g_1, g_2, \dots, g_n) \in \mathcal{G}^n \mid \boldsymbol{d}(g_i) = \boldsymbol{r}(g_{i+1}) \text{ for all } i = 1, 2, \dots, n-1\}.$$

For $i = 0, 1, \dots, n$, with $n \geq 2$ we let $d_i : \mathcal{G}^{(n)} \to \mathcal{G}^{(n-1)}$ be a map defined by

$$d_i(g_1, g_2, \dots, g_n) = \begin{cases} (g_2, g_3, \dots, g_n) & i = 0 \\ (g_1, \dots, g_i g_{i+1}, \dots, g_n) & 1 \leq i \leq n-1 \\ (g_1, g_2, \dots, g_{n-1}) & i = n. \end{cases}$$

When $n = 1$, we let $d_0, d_1 : \mathcal{G}^{(1)} \to \mathcal{G}^{(0)}$ be the source map and the range map, respectively (i.e., the $\boldsymbol{d}$ and $\boldsymbol{r}$ maps). Clearly, the maps d_i are local homeomorphisms.

Define the homomorphisms $\partial_n : C_c(\mathcal{G}^{(n)}, R) \to C_c(\mathcal{G}^{(n-1)}, R)$ by

$$\partial_1 = \boldsymbol{d}_* - \boldsymbol{r}_* \quad \text{and} \quad \partial_n = \sum_{i=0}^{n} (-1)^i d_{i*}. \tag{3.16}$$

One can check that the sequence

$$0 \xleftarrow{\partial_0} C_c(\mathcal{G}^{(0)}, R) \xleftarrow{\partial_1} C_c(\mathcal{G}^{(1)}, R) \xleftarrow{\partial_2} C_c(\mathcal{G}^{(2)}, R) \xleftarrow{\partial_3} \cdots \tag{3.17}$$

is a chain complex of abelian groups.

The following definition comes from [26, 41].

Definition 3.3 (*Homology groups of a groupoid* $\mathcal{G}$) Let $\mathcal{G}$ be an étale groupoid. Define the homology groups of $\mathcal{G}$ with coefficients R, $H_n(\mathcal{G}, R)$, $n \geq 0$, to be the homology groups of the Moore complex (3.17), i.e., $H_n(\mathcal{G}, R) = \ker \partial_n / \operatorname{Im} \partial_{n+1}$. When $R = \mathbb{Z}$, we simply write $H_n(\mathcal{G}) = H_n(\mathcal{G}, \mathbb{Z})$. In addition, we define

$$H_0(\mathcal{G})^+ = \{[f] \in H_0(\mathcal{G}) \mid f(x) \geq 0 \text{ for all } x \in \mathcal{G}^{(0)}\},$$

where $[f]$ denotes the equivalence class of $f \in C_c(\mathcal{G}^{(0)}, \mathbb{Z})$.

Extending Matui's result (3.15), in [33] using the description of monoid of Leavitt path algebras, it could be proved that for any graph (with sinks, source and infinite emitters), we have

$$H_0(\mathcal{G}_E) \cong K_0(A(\mathcal{G}_E)) \cong K_0(L(E)) \cong K_0(C^*(E)) \cong K_0(C^*(\mathcal{G}_E)).$$

Before stating Matui's conjecture (Conjecture 3.1) we also state a class of groupoids that the zeroth homology H_0 coincides with Grothendieck group K_0 and their algebras fall into Elliott's class of algebras that can be classified by K_0-groups.

Let $\mathcal{G}$ be a second countable étale groupoid whose unit space is compact and totally disconnected. Then the subgroupoid $\mathcal{H} \subseteq \mathcal{G}$ is an *elementary subgroupoid* if $\mathcal{H}$ is a compact open principal subgroupoid of $\mathcal{G}$ such that $\mathcal{H}^{(0)} = \mathcal{G}^{(0)}$. The groupoid $\mathcal{G}$ is called an *AF groupoid* if it can be written as an increasing union of elementary subgroupoids.

If $\mathcal{G}$ is an AF groupoid, then Steinberg algebra $A_R(\mathcal{G})$, for a field R, is an ultramatricial algebra (or the reduced groupoid C^*-algebra $C^*(\mathcal{G})$ is an AF algebra and thus the terminology). For such groupoids, there is an order-preserving isomorphism

$$\begin{aligned}\pi : H_0(\mathcal{G}) &\longrightarrow K_0(A_R(\mathcal{G})),\\ [1_{\mathcal{G}^{(0)}}] &\longmapsto [1_{A_R(\mathcal{G}^{(0)})}].\end{aligned}$$

We have then the following theorem.

Theorem 3.8 *Let R be a field and $\mathcal{G}$ and $\mathcal{H}$ are AF groupoids. Then the following are equivalent.*

(1) *$\mathcal{G}$ and $\mathcal{H}$ are isomorphic;*
(2) *There is an order-preserving isomorphism $H_0(\mathcal{G}) \cong H_0(\mathcal{H})$ which sends $[1_{\mathcal{G}^{(0)}}]$ to $[1_{\mathcal{H}^{(0)}}]$;*
(3) *there exists a R-algebra isomorphism between the Steinberg algebras $A_R(\mathcal{G})$ and $A_R(\mathcal{H})$.*

This theorem points to a direction which is gaining evermore importance of finding a class of étale groupoids that a variant of K-theory and homology theory can be a complete invariant.

Recall that an étale groupoid $\mathcal{G}$ is said to be effective if the interior of its isotropy coincides with its unit space $\mathcal{G}^{(0)}$ and minimal if every orbit is dense.

The following conjecture of Matui [43, Conjecture 2.6] expresses the K-theory of a groupoid C^*-algebra as a direct sum of homology groups of the associated groupoid. For one thing, this indicates that the homology groups provide much finer invariants than the K-groups.

Conjecture 3.1 (Matui's HK Conjecture) Let $\mathcal{G}$ be a locally compact Hausdorff étale groupoid such that $\mathcal{G}^{(0)}$ is a Cantor set. Suppose that $\mathcal{G}$ is both effective and minimal. Then

$$K_0(C_r^*(\mathcal{G})) \cong \bigoplus_{i=0}^{\infty} H_{2i}(\mathcal{G}) \tag{3.18}$$

$$K_1(C_r^*(\mathcal{G})) \cong \bigoplus_{i=0}^{\infty} H_{2i+1}(\mathcal{G}) \tag{3.19}$$

Apart from (arbitrary graphs) Matui proved this conjecture for AF groupoids with compact unit space and in [43] for all finite Cartesian products of groupoids associated with shifts of finite type. Ortega also showed in [45] that the conjecture is valid for the Katsura-Exel-Pardo groupoid $\mathcal{G}_{A,B}$ associated to square integer matrices with $A \geq 0$.

3.6 Partial Crossed Product Rings

In this section, we consider the 'partially' group ring-like algebras and relate them to Steinberg algebras. This is also another demonstration of how the inverse semigroups and algebras arising from them are related to groupoids and algebras coming from them.

Let $\pi = (\pi_s, A_s, A)_{s\in\mathcal{S}}$ be a partial action of the inverse semigroup $\mathcal{S}$ on an algebra A. Here $A_s \subseteq A$, $s \in \mathcal{S}$ is an ideal of A and $\pi_s : A_{s^*} \to A_s$ an isomorphism such that for all $s, t \in \mathcal{S}$

(i) $\pi_s^{-1} = \pi_{s^*}$;
(ii) $\pi_s(A_{s^*} \cap A_t) \subseteq A_{st}$;
(iii) if $s \leq t$, then $A_s \subseteq A_t$;
(iv) For every $x \in A_{t^*} \cap A_{t^*s^*}$, $\pi_s(\pi_t(x)) = \pi_{st}(x)$.

This is a generalisation of the concept of partial group actions (see Sect. 3.5). Define $\mathcal{L}$ as the set of all formal forms $\sum_{s\in\mathcal{S}} a_s\delta_s$ (with finitely many a_s nonzero), where $a_s \in A_s$ and δ_s are symbols, with addition defined in the obvious way and multiplication being the linear extension of

$$(a_s\delta_s)(a_t\delta_t) = \pi_s\big(\pi_{s^{-1}}(a_s)a_t\big)\delta_{st}.$$

Then $\mathcal{L}$ is an algebra which is possibly not associative. Exel and Vieira proved under which condition $\mathcal{L}$ is associative (see [31, Theorem 3.4]). In particular, if each ideal A_s is idempotent or non-degenerate, then $\mathcal{L}$ is associative (see [31, Theorem 3.4] and [28, Proposition 2.5]). This algebra is too large for us and we need to consider this ring modulo idempotents, as follows. Consider $\mathcal{N} = \langle a\delta_s - a\delta_t : a \in A_s, s \leq t\rangle$, which is the ideal generated by $a\delta_s - a\delta_t$. The *partial skew inverse semigroup ring* $A \rtimes_\pi \mathcal{S}$ is defined as the quotient ring $\mathcal{L}/\mathcal{N}$.

Next, we equip these algebras with a graded structure. Suppose $\mathcal{S}$ is a Γ-graded inverse semigroup (see Sect. 3.2.1). Observe that the algebra $\mathcal{L}$ is a G-graded algebra with elements $a_s\delta_s \in \mathcal{L}$ with $a_s \in A_s$ are homogeneous elements of degree $w(s)$. Furthermore, if $s \leq t$, then $s = ts^*s$. It follows that $w(s) = w(t)w(s^*)w(s) = w(t)$. Hence $a\delta_s - a\delta_t$ with $s \leq t$ and $a \in A_s$ is a homogeneous element in $\mathcal{L}$. Thus, the ideal $\mathcal{N}$ generated by homogeneous elements is a graded ideal and, therefore, the quotient algebra $A \rtimes_\pi \mathcal{S} = \mathcal{L}/\mathcal{N}$ is Γ-graded.

Let X be a Hausdorff topological space and R a unital commutative ring with a discrete topology. Let $C_R(X)$ be the set of R-valued continuous function (i.e. locally constant) with compact support (see also Sect. 3.4.2). If D is a compact open subset of X, the characteristic function of D, denoted by 1_D, is clearly an element of $C_R(X)$. In fact, every f in $C_R(X)$ may be written as

$$f = \sum_{i=1}^{n} r_i 1_{D_i}, \tag{3.20}$$

where $r_i \in R$ and the D_i are compact open, pairwise disjoint subsets of X. $C_R(X)$ is a commutative R-algebra with pointwise multiplication. The support of f, defined by $\mathrm{supp}(f) = \{x \in X \mid f(x) \neq 0\}$, is clearly a compact open subset.

We observe that $C_R(X)$ is an idempotent ring. We have

$$\sum_{i=1}^{n} 1_{D_i} \cdot f = f \cdot \sum_{i=1}^{n} 1_{D_i} = \sum_{i=1}^{n} r_i 1_{D_i} = f \tag{3.21}$$

for any $f \in C_R(X)$ which is written as (3.20). So $C_R(X)$ is a ring with local units and thus an idempotent ring.

For Γ-graded Hausdorff ample groupoid $\mathcal{G}$, recall the inverse semigroup $\mathcal{G}^h$ from (3.6) and the action of $\mathcal{G}^h$ on $\mathcal{G}^{(0)}$ from (3.7). There is an induced action $(\pi_B, C_R(BB^{-1}), C_R(\mathcal{G}^{(0)}))_{B\in\mathcal{G}^h}$ of $\mathcal{G}^h$ on an algebra $C_R(\mathcal{G}^{(0)})$. Here the map $\pi_B : C_R(B^{-1}B) \to C_R(BB^{-1})$ is given by $\pi_B(f) = f \circ \pi_B^{-1}$. We still denote the induced action by π. In this case, $\mathcal{L} = \left\{ \sum_{B\in\mathcal{G}^h} a_B\delta_B \mid a_B \in C_R(BB^{-1})\right\}$ is associative, since each ideal $C_R(BB^{-1})$ is idempotent. Since $\mathcal{G}^h$ is Γ-graded, $C_R(\mathcal{G}^{(0)}) \rtimes_\pi \mathcal{G}^h$ is a Γ-graded algebra.

We are in a position to relate partial skew inverse semigroup rings to Steinberg algebras.

Theorem 3.9 *Let $\mathcal{G}$ be a Γ-graded Hausdorff ample groupoid and*

$$\pi = \Big(\pi_B, C_R(BB^{-1}), C_R(\mathcal{G}^{(0)})\Big)_{B\in\mathcal{G}^h},$$

the induced action of $\mathcal{G}^h$ on $C_R(\mathcal{G}^{(0)})$. Then there is a Γ-graded isomorphism of R-algebras

$$A_R(\mathcal{G}) \cong_{\mathrm{gr}} C_R(\mathcal{G}^{(0)}) \rtimes_\pi \mathcal{G}^h. \tag{3.22}$$

Proof For each $D \in \mathcal{G}^{(h)}$, define

$$t_D = 1_{r(D)}\delta_D \in C_R(\mathcal{G}^{(0)}) \rtimes_\pi \mathcal{G}^h.$$

One can check that the set $\{t_D \mid D \in \mathcal{G}^{(h)}\}$ satisfies (R1), (R2) and (R3) relations in the Definition 3.2 of Steinberg algebras. Thus, we obtain a homomorphism

$$f : A_R(\mathcal{G}) \to C_R(\mathcal{G}^{(0)}) \rtimes_\pi \mathcal{G}^h.$$

Next define a map $g : C_R(\mathcal{G}^{(0)}) \rtimes_\pi \mathcal{G}^{(h)} \to A_R(\mathcal{G})$. For each $B \in \mathcal{G}^{(h)}$ and $a_B\delta_B \in \mathcal{L}$, we define

$$g(a_B\delta_B) = \begin{cases} a_B(r(x)), & \text{if } x \in B, \\ 0, & \text{otherwise.} \end{cases}$$

One can check that the map g is well-defined and $gf = \mathrm{id}_{A_R(\mathcal{G})}$ and $fg = 1_{C_R(\mathcal{G}^{(0)})\rtimes_\pi \mathcal{G}^h}$. □

We refer the reader to [9, 10, 33] for more results relating the partial inverse semigroup algebras to Steinberg algebras and [30] for the C^*-versions of these results.

3.7 Non-Hausdorff Ample Groupoids

We finish this paper with a brief discussion about non-Hausdorff groupoids. When relaxing the Hausdorff assumption on $\mathcal{G}$, we still insist that the unit space $\mathcal{G}^{(0)}$ be Hausdorff so that, in the setting of étale and ample groupoids, they are locally Hausdorff. With this weakened hypothesis the universally defined Steinberg algebra of Definition 3.2 no longer works. However, we can still study the Steinberg algebra of such a groupoid using one of the other characterisations, such as

$$A_R(\mathcal{G}) = \mathrm{span}\{1_B : B \in \mathcal{G}^h\}.$$

Since $\mathcal{G}$ is not Hausdorff, compact open bisections are no longer closed and hence characteristic functions might not be continuous.

Other fundamental results in the theory of Steinberg algebras fail for non-Hausdorff groupoids, for example, the Uniqueness theorems (see [24, Example 3.5]). Still, progress is slowly being made to develop a theory. Recall that in an ample groupoid $\mathcal{G}^{(0)}$ is always open in $\mathcal{G}$. It turns out that $\mathcal{G}^{(0)}$ is closed in $\mathcal{G}$ if and only if $\mathcal{G}$ is Hausdorff. So an important step for understanding non-Hausdorff groupoids is to understand the closure of $\mathcal{G}^{(0)}$.

3.7.1 Non-Hausdorff Simplicity

For non-Hausdorff groupoids, necessary and sufficient groupoid conditions that ensure the Steinberg algebra is simple are not known. The forward implication of Theorem 3.4 does hold in the non-Hausdorff setting (see [55, Theorem 3.5]). But the reverse implication uses Theorem 3.2 which fails for non-Hausdorff groupoids. Here is why it fails: The proof of Theorem 3.2 assumes that for every function $f \in A_k(\mathcal{G})$, the set

$$\mathrm{supp}(f) := \{\gamma \in \mathcal{G} : f(\gamma) \neq 0\} = f^{-1}(0)$$

has non-empty interior. This is clearly true if f is continuous but can fail when $\mathcal{G}$ is not Hausdorff. We call a function f such that $\mathrm{supp}(f)$ has empty interior a *singular function*. The collection of all singular functions forms an ideal in $A_k(\mathcal{G})$. It turns out this is the only obstruction to simplicity.

Theorem 3.10 [19, Theorem 3.14] *Let $\mathcal{G}$ be a second countable, ample groupoid such that $\mathcal{G}^{(0)}$ is Hausdorff and let k be a field. Then $A_k(\mathcal{G})$ is simple if and only if the following three conditions are satisfied:*

1. *$\mathcal{G}$ is minimal,*
2. *$\mathcal{G}$ is effective, and*
3. *for every nonzero $f \in A_k(\Gamma)$, $\operatorname{supp}(f)$ has non-empty interior.*

It is not known whether condition (3) is an automatic given conditions (1) and (2).

Acknowledgements The authors would like to acknowledge the grant DP160101481 from the Australian Research Council and Marsden grant VUW1514 from the Royal Society of New Zealand.

References

1. G. Abrams, P. Ara, M. Siles Molina, *Leavitt Path Algebras. Lecture Notes in Mathematics*, vol. 2191 (Springer, London, 2017)
2. G. Abrams, G. Aranda Pino, The Leavitt path algebra of a graph. J. Algebra **293**, 319–334 (2005)
3. P. Ara, M.A. Moreno, E. Pardo, Nonstable K-theory for graph algebras. Algebr. Represent. Theory **10**, 157–178 (2007)
4. P. Ara, J. Bosa, R. Hazrat, A. Sims, Reconstruction of graded groupoids from graded Steinberg algebras. Forum Math. **29**(5), 1023–1037 (2017)
5. C.J. Ash, T.E. Hall, Inverse semigroups on graphs. Semigroup Forum **11**, 140–145 (1975)
6. T. Bates, D. Pask, I. Raeburn, W. Szymański, The C^*-algebras of row-finite graphs. N. Y. J. Math. **6**(307), 324 (2000)
7. E. Batista, Partial actions: what they are and why we care. Bull. Belg. Math. Soc. Simon Stevin **24**(1), 35–71 (2017)
8. S. Berberian, Baer $*$-rings. Die Grundlehren der mathematischen Wissenschaften, Band 195 (Springer, New York, 1972)
9. V.M. Beuter, L. Cordeiro, The dynamics of partial inverse semigroup actions, arXiv:1804.00396
10. V.M. Beuter, D. Gonçalves, The interplay between Steinberg algebras and partial skew rings, arXiv:1706.00127v1
11. H. Brandt, Über eine Verallgemeinerung des Gruppenbegriffes. Math. Ann. **96**, 360–366 (1926)
12. R. Brown, *Topology and Groupoids. Elements of Modern Topology*, 3rd edn. (BookSurge LLC, Charleston, 2006)
13. R. Brown, From groups to groupoids: a brief survey. Bull. Lond. Math. Soc. **19**, 113–134 (1987)
14. J. Brown, L.O. Clark, C. Farthing, A. Sims, Simplicity of algebras associated to étale groupoids. Semigroup Forum **88**(2), 433–452 (2014)
15. N. Brownlowe, A.P.W. Sørensen, $L_{2,\mathbb{Z}} \otimes L_{2,\mathbb{Z}}$ does not embed in $L_{2,\mathbb{Z}}$. J. Algebra **456**, 1–22 (2016)
16. A. Buss, R. Exel, Inverse semigroup expansions and their actions on C^*-algebras. Ill. J. Math. **56**, 1185–1212 (2014)
17. T.M. Carlsen, J. Rout, Diagonal-preserving graded isomorphisms of Steinberg algebras. Commun. Contemp. Math. **20**(6), 1750064 (2018)
18. L.O. Clark, C. Edie-Michell, A. an Huef, A. Sims, Ideals of Steinberg algebras of strongly effective groupoids, with applications to Leavitt path algebras. Trans. Am. Math. Soc. **371**(8), 5461–5486. arXiv:1601.07238

19. L.O. Clark, R. Exel, E. Pardo, A. Sims, C. Starling, Simplicity of algebras associated to non-Hausdorff groupoids. Trans. Am. Math. Soc. (2018). arXiv:1806.04362
20. L.O. Clark, R. Hazrat, S.W. Rigby, Strongly graded groupoids and strongly graded Steinberg algebras. J. Algebra. **530**, 34–68 (2019). arXiv:1711.04904
21. L.O. Clark, C. Edie-Michell, Uniqueness theorems for Steinberg algebras. Algebr. Represent. Theory **18**(4), 907–916 (2015)
22. L.O. Clark, A. Sims, Equivalent groupoids have Morita equivalent Steinberg algebras. J. Pure Appl. Algebra **219**, 2062–2075 (2015)
23. L.O. Clark, C. Farthing, A. Sims, M. Tomforde, A groupoid generalisation of Leavitt path algebras. Semigroup Forum **89**, 501–517 (2014)
24. L.O. Clark, R. Exel, E. Pardo, A generalised uniqueness theorem and the graded ideal structure of Steinberg algebras. Forum Math. **30**(3), 533–552 (2018)
25. A. Connes, *Noncommutative Geometry* (Academic Press, San Diego, 1994)
26. M. Crainic, I. Moerdijk, A homology theory for étale groupoids. J. Reine Angew. Math. **521**, 25–46 (2000)
27. J. Cuntz, Simple C^*-algebras generated by isometries. Comm. Math. Phys. **57**(2), 173–185 (1977)
28. M. Dokuchaev, R. Exel, Associativity of crossed products by partial actions, enveloping actions and partial representations. Trans. Am. Math. Soc. **357**(5), 1931–1952 (2005)
29. R. Exel, Partial actions of groups and actions of semigroups. Proc. Am. Math. Soc. **126**(12), 3481–3494 (1998)
30. R. Exel, Inverse semigroups and combinatorial C^*-algebras. Bull. Braz. Math. Soc. New Ser. **39**(2), 191–313 (2008)
31. R. Exel, F. Vieira, Actions of inverse semigroups arising from partial actions of groups. J. Math. Anal. Appl. **363**, 86–96 (2010)
32. T. Giordano, I.F. Putnam, C.F. Skau, Full groups of Cantor minimal systems. Isr. J. Math. **111**, 285–320 (1999)
33. R. Hazrat, H. Li, Homology of étale groupoids, a graded approach, arXiv:1806.03398 [math.KT]
34. P.J. Higgins, Categories and groupoids. Repr. Theory Appl. Categ. **7**, 1–195 (2005)
35. I. Kaplansky, *Rings of Operators* (W. A. Benjamin Inc., New York, 1968)
36. A. Kumjian, D. Pask, C^*-algebras of directed graphs and group actions. Ergod. Theory Dyn. Syst. **19**, 1503–1519 (1999)
37. A. Kumjian, D. Pask, Higher rank graph C^*-algebras. N. Y. J. Math. **6**(1), 1–20 (2000)
38. A. Kumjian, D. Pask, I. Raeburn, J. Renault, Graphs, groupoids, and Cuntz–Krieger algebras. J. Funct. Anal. **144**(2), 505–541 (1997)
39. M.V. Lawson, *Inverse Semigroups: The Theory of Partial Symmetries* (World Scientific Publishing Co. Inc., River Edge, NJ, 1998)
40. W.G. Leavitt, Modules without invariant basis number. Proc. Am. Math. Soc. **8**, 322–328 (1957)
41. H. Matui, Homology and topological full groups of étale groupoids on totally disconnected spaces. Proc. Lond. Math. Soc. **104**(3), 27–56 (2012)
42. H. Matui, Topological full groups of one-sided shifts of finite type. J. Reine Angew. Math. **705**, 35–84 (2015)
43. H. Matui, Étale groupoids arising from products of shifts of finite type. Adv. Math. **303**, 502–548 (2016)
44. Z. Mesyan, J. Mitchell, The structure of a graph inverse semigroup. Semigroup Forum **93**, 111–130 (2016)
45. E. Ortega, Homology of the Katsura-Exel-Pardo groupoid (2018). arXiv:1806.09297
46. E. Pardo, The isomorphism problem for Higman–Thompson groups. J. Algebra **344**, 172–183 (2011)
47. A.L.T. Paterson, *Groupoids, Inverse Semigroups, and Their Operator Algebras*, vol. 170 (Birkhäuser, Basel, 1999)
48. A.L.T. Paterson, Graph inverse semigroups, groupoids and their C^*-algebras. J. Oper. Theory **48**, 645–662 (2002)

49. D. Quillen, *Higher Algebraic K-Theory. I. Algebraic K-Theory, I: Higher K-Theories (Proceedings of the Conference on Battelle Memorial Institute, Seattle, WA, 1972)*. Lecture Notes in Mathematics, vol. 341 (Springer, Berlin, 1973), pp. 85–147
50. J. Renault, *A Groupoid Approach to C^*-Algebras*. Lecture Notes in Mathematics, vol. 793 (Springer, Berlin, 1980)
51. J. Renault, Cuntz-like algebras. in *Operator Theoretical Methods, Timioara, 1998* (Theta Foundation, Bucharest, 2000), pp. 371–386
52. J. Renault, Cartan subalgebras in C^*-algebras. Ir. Math. Soc. Bull. **61**, 29–63 (2008)
53. S.W. Rigby, The groupoid approach to Leavitt path algebras. arXiv:1811.02269
54. B. Steinberg, A groupoid approach to discrete inverse semigroup algebras. Adv. Math. **223**, 689–727 (2010)
55. B. Steinberg, Simplicity, primitivity and semiprimitivity of étale groupoid algebras with applications to inverse semigroup algebras. J. Pure Appl. Algebra **220**(3), 1035–1054 (2016)
56. B. Steinberg, Chain conditions on étale groupoid algebras with applications to Leavitt path algebras and inverse semigroup algebras. J. Aust. Math. Soc. **104**, 403–411 (2018)
57. B. Steinberg, Prime étale groupoid algebras with applications to inverse semigroup and Leavitt path algebras. J. Pure Appl. Algebra **223**(6), 2474–2488 (2019)
58. B. Steinberg, Diagonal-preserving isomorphisms of étale groupoid algebras. J. Algebra **518**, 412–439 (2019)
59. S.B.G. Webster, The path space of a directed graph. Proc. Am. Math. Soc. **142**, 213–225 (2014)
60. A. Weinstein, Groupoids: unifying internal and external symmetry. A tour through some examples. Not. Am. Math. Soc. **43**(7), 744–752 (1996)

Chapter 4
The Injective and Projective Leavitt Complexes

Huanhuan Li

4.1 Introduction

Let A be a finite dimensional algebra over a field k. We denote by $\mathbf{K}_{\mathrm{ac}}(A\text{-Inj})$ the homotopy category of acyclic complexes of injective A-modules which are called the stable derived category of A in [16]. This category is a compactly generated triangulated category such that its subcategory of compact objects is triangle equivalent to the singularity category [4, 22] of A.

In general, it seems very difficult to give an explicit compact generator for the stable derived category of an algebra. An explicit compact generator called the *injective Leavitt complex*, for the homotopy category $\mathbf{K}_{\mathrm{ac}}(A\text{-Inj})$ in the case that the algebra A is with radical square zero was constructed in [18]. This terminology is justified by the following result: the differential graded endomorphism algebra of the injective Leavitt complex is quasi-isomorphic to the Leavitt path algebra in the sense of [2, 3]. Here, the Leavitt path algebra is naturally $\mathbb{Z}$-graded and viewed as a differential graded algebra with trivial differential.

We denote by $\mathbf{K}_{\mathrm{ac}}(A\text{-Proj})$ the homotopy category of acyclic complexes of projective A-modules. This category is a compactly generated triangulated category whose subcategory of compact objects is triangle equivalent to the opposite category of the singularity category of the opposite algebra A^{op}. An explicit compact generator, called the *projective Leavitt complex*, for the homotopy category $\mathbf{K}_{\mathrm{ac}}(A\text{-Proj})$ in the case that A is an algebra with radical square zero was constructed in [19]. It is shown that the opposite differential graded endomorphism algebra of the projective Leavitt

H. Li (✉)
School of Mathematical Sciences, Nanjing Normal University, Nanjing, China

Centre for Research in Mathematics, Western Sydney University,
Sydney, NSW, Australia
e-mail: lihuanhuan2019@gmail.com

A. A. Ambily et al. (eds.), *Leavitt Path Algebras and Classical* K*-Theory*,
Indian Statistical Institute Series,
https://doi.org/10.1007/978-981-15-1611-5_4

complex of a finite quiver without sources is quasi-isomorphic to the Leavitt path algebra of the opposite graph [19].

We recall the constructions of the injective and projective Leavitt complexes. We overview the connection between the injective and projective Leavitt complexes and the Leavitt path algebras of the given graphs. A differential graded bimodule structure, which is right quasi-balanced, is endowed to the injective and projective Leavitt complex in [18, 19]. We prove that neither the injective nor the projective Leavitt complex is not left quasi-balanced.

4.2 Preliminaries

In this section, we recall some notation on compactly generated triangulated categories and differential graded algebras.

4.2.1 Triangulated Categories

The construction of derived categories for an arbitrary abelian category goes back to the inspiration of Grothendieck in the early sixties in order to formulate Grothendieck's duality theory for schemes [10]. The formulation in terms of triangulated categories was achieved by Verdier in the sixties and an account of this is published in Verdier [29]. Happel [11, 12] investigated the derived category of bounded complexes of the category of modules over a finite dimensional algebra. Rickard introduced the concept of tilting complex [24–26] in order to investigate the derived category. Now triangulated categories and derived categories have become important tools in many branches of algebraic geometry, in algebraic analysis, non-commutative algebraic geometry, representation theory, and so on.

For the definition of a triangulated category, refer to [12, Sect. 1.1], [21, Sect. 1.3], etc. Homotopy categories and derived categories over an abelian category are classical examples of triangulated categories (refer to [12, Sect. 1.3] etc.).

We recall the triangulated structure for the homotopy category. We first recall that a complex $X^\bullet = (X^i, d_X^i)_{i\in\mathbb{Z}}$ over an abelian category $\mathcal{A}$ is by definition a collection of objects X^i in $\mathcal{A}$ and morphisms $d_X^i : X^i \to X^{i+1}$ in $\mathcal{A}$ such that $d_X^{i+1} \circ d_X^i = 0$ for all $i \in \mathbb{Z}$. Usually we write a complex $X^\bullet = (X^i, d_X^i)_{i\in\mathbb{Z}}$ as

$$\cdots \longrightarrow X^{i-1} \xrightarrow{d_X^{i-1}} X^i \xrightarrow{d_X^i} X^{i+1} \longrightarrow \cdots .$$

Let A be a finite dimensional algebra over a field k. Denote by A-Mod the category of left A-modules and $\mathbf{K}(A\text{-Mod})$ its homotopy category. For a complex $X^\bullet = (X^i, d_X^i)_{i\in\mathbb{Z}}$ of A-modules, the complex $X^\bullet[1]$ is given by $(X^\bullet[1])^i = X^{i+1}$ and $d_{X[1]}^i = -d_X^{i+1}$ for $i \in \mathbb{Z}$. For a chain map $f^\bullet : X^\bullet \to Y^\bullet$, its *mapping cone* $\mathrm{Con}(f^\bullet)$ is a complex such that $\mathrm{Con}(f^\bullet) = X^\bullet[1] \oplus Y^\bullet$ with the differential $d^i_{\mathrm{Con}(f^\bullet)} = \begin{pmatrix} -d_X^{i+1} & 0 \\ f^{i+1} & d_Y^i \end{pmatrix}$. Each triangle in $\mathbf{K}(A\text{-Mod})$ is isomorphic to

$$X^\bullet \xrightarrow{f^\bullet} Y^\bullet \xrightarrow{\binom{0}{1}} \mathrm{Con}(f^\bullet) \xrightarrow{(1\ 0)} X^\bullet[1]$$

for some chain map $f^\bullet$.

4.2.2 Compactly Generated Triangulated Categories

For a triangulated category $\mathcal{T}$, a *thick* subcategory of $\mathcal{T}$ is a triangulated subcategory of $\mathcal{T}$ which is closed under direct summands. Let $\mathcal{S}$ be a class of objects in $\mathcal{T}$. We denote by $\mathrm{thick}\langle\mathcal{S}\rangle$ the smallest thick subcategory of $\mathcal{T}$ containing $\mathcal{S}$. If $\mathcal{T}$ has arbitrary coproducts, we denote by $\mathrm{Loc}\langle\mathcal{S}\rangle$ the smallest triangulated subcategory of $\mathcal{T}$ which contains $\mathcal{S}$ and is closed under arbitrary coproducts. By [6, Proposition 3.2] we have that $\mathrm{thick}\langle\mathcal{S}\rangle \subseteq \mathrm{Loc}\langle\mathcal{S}\rangle$.

For a triangulated category $\mathcal{T}$ with arbitrary coproducts, an object M in $\mathcal{T}$ is *compact* if the functor $\mathrm{Hom}_{\mathcal{T}}(M, -)$ commutes with arbitrary coproducts. Denote by $\mathcal{T}^c$ the full subcategory consisting of compact objects; it is a thick subcategory.

A triangulated category $\mathcal{T}$ with arbitrary coproducts is *compactly generated* [14, 20] if there exists a set $\mathcal{S}$ of compact objects such that any nonzero object T satisfies that $\mathrm{Hom}_{\mathcal{T}}(S, T[n]) \neq 0$ for some $S \in \mathcal{S}$ and $n \in \mathbb{Z}$. Here, $[n]$ is the n-th power of the shift function of $\mathcal{T}$. This is equivalent to the condition that $\mathcal{T} = \mathrm{Loc}\langle\mathcal{S}\rangle$, in which case we have $\mathcal{T}^c = \mathrm{thick}\langle\mathcal{S}\rangle$; see [20, Lemma 3.2]. If the above set $\mathcal{S}$ consists of a single object S, we call S a *compact generator* of $\mathcal{T}$.

Let A-Inj be the category of injective A-modules. Denote by $\mathbf{K}(A\text{-Inj})$ the homotopy category of complexes of injective A-modules, which is a triangulated subcategory of $\mathbf{K}(A\text{-Mod})$ that is closed under coproducts. By [16, Proposition 2.3(1)] $\mathbf{K}(A\text{-Inj})$ is a compactly generated triangulated category.

Denote by $\mathbf{K}_{\mathrm{ac}}(A\text{-Inj})$ the full subcategory of $\mathbf{K}(A\text{-Inj})$ formed by acyclic complexes of injective A-modules. The homotopy category $\mathbf{K}_{\mathrm{ac}}(A\text{-Inj})$ is called the stable derived category of A in [16]. This category is a compactly generated triangulated category such that its subcategory of compact objects is triangle equivalent to the singularity category [4, 22] of A.

We denote by $\mathbf{K}_{\mathrm{ac}}(A\text{-Proj})$ the homotopy category of acyclic complexes of projective A-modules. This category is a compactly generated triangulated category whose subcategory of compact objects is triangle equivalent to the opposite category of the singularity category of the opposite algebra A^{op}.

In the last decade, Leavitt path algebras of directed graphs [2, 3] were introduced as an algebraisation of graph C^*-algebras [17, 23] and in particular Cuntz–Krieger algebras [8].

For a finite directed graph E, Smith [28] described the quotient category

$$\mathrm{QGr}(kE) := \mathrm{Gr}(kE)/\mathrm{Fdim}(kE)$$

of graded kE-modules modulo those that are the sum of their finite dimensional submodules in terms of the category of graded modules over the Leavitt path algebra of E^o over a field k. Here, kE is the path algebra of E and E^o is the graph without sources or sinks that are obtained by repeatedly removing all sinks and sources from E. The full subcategory qgr(kE) of finitely presented objects in QGr(kE) is triangulated equivalent to the singularity category of the radical square zero algebra $kE/kE_{\geq 2}$; see [28, Theorem 7.2].

The homotopy categories $\mathbf{K}_{\mathrm{ac}}(A\text{-Inj})$ and $\mathbf{K}_{\mathrm{ac}}(A\text{-Proj})$ were described as derived categories of Leavitt path algebras, in the case that A is an algebra with radical square zero associated to a certain finite directed graph; see [7, Theorem 6.1] and [7, Theorem 6.2].

In general, it seems very difficult to give an explicit compact generator for the stable derived category of algebra or the homotopy category of acyclic complexes of projective modules over an algebra. An explicit compact generator for the homotopy categories $\mathbf{K}_{\mathrm{ac}}(A\text{-Inj})$ and $\mathbf{K}_{\mathrm{ac}}(A\text{-Proj})$, were constructed in [18, 19] respectively, in the case that the algebra $A = kE/kE_{\geq 2}$ is with radical square zero, where E is a finite directed graph without sources or sinks.

4.2.3 Differential Graded Algebras

Differential graded (dg for short) algebras appeared in [15]. They found applications in the representation theory of finite dimensional algebras in the seventies; see [9, 27]. The idea to use dg categories to 'enhance' triangulated categories goes back at least to Bondal-Kapranov [5], who were motivated by the study of exceptional collections of coherent sheaves on projective varieties.

We recall from [14] some notation on differential graded modules. Let $A = \bigoplus_{n\in\mathbb{Z}} A^n$ be a $\mathbb{Z}$-graded algebra. For a (left) graded A-module $M = \bigoplus_{n\in\mathbb{Z}} M^n$, elements m in M^n are said to be homogeneous of degree n, denoted by $|m| = n$.

A *differential graded algebra* (dg algebra for short) is a $\mathbb{Z}$-graded algebra A with a differential $d : A \to A$ of degree one such that $d(ab) = d(a)b + (-1)^{|a|}ad(b)$ for homogenous elements $a, b \in A$.

A *(left) differential graded* A-module (dg A-module for short) M is a graded A-module $M = \bigoplus_{n\in\mathbb{Z}} M^n$ with a differential $d_M : M \to M$ of degree one such that $d_M(a{\cdot}m) = d(a){\cdot}m + (-1)^{|a|}a{\cdot}d_M(m)$ for homogenous elements $a \in A$ and $m \in M$. A morphism of dg A-modules is a morphism of A-modules preserving degrees and commuting with differentials. A *right differential graded* A-module (right dg A-module for short) N is a right graded A-module $N = \bigoplus_{n\in\mathbb{Z}} N^n$ with a differential $d_N : N \to N$ of degree one such that $d_N(m \cdot a) = d_N(m) \cdot a + (-1)^{|m|}m \cdot$

$d(a)$ for homogenous elements $a \in A$ and $m \in N$. Here, we use central dots to denote the A-module action.

Let B be another dg algebra. Recall that a *dg A-B-bimodule M* is a left dg A-module as well as a right dg B-module such that $(a \cdot m) \cdot b = a \cdot (m \cdot b)$ for $a \in A$, $m \in M$ and $b \in B$.

Let M, N be (left) dg A-modules. We have a $\mathbb{Z}$-graded vector space

$$\mathrm{Hom}_A(M, N) = \bigoplus_{n \in \mathbb{Z}} \mathrm{Hom}_A(M, N)^n$$

such that each component $\mathrm{Hom}_A(M, N)^n$ consists of k-linear maps $f : M \to N$ satisfying $f(M^i) \subseteq N^{i+n}$ for all $i \in \mathbb{Z}$ and $f(a \cdot m) = (-1)^{n|a|} a \cdot f(m)$ for all homogenous elements $a \in A$. The differential on $\mathrm{Hom}_A(M, N)$ sends $f \in \mathrm{Hom}_A(M, N)^n$ to $d_N \circ f - (-1)^n f \circ d_M \in \mathrm{Hom}_A(M, N)^{n+1}$. Furthermore, $\mathrm{End}_A(M) := \mathrm{Hom}_A(M, M)$ becomes a dg algebra with this differential and the usual composition as multiplication. The dg algebra $\mathrm{End}_A(M)$ is usually called the *dg endomorphism algebra* of M.

We denote by A^{opp} the *opposite dg algebra* of a dg algebra A. More precisely, $A^{\mathrm{opp}} = A$ as graded spaces with the same differential, and the multiplication '$\circ$' on A^{opp} is given by $a \circ b = (-1)^{|a||b|} ba$.

Let B be another dg algebra. Recall that a right dg B-module is a left dg B^{opp}-module via $bm = (-1)^{|b||m|} m \cdot b$ for homogenous elements $b \in B$, $m \in M$. For a dg A-B-bimodule M, the canonical map $A \to \mathrm{End}_{B^{\mathrm{opp}}}(M)$ is a homomorphism of dg algebras, sending a to l_a with $l_a(m) = a \cdot m$ for $a \in A$ and $m \in M$. Similarly, the canonical map $B \to \mathrm{End}_A(M)^{\mathrm{opp}}$ is a homomorphism of dg algebras, sending b to r_b with $r_b(m) = (-1)^{|b||m|} m \cdot b$ for homogenous elements $b \in B$ and $m \in M$.

A dg A-B-bimodule M is called *right quasi-balanced* provided that the canonical homomorphism $B \to \mathrm{End}_A(M)^{\mathrm{opp}}$ of dg algebras is a quasi-isomorphism; see [7, 2.2]. Dually a dg A-B-bimodule M is called *left quasi-balanced* provided that the canonical homomorphism $A \to \mathrm{End}_{B^{\mathrm{opp}}}(M)$ of dg algebras is a quasi-isomorphism.

Denote by $\mathbf{K}(A)$ the homotopy category and by $\mathbb{D}(A)$ the derived category of left dg A-modules; they are triangulated categories with arbitrary coproducts. For a dg A-B-bimodule M and a left dg A-module N, $\mathrm{Hom}_A(M, N)$ has a natural structure of left dg B-module.

Recall that $\mathrm{Loc}\langle M \rangle \subseteq \mathbf{K}(A)$ is the smallest triangulated subcategory of $\mathbf{K}(A)$ which contains M and is closed under arbitrary coproducts. If M is a compact object in $\mathrm{Loc}\langle M \rangle$ and a dg A-B-bimodule which is right quasi-balanced, then we have a triangle equivalence

$$\mathrm{Hom}_A(M, -) : \mathrm{Loc}\langle M \rangle \xrightarrow{\sim} \mathbb{D}(B),$$

see [7, Proposition 2.2]; compare [14, 4.3] and [16, Appendix A].

4.3 The Injective Leavitt Complex of a Finite Graph Without Sinks

In this section, we recall the construction of the injective Leavitt complex of a finite graph without sinks and prove that the injective Leavitt complex is not left quasi-balanced.

4.3.1 *The Injective Leavitt Complex*

Let $E = (E^0, E^1; s, t)$ be a finite (directed) graph. A path in the graph E is a sequence $p = \alpha_n \cdots \alpha_2\alpha_1$ of edges with $t(\alpha_j) = s(\alpha_{j+1})$ for $1 \leq j \leq n-1$. The length of p, denoted by $l(p)$, is n. The starting vertex of p, denoted by $s(p)$, is $s(\alpha_1)$. The terminating vertex of p, denoted by $t(p)$, is $t(\alpha_n)$. We identify an edge with a path of length one. We associate to each vertex $i \in E^0$ a trivial path e_i of length zero. Set $s(e_i) = i = t(e_i)$. Denote by E^n the set of all paths in E of length n for each $n \geq 0$. Recall that a vertex of E is a sink if there is no edge starting at it and a vertex of E is a source if there is no edge terminating at it.

Let E be a finite graph without sinks. For any vertex $i \in E^0$, fix an edge γ with $s(\gamma) = i$. We call the fixed edge the *special edge* starting at i. For a special edge α, we set

$$S(\alpha) = \{\beta \in E^1 \mid s(\beta) = s(\alpha), \beta \neq \alpha\}. \tag{4.1}$$

We mention that the terminology 'special edge' is taken from [1].

The following notion is inspired by a basis for Leavitt path algebra (refer to [1, 13] for the basis).

Definition 4.1 For two paths $p = \alpha_m \cdots \alpha_1$ and $q = \beta_n \cdots \beta_1$ in E with $m, n \geq 1$, we call the pair (p, q) an *admissible pair* in E if $t(p) = t(q)$, and either $\alpha_m \neq \beta_n$, or $\alpha_m = \beta_n$ is not special. For each path r in E, we define two additional *admissible pairs* $(r, e_{t(r)})$ and $(e_{t(r)}, r)$ in E. □

For each vertex $i \in E^0$ and $l \in \mathbb{Z}$, set

$$\mathbf{B}_i^l = \{(p, q) \mid (p, q) \text{ is an admissible pair with } l(q) - l(p) = l \text{ and } s(q) = i\}. \tag{4.2}$$

The above set $\mathbf{B}_i^l$ is not empty for each vertex i and each integer l; see [18, Lemma 1.2].

Let E be a finite graph without sinks. Set $A = kE/kE^{\geq 2}$ to be the corresponding finite dimensional algebra with radical square zero. Indeed, $A = kE^0 \oplus kE^1$ as a k-vector space and its Jacobson radical $\operatorname{rad} A = kE^1$ satisfying $(\operatorname{rad} A)^2 = 0$.

Denote by $I_i = D(e_i A)$ the injective left A-module for each $i \in E^0$, where $(e_i A)_A$ is the indecomposable projective right A-module and $D = \operatorname{Hom}_k(-, k)$ denotes the standard k-duality. Denote by $\{e_i^\sharp\} \cup \{\alpha^\sharp | \alpha \in E^1, t(\alpha) = i\}$ the basis of I_i, which is dual to the basis $\{e_i\} \cup \{\alpha | \alpha \in E^1, t(\alpha) = i\}$ of $e_i A$.

For a set X and an A-module M, the coproduct $M^{(X)}$ will be understood as $\bigoplus_{x \in X} M\zeta_x$, where each component $M\zeta_x$ is M. For an element $m \in M$, we use $m\zeta_x$ to denote the corresponding element in $M\zeta_x$.

For a path $p = \alpha_n \cdots \alpha_2\alpha_1$ in E of length $n \geq 2$, we denote by $\widehat{p} = \alpha_{n-1} \cdots \alpha_1$ and $\widetilde{p} = \alpha_n \cdots \alpha_2$ the two *truncations* of p. For an edge α, denote by $\widehat{\alpha} = e_{s(\alpha)}$ and $\widetilde{\alpha} = e_{t(\alpha)}$.

Definition 4.2 Let E be a finite graph without sinks. The *injective Leavitt complex* $\mathcal{I}^\bullet = (\mathcal{I}^l, \partial^l)_{l \in \mathbb{Z}}$ of E is defined as follows:

(1) the l-th component $\mathcal{I}^l = \bigoplus_{i \in E^0} I_i{}^{(\mathbf{B}_i^l)}$;
(2) the differential $\partial^l : \mathcal{I}^l \to \mathcal{I}^{l+1}$ is given by $\partial^l(e_i^\sharp \zeta_{(p,q)}) = 0$ and

$$\partial^l(\alpha^\sharp \zeta_{(p,q)}) = \begin{cases} e^\sharp_{s(\alpha)}\zeta_{(\widehat{p}, e_{s(\alpha)})} - \sum\limits_{\beta \in S(\alpha)} e^\sharp_{s(\alpha)}\zeta_{(\beta\widehat{p}, \beta)}, & \text{if } q = e_i,\ \ p = \alpha\widehat{p} \\ & \text{and } \alpha \text{ is special;} \\ e^\sharp_{s(\alpha)}\zeta_{(p, q\alpha)}, & \text{otherwise,} \end{cases}$$

for any $i \in E^0$, $(p, q) \in \mathbf{B}_i^l$ and $\alpha \in E^1$ with $t(\alpha) = i$. Here, the set $S(\alpha)$ is defined in (4.1). □

Each component $\mathcal{I}^l$ is an injective A-module. The differentials ∂^l are A-module morphisms. Indeed, $\mathcal{I}^\bullet$ is an acyclic complex of injective A-modules; see [18, Proposition 1.9].

Example 4.1 Let E be the following graph with one vertex and n loops with $n \geq 2$.

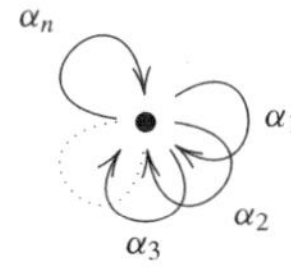

We choose α_1 to be the special edge. Let e be the trivial path corresponding the unique vertex. Set $\mathbf{B}^l$ to be the set of admissible pairs (p, q) in E with $l(q) - l(p) = l$ for each $l \in \mathbb{Z}$. A pair (p, q) of paths lies in $\mathbf{B}^l$ if and only if $l(q) - l(p) = l$ and p, q do not end with α_1 simultaneously. In particular, the set $\mathbf{B}^l$ is infinite.

The corresponding algebra A with radical square zero has a k-basis $\{e, \alpha_1, \ldots, \alpha_n\}$. Set $I = D(A_A)$. Then the injective Leavitt complex $\mathcal{I}^\bullet$ of E is as follows.

$$\cdots \longrightarrow I^{(\mathbf{B}^{-1})} \xrightarrow{\partial^{-1}} I^{(\mathbf{B}^0)} \xrightarrow{\partial^0} I^{(\mathbf{B}^1)} \longrightarrow \cdots$$

We write the differential ∂^{-1} explicitly: $\partial^{-1}(e^\sharp\zeta_{(p,q)}) = 0$, $\partial^{-1}(\alpha_i^\sharp\zeta_{(p,q)}) = e^\sharp\zeta_{(p,q\alpha_i)}$ for $2 \leq i \leq n$ and

$$\partial^{-1}(\alpha_1^\sharp \zeta_{(p,q)}) = \begin{cases} e^\sharp \zeta_{(e,e)} - \sum_{i=2}^n e^\sharp \zeta_{(\alpha_i,\alpha_i)}, & \text{if } q = e \text{ and } p = \alpha_1; \\ e^\sharp \zeta_{(p,q\alpha_1)}, & \text{otherwise,} \end{cases}$$

for $(p, q) \in \mathbf{B}^{-1}$.

Recall that the Leavitt path algebra $L_k(E)$ of a finite graph E is the k-algebra generated by the set $\{v \mid v \in E^0\} \cup \{e \mid e \in E^1\} \cup \{e^* \mid e \in E^1\}$ subject to the following relations:

(0) $vw = \delta_{v,w} v$ for every $v, w \in E^0$;
(1) $t(e)e = es(e) = e$ for all $e \in E^1$;
(2) $e^* t(e) = s(e)e^* = e^*$ for all $e \in E^1$;
(3) $ef^* = \delta_{e,f} t(e)$ for all $e, f \in E^1$;
(4) $\sum_{\{e \in E^1 \mid s(e)=v\}} e^* e = v$ for every $v \in E^0$ which is not a sink.

Here, δ is the Kronecker symbol. The relations (3) and (4) are called *Cuntz–Krieger relations*. The relation (3) is called (CK1)-relation and (4) is called (CK2)-relation. The elements α^* for $\alpha \in E^1$ are called *ghost arrows*.

If $p = e_n \cdots e_2 e_1$ is a path in E of length $n \geq 1$, we define $p^* = e_1^* e_2^* \cdots e_n^*$. For convention, we set $v^* = v$ for $v \in E^0$. We observe by (2) that for paths p, q in E, $p^* q = 0$ for $t(p) \neq t(q)$.

Recall that the Leavitt path algebra is naturally $\mathbb{Z}$-graded by the length of paths. In what follows, we write $B = L_k(E)^{\text{op}}$, which is the opposite algebra of $L_k(E)$. Then B is a $\mathbb{Z}$-graded algebra. We view B as a dg algebra with trivial differential.

Consider $A = kE/kE^{\geq 2}$ as a dg algebra concentrated on degree zero. Recall the injective Leavitt complex $\mathcal{I}^\bullet = \bigoplus_{l \in \mathbb{Z}} \mathcal{I}^l$, which is a left dg A-module. By [18, Proposition 3.6], $\mathcal{I}^\bullet$ is a right dg B-module and a dg A-B-bimodule.

The following theorem demonstrates the role of the injective Leavitt complex in the stable category and establishes a connection between the injective Leavitt complex and the Leavitt path algebra, which justifies the terminology.

Theorem 4.1 *Let E be a finite graph without sinks, and let $A = kE/kE^{\geq 2}$ be the corresponding algebra with radical square zero.*

(1) *The injective Leavitt complex $\mathcal{I}^\bullet$ of E is a compact generator for the homotopy category* $\mathbf{K}_{\text{ac}}(A\text{-Inj})$.
(2) *The dg A-B-bimodule $\mathcal{I}^\bullet$ is right quasi-balanced. In particular, the dg endomorphism algebra* $\text{End}_A(\mathcal{I}^\bullet)$ *is quasi-isomorphic to the Leavitt path algebra $L_k(E)$. Here, $L_k(E)$ is naturally $\mathbb{Z}$-graded and viewed as a dg algebra with trivial differential.*

For the proof of Theorem 4.1, refer to [18, Theorem 2.13] and [18, Theorem 4.2].

There is a unique right B-module morphism $\psi : B \to \mathcal{I}^\bullet$ with

$$\psi(1) = \sum_{i \in E^0} e_i^\sharp \zeta_{(e_i,e_i)}.$$

Here, 1 is the unit of B. For each edge $\beta \in E^1$, there is a unique right B-module morphism $\psi_\beta : B \to \mathcal{I}^\bullet$ with $\psi_\beta(1) = \beta^\sharp \zeta_{(e_{t(\beta)}, e_{t(\beta)})}$. Let (p, q) be an admissible pair in E. We have $\sum_{i \in E^0} e_i^\sharp \zeta_{(e_i, e_i)} \cdot p^* q = e_{s(q)}^\sharp \zeta_{(p,q)}$ and $\beta^\sharp \zeta_{(e_i, e_i)} \cdot p^* q = \delta_{i, s(q)} \beta^\sharp \zeta_{(p,q)}$ for each edge $\beta \in E^1$ with $t(\beta) = i$. It follows that

$$\psi(p^*q) = e_{s(q)}^\sharp \zeta_{(p,q)} \quad \text{and} \quad \psi_\beta(p^*q) = \delta_{s(q), t(\beta)} \beta^\sharp \zeta_{(p,q)}. \tag{4.3}$$

It follows that ψ is injective and $\mathrm{Im}\psi_\beta \cong \mathrm{e}_{\mathrm{t}(\beta)}\mathrm{B}$ for each $\beta \in E^1$. Consider the gradings of B and $\mathcal{I}^\bullet$. We have that ψ and ψ_β for $\beta \in E^1$ are right graded B-module morphisms.

Set $\Psi = \big(\psi \ (\psi_\beta)_{\beta \in E^1}\big) : B \oplus B^{(E^1)} \to \mathcal{I}^\bullet$. The map Ψ is a graded right B-module morphism.

Lemma 4.1 *The above morphism Ψ is surjective and the injective Leavitt complex $\mathcal{I}^\bullet$ of E is a graded projective right B-module.*

Proof For each $i \in E^0$, $l \in \mathbb{Z}$ and $(p, q) \in \mathbf{B}_i^l$, we have that $e_i^\sharp \zeta_{(p,q)} = \psi(p^*q)$ and $\alpha^\sharp \zeta_{(p,q)} = \psi_\alpha(p^*q)$ for each edge $\alpha \in E^1$ with $t(\alpha) = i$. Then Ψ is surjective and $\mathcal{I}^\bullet = \mathrm{Im}\Psi$. Observe that $\mathrm{Im}\Psi = \mathrm{Im}\psi + \sum_{\beta \in \mathrm{E}^1} \mathrm{Im}\psi_\beta = \mathrm{Im}\psi \oplus (\oplus_{\beta \in \mathrm{E}^1} \mathrm{Im}\psi_\beta)$. Then $\mathrm{Im}\Psi \cong \mathrm{B} \oplus (\oplus_{\beta \in \mathrm{E}^1} \mathrm{e}_{\mathrm{t}(\beta)}\mathrm{B})$ and we are done. □

The following observation implies that the dg A-B-bimodule $\mathcal{I}^\bullet$ is not left quasi-balanced. In other words, the canonical dg algebra homomorphism $A \to \mathrm{End}_{B^{\mathrm{opp}}}(\mathcal{I}^\bullet)$ is not a quasi-isomorphism. Indeed, we infer from the observation that the dg endomorphism algebra $\mathrm{End}_{B^{\mathrm{opp}}}(\mathcal{I}^\bullet)$ is acyclic.

Proposition 4.1 *We have that $\mathcal{I}^\bullet = 0$ in the homotopy category of right dg B-modules.*

Proof Recall from (4.3) the right B-module morphisms ψ and ψ_β for $\beta \in E^1$. For each $l \in \mathbb{Z}$, the set $\{e_i^\sharp \zeta_{(p,q)}, \alpha^\sharp \zeta_{(p,q)} \mid i \in E^0, (p, q) \in \mathbf{B}_i^l \text{ and } \alpha \in E^1 \text{ with } t(\alpha) = i\}$ is a k-basis of $\mathcal{I}^l$. We define a k-linear map $h^l : \mathcal{I}^l \to \mathcal{I}^{l-1}$ such that

$$\begin{cases} h^l(\alpha^\sharp \zeta_{(p,q)}) = 0; \\ h^l(e_i^\sharp \zeta_{(p,q)}) = \sum_{\{\beta \in E^1 \mid s(\beta) = i\}} \psi_\beta(p^* q \beta^*), \end{cases}$$

for each $i \in E^0$, $l \in \mathbb{Z}$, $(p, q) \in \mathbf{B}_i^l$, and $\alpha \in E^1$ with $t(\alpha) = i$. Here, $p^*q\beta^*$ is the multiplication of p^*q and β^* in $L_k(E)$. For any element $b \in L_k(E)^l e_i$, we have

$$h^l(\psi(b)) = \sum_{\{\beta \in E^1 \mid s(\beta) = i\}} \psi_\beta(b\beta^*).$$

Recall that ψ and ψ_β for $\beta \in E^1$ are right B-module morphisms. Then we have that $h = (h^l)_{l \in \mathbb{Z}}$ is a right graded B-module morphism of degree -1. In other words, we have $h \in \mathrm{End}_{B^{\mathrm{opp}}}(\mathcal{I}^\bullet)^{-1}$.

It suffices to prove that

$$\partial^{l-1} \circ h^l + h^{l+1} \circ \partial^l = \mathrm{Id}_{\mathcal{I}^l} \text{ for each } l \in \mathbb{Z}. \tag{4.4}$$

For each $i \in E^0$, $l \in \mathbb{Z}$ and $(p, q) \in \mathbf{B}_i^l$, we have that

$$\begin{aligned}
&(\partial^{l-1} \circ h^l + h^{l+1} \circ \partial^l)(e_i^\sharp \zeta_{(p,q)}) \\
&= \sum\nolimits_{\{\beta \in E^1 \mid s(\beta)=i\}} (\partial^{l-1} \circ \psi_\beta)(p^* q \beta^*) \\
&= \sum\nolimits_{\{\beta \in E^1 \mid s(\beta)=i\}} \psi(p^* q \beta^* \beta) \\
&= e_i^\sharp \zeta_{(p,q)},
\end{aligned}$$

where the second equality uses [18, Lemma 3.7] and the last equality uses the (CK2)-relation for the Leavitt path algebra. Similarly, we have

$$\begin{aligned}
&(\partial^{l-1} \circ h^l + h^{l+1} \circ \partial^l)(\alpha^\sharp \zeta_{(p,q)}) \\
&= h^{l+1}((\partial^l \circ \psi_\alpha)(p^* q)) \\
&= h^{l+1}(\psi(p^* q \alpha)) \\
&= \sum\nolimits_{\{\beta \in E^1 \mid s(\beta)=s(\alpha)\}} \psi_\beta(p^* q \alpha \beta^*) \\
&= \alpha^\sharp \zeta_{(p,q)}
\end{aligned}$$

for $\alpha \in E^1$ with $t(\alpha) = i$. Here, the last equality uses the (CK1)-relation for the Leavitt path algebra. □

Remark 4.1 Note that the Eq. (4.4) implies that the injective Leavitt complex I is acyclic.

4.3.2 *The Independence of the Injective Leavitt Complex*

We show that the definition of the injective Leavitt complex of E is independent of the choice of special edges in E.

Denote by S and S' two different sets of special edges of E. For each $i \in E^0$ and $l \in \mathbb{Z}$, let $\mathbf{B}_i^l$ and $(\mathbf{B}_i^l)'$ be the corresponding sets of admissible pairs with respect to S and S' respectively; see (4.6). Define a map $\Phi_{li} : \mathbf{B}_i^l \to (\mathbf{B}_i^l)'$ such that for $(p, q) \in \mathbf{B}_i^l$

$$\Phi_{li}((p,q)) = \begin{cases} (\alpha \widehat{p}, \alpha \widehat{q}), & \text{if } p = \alpha' \widehat{p}, q = \alpha' \widehat{q} \text{ and } \alpha' \in S'; \\ (p,q), & \text{otherwise}, \end{cases}$$

where $\alpha \in S$ with $s(\alpha) = s(\alpha')$. The map $\Phi_{li} : \mathbf{B}_i^l \to (\mathbf{B}_i^l)'$ is a bijection. In fact, the inverse map of Φ_{li} can be defined symmetrically.

Denote by $\mathcal{I}^\bullet = (\mathcal{I}^l, \partial^l)_{l\in\mathbb{Z}}$ and $\mathcal{I}^{\bullet\prime} = (\mathcal{I}^{l\prime}, (\partial^l)')_{l\in\mathbb{Z}}$ the injective Leavitt complexes of E with S and S' sets of special edges, respectively. Let Λ' be the k-basis of $L_k(E)$ given by [1, Theorem 1] and [13, Corollary 17] with S' the set of special edges.

Recall that $B = L_k(E)^{\mathrm{op}}$. The maps $\psi' : B \to \mathcal{I}^{\bullet\prime}$, $p^*q \mapsto e^\sharp_{s(q)}\zeta_{(p,q)}$ and $\psi'_\beta : B \to \mathcal{I}^{\bullet\prime}$, $p^*q \mapsto \delta_{s(q),t(\beta)}\beta^\sharp\zeta_{(p,q)}$ for $p^*q \in \Lambda'$ and $\beta \in E^1$ are graded right B-module morphisms.

For each $l \in \mathbb{Z}$, define a k-linear map $\omega^l : \mathcal{I}^l \to (\mathcal{I}^l)'$ such that $\omega^l(e_i^\sharp\zeta_{(p,q)}) = \psi'(p^*q)$ and $\omega^l(\alpha^\sharp\zeta_{(p,q)}) = \psi'_\alpha(p^*q)$ for $i \in E^0$ and $(p, q) \in \mathbf{B}_i^l$ and $\alpha \in E^1$ with $t(\alpha) = i$. Let $\omega^\bullet = (\omega^l)_{l\in\mathbb{Z}} : \mathcal{I}^\bullet \to \mathcal{I}^{\bullet\prime}$. By definitions we have $\omega^\bullet \circ \psi = \psi'$ and $\omega^\bullet \circ \psi_\beta = \psi'_\beta$ for each edge $\beta \in E^1$. Then we have $\omega^\bullet \circ \Psi = \Psi'$ with $\Psi' = \big(\psi'\ (\psi'_\beta)_{\beta\in E^1}\big) : B \oplus B^{(E^1)} \to \mathcal{I}^{\bullet\prime}$ a morphism of graded right B-modules.

Proposition 4.2 (1) *The above map $\omega^\bullet : \mathcal{I}^\bullet \to \mathcal{I}^{\bullet\prime}$ is an isomorphism of complexes of A-modules.*
(2) *The above map $\omega^\bullet : \mathcal{I}^\bullet \to \mathcal{I}^{\bullet\prime}$ is an isomorphism of right dg B-modules.*

Proof (1) We can directly check that ω^l is an A-module map for each $l \in \mathbb{Z}$. The inverse map of ω^l can be defined symmetrically. We have that ω^l is an isomorphism of A-modules for each $l \in \mathbb{Z}$. It remains to prove that $\omega^\bullet$ is a chain map of complexes. For each $i \in E^0$, $l \in \mathbb{Z}$ and $(p, q) \in \mathbf{B}_i^l$, since $((\partial^l)' \circ \omega^i)(e_i^\sharp\zeta_{(p,q)}) = 0 = (\omega^{l+1} \circ \partial^l)(e_i^\sharp\zeta_{(p,q)})$, it suffices to prove that $((\partial^l)' \circ \omega^l)(\alpha^\sharp\zeta_{(p,q)}) = (\omega^{l+1} \circ \partial^l)(\alpha^\sharp\zeta_{(p,q)})$ for $\alpha \in E^1$ with $t(\alpha) = i$. By [18, Lemma 3.7], we have that $((\partial^l)' \circ \omega^l)(\alpha^\sharp\zeta_{(p,q)}) = (\partial^l)'(\psi'_\alpha(p^*q)) = \psi'(p^*q\alpha)$ and $(\omega^{l+1} \circ \partial^l)(\alpha^\sharp\zeta_{(p,q)}) = \omega^{l+1}(\psi(p^*q\alpha)) = \psi'(p^*q\alpha)$. Then we are done.

(2) It remains to prove that $\omega^\bullet$ is a graded right B-module morphism. By Lemma 4.1, there exists a graded right B-module morphism $\Xi : \mathcal{I}^\bullet \longrightarrow B \oplus B^{(E^1)}$ such that $\Psi \circ \Xi = \mathrm{Id}_{\mathcal{I}^\bullet}$. Since $\omega^\bullet \circ \Psi = \Psi'$, we have that $\omega^\bullet = \omega^\bullet \circ (\Psi \circ \Xi) = \Psi' \circ \Xi$ is a composition of graded right B-module morphisms. □

4.4 The Projective Leavitt Complex of a Finite Graph Without Sources

In this section, we recall the construction of the projective Leavitt complex of a finite graph without sources and prove that the projective Leavitt complex is not left quasi-balanced.

4.4.1 The Projective Leavitt Complex

Let E be a finite graph without sources. For a vertex $i \in E^0$, fix an edge γ with $t(\gamma) = i$. We call the fixed edge the *associated edge* terminating at i. For an associated edge α, we set

$$T(\alpha) = \{\beta \in E^1 \mid t(\beta) = t(\alpha), \beta \neq \alpha\}. \tag{4.5}$$

Definition 4.3 For two paths $p = \alpha_m \cdots \alpha_2\alpha_1$ and $q = \beta_n \cdots \beta_2\beta_1$ with $m, n \geq 1$, we call the pair (p, q) an *associated pair* in E if $s(p) = s(q)$, and either $\alpha_1 \neq \beta_1$, or $\alpha_1 = \beta_1$ is not associated. In addition, we call $(p, e_{s(p)})$ and $(e_{s(p)}, p)$ *associated pairs* in E for each path p in E. □

For each vertex $i \in E^0$ and $l \in \mathbb{Z}$, set

$$\mathbf{\Lambda}_i^l = \{(p, q) \mid (p, q) \text{ is an associated pair with } l(q) - l(p) = l \text{ and } t(p) = i\}. \tag{4.6}$$

The set $\mathbf{\Lambda}_i^l$ is not empty for each vertex i and each integer l; see [19, Lemma 2.2].

Recall that $A = kE/kE^{\geq 2}$ is a finite dimensional algebra with radical square zero. Denote by $P_i = Ae_i$ the indecomposable projective left A-module for $i \in E^0$.

Definition 4.4 Let E be a finite graph without sources. The *projective Leavitt complex* $\mathcal{P}^\bullet = (\mathcal{P}^l, \delta^l)_{l\in\mathbb{Z}}$ of E is defined as follows:
(1) the l-th component $\mathcal{P}^l = \bigoplus_{i\in E^0} P_i^{(\Lambda_i^l)}$.
(2) the differential $\delta^l : \mathcal{P}^l \longrightarrow \mathcal{P}^{l+1}$ is given by $\delta^l(\alpha\zeta_{(p,q)}) = 0$ and

$$\delta^l(e_i\zeta_{(p,q)}) = \begin{cases} \beta\zeta_{(\widehat{p},q)}, & \text{if } p = \beta\widehat{p}; \\ \sum_{\{\beta\in E^1 \mid t(\beta)=i\}} \beta\zeta_{(e_{s(\beta)},q\beta)}, & \text{if } l(p) = 0, \end{cases}$$

for any $i \in E^0$, $(p, q) \in \Lambda_i^l$ and $\alpha \in E^1$ with $s(\alpha) = i$. □

Each component $\mathcal{P}^l$ is a projective A-module and the differential δ^l is an A-module morphism. $\mathcal{P}^\bullet$ is an acyclic complex of projective A-modules; see [19, Proposition 2.7].

Example 4.2 Let E be the following graph with one vertex and n loops with $n \geq 2$.

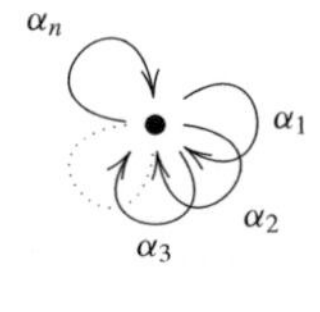

We choose α_1 to be the associated edge. Let e be the trivial path corresponding the unique vertex. Set $\mathbf{\Lambda}^l$ to be the set of associated pairs (p, q) in E with $l(q) - l(p) = l$ for each $l \in \mathbb{Z}$. A pair (p, q) of paths lies in $\mathbf{\Lambda}^l$ if and only if $l(q) - l(p) = l$ and p, q do not begin with α_1 simultaneously. In particular, the set $\mathbf{\Lambda}^l$ is infinite.

The corresponding algebra A with radical square zero has a k-basis $\{e, \alpha_1, \dots, \alpha_n\}$. Set $P = A$. Then the projective Leavitt complex $\mathcal{P}^\bullet$ of E is as follows.

$$\cdots \longrightarrow P^{(\mathbf{\Lambda}^0)} \stackrel{\delta^0}{\longrightarrow} P^{(\mathbf{\Lambda}^1)} \stackrel{\delta^1}{\longrightarrow} P^{(\mathbf{\Lambda}^2)} \longrightarrow \cdots$$

We write the differential δ^1 explicitly: $\delta^1(e\zeta_{(p,q)}) = \begin{cases} \beta\zeta_{(\widehat{p},q)}, & \text{if } p = \beta\widehat{p}; \\ \sum_{i=1}^n \alpha_i\zeta_{(e,q\alpha_i)}, & \text{if } l(p) = 0, \end{cases}$

$\delta^1(\alpha_j\zeta_{(p,q)}) = 0$ for $1 \leq j \leq n$ and $(p, q) \in \mathbf{\Lambda}^1$.

For notation, $E^{\rm op}$ is the opposite graph of E. For a path p in E, denote by $p^{\rm op}$ the corresponding path in $E^{\rm op}$. The starting and terminating vertices of $p^{\rm op}$ are $t(p)$ and $s(p)$, respectively. For convention, $e_j^{\rm op} = e_j$ for each vertex $j \in E^0$.

In what follows, we write $B = L_k(E^{\rm op})$, which is a $\mathbb{Z}$-graded algebra. We view B as a dg algebra with trivial differential.

Consider $A = kE/kE^{\geq 2}$ as a dg algebra concentrated on degree zero. The projective Leavitt complex $\mathcal{P}^\bullet = \bigoplus_{l\in\mathbb{Z}} \mathcal{P}^l$ is a left dg A-module. By [19, Proposition 4.6], $\mathcal{P}^\bullet$ is a right dg B-module and a dg A-B-bimodule.

The following theorem establishes a connection between the projective Leavitt complex and the Leavitt path algebra, which justifies the terminology.

Theorem 4.2 *Let E be a finite graph without sources, and let $A = kE/kE^{\geq 2}$ be the corresponding algebra with radical square zero.*

(1) *The projective Leavitt complex $\mathcal{P}^\bullet$ of E is a compact generator for the homotopy category $\mathbf{K}_{\rm ac}(A\text{-Proj})$.*
(2) *The dg A-B-bimodule $\mathcal{P}^\bullet$ is right quasi-balanced. In particular, the opposite dg endomorphism algebra of the projective Leavitt complex of E is quasi-isomorphic to the Leavitt path algebra $L_k(E^{\rm op})$. Here, $E^{\rm op}$ is the opposite graph of E; $L_k(E^{\rm op})$ is naturally $\mathbb{Z}$-graded and viewed as a dg algebra with trivial differential.*

For the proof of Theorem II, refer to [19, Theorem 3.7] and [19, Theorem 5.2].

There is a unique right B-module morphism $\phi : B \to \mathcal{P}^\bullet$ with

$$\phi(1) = \sum_{i\in E^0} e_i\zeta_{(e_i,e_i)}.$$

Here, 1 is the unit of B. For each edge $\beta \in E^1$, there is a unique right B-module morphism $\phi_\beta : B \to \mathcal{P}^\bullet$ with $\phi_\beta(1) = \beta\zeta_{(e_{s(\beta)},e_{s(\beta)})}$. Let (p, q) be an associated pair in E. We have $\sum_{i\in E^0} e_i\zeta_{(e_i,e_i)} \cdot (p^{\rm op})^* q^{\rm op} = e_{t(p)}\zeta_{(p,q)}$; and $\beta\zeta_{(e_{s(\beta)},e_{s(\beta)})} \cdot (p^{\rm op})^* q^{\rm op} = \delta_{s(\beta),t(p)}\beta\zeta_{(p,q)}$ for each edge $\beta \in E^1$; see [19, Lemma 4.5]. It follows that

$$\phi((p^{\rm op})^* q^{\rm op}) = e_{t(p)}\zeta_{(p,q)} \quad \text{and} \quad \phi_\beta((p^{\rm op})^* q^{\rm op}) = \delta_{t(p),s(\beta)}\beta\zeta_{(p,q)}. \tag{4.7}$$

It follows that ϕ is injective and $\mathrm{Im}\phi_\beta \cong \mathrm{e}_{\mathrm{s}(\beta)}\mathrm{B}$ for each $\beta \in E^1$. Consider the gradings of B and $\mathcal{P}^\bullet$. We have that ϕ and ϕ_β for $\beta \in E^1$ are right graded B-module morphisms.

Set $\Phi = \big(\phi\ (\phi_\beta)_{\beta\in E^1}\big) : B \oplus B^{(E^1)} \to \mathcal{P}^\bullet$. The map Φ is a graded right B-module morphism.

Lemma 4.2 *The above morphism Φ is surjective and the projective Leavitt complex $\mathcal{P}^\bullet$ of E is a graded projective right B-module.*

Proof For each $i \in E^0, l \in \mathbb{Z}$ and $(p, q) \in \mathbf{\Lambda}_i^l$, we have that $e_i\zeta_{(p,q)} = \phi((p^{\mathrm{op}})^* q^{\mathrm{op}})$ and $\alpha\zeta_{(p,q)} = \phi_\alpha((p^{\mathrm{op}})^* q^{\mathrm{op}})$ for each edge $\alpha \in E^1$ with $s(\alpha) = i$. Then Φ is surjective and $\mathcal{P}^\bullet = \mathrm{Im}\Phi$. Observe that $\mathrm{Im}\Phi = \mathrm{Im}\phi + \sum_{\beta\in \mathrm{E}^1} \mathrm{Im}\phi_\beta = \mathrm{Im}\phi \oplus (\oplus_{\beta\in\mathrm{E}^1}\mathrm{Im}\phi_\beta)$. Then $\mathrm{Im}\Phi \cong \mathrm{B} \oplus (\oplus_{\beta\in\mathrm{E}^1}\mathrm{e}_{\mathrm{s}(\beta)}\mathrm{B})$ and we are done. □

We prove that the dg A-B-bimodule $\mathcal{P}^\bullet$ is not left quasi-balanced. In other words, the canonical dg algebra homomorphism $A \to \mathrm{End}_{B^{\mathrm{opp}}}(\mathcal{P}^\bullet)$ is not a quasi-isomorphism. Indeed, we infer from the observation that the dg endomorphism algebra $\mathrm{End}_{B^{\mathrm{opp}}}(\mathcal{P}^\bullet)$ is acyclic.

Proposition 4.3 *We have that $\mathcal{P}^\bullet = 0$ in the homotopy category of right dg B-modules.*

Proof Recall from (4.7) the right B-module morphisms ϕ and ϕ_β for $\beta \in E^1$. For each $l \in \mathbb{Z}$, the set $\{e_i\zeta_{(p,q)}, \alpha\zeta_{(p,q)} \mid i \in E^0, (p,q) \in \mathbf{\Lambda}_i^l$ and $\alpha \in E^1$ with $s(\alpha) = i\}$ is a k-basis of $\mathcal{P}^l$. We define a k-linear map $h^l : \mathcal{P}^l \to \mathcal{P}^{l-1}$ such that $h^l(e_i\zeta_{(p,q)}) = 0$ and

$$h^l(\alpha\zeta_{(p,q)}) = \begin{cases} e_{t(\alpha)}\zeta_{(e_{t(\alpha)},\widetilde{q})} - \sum_{\beta\in T(\alpha)} e_{t(\alpha)}\zeta_{(\beta,\widetilde{q}\beta)}; & \text{if } l(p) = 0,\ \ q = \widetilde{q}\alpha \text{ and } \alpha \text{ is associated;} \\ e_{t(\alpha)}\zeta_{(\alpha p,q)}, & \text{otherwise.} \end{cases}$$

for each $i \in E^0$, $l \in \mathbb{Z}$, $(p, q) \in \mathbf{\Lambda}_i^l$, and $\alpha \in E^1$ with $s(\alpha) = i$. It follows that $h^l(\phi_\alpha((p^{\mathrm{op}})^* q^{\mathrm{op}})) = \phi((\alpha^{\mathrm{op}})^*(p^{\mathrm{op}})^* q^{\mathrm{op}})$. For any element $b \in e_i L_k(E)^l$, we have

$$h^l(\phi_\alpha(b) = \phi((\alpha^{\mathrm{op}})^* b). \tag{4.8}$$

Here, $(\alpha^{\mathrm{op}})^* b$ is the multiplication of b and $(\alpha^{\mathrm{op}})^*$ in $B = L_k(E^{\mathrm{op}})$. Recall that ϕ and ϕ_α for $\alpha \in E^1$ are right B-module morphisms. Then we have that $h = (h^l)_{l\in\mathbb{Z}}$ is right graded B-module morphisms of degree -1. In other words, we have $h \in \mathrm{End}_{B^{\mathrm{opp}}}(\mathcal{P}^\bullet)^{-1}$.

It suffices to prove that

$$\delta^{l-1} \circ h^l + h^{l+1} \circ \delta^l = \mathrm{Id}_{\mathcal{P}^l} \text{ for each } l \in \mathbb{Z}. \tag{4.9}$$

For each $i \in E^0, l \in \mathbb{Z}$ and $(p, q) \in \mathbf{\Lambda}_i^l$, we have that

$$\begin{aligned}
&(\delta^{l-1} \circ h^{l} + h^{l+1} \circ \delta^{l})(e_i \zeta_{(p,q)}) \\
&= 0 + h^{l+1}(\delta^{l}(e_i \zeta_{(p,q)})) \\
&= \begin{cases} h^{l+1}(\beta \zeta_{(\widehat{p},q)}); & \text{if } p = \beta \widehat{p}; \\ h^{l+1}(\sum_{\{\gamma \in E^1 \mid t(\gamma)=i\}} \gamma \zeta_{(e_{s(\gamma)}, q\gamma)}), & \text{if } l(p) = 0. \end{cases} \\
&= e_i \zeta_{(p,q)}
\end{aligned}$$

where the last equality uses the definition of h^{l+1}. Similarly, we have

$$\begin{aligned}
&(\delta^{l-1} \circ h^{l} + h^{l+1} \circ \delta^{l})(\alpha \zeta_{(p,q)}) \\
&= (\delta^{l-1} \circ h^{l})(\alpha \zeta_{(p,q)}) \\
&= \delta^{l-1}\big((h^{l} \circ \phi_{\alpha})((p^{\mathrm{op}})^{*} q^{\mathrm{op}})\big) \\
&= \delta^{l-1}(\phi((\alpha^{\mathrm{op}})^{*}(p^{\mathrm{op}})^{*} q^{\mathrm{op}})) \\
&= \sum_{\beta \in E^1 \mid t(\beta)=t(\alpha)} \phi_{\beta}(\beta^{\mathrm{op}}(\alpha^{\mathrm{op}})^{*}(p^{\mathrm{op}})^{*} q^{\mathrm{op}}) \\
&= \alpha \zeta_{(p,q)}
\end{aligned}$$

for $\alpha \in E^1$ with $s(\alpha) = i$. Here, the third equality uses (4.8), the fourth equality uses [19, Lemma 4.7], and the second last equality uses the (CK1)-relation for the Leavitt path algebra $L_k(E^{\mathrm{op}})$. □

Remark 4.2 Note that Eq. (4.9) implies that the projective Leavitt complex $\mathcal{P}^{\bullet}$ is acyclic.

4.4.2 The Independence of the Projective Leavitt Complex

We show that the definition of the projective Leavitt complex of E is independent of the choice of associated edges in E.

Denote by H and H' two different sets of associated edges of E. For each $i \in E^0$ and $l \in \mathbb{Z}$, let $\mathbf{\Lambda}_i^l$ and $\mathbf{\Lambda}_i^{l'}$ be the corresponding sets of associated pairs with respect to H and H' respectively; see (4.6). Define a map $\tau_{li} : \mathbf{\Lambda}_i^l \to \mathbf{\Lambda}_i^{l'}$ such that for $(p, q) \in \mathbf{\Lambda}_i^l$

$$\tau_{li}((p,q)) = \begin{cases} (\widetilde{p}\alpha, \widetilde{q}\alpha), & \text{if } p = \widetilde{p}\alpha', q = \widetilde{q}\alpha' \text{ and } \alpha' \in H'; \\ (p,q), & \text{otherwise,} \end{cases}$$

where $\alpha \in H$ with $t(\alpha) = t(\alpha')$. The map $\tau_{li} : \mathbf{\Lambda}_i^l \to \mathbf{\Lambda}_i^{l'}$ is a bijection. In fact, the inverse map of τ_{li} can be defined symmetrically.

Denote by $\mathcal{P}^{\bullet} = (\mathcal{P}^l, \delta^l)_{l \in \mathbb{Z}}$ and $\mathcal{P}^{\bullet\prime} = ((\mathcal{P}^l)', (\delta^l)')_{l \in \mathbb{Z}}$ the projective Leavitt complexes of E with H and H' sets of associated edges, respectively.

Recall that $B = L_k(E^{\text{op}})$. The maps $\phi' : B \to \mathcal{P}^{\bullet\prime}$, $(p^{\text{op}})^* q^{\text{op}} \mapsto e_{t(p)}\zeta_{(p,q)}$ and $\phi'_\beta : B \to \mathcal{P}^{\bullet\prime}$, $(p^{\text{op}})^* q^{\text{op}} \mapsto \delta_{t(p),s(\beta)}\beta\zeta_{(p,q)}$ with (p,q) an associated pair and $\beta \in E^1$ are graded right B-module morphisms.

For each $l \in \mathbb{Z}$, define a k-linear map $\theta^l : \mathcal{P}^l \to (\mathcal{P}^l)'$ such that $\theta^l(e_i\zeta_{(p,q)}) = \phi'((p^{\text{op}})^* q^{\text{op}})$ and $\theta^l(\alpha\zeta_{(p,q)}) = \phi'_\alpha((p^{\text{op}})^* q^{\text{op}})$ for $i \in E^0$ and $(p,q) \in \mathbf{\Lambda}_i^l$ and $\alpha \in E^1$ with $s(\alpha) = i$. Let $\theta^\bullet = (\theta^l)_{l\in\mathbb{Z}} : \mathcal{P}^\bullet \to \mathcal{P}^{\bullet\prime}$. By definitions we have $\theta^\bullet \circ \phi = \phi'$ and $\theta^\bullet \circ \phi_\beta = \phi'_\beta$ for each edge $\beta \in E^1$. Then we have $\theta^\bullet \circ \Phi = \Phi'$ with $\Phi' = \left(\phi' \ (\phi'_\beta)_{\beta\in E^1}\right) : B \oplus B^{(E^1)} \to \mathcal{P}^{\bullet\prime}$ a morphism of graded right B-modules.

Proposition 4.4 (1) *The above map $\theta^\bullet : \mathcal{P}^\bullet \to \mathcal{P}^{\bullet\prime}$ is an isomorphism of complexes of A-modules.*

(2) *The above map $\theta^\bullet : \mathcal{P}^\bullet \to \mathcal{P}^{\bullet\prime}$ is an isomorphism of right dg B-modules.*

Proof (1) We can directly check that θ^l is an A-module map for each $l \in \mathbb{Z}$. The inverse map of θ^l can be defined symmetrically. We have that θ^l is an isomorphism of A-modules for each $l \in \mathbb{Z}$. It remains to prove that $\theta^\bullet$ is a chain map of complexes. For each $i \in E^0$, $l \in \mathbb{Z}$ and $(p,q) \in \mathbf{\Lambda}_i^l$, since $((\delta^l)' \circ \theta^i)(\alpha\zeta_{(p,q)}) = 0 = (\theta^{l+1} \circ \delta^l)(\alpha\zeta_{(p,q)})$, it suffices to prove that $((\delta^l)' \circ \theta^l)(e_i\zeta_{(p,q)}) = (\theta^{l+1} \circ \delta^l)(e_i\zeta_{(p,q)})$ for $\alpha \in E^1$ with $s(\alpha) = i$. By [19, Lemma 4.7], we have that

$$((\delta^l)' \circ \theta^l)(e_i\zeta_{(p,q)}) = (\delta^l)'(\phi'((p^{\text{op}})^* q^{\text{op}})) = \sum_{\{\beta\in E^1 \mid t(\beta)=i\}} \phi'_\beta(\alpha^{\text{op}}(p^{\text{op}})^* q^{\text{op}}))$$

and

$$\begin{aligned}(\theta^{l+1} \circ \delta^l)(e_i\zeta_{(p,q)}) &= \begin{cases} \phi'_{\alpha_1}(\alpha_1^{\text{op}}(p^{\text{op}})^* q^{\text{op}})), & \text{if } p = \alpha_1\widehat{p}; \\ \sum_{\{\beta\in E^1 \mid t(\beta)=i\}} \phi'_\beta(\alpha^{\text{op}}(p^{\text{op}})^* q^{\text{op}})), & \text{if } l(p) = 0. \end{cases} \\ &= \sum_{\{\beta\in E^1 \mid t(\beta)=i\}} \phi'_\beta(\alpha^{\text{op}}(p^{\text{op}})^* q^{\text{op}})).\end{aligned}$$

Then we are done.

(2) It remains to prove that $\theta^\bullet$ is a graded right B-module morphism. By Lemma 4.2, there exists a graded right B-module morphism $\Omega : \mathcal{P}^\bullet \longrightarrow B \oplus B^{(E^1)}$ such that $\Phi \circ \Omega = \text{Id}_{\mathcal{P}^\bullet}$. Since $\theta^\bullet \circ \Phi = \Phi'$, we have that $\theta^\bullet = \theta^\bullet \circ (\Phi \circ \Omega) = \Phi' \circ \Omega$ is a composition of graded right B-module morphisms. □

Acknowledgements The author thanks Xiao-Wu Chen for many helpful discussions and encouragement. This project was supported by the National Natural Science Foundation of China (No.s 11522113 and 11571329). The author thanks Dr. Ambily Ambattu Asokan for hospitality when the author was in Cochin University of Science and Technology. The author thanks the support from CIMPA and the support from Centre for Research in Mathematics and Data Science. The author thanks Roozbeh Hazrat for many helpful discussions and encouragement. The author also would like to acknowledge the support of the Australian Research Council grant DP160101481. The author thanks Pere Ara, Jie Du, and Steffen Koenig for their support and help. The author thanks the anonymous referees for their very helpful suggestions to improve this paper.

References

1. A. Alahmadi, H. Alsulami, S.K. Jain, E. Zelmanov, Leavitt path algebras of finite Gelfand–Kirillov dimension. J. Algebra Appl. **11**(6), 6 (2012)
2. G. Abrams, G. Aranda Pino, The Leavitt path algebra of a graph. J. Algebra **293**(2), 319–334 (2005)
3. P. Ara, M.A. Moreno, E. Pardo, Nonstable **K**-theory for graph algebras. Algebr. Represent. Theory **10**(2), 157–178 (2007)
4. R.O. Buchweitz, *Maximal Cohen-Macaulay Modules and Tate-Cohomology Over Gorenstein Rings*, (unpublished manuscript) (1987). http://hdl.handle.net/1807/16682
5. A.I. Bondal, M.M. Kapranov, Enhanced triangulated categories. Mat. Sb. **181**(5), 669–683 (1990); (English translation Math. USSR-Sb. **70**(1), 93–107 (1990))
6. M. Bökstedt, A. Neeman, Homotopy limits in triangulated categories. Compos. Math. **86**, 209–234 (1993)
7. X.W. Chen, D. Yang, Homotopy categories, Leavitt path algebras, and Gorenstein projective modules. Int. Math. Res. Not. **10**, 2597–2633 (2015)
8. J. Cuntz, W. Krieger, A class of C^*-algebras and topological Markov chains. Invent. Math. **63**, 25–40 (1981)
9. J.A. Drozd, Tame and wild matrix problems, in *Representation theory, II (Proceedings of the Second International Conference on Carleton University, Ottawa, 1979)*. Lecture Notes in Mathematics, vol. 832. (Springer, Berlin, 242–258, 1980)
10. A. Grothendieck, The cohomology theory of abstract algebraic varieties, in *Proceedings of the International Congress on Mathematics (Edinburgh, 1958)* (Cambridge University Press, New York, 1960), pp. 103–118
11. D. Happel, On the derived category of a finite-dimensional algebra. Comment. Math. Helv. **62**, 339–389 (1987)
12. D. Happel, *Triangulated Categories in the Representation Theory of Finite-Dimensional Algebras*. London Mathematical Society Lecture Note Series, vol. 119 (Cambridge University Press, Cambridge, 1988)
13. R. Hazrat, R. Preusser, Applications of normal forms for weighted Leavitt path algebras: simple rings and domains. Algebr. Represent. Theory, https://doi.org/10.1007/s10468-017-9674-3
14. B. Keller, Deriving DG categories. Ann. Sci. Éc. Norm. Supér. (4) **27**(1), 63–102 (1994)
15. G.M. Kelly, Chain maps inducing zero homology maps. Proc. Camb. Philos. Soc. **61**, 847–854 (1965)
16. H. Krause, The stable derived category of a Noetherian scheme. Compos. Math. **141**, 1128–1162 (2005)
17. A. Kumjian, D. Pask, I. Raeburn, J. Renault, Graphs, groupoids, and Cuntz-Krieger algebras. J. Funct. Anal. **144**, 505–541 (1997)
18. H. Li, The injective Leavitt complex. Algebr. Represent Theor. **21**, 833–858 (2018)
19. H. Li, The projective Leavitt complex. Proc. Edinb. Math. Soc. **61**, 1155–1177 (2018)
20. A. Neeman, The Grothendieck duality theorem via Bousfield's techniques and Brown representability. J. Am. Math. Soc. **9**, 205–236 (1996)
21. A. Neeman, *Triangulated Categories*, Annals of Mathematics Studies, vol. 148 (Princeton University Press, Princeton, NJ, 2001)
22. D.O. Orlov, Triangulated categories of sigularities and D-branes in Landau–Ginzburg models. Proc. Steklov Inst. Math. **246**(3), 227–248 (2004)
23. I. Raeburn, *Graph Algebras*. CBMS Regional Conference Series in Mathematics, vol. 103 (The American Mathematical Society, Providence, RI, 2005)
24. J. Rickard, Derived categories and stable equivalence. J. Pure Appl. Algebra **61**, 303–317 (1989)
25. J. Rickard, Morita theory for derived categories. J. Lond. Math. Soc. **39**(2), 436–456 (1989)
26. J. Rickard, Derived equivalences as derived functors. J. Lond. Math. Soc. **43**(2), 37–48 (1991)
27. A.V. Roiter, Matrix problems, in *Proceedings of the International Congress of Mathematicians (Helsinki, 1978)*, vol. 1 (Academia Scientiarum Fennica, Helsinki 1980), pp. 319–322

28. S.P. Smith, Category equivalences involving graded modules over path algebras of quivers. Adv. Math. **230**, 1780–1810 (2012)
29. J.-L. Verdier, *Catégories dérivées, in SGA* $4\frac{1}{2}$. Lecture Notes in Mathematics, vol. 569 (Springer, Berlin, 1977), pp. 262–311

Chapter 5
A Survey on the Ideal Structure of Leavitt Path Algebras

Müge Kanuni and Suat Sert

5.1 Introduction

Leavitt path algebras are introduced independently by Abrams and Aranda Pino in [3] and by Ara, Moreno and Pardo in [6] around 2005. When the Leavitt path algebra is defined over the complex field it is the dense subalgebra of the graph C^*-algebra. (For a comprehensive survey on the graph C^*-algebras by Raeburn, see [11]). This close connection between algebra and analysis flourished with many similar results on the algebraic and analytic structures. A survey article by Abrams [1] summarized this interaction, also listed the similarities/differences of algebraic and analytic results giving an extensive list of references. This topic attracted the interest of many mathematicians immediately as the structure reveals itself in the graph properties on which it is constructed. Leavitt path algebras produced examples to answer some well-known open problems. Hence, hundreds of papers are published within a decade.

For a detailed discussion on Leavitt path algebras, interactions with various topics, we refer the interested reader to a well-written introductory level book published in 2017 by Abrams, Ara, and Siles Molina [2] which covers most of the literature.

Our main aim in this article is to focus only on the prime, primitive, and maximal two-sided ideals of Leavitt path algebras over a field, we gather and cite the known results that are either included in the book [2] or some recent *to appear* results [9, 14]. To keep the survey short and to avoid overlap with other expository papers, we did not include many other important and interesting topics in the ideal structure of Leavitt path algebras. We also did not extend the discussion to the results on Leavitt path algebras over commutative rings.

M. Kanuni (✉) · S. Sert
Department of Mathematics, Düzce University,
Konuralp 81620 Düzce, Turkey
e-mail: mugekanuni@duzce.edu.tr

S. Sert
e-mail: suatsert@gmail.com

A. A. Ambily et al. (eds.), *Leavitt Path Algebras and Classical* K-*Theory*,
Indian Statistical Institute Series,
https://doi.org/10.1007/978-981-15-1611-5_5

5.2 Preliminaries

The first section consists of preliminary definitions all of which can be found in the book [2].

5.2.1 *Graph Theory*

We first start with the basic definitions on graphs that is the main discrete structure of our interest. In this paper, $E = (E^0, E^1, s, r)$ will denote a directed graph with vertex set E^0, edge set E^1, source function s, and range function r. In particular, the source vertex of an edge e is denoted by $s(e)$, and the range vertex by $r(e)$. The graph E is called *finite* if both E^0 and E^1 are finite sets, and called *row-finite* if every vertex emits only finitely many edges. A vertex which emits infinitely many edges is called an *infinite emitter*. A *sink* is a vertex v for which the set $s^{-1}(v) = \{e \in E^1 \mid s(e) = v\}$ is empty, i.e., emits no edges. A vertex is a *regular* vertex if it is neither a sink nor an infinite emitter.

A proper path μ is a sequence of edges $\mu = e_1e_2...e_n$ such that $s(e_i) = r(e_{i-1})$ for $i = 2, ..., n$. Any vertex is considered to be a trivial path of length zero. The *length of a path* μ is the number of edges forming the path, i.e. $l(\mu) = n$ and the set of all paths is denoted by Path(E). If $n = l(\mu) \geq 1$, and $v = s(\mu) = r(\mu)$, then μ is called a *closed path based at* v. Again, μ is a *closed simple path based at* v if $s(e_j) \neq v$ for every $j > 1$. If $\mu = e_1e_2...e_n$ is a closed path based at v and $s(e_i) \neq s(e_j)$ for every $i \neq j$, then μ is called a *cycle based at* v. An *exit* for a path $\mu = e_1 \dots e_n$ is an edge e such that $s(e) = s(e_i)$ for some i and $e \neq e_i$. A cycle of length 1 is called a *loop*. A graph E is said to be *acyclic* in case it does not have any closed paths based at any vertex of E.

There are some graph properties that deserve to be named which will be used in the sequel.

Definition 5.1 For $v, w \in E^0$, we write $v \geq w$ in case there is a path $\mu \in Path(E)$ such that $s(\mu) = v$ and $r(\mu) = w$.

If $v \in E^0$ then the *tree of* v, denoted $T(v)$, is the set

$$T(v) = \{w \in E^0 \mid v \geq w\}.$$

Also, define $M(v) = \{w \in E^0 : w \geq v\}$.

Definition 5.2 A graph E satisfies *Condition* (K) if for each $v \in E^0$ which lies on a closed simple path, there exist at least two distinct closed simple paths α, β based at v.

A graph E satisfies *Condition* (L) if every cycle in E has an exit.

A cycle c in a graph E is called a *cycle without* K, if no vertex on c is the base of another distinct cycle in E (where distinct cycles possess different sets of edges).

A graph E satisfies the *Countable Separation Property*, if there exists a countable set S of vertices in E such that, for each vertex $u \in E$, there exists $w \in S$ for which $u \geq w$.

A graph E is said to be *countably directed* if there is a non-empty at most countable subset S of E^0 such that, for any two $u, v \in E^0$, there is a $w \in S$ such that $u \geq w$ and $v \geq w$.

Definition 5.3 Let E be a graph, and $H \subseteq E^0$. H is *hereditary* if whenever $v \in H$ and $w \in E^0$ for which $v \geq w$, then $w \in H$.

H is *saturated* if whenever a regular vertex v has the property that $\{r(e) | e \in E^1, s(e) = v\} \subseteq H$, then $v \in H$.

We denote $\mathcal{H}_E$ the set of those subsets of E^0 which are both hereditary and saturated.

For a given graph, there are many different new graph constructions that play a role in the ideal theory of Leavitt path algebras.

Definition 5.4 (*The restriction graph E_H*) Let E be an arbitrary graph, and let H be a hereditary subset of E^0. We denote by E_H the restriction graph:

$$E_H^0 := H, \quad E_H^1 := \{e \in E^1 | s(e) \in H\},$$

and the source and range functions in E_H are the source and range functions in E, restricted to H.

(The quotient graph by a hereditary subset E/H) Let E be an arbitrary graph, and let H be a hereditary subset of E^0. We denote by E/H the *quotient graph of E by H*, defined as follows:

$$(E/H)^0 = E^0 \backslash H, \text{ and } (E/H)^1 = \{e \in E^1 | r(e) \notin H\}.$$

The range and source functions for E/H are defined by restricting the range and source functions of E to (E/H).

(The hedgehog graph for a hereditary subset $F_E(H)$) Let E be an arbitrary graph. Let H be a non-empty hereditary subset of E^0. We denote by $F_E(H)$ the set

$$F_E(H) = \{\alpha \in Path(E) | \alpha = e_1....e_n, \text{ with } s(e_1) \in E^0 \backslash H, r(e_i) \in E^0 \backslash H \text{ for all } 1 \leq i < n, \text{ and } r(e_n) \in H\}$$

We denote by $\overline{F}_E(H)$ another copy of $F_E(H)$. If $\alpha \in F_E(H)$, we will write $\overline{\alpha}$ to refer to a copy of α in $\overline{F}_E(H)$. We define the graph ${}_HE = ({}_HE^0, {}_HE^1, s', r')$ as follows:

$${}_HE^0 = H \cup F_E(H), \quad \text{and} \quad {}_HE^1 = \{e \in E^1 | s(e) \in H\} \cup \overline{F}_E(H).$$

The source and range functions s' and r' are defined by setting $s'(e) = s(e)$ and $r'(e) = r(e)$ for every $e \in E^1$ such that $s(e) \in H$; and by setting $s'(\overline{\alpha}) = \alpha$ and $r'(\overline{\alpha}) = r(\alpha)$ for all $\overline{\alpha} \in \overline{F}_E(H)$.

Intuitively, $F_E(H)$ can be viewed as H, together with a new vertex corresponding to each path in E which ends at a vertex in H, but for which none of the previous edges in the path ends at a vertex in H. For every such new vertex, a new edge is added going into H. In $F_E(H)$, the only paths entering the subgraph H have common length 1; (the new graph looks like a hedgehog where the body is H and the quills are the edges into H).

Example 5.1 Consider the graph E below and take the hereditary saturated subset $H = \{v, w\}$,

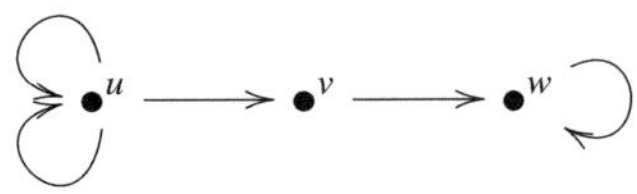

The restriction graph is

E_H

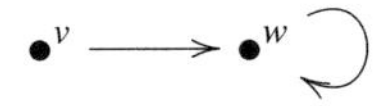

The quotient graph E/H is

E/H

Example 5.2 Consider the graph E below and take the hereditary saturated subset $H = \{v, w\}$,

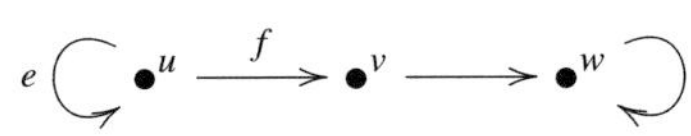

The hedgehog graph $_HE$ is

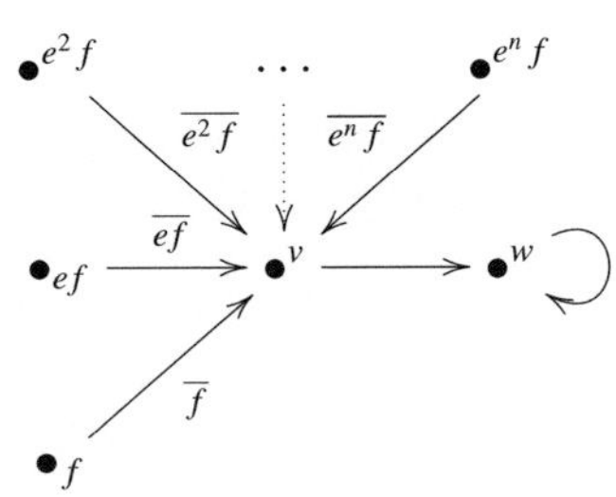

When we have infinite emitters in a graph, the graph is not row-finite and we need to introduce the notion of breaking vertices.

Definition 5.5 Let E be an arbitrary graph and K be any field. Let H be a hereditary subset of E^0, and let $v \in E^0$. We say that v is a *breaking vertex of* H if v belongs to the set

$$B_H := \{v \in E^0 \backslash H \mid \text{v is an infinite emitter and } 0 < |s^{-1}(v) \cap r^{-1}(E^0 \backslash H)| < \infty\}.$$

In words, B_H consists of those vertices of E which are infinite emitters, which do not belong to H, and for which the ranges of the edges they emit are all, except for a finite (but non-zero) number, inside H. For $v \in B_H$, we define the element v^H of $L_K(E)$ by setting

$$v^H := v - \sum_{e \in s^{-1}(v) \cap r^{-1}(E^0 \setminus H)} ee^*.$$

We note that any such v^H is homogeneous of degree 0 in the standard $\mathbb{Z}$-grading on $L_K(E)$. For any subset $S \subseteq B_H$, we define $S^H \subseteq L_K(E)$ by setting $S^H = \{v^H | v \in S\}$. Given a hereditary saturated subset H and a subset $S \subset B_H$, (H, S) is called an *admissible pair*. Given an admissible pair (H, S), the ideal generated by $H \cup S^H$ is denoted by $I(H, S)$.

Now, the new graph constructions that we defined in definition 5.4, can be extended to graphs with infinite emitters.

Definition 5.6 (*The quotient graph* $E/(H, S)$) Let E be an arbitrary graph, $H \in \mathcal{H}_E$, and $S \subseteq B_H$. We denote by $E/(H, S)$ the *quotient graph of* E *by* (H, S), defined as follows:

$$(E/(H, S))^0 = (E^0 \setminus H) \cup \{v' | v \in B_H \setminus S\},$$

$$(E/(H, S))^1 = \{e \in E^1 | r(e) \notin H\} \cup \{e' | e \in E^1 \text{ and } r(e) \in B_H \setminus S\},$$

and range and source maps in $E/(H, S)$ are defined by extending the range and source maps in E when appropriate, and in addition setting $s(e') = s(e)$ and $r(e') = r(e)'$.

(The generalized hedgehog graph construction ${}_{(H,S)}E$) Let E be an arbitrary graph, H a non-empty hereditary subset of E, and $S \subseteq B_H$. We define

$$F_1(H, S) := \{\alpha \in Path(E) | \alpha = e_1 \ldots e_n, r(e_n) \in H \text{ and } s(e_n) \notin H \cup S\}, \text{ and}$$

$$F_2(H, S) := \{\alpha \in Path(E) | \quad |\alpha| \geq 1 \text{ and } r(\alpha) \in S\}.$$

For $i = 1, 2$ we denote a copy of $F_i(H, S)$ by $\overline{F}_i(H, S)$. We define the graph ${}_{(H,S)}E$ as follows:

$${}_{(H,S)}E^0 := H \cup S \cup F_1(H, S) \cup F_2(H, S), \text{ and}$$

$${}_{(H,S)}E^1 := \{e \in E^1 | s(e) \in H\} \cup \{e \in E^1 | s(e) \in S \text{ and } r(e) \in H\} \cup \overline{F}_1(H, S) \cup \overline{F}_2(H, S).$$

The range and source map for ${}_{(H,S)}E$ are described by extending r and s to ${}_{(H,S)}E^1$, and by defining $r(\overline{\alpha}) = \alpha$ and $s(\overline{\alpha}) = \alpha$ for all $\overline{\alpha} \in \overline{F}_1(H, S) \cup \overline{F}_2(H, S)$.

Definition 5.7 A graph F is a *subgraph* of a graph E, if $F^0 \subset E^0$ and $F^1 \subset E^1$ where for any $f \in F^1$, $s(f), r(f) \in F^0$.

A subgraph F of a graph E is called *full* in case for each $v, w \in F^0$,

$$\{f \in F^1 | s(f) = v, r(f) = w\} = \{e \in E^1 | s(e) = v, r(e) = w\}.$$

In other words, the subgraph F is full in case whenever two vertices of E are in the subgraph, then all of the edges connecting those two vertices in E are also in F.

A non-empty full subgraph M of E is a *maximal tail* if it satisfies the following properties:

(MT – 1) If $v \in E^0$, $w \in M^0$ and $v \geq w$, then $v \in M^0$;
(MT – 2) If $v \in M^0$ and $s_E^{-1}(v) \neq \emptyset$, then there exists $e \in E^1$ such that $s(e) = v$ and $r(e) \in M^0$; and
(MT – 3) If $v, w \in M^0$, then there exists $y \in M^0$ such that $v \geq y$ and $w \geq y$.

Condition $MT - 3$ is now more commonly called *downward directedness* in literature, however we will use the term $MT - 3$ for consistency throughout the text.

5.2.2 Leavitt Path Algebra

Definition 5.8 Given an arbitrary graph E and a field K, the *Leavitt path algebra* $L_K(E)$ is defined to be the K-algebra generated by a set $\{v : v \in E^0\}$ of pair-wise orthogonal idempotents together with a set of variables $\{e, e^* : e \in E^1\}$ which satisfy the following conditions:

(1) $s(e)e = e = er(e)$ for all $e \in E^1$.
(2) $r(e)e^* = e^* = e^*s(e)$ for all $e \in E^1$.
(3) (CK-1 relations) For all $e, f \in E^1$, $e^*e = r(e)$ and $e^*f = 0$ if $e \neq f$.
(4) (CK-2 relations) For every regular vertex $v \in E^0$,

$$v = \sum_{e \in E^1,\ s(e)=v} ee^*.$$

The Leavitt path algebra is spanned as a K-vector space by the set of monomials

$$\{\gamma\lambda^* | \gamma, \lambda \in Path(E) \text{ such that } r(\gamma) = r(\lambda)\}$$

That is, any $x \in L_K(E)$,

$$x = \sum_{i=1}^{n} k_i \gamma_i \lambda_i^* \quad \text{for any} \quad k_i \in K, \gamma_i, \lambda_i \in Path(E).$$

Some familiar rings appear as examples of Leavitt path algebras, for instance:

Example 5.3 Take the graph E as

$L_K(E) \cong M_n(K)$.

Example 5.4 Take the graph R_1 as

In this case, $L_K(R_1) \cong K[x, x^{-1}]$ via $v \mapsto 1, e \mapsto x, e^* \mapsto x^{-1}$.

Example 5.5 For $n \geq 2$, consider the graph

$R_n =$

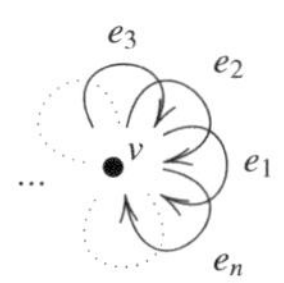

Then $L_K(R_n) \cong L_K(1, n)$ which is Leavitt algebra of type $(1, n)$.

Recall that a ring R is said to have *a set of local units* F, where F is a set of idempotents in R having the property that, for each finite subset $r_1, \ldots, r_n$ of R, there exists $f \in F$ with $fr_if = r_i$ for all $1 \leq i \leq n$. A ring R with unit 1 is, clearly, a ring with a set of local units where $F = \{1\}$.

In the case of Leavitt path algebras, for each $x \in L_K(E)$ there exists a finite set of distinct vertices $V(x)$ for which $x = fxf$, where $f = \sum_{v \in V(x)} v$. When E^0 is finite, $L_K(E)$ is a ring with unit element $1 = \sum\limits_{v \in E^0} v$. Otherwise, $L_K(E)$ is not a unital ring, but is a ring with local units consisting of sums of distinct elements of E^0.

One of the most important properties of the class of Leavitt path algebras is that each $L_K(E)$ is a $\mathbb{Z}$-graded K-algebra. that is, $L_K(E) = \bigoplus\limits_{n \in \mathbb{Z}} L_n$ induced by defining, for all $v \in E^0$ and $e \in E^1$, $\deg(v) = 0$, $\deg(e) = 1$, $\deg(e^*) = -1$. Further, for each $n \in \mathbb{Z}$, the homogeneous component L_n is given by

$$L_n = \left\{ \sum k_i \alpha_i \beta_i^* \in L : l(\alpha_i) - l(\beta_i) = n, \ k_i \in K, \ \alpha_i, \beta_i \in Path(E) \right\}.$$

An ideal I of $L_K(E)$ is said to be a *graded ideal* if $I = \bigoplus\limits_{n \in \mathbb{Z}} (I \cap L_n)$. In the sequel, all ideals of our concern will be two-sided.

5.3 Ideals in Leavitt Path Algebras

Recall that an (not necessarily unital) algebra R is called *simple*, if $R^2 \neq 0$ and R has no proper non-trivial ideals. Simple Leavitt path algebras are characterized in [3] by Abrams and Aranda Pino.

Theorem 5.1 *Let E be an arbitrary graph and K be any field. Then $L_K(E)$ is simple if and only if E has Condition (L) and the only hereditary saturated subsets of E^0 are $\emptyset$ and E^0.*

In a Leavitt path algebra, the intersection of any ideal with the set of vertices is always a hereditary set.

Lemma 5.1 ([3, Lemma 3.9]) *Let E be an arbitrary graph, K be any field and N be an ideal of $L_K(E)$. Then $N \cap E^0 \in \mathcal{H}_E$.*

$N \cap E^0$ may very well be the empty set, however if the Leavitt path algebra is over a graph that satisfies Condition (L) then N definitely contains a vertex (an idempotent).

Proposition 5.1 ([3, Corollary 3.8]) *Let E be an arbitrary graph satisfying Condition (L) and K be any field. Then every non-zero ideal of $L_K(E)$ contains a vertex.*

Proposition 5.2 *Let E be an arbitrary graph, K be any field and H be a hereditary subset of E^0. Then there is a $\mathbb{Z}$-graded monomorphism φ from $L_K(E_H)$ into $L_K(E)$ via $v \mapsto v$, $e \mapsto e$, $e^* \mapsto e^*$ for all $v \in E_H^0$, $e \in E_H^1$.*

We give a description of the elements in the ideal generated by a hereditary subset of vertices.

Lemma 5.2 ([15, Lemma 5.6]) *Let E be an arbitrary graph and K be any field.*
(i) *Let H be a hereditary subset of E^0. Then the ideal $I(H)$ is*

$$I(H) = span_K(\{\gamma\lambda^* | \gamma, \lambda \in Path(E) \text{ such that } r(\gamma) = r(\lambda) \in H\})$$

$$= \left\{ \sum_{i=1}^{n} k_i \gamma_i \lambda_i^* | n \geq 1, k_i \in K, \gamma_i, \lambda_i \in Path(E) \text{ such that } r(\gamma_i) = r(\lambda_i) \in H \right\}$$

(ii) *Let H be a hereditary subset of E^0 and S a subset of B_H. Then the ideal*

$$I(H, S) = span_K(\{\gamma\lambda^* | \gamma, \lambda \in Path(E) \text{ such that } r(\gamma) = r(\lambda) \in H\})$$
$$+ span_K(\{\alpha v^H \beta^* | \alpha, \beta \in Path(E) \text{ and } v \in S\}).$$

5.3.1 Graded Ideals

First, we mention the result on graded simplicity, that is when $L_K(E)$ has no non-trivial graded ideals. As stated in [2, Corollary 2.5.15], $L_K(E)$ is graded simple if and only if the only hereditary saturated subsets of E^0 are $\emptyset$ and E^0. A typical example of a graded simple Leavitt path algebra is $K[x, x^{-1}]$, see Example 5.4. However, since $\langle 1 + x \rangle$ is a (non-graded) ideal, $K[x, x^{-1}]$ is not simple. Hence, it is possible to have non-trivial non-graded ideals in a graded simple ring.

Now, we are ready to describe the graded ideals in Leavitt path algebras which is in [7, Remark 2.2].

Theorem 5.2 *Let E be an arbitrary graph and K be any field. Then every graded ideal N of $L_K(E)$ is generated by $H \cup S^H$, where $H = N \cap E^0 \in \mathcal{H}_E$, and $S = \{v \in B_H | v^H \in N\}$, i.e. $N = I(H, S)$.*

In particular, every graded ideal of $L_K(E)$ is generated by a set of homogeneous idempotents.

Observe that if $N = I(H, S)$ is a graded ideal, so that $N = \langle H, v^H : v \in S \rangle$, the generators u in H and v^H are all idempotents. So they all belong to N^2, that is if N is a graded ideal, then $N = N^2$. Conversely, if N is an ideal such that $N = N^2$, we use a result from [10]. In [10, Theorem 3.6], it was shown that for any ideal N, the intersection of $\{N^n : n > 0\}$ is a graded ideal. So, if $N^2 = N$, then $N = \cap\{N^n : n > 0\}$ is a graded ideal. Thus, we obtain the following characterization of graded ideals of a Leavitt path algebra (which also appears in [2, Corollary 2.9.11] via a different proof.)

Theorem 5.3 *Let E be an arbitrary graph and K be any field. Then, an ideal N of $L_K(E)$ is graded if and only if $N^2 = N$.*

The correspondence between the quotient Leavitt path algebra and the Leavitt path algebra of the quotient graph is noteworthy to state at this point. Part (i) of the following theorem appears as [7, Lemma 2.3] and part (ii) appears in [15, Theorem 5.7].

Theorem 5.4 *Let K be any field,*

(i) *E be a row-finite graph, and $H \in \mathcal{H}_E$. Then $L_K(E)/I(H) \cong L_K(E/H)$ as $\mathbb{Z}$-graded K-algebras.*
(ii) *E be an arbitrary graph, $H \in \mathcal{H}_E$ and $S \subset B_H$. Then $L_K(E)/I(H, S) \cong L_K(E/(H, S))$ as $\mathbb{Z}$-graded K-algebras.*

5.3.2 The Structure Theorem of Graded Ideals

Now, we are ready to give a complete description of the lattice of graded ideals of a Leavitt path algebra in terms of specified subsets of E^0, that is the Structure Theorem for Graded Ideals. The results in this section first appeared for row-finite graphs in [6] and for arbitrary graphs in [15].

Definition 5.9 Let E be an arbitrary graph and K be any field. Denote $\mathcal{L}_{gr}(L_K(E))$ the lattice of graded ideals of $L_K(E)$, whose order is inclusion, also supremum and infimum are the usual operations of ideal sum and intersection.

Remark 5.1 Let E be an arbitrary graph. We define in $\mathcal{H}_E$ a partial order by setting $H \leq H'$ in case $H \subseteq H'$. So, $\mathcal{H}_E$ is a complete lattice, with supremum $\vee$ and infimum $\wedge$ in $\mathcal{H}_E$ given by setting $\vee_{i\in\Gamma} H_i := \overline{\cup_{i\in\Gamma} H_i}$ and $\wedge_{i\in\Gamma} H_i := \cap_{i\in\Gamma} H_i$ respectively.

Definition 5.10 Let E be an arbitrary graph. We set

$$\mathcal{S} = \bigcup_{H\in\mathcal{H}_E} \mathcal{P}(B_H),$$

where $\mathcal{P}(B_H)$ denotes the set of all subsets of B_H.

We denote by $\mathcal{T}_E$ the subset of $\mathcal{H}_E \times \mathcal{S}$ consisting of pairs of the form (H, S), where $S \in \mathcal{P}(B_H)$. We define in $\mathcal{T}_E$ the following relation:

$$(H_1, S_1) \leq (H_2, S_2) \text{ if and only if } H_1 \subseteq H_2 \text{ and } S_1 \subseteq H_2 \cup S_2.$$

Proposition 5.3 *Let E be an arbitrary graph. For* $(H_1, S_1), (H_2, S_2) \in \mathcal{T}_E$, *we have*

$$(H_1, S_1) \leq (H_2, S_2) \Longleftrightarrow I(H_1, S_1) \subseteq I(H_2, S_2).$$

In particular, $\leq$ *is a partial order on* $\mathcal{T}_E$.

For more details on the lattice structure of $\mathcal{T}_E$, see [2].

Theorem 5.5 ([15, Theorem 5.7]) *Let E be an arbitrary graph and K be any field. Then the map* φ *given here provides a lattice isomorphism:*

$$\varphi : \mathcal{L}_{gr}(L_K(E)) \to \mathcal{T}_E \quad via \quad I \mapsto (I \cap E^0, S).$$

where $S = \{v \in B_H | v^H \in I\}$ *for* $H = I \cap E^0$. *The inverse* φ' *of* φ *is given by:*

$$\varphi' : \mathcal{T}_E \to \mathcal{L}_{gr}(L_K(E)) \quad via \quad (H, S) \mapsto I(H \cup S^H).$$

Theorem 5.6 ([6, Theorem 5.3]) *Let E be a row-finite graph and K be any field. The following map* φ *provides a lattice isomorphism:*

$$\varphi : \mathcal{L}_{gr}(L_K(E)) \to \mathcal{H}_E \quad via \quad \varphi(I) = I \cap E^0,$$

with inverse given by

$$\varphi' : \mathcal{H}_E \to \mathcal{L}_{gr}(L_K(E)) \quad via \quad \varphi'(H) = I(H).$$

Let E be an arbitrary graph and K be any field. Then every graded ideal of $L_K(E)$ is K-algebra isomorphic to a Leavitt path algebra. Part (i) of the following theorem first appears in [7, Lemma 5.2] under the hypothesis that graph E_H satisfies Condition (L).

Theorem 5.7 *Let E be an arbitrary graph and K be any field. Let H be a non-empty hereditary subset of E and $S \subseteq B_H$. Then* (i) *$I(H)$ is K-algebra isomorphic to $L_K({}_HE)$;*
(ii) *$I(H, S)$ is isomorphic as K-algebras to $L_K({}_{(H,S)}E)$.*

5.3.3 Structure of Two-Sided Ideals

The generators of an ideal are studied in [13] and gives a useful characterization of the graded and non-graded part of an ideal. The following results are due to Rangaswamy and finally achieving that in a Leavitt path algebra, any finitely generated ideal is principal [13].

Theorem 5.8 *Let E be an arbitrary graph and K be any field. Then any non-zero ideal of the $L_K(E)$ is generated by elements of the form*

$$\left(u + \sum_{i=1}^{k} k_i g^{ri}\right)\left(u - \sum_{e \in X} ee^*\right)$$

where $u \in E^0$, $k_i \in K$, r_i are positive integers, X is a finite (possibly empty) proper subset of $s^{-1}(u)$ and, whenever $k_i \neq 0$ for some i, then g is a unique cycle based at u.

The main result of [13] is the following theorem:

Theorem 5.9 *Let I be an arbitrary non-zero ideal of $L_K(E)$ with $I \cap E^0 = H$ and $S = \{v \in B_H : v^H \in I\}$. Then I is generated by $H \cup \{v^H : v \in S\} \cup Y$ where Y is a set of mutually orthogonal elements of the form $(u + \sum_{i=1}^{n} k_i g^{r_i})$ in which the following statements hold:*

(i) *g is a (unique) cycle with no exits in $E^0 \backslash H$ based at a vertex u in $E^0 \backslash H$; and*
(ii) *$k_i \in K$ with at least one $k_i \neq 0$.*

If I is non-graded, then Y is non-empty.

Corollary 5.1 *Every finitely generated ideal of $L_K(E)$ is a principal ideal. Moreover, if E is a finite graph, then every ideal is principal.*

5.3.4 Prime and Primitive Ideals

The structure of prime ideals has played a key role in ring theory. In the Leavitt path algebra setting the first paper to focus on the prime and primitive ideals of Leavitt path algebras on row-finite graphs have been [8]. Later the prime ideal structure on an arbitrary graph was studied in [12], while the primitive Leavitt path algebras are described in [4]. The primitive algebras have also been important as a consequence of Kaplansky's question: "Is a regular prime ring necessarily primitive?"

We recall a few ring-theoretic definitions. A two-sided ideal P of a ring R is *prime* in case $P \neq R$ and P has the property that for any two-sided ideals I, J of R, if $IJ \subseteq P$ then either $I \subseteq P$ of $J \subseteq P$. The ring R is called *prime* in case $\{0\}$ is a prime ideal of R. It is easily shown that P is a prime ideal of R if and only if R/P is a prime ring. The set of all prime ideals of R is denoted by Spec(R), call the prime spectrum of R. A ring R is called *left primitive* if R admits a simple faithful left R-module. It is easy to show that any primitive ring is prime.

A ring is *von Neumann regular* (or regular) in case for each $a \in R$ there exists $x \in R$ for which $a = axa$. In the theory of Leavitt path algebras, the necessary and sufficient condition for $L_K(E)$ to be regular is given by Abrams and Rangaswamy [5].

Theorem 5.10 *Let E be an arbitrary graph and K be any field. $L_K(E)$ is von Neumann regular if and only if E is acyclic.*

Recall the one vertex, one loop graph R_1 of the Example 5.4. The prime ideals of the principal ideal domain $K[x, x^{-1}] \cong L_K(R_1)$ provides a model for the prime spectra of general Leavitt path algebras. The key property of R_1 in this setting is that it contains a unique cycle without exits. Specifically, Spec($K[x, x^{-1}]$) consists of the ideal $\{0\}$, together with ideals generated by the irreducible polynomials of $K[x, x^{-1}]$. The irreducible polynomials are of the form $x^n f(x)$, where $f(x)$ is an irreducible polynomial in the standard polynomial ring $K[x]$, and $n \in \mathbb{Z}$. In particular, there is exactly one graded prime ideal (namely,$\{0\}$) in $L_K(R_1)$. All the remaining prime ideals of $L_K(R_1)$ are non-graded corresponding to irreducible polynomials in $K[x, x^{-1}]$.

The prime ideals of a Leavitt path algebra are completely characterized in the following theorem. Recall that $M(u)$ is defined in Definition 5.1.

Theorem 5.11 ([12, Theorem 3.12]) *Let E be an arbitrary graph and K be any field. Let P be an ideal of $L_K(E)$ with $P \cap E^0 = H$. Then P is a prime ideal of $L_K(E)$ if and only if P satisfies one of the following conditions:*

(i) $P = \langle H, \{v^H : v \in B_H\}\rangle$ *and $E^0 \backslash H$ satisfies the $MT-3$ condition;*
(ii) $P = \langle H, \{v^H : v \in B_H \backslash \{u\}\}\rangle$ *for some $u \in B_H$ and $E^0 \backslash H = M(u)$;*
(iii) $P = \langle H, \{v^H : v \in B_H\}, f(c)\rangle$ *where c is a cycle without K in E based at a vertex u, $E^0 \backslash H = M(u)$ and $f(x)$ is an irreducible polynomial in $K[x, x^{-1}]$.*

Recall that a ring R is prime if $\{0\}$ is a prime ideal, hence the immediate corollary to Theorem 5.11 follows.

Corollary 5.2 *Let E be an arbitrary graph and K any field. Then $L_K(E)$ is prime if and only if E is $MT-3$.*

When E is row-finite, the characterization of a primitive $L_K(E)$ is given in [8].

Theorem 5.12 *Let E be a row-finite graph and K be any field. Then $L_K(E)$ is primitive if and only if E has $MT-3$ and Condition(L).*

When E is an arbitrary graph, the result requires a new condition on the graph [4].

Theorem 5.13 *Let E be any graph and K be any field. Then $L_K(E)$ is primitive if and only if E has $MT-3$, Condition(L) and Countable Separation Property.*

We pause here to construct a Leavitt path algebra which is a counter example to Kaplansky's question "Is a regular prime ring necessarily primitive?", (see [4] for details).

Example 5.6 Let X be an uncountable set and S be the set of finite subsets of X. Define the graph E with

(1) Vertices indexed by S, and
(2) Edges induced by proper subset relationship.

Then $L_K(E)$ is a regular, prime and not primitive Leavitt path algebra.

The following results are from [12].

Lemma 5.3 ([12, Lemma 3.8]) *Let P be a prime ideal of $L_K(E)$ with $H = P \cap E^0$ and let $S = \{v \in B_H : v^H \in P\}$. Then the ideal $I(H, S)$ is also a prime ideal of $L_K(E)$.*

Corollary 5.3 ([12, Corollary 3.9]) *Let E be an arbitrary graph and K be any field. Then the Leavitt path algebra $L_K(E)$ is a prime ring if and only if there is a prime ideal of $L_K(E)$ which does not contain any vertices.*

A natural question that arose is to answer the graded version of Kaplansky's question, namely whether every graded prime von Neumann regular Leavitt path algebra is graded primitive. This question is solved by the recent unpublished work of Rangaswamy [14].

Theorem 5.14 *For any arbitrary graph E given, the following are equivalent*

(i) *$L_K(E)$ is graded primitive;*
(ii) *E^0 is countably directed;*
(iii) *$L_K(E)$ is graded prime and, for some vertex $v \in E^0$, the tree $T(v)$ satisfies the Countable Separation Property.*

The author in [14], provides many examples of graded von Neumann regular rings which are graded prime but not graded primitive.

5.3.5 *Maximal Ideals*

This section is quoted from [9] by Esin and the first named author.

In a unital ring, any maximal ideal is also a prime ideal. However, this is not necessarily true for a non-unital ring. Consider, for instance, the non-unital ring $2\mathbb{Z}$, and it's ideal $4\mathbb{Z}$. Notice that $4\mathbb{Z}$ is a maximal ideal, but not prime ideal in $2\mathbb{Z}$. The

Leavitt path algebra is a unital ring, only if E^0 is finite. So it is worthwhile to study the maximal ideals in a non-unital setting. The following argument on maximal and prime ideals in non-unital Leavitt path algebras appears in [12, pp. 86–87].

Proposition 5.4 *In a ring R satisfying $R^2 = R$, any maximal ideal is a prime ideal. Hence, in any Leavitt path algebra, any maximal ideal is a prime ideal.*

Proof Suppose $R^2 = R$, and let M be a maximal ideal of R such that $A \subsetneq M$ and $B \subsetneq M$ for some ideals A, B of R. Then $R = R^2 = (M + A)(M + B) = M^2 + AM + MB + AB \subseteq M + AB$. Then $M + AB = R$, and $AB \subsetneq M$. Thus M is a prime ideal. Now, since any Leavitt path algebra, R is a ring with local units, $R^2 = R$ is satisfied and the result holds.

As stated in [12, Lemma 3.6], in a Leavitt path algebra $L_K(E)$, the largest graded ideal contained in any ideal N (which is denoted by $gr(N)$) is the ideal generated by the admissible pair (H, S) where $H = N \cap E^0$, and $S = \{v \in B_H | v^H \in N\}$, i.e. $gr(N) = I(H, S)$. One useful observation is that: if a non-graded ideal N is a maximal element in $\mathcal{L}(L_K(E))$, the lattice of all two-sided ideals of a Leavitt path algebra, then $gr(N)$ is a maximal element in $\mathcal{L}_{gr}(L_K(E))$, the lattice of all two-sided graded ideals of this Leavitt path algebra (e.g. Example 5.10).

Maximal ideals always exist in a unital ring; however, this is not always true in a non-unital ring. Consider the Leavitt path algebra of the next example:

Example 5.7 Let E be the row-finite graph with $E^0 = \{v_i : i = 1, 2, \ldots\}$ and for each i, there is an edge e_i with $r(e_i) = v_i$, $s(e_i) = v_{i+1}$, also at each v_i there are two loops f_i, g_i so that $v_i = s(f_i) = r(f_i) = s(g_i) = r(g_i)$:

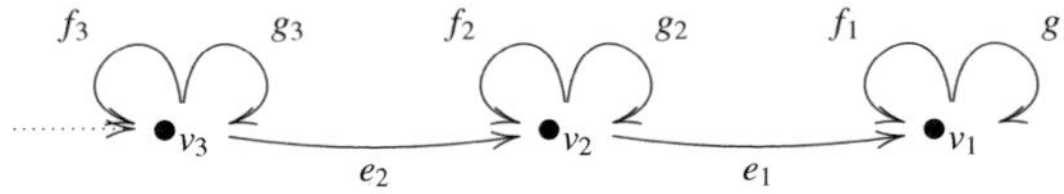

The non-empty proper hereditary saturated subsets of vertices in E are the sets $H_n = \{v_1, \ldots, v_n\}$ for some $n \geq 1$ and they form an infinite chain under set inclusion. Graph E satisfies Condition (K), so all ideals are graded, generated by H_n for some n and they form a chain under set inclusion. As the chain of ideals does not terminate, $L_K(E)$ does not contain any maximal ideals. Note also that, $E^0 \backslash (H_n, \emptyset)$ is $MT - 3$ for each n, thus all ideals are prime ideals.

A well-established question is to find out when a maximal ideal exists in a non-unital Leavitt path algebra. The necessary and sufficient condition depends on the existence of a maximal hereditary and saturated subset of E^0 as proved in [9].

Theorem 5.15 (Existence Theorem) $L_K(E)$ *has a maximal ideal if and only if* $\mathcal{H}_E$ *has a maximal element.*

Proof (Sketch: see [9] for details) Assume $L_K(E)$ has a maximal ideal M, then there are two cases:

If M is a graded ideal, then $M = I(H, S)$ for some $H \in \mathcal{H}_E$ and $S = \{v \in B_H | v^H \in M\}$. However, $M = I(H, S) \leq I(H, B_H)$, and as M is a maximal ideal, $S = B_H$. Then it can be shown that: $I(H, B_H)$ is a maximal ideal in $L_K(E)$ if and only if H is a maximal element in $\mathcal{H}_E$ and the quotient graph $E\backslash(H, B_H)$ has Condition(L).

If M is a non-graded maximal ideal, then $gr(M) = I(H, S)$ is a maximal graded ideal where $H = M \cap E^0$, and $S = \{v \in B_H | v^H \in M\}$. Similarly since $gr(M)$ is maximal, $S = B_H$. Again, it can be shown that: H is a maximal element in $\mathcal{H}_E$ with $E\backslash(H, B_H)$ not satisfying Condition(L), if and only if there is a maximal non-graded ideal M containing $I(H, B_H)$ with $H = M \cap E^0$.

This completes the proof.

Moreover, the poset structure of $\mathcal{H}_E$ determines whether every ideal of the Leavitt path algebra is contained in a maximal ideal.

Theorem 5.16 *The following assertions are equivalent:*

(i) *Every element $X \in \mathcal{H}_E$ is contained in a maximal element $Z \in \mathcal{H}_E$.*
(ii) *Every ideal of $L_K(E)$ is contained in a maximal ideal.*

Example 5.8 Let E be the graph

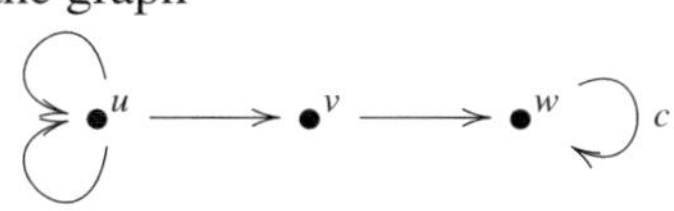

Then E does not satisfy Condition (K), so the Leavitt path algebra on E has both graded and non-graded ideals. Let Q be the graded ideal generated by the hereditary saturated set $H = \{v, w\}$. Q is a maximal ideal as L/Q is isomorphic to $L_K(E\backslash H)$ which is also isomorphic to the simple Leavitt algebra $L(1, 2)$ (See Example 5.1). By using Theorem 5.11, we classify the prime ideals in L. There are infinitely many non-graded prime ideals each generated by $f(c)$ where $f(x)$ is an irreducible polynomial in $K[x, x^{-1}]$ which are all contained in Q. Also, the trivial ideal $\{0\}$ is prime as E satisfies condition $MT - 3$ and $L_K(E)$ has a unique maximal element Q.

We now give an example of a graph with infinitely many hereditary saturated sets and the corresponding Leavitt path algebra has a unique maximal ideal which is graded.

Example 5.9 Let E be a graph with $E^0 = \{v_i : i = 1, 2, \ldots\}$. For each i, there is an edge e_i with $s(e_i) = v_i$ and $r(e_i) = v_{i+1}$ and at each v_i there are two loops f_i, g_i so that $v_i = s(f_i) = r(f_i) = s(g_i) = r(g_i)$. Thus E is the graph

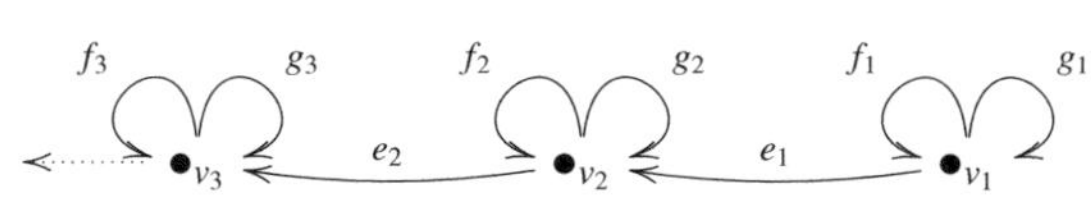

Now E is a row-finite graph and the non-empty proper hereditary saturated subsets of vertices in E are the sets $H_n = \{v_n, v_{n+1}, \ldots\}$ for some $n \geq 2$ and $H_{n+1} \subsetneq H_n$ form

an infinite chain under set inclusion and $H_2 = \{v_2, v_3, \ldots\}$ is the maximal element in $\mathcal{H}_E$. The graph E satisfies Condition (K), so all ideals are graded, generated by H_n for some n. So $L_K(E)$ contains a unique maximal ideal $I(H_2)$. Note also that, $E^0 \backslash H_n$ is $MT-3$ for each n, thus all ideals of L are prime ideals.

In a Leavitt path algebra, if a unique maximal ideal exists, then it is a graded ideal. Also, every maximal ideal is graded in $L_K(E)$ if and only if for every maximal element H in $\mathcal{H}_E$, $E\backslash(H, B_H)$ satisfies Condition(L). Note that there are Leavitt path algebras with both graded and non-graded maximal ideals as the following example illustrates.

Example 5.10 Let E be the graph

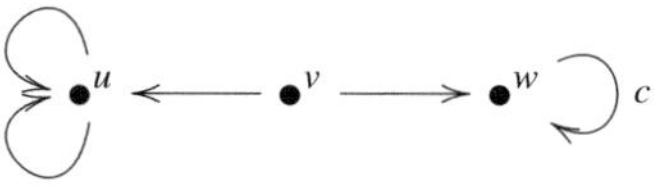

Then the Leavitt path algebra on E has both graded and non-graded maximal ideals. The set $\mathcal{H}_E$ is finite and hence any ideal is contained in a maximal ideal. The trivial ideal $\{0\}$ which is a graded ideal generated by the empty set, is not prime as E does not satisfy condition $MT-3$. There are infinitely many non-graded prime ideals each generated by $f(c)$ where $f(x)$ is an irreducible polynomial in $K[x, x^{-1}]$ which all contain $\{0\}$. Let N be the graded ideal generated by the hereditary saturated set $H = \{u\}$ and in this case, the quotient graph $E\backslash H$ does not satisfy condition (L). Then there are infinitely many maximal non-graded ideals each generated by $f(c)$ where $f(x)$ is an irreducible polynomial in $K[x, x^{-1}]$ which all contain N. Also, let Q be the graded ideal generated by the hereditary saturated set $H = \{w\}$. In this case, the quotient graph $E\backslash H$ satisfy condition (L). Hence, Q is a maximal ideal.

$L_K(E)$ has infinitely many maximal ideals, one of them is graded, namely Q and infinitely many are non-graded ideals whose graded part is N.

It is an interesting question to answer when all non-zero prime ideals are maximal, as these rings are called rings with Krull dimension zero. In fact, Leavitt path algebras with prescribed Krull dimension are studied in [12]. We conclude this article with two results from [12].

Theorem 5.17 ([12, Theorem 6.1]) *Let E be an arbitrary graph and K be any field. Then every non-zero prime ideal of the Leavitt path algebra $L_K(E)$ is maximal if and only if E satisfies one of the following two conditions:*

Condition I: (i) *E^0 is a maximal tail;* (ii) *The only hereditary saturated subsets of E^0 are E^0 and $\emptyset$;* (iii) *E does not satisfy the Condition(K).*

Condition II: (a) *E satisfies the Condition(K);* (b) *For each maximal tail M, the restricted graph E_M contains no proper non-empty hereditary saturated subsets;* (c) *If H is a hereditary saturated subset of E^0, then for each $u \in B_H$, $M(u) \subsetneqq E^0 \backslash H$*

When E is finite, the answer is much simpler.

Corollary 5.4 *Let E be a finite graph. Then every non-zero prime ideal of $L_K(E)$ is maximal if and only if either $L_K(E) \cong M_n(K[x, x^{-1}])$ for some positive integer n*

or E satisfies the Condition(K) and, for each maximal tail M, the restricted graph E_M contains no proper non-empty hereditary saturated subsets of vertices.

Acknowledgements The authors would like to thank the referees for a very careful, thorough reading of the paper and well-worthy comments to improve both the text and the references.

References

1. G. Abrams, Leavitt path algebras: the first decade. Bull. Math. Sci. **5**, 59–120 (2015)
2. G. Abrams, P. Ara, M. Siles Molina, *Leavitt Path Algebras*. Lecture Notes in Mathematics (Springer, London, 2017)
3. G. Abrams, G. Aranda Pino, The Leavitt path algebra of a graph. J. Algebra **293**, 319–334 (2005)
4. G. Abrams, P.J. Bell, K.M. Rangaswamy, On prime nonprimitive von Neumann regular algebras. Trans. Am. Math. Soc. **366**(5), 2375–2392 (2014)
5. G. Abrams, K.M. Rangaswamy, Regularity conditions for arbitrary Leavitt path algebras. Algebra Represent. Theory **13**(3), 319–334 (2010)
6. P. Ara, M.A. Moreno, E. Pardo, Nonstable K-theory for graph algebras. Algebra Represent. Theory **10**, 157–178 (2007)
7. G. Aranda Pino, E. Pardo, M. Siles Molina, Exchange Leavitt path algebras and stable rank. J. Algebra **305**, 912–936 (2006)
8. G. Aranda Pino, E. Pardo, M. Siles Molina, Prime spectrum and primitive Leavitt path algebras. Indiana Univ. Math. J. **58**(2), 869–890 (2009)
9. S. Esin, M. Kanuni, Existence of maximal ideals in Leavitt path algebras. Turkish J. Math. **42**, 2081–2090 (2018). https://doi.org/10.3906/mat-1704-116
10. S. Esin S, M. Kanuni, K.M. Rangaswamy, On the intersections of two-sided ideals of Leavitt path algebras. J. Pure Appl. Algebra **221**, 632–644 (2017)
11. I. Raeburn, *Graph Algebras*, vol. 103, CBMS Regional Conference Series in Mathematics (American Mathematical Society, Providence, 2005)
12. K.M. Rangaswamy, The theory of prime ideals of Leavitt path algebras over arbitrary graphs. J. Algebra **375**, 73–96 (2013)
13. K.M. Rangaswamy, On generators of two-sided ideals of Leavitt path algebras over arbitrary graphs. Commun. Algebra **42**, 2859–2868 (2014)
14. K.M. Rangaswamy, *On Graded Primitive Leavitt Path Algebras*, Preprint, Personal Correspondence
15. T. Tomforde, Uniqueness theorems and ideal structure of Leavitt path algebras. J. Algebra **318**, 270–299 (2007)

Chapter 6
Gröbner Bases and Dimension Formulas for Ternary Partially Associative Operads

Fatemeh Bagherzadeh and Murray Bremner

6.1 Introduction

We consider nonsymmetric operads in the category of $\mathbb{Z}$-graded vector spaces over a field of characteristic 0. The product is the tensor product (with Koszul signs) and the coproduct is the direct sum. Gröbner bases for operads were introduced by Dotsenko, Khoroshkin and Vallette [5, 6]; see also [2].

Let $\mathcal{LT}$ be the free nonsymmetric operad with one ternary operation $t = (***)$. Let α denote ternary partial associativity, which may be written as a tree polynomial, using partial compositions or as a nonassociative polynomial:

$\alpha =$ 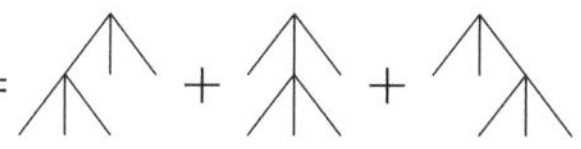

$$t \circ_1 t + t \circ_2 t + t \circ_3 t, \qquad ((***)**) + (*(***)*) + (**(***)). \tag{6.1}$$

We compute a Gröbner basis for the ideal $\langle \alpha \rangle$ when t has even (homological) degree so that Koszul signs are irrelevant, and when t has odd degree so that Koszul signs are essential. We include details of the calculations to clarify the Gröbner basis algorithm for nonsymmetric operads. As an application, we calculate dimension formulas for the quotient operads. Similar results have been obtained independently in unpublished work of Vladimir Dotsenko.

For earlier work on partial associativity and its applications, see [1, 3, 7, 9–11, 13–15]. Recent results of Dotsenko, Shadrin and Vallette [8] have shown that

F. Bagherzadeh · M. Bremner (✉)
Department of Mathematics and Statistics, University of Saskatchewan, Saskatoon, Canada
e-mail: bremner@math.usask.ca

F. Bagherzadeh
e-mail: bagherzadeh@math.usask.ca

A. A. Ambily et al. (eds.), *Leavitt Path Algebras and Classical K-Theory*, Indian Statistical Institute Series,
https://doi.org/10.1007/978-981-15-1611-5_6

the ternary partially associative operad with an odd generator arises naturally in the homology of the poset of interval partitions into intervals of odd length and in certain De Concini–Procesi models of subspace arrangements [4] over the real numbers.

6.2 Preliminaries

Definition 6.1 An *m-ary tree* is a rooted plane tree p in which every node has either no children (*leaf*) or m children (*internal node*). The *weight* $w(p)$ counts internal nodes; the *arity* $\ell(p) = 1 + w(p)(m-1)$ counts leaves indexed $1, \ldots, \ell(p)$ from left to right. The *basic tree* t is the m-ary tree of weight 1. Set $[n] = \{1, \ldots, n\}$.

Definition 6.2 If $n \equiv 1 \pmod{m-1}$ then $\mathcal{T}(n)$ is the set of m-ary trees of arity n, and $\mathcal{T}$ is the disjoint union of the $\mathcal{T}(n)$ for $n \geq 1$.

Definition 6.3 If $p, q \in \mathcal{T}$ then for $i \in [\ell(p)]$ the *partial composition* $p \circ_i q \in \mathcal{T}$ is obtained by identifying leaf i of p with the root of q.

Lemma 6.1 *Starting with t, every m-ary tree of weight w can be obtained by a sequence of $w-1$ partial compositions.*

Lemma 6.2 *Let p, q, r be m-ary trees. Partial composition satisfies* [2, p. 72]*:*

$$(p \circ_i q) \circ_j r = \begin{cases} p \circ_i (q \circ_{j-i+1} r), & i \leq j \leq i+\ell(q)-1; \\ (p \circ_{j-\ell(q)+1} r) \circ_i q, & i+\ell(q) \leq j \leq \ell(p)+\ell(q)-1; \\ (p \circ_j r) \circ_{i+\ell(r)-1} q, & 1 \leq j \leq i-1. \end{cases}$$

Lemma 6.3 *The set $\mathcal{T}$ with partial compositions is isomorphic to the free nonsymmetric (set) operad with one m-ary operation t.*

Definition 6.4 If $n \equiv 1 \pmod{m-1}$ then $\mathcal{LT}(n)$ is the vector space with basis $\mathcal{T}(n)$, and $\mathcal{LT}$ is the direct sum of $\mathcal{LT}(n)$ for $n \geq 1$. A *tree polynomial* of arity n is an element of $\mathcal{LT}(n)$. Partial composition in $\mathcal{T}$ extends bilinearly to $\mathcal{LT}$.

Lemma 6.4 *The vector space $\mathcal{LT}$ with partial compositions is isomorphic to the free nonsymmetric (vector) operad with one m-ary operation t.*

Definition 6.5 A *relation* of arity n is an element of $\mathcal{LT}(n) \setminus 0$. The *operad ideal* $\mathcal{I} = \langle f_1, \ldots, f_k \rangle$ generated by relations $f_1, \ldots, f_k$ is the intersection of all homogeneous subspaces $\mathcal{S} \subseteq \mathcal{LT}$ such that $f_1, \ldots, f_k \in \mathcal{S}$, and for all $f \in \mathcal{S}(m)$, $g \in \mathcal{LT}(n)$ we have $f \circ_i g, g \circ_j f \in \mathcal{S}$ $(i \in [m], j \in [n])$.

The following results come from [2, Sect. 3.4] and [6, Sects. 2.4, 3.1] with minor changes.

Definition 6.6 The *path sequence* of $p \in \mathcal{T}(n)$ is $\text{path}(p) = (a_1, \ldots, a_n)$, where a_i is the length of the path from the root to the leaf i.

Lemma 6.5 *If $p, q \in \mathcal{T}$ then $p = q$ if and only if* $\text{path}(p) = \text{path}(q)$.

Definition 6.7 For $p, q \in \mathcal{T}(n)$ we write $p \prec q$ and say *p precedes q* in *path-lex order* if and only if $\text{path}(p) \prec \text{path}(q)$ in lex order on n-tuples of positive integers. If $f \in \mathcal{LT}(n)$ then its *leading monomial $\ell m(f) \in \mathcal{T}(n)$* is the greatest monomial in path-lex order, and its *leading coefficient $\ell c(f)$* is the coefficient of $\ell m(f)$.

Definition 6.8 If $p, q \in \mathcal{T}$ then q is *divisible* by p (written $p \mid q$) if p is a subtree of q: that is, $q = \cdots p \cdots$ where the dots denote sequences of partial compositions with parentheses. If $p \in \mathcal{T}(m)$, $q \in \mathcal{T}(n)$, $p \mid q$, and $f \in \mathcal{LT}(m)$ then we may replace p by f in q and use linearity and the same partial compositions to obtain the *substitution* of f for p in q:

$$M(q, p, f) = \cdots f \cdots \in \mathcal{LT}(n).$$

Definition 6.9 If $f, g \in \mathcal{LT}$ and $\ell m(g) \mid \ell m(f)$ then the *reduction of f by g* (which eliminates the leading term of f) is

$$R(f, g) = f - \frac{\ell c(f)}{\ell c(g)} M\big(\ell m(f), \ell m(g), g\big).$$

This extends to reduction of f by $g_1, \ldots, g_k$; see [2, Algorithm 3.4.2.16].

Definition 6.10 If $p, q, r \in \mathcal{T}$ then we call p a *small common multiple* (SCM) of q and r if $q \mid p$, $r \mid p$, every node of p is a node of q or r (or both), and $\ell(p) < \ell(q) + \ell(r)$.

Definition 6.11 If f, g, h are monic tree polynomials and $\ell m(f)$ is an SCM of $\ell m(g)$, $\ell m(h)$ then the resulting *S-polynomial* is

$$S(f, g, h) = M\big(\ell m(f), \ell m(g), g\big) - M\big(\ell m(f), \ell m(h), h\big).$$

Definition 6.12 Let G be a finite set of relations and let $I = \langle G \rangle$. If for all $f \in I$ there exists $g \in G$ such that $\ell m(g) \mid \ell m(f)$ then we call G a *Gröbner basis* for I. We say G is *reduced* if $\ell m(g)$ is not divisible by $\ell m(h)$ for all $g, h \in G$.

Lemma 6.6 *Every operad ideal has a unique reduced Gröbner basis.*

Theorem 6.1 *If $I = \langle G \rangle$ then G is a Gröbner basis for I if and only if for every SCM f of elements $g, h \in G$ the reduction of $S(f, g, h)$ by G is 0.*

6.3 Gröbner Bases and Dimension Formulas

In the rest of this paper, we consider a ternary operation ($m = 3$). We usually indicate the leading monomial of a tree polynomial by a bullet at the root, and write the terms of a tree polynomial from left to right in reverse path-lex order. The partially associative relation α corresponds to this rewrite rule:

$t \circ_1 t =$ 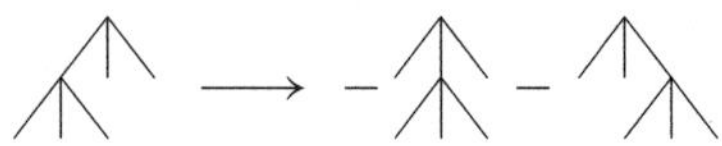 $= -\, t \circ_2 t - t \circ_3 t$ (6.2)

Theorem 6.2 *For the path-lex monomial order, the following tree polynomials form the reduced Gröbner basis for $\langle \alpha \rangle$ with an operation of even degree:*

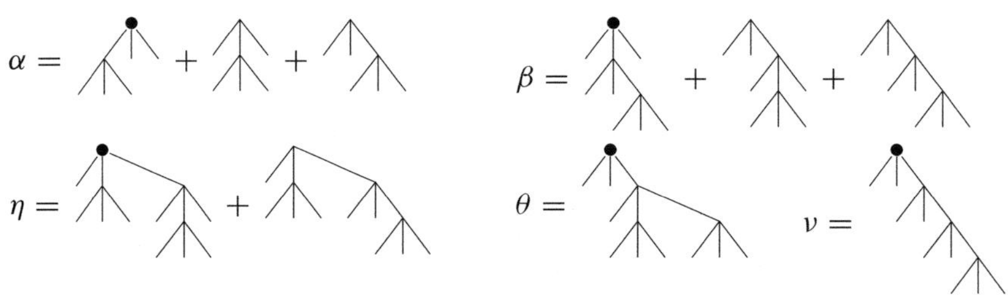

Proof The proof consists of Lemmas 6.7 to 6.13. □

Remark 6.1 As nonassociative polynomials, the relations of Theorem 6.2 are

$$((***)**) + (*(***)*) + (**(***)),$$
$$(*(**(***))*) + (**(*(***)*)) + (**(**(***))),$$
$$(*(***)(*(***)*)) + (*(***)(**(***))),\ (**(*(***)(***))),\ (**(**(**(***)))).$$

Lemma 6.7 *There is only one SCM of $\ell m(\alpha)$ with itself; this produces reduced S-polynomial β, and the set $\{\alpha, \beta\}$ is self-reduced:*

$\beta =$ 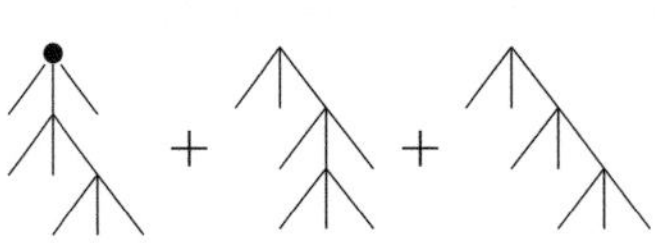 $= t \circ_2 (t \circ_3 t) + t \circ_3 (t \circ_2 t) + t \circ_3 (t \circ_3 t).$

Proof We have $\ell m(\alpha) = t \circ_1 t$ and hence

$\ell m(\alpha) \circ_1 t =$ 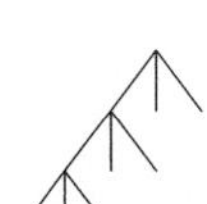 $= t \circ_1 \ell m(\alpha).$

From this, we obtain these tree polynomials by substitution (Definition 6.8):

$$\alpha \circ_1 t = (t \circ_1 t) \circ_1 t + (t \circ_2 t) \circ_1 t + (t \circ_3 t) \circ_1 t$$

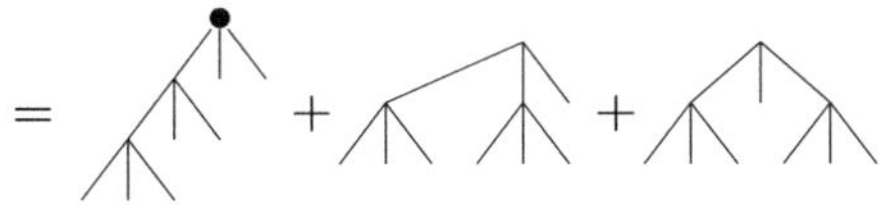

$$t \circ_1 \alpha = t \circ_1 (t \circ_1 t) + t \circ_1 (t \circ_2 t) + t \circ_1 (t \circ_3 t)$$

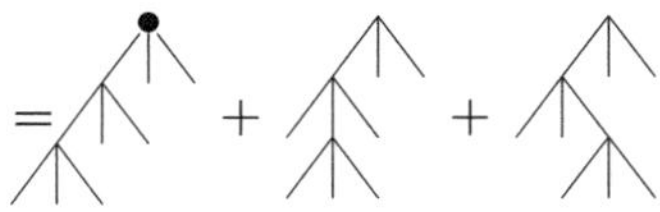

The difference is this (non-reduced) S-polynomial:

$$\alpha \circ_1 t - t \circ_1 \alpha = (t \circ_2 t) \circ_1 t + (t \circ_3 t) \circ_1 t - t \circ_1 (t \circ_2 t) - t \circ_1 (t \circ_3 t)$$

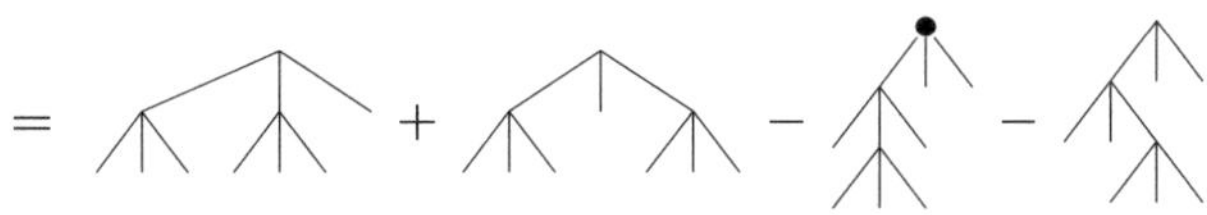

$$= (t \circ_1 t) \circ_4 t + (t \circ_1 t) \circ_5 t - (t \circ_1 t) \circ_2 t - (t \circ_1 t) \circ_3 t.$$

We have rewritten the partial compositions (Lemma 6.2). We apply rewrite rule (6.2) to the top subtree $\ell m(\alpha) = t \circ_1 t$ of each monomial (reduce using α):

$$\begin{aligned} &- (t \circ_2 t) \circ_4 t - (t \circ_3 t) \circ_4 t - (t \circ_2 t) \circ_5 t - (t \circ_3 t) \circ_5 t \\ &+ (t \circ_2 t) \circ_2 t + (t \circ_3 t) \circ_2 t + (t \circ_2 t) \circ_3 t + (t \circ_3 t) \circ_3 t. \end{aligned}$$

Terms 3 and 6 cancel since both monomials represent the same tree:

$$(t \circ_2 t) \circ_5 t \;=\; (t \circ_3 t) \circ_2 t \;=$$

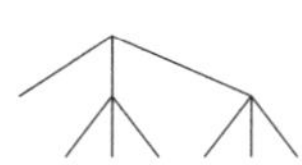

Six terms remain:

$$\begin{aligned} &- (t \circ_2 t) \circ_4 t - (t \circ_3 t) \circ_4 t - (t \circ_3 t) \circ_5 t \\ &\quad + (t \circ_2 t) \circ_2 t + (t \circ_2 t) \circ_3 t + (t \circ_3 t) \circ_3 t \end{aligned}$$

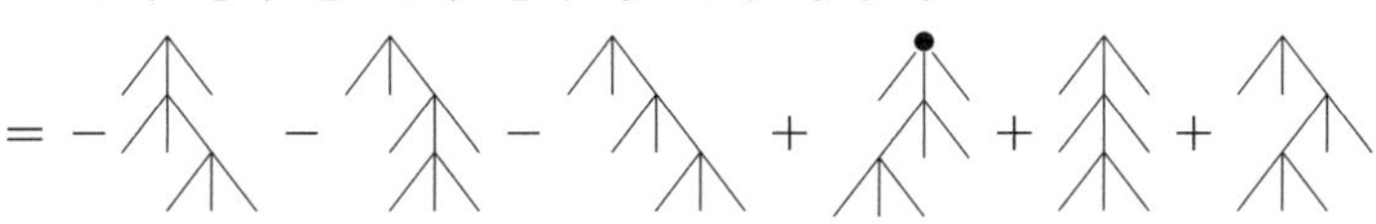

$$\begin{aligned} &= - t \circ_2 (t \circ_3 t) - t \circ_3 (t \circ_2 t) - t \circ_3 (t \circ_3 t) \\ &\quad + t \circ_2 (t \circ_1 t) + t \circ_2 (t \circ_2 t) + t \circ_3 (t \circ_1 t). \end{aligned}$$

In terms 4 and 6, we reduce the bottom subtree $\ell m(\alpha) = t \circ_1 t$ using α:

$$\begin{aligned} &- t \circ_2 (t \circ_3 t) - t \circ_3 (t \circ_2 t) - t \circ_3 (t \circ_3 t) - t \circ_2 (t \circ_2 t) \\ &- t \circ_2 (t \circ_3 t) + t \circ_2 (t \circ_2 t) - t \circ_3 (t \circ_2 t) - t \circ_3 (t \circ_3 t). \end{aligned}$$

Terms 4 and 6 cancel and the others combine in pairs:

$$-2\big(t \circ_2 (t \circ_3 t) + t \circ_3 (t \circ_2 t) + t \circ_3 (t \circ_3 t)\big) = -2$$ 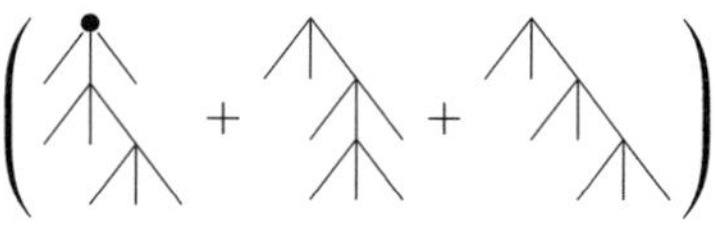

No further reduction is possible. The monic form of this polynomial is β. □

The relation β corresponds to this rewrite rule:

$$t \circ_2 (t \circ_3 t) = -t \circ_3 (t \circ_2 t) - t \circ_3 (t \circ_3 t)$$

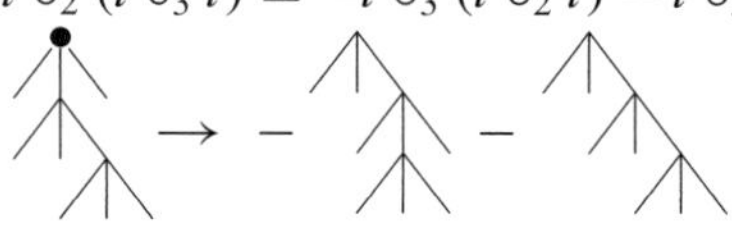 (6.3)

We consider separately the four SCMs of $\ell m(\alpha) = t \circ_1 t$ and $\ell m(\beta) = t \circ_2 (t \circ_3 t)$.

Lemma 6.8 *Identifying the second t of* $\ell m(\alpha) = t \circ_1 t$ *with the first t of* $\ell m(\beta) = t \circ_2 (t \circ_3 t)$ *produces the reduced S-polynomial* γ*, and* $\{\alpha, \beta, \gamma\}$ *is self-reduced:*

$\gamma = 2$ 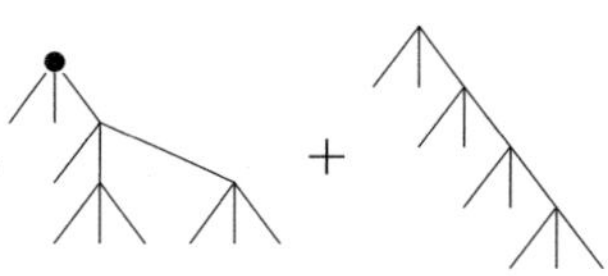 $= 2(t \circ_3 (t \circ_2 t)) \circ_7 t + t \circ_3 (t \circ_3 (t \circ_3 t)).$

Proof We have the following equations:

$\ell m(\alpha) \circ_2 (t \circ_3 t) = (t \circ_1 t) \circ_2 (t \circ_3 t) =$ 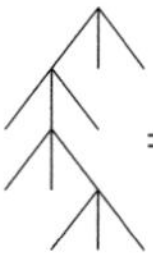 $= t \circ_1 (t \circ_2 (t \circ_3 t)) = t \circ_1 \ell m(\beta).$

We apply the same partial compositions to α and β:

$$\begin{aligned} \alpha \circ_2 (t \circ_3 t) &= (t \circ_1 t) \circ_2 (t \circ_3 t) + (t \circ_2 t) \circ_2 (t \circ_3 t) + (t \circ_3 t) \circ_2 (t \circ_3 t), \\ t \circ_1 \beta &= t \circ_1 (t \circ_2 (t \circ_3 t)) + t \circ_1 (t \circ_3 (t \circ_2 t)) + t \circ_1 (t \circ_3 (t \circ_3 t)). \end{aligned}$$

Taking the difference, we obtain this (non-reduced) S-polynomial:

$$\begin{aligned} &(t \circ_1 t) \circ_2 (t \circ_3 t) + (t \circ_2 t) \circ_2 (t \circ_3 t) + (t \circ_3 t) \circ_2 (t \circ_3 t) \\ &- t \circ_1 (t \circ_2 (t \circ_3 t)) - t \circ_1 (t \circ_3 (t \circ_2 t)) - t \circ_1 (t \circ_3 (t \circ_3 t)). \end{aligned}$$

Terms 1 and 4 cancel, leaving

$$(t \circ_2 t) \circ_2 (t \circ_3 t) + (t \circ_3 t) \circ_2 (t \circ_3 t) \\ - t \circ_1 (t \circ_3 (t \circ_2 t)) - t \circ_1 (t \circ_3 (t \circ_3 t))$$

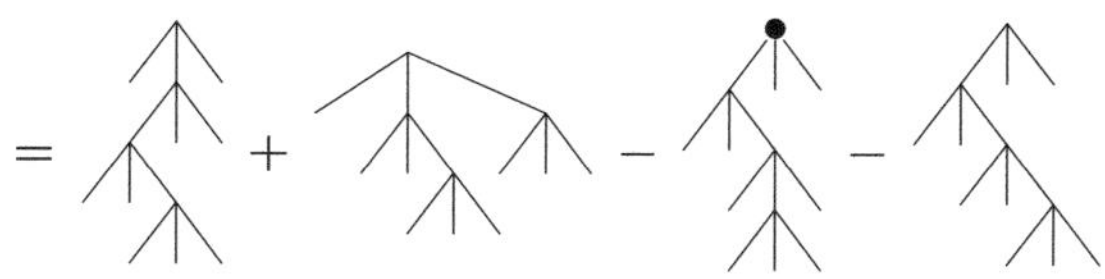

$$= t \circ_2 ((t \circ_1 t) \circ_3 t) + (t \circ_3 t) \circ_2 (t \circ_3 t) \\ - (t \circ_1 t) \circ_3 (t \circ_2 t) - (t \circ_1 t) \circ_3 (t \circ_3 t).$$

Terms 1, 3, 4 contain the subtree $\ell m(\alpha) = t \circ_1 t$, so we reduce them using α:

$$- t \circ_2 ((t \circ_2 t) \circ_3 t) - t \circ_2 ((t \circ_3 t) \circ_3 t) + (t \circ_3 t) \circ_2 (t \circ_3 t) \\ + (t \circ_2 t) \circ_3 (t \circ_2 t) + (t \circ_3 t) \circ_3 (t \circ_2 t) + (t \circ_2 t) \circ_3 (t \circ_3 t) \\ + (t \circ_3 t) \circ_3 (t \circ_3 t).$$

We write this polynomial in terms of trees:

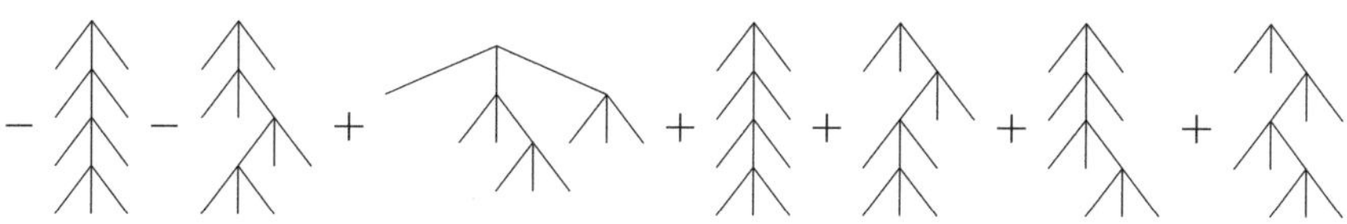

Terms 1 and 4 cancel, leaving

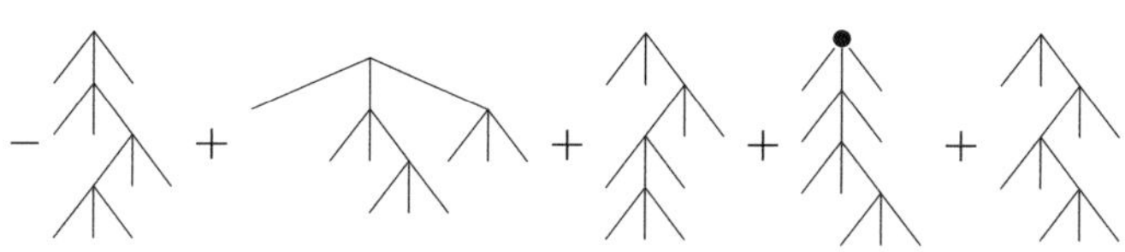

The leading monomial is divisible by $\ell m(\beta)$ but not $\ell m(\alpha)$; we reduce using β:

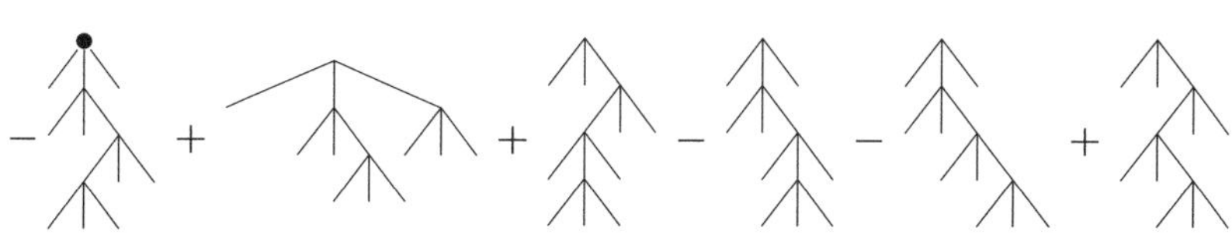

The leading monomial is divisible by α (bottom) and β (top). Using α gives

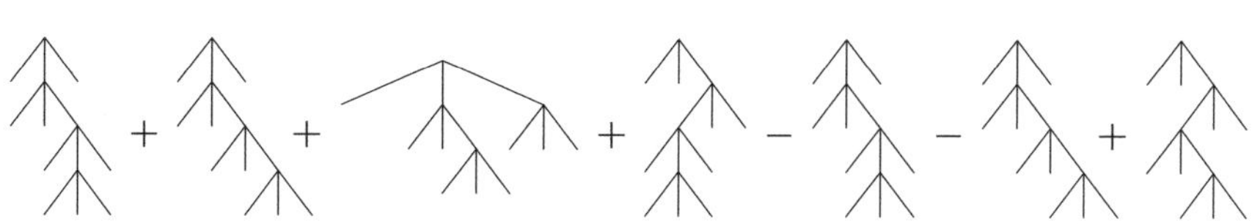

Terms 1, 5 and terms 2, 6 cancel, leaving

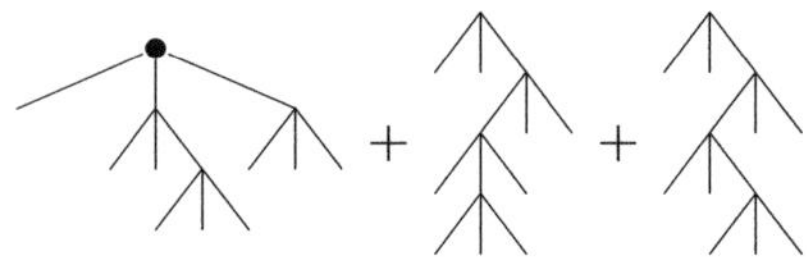

We reduce the leading monomial using β:

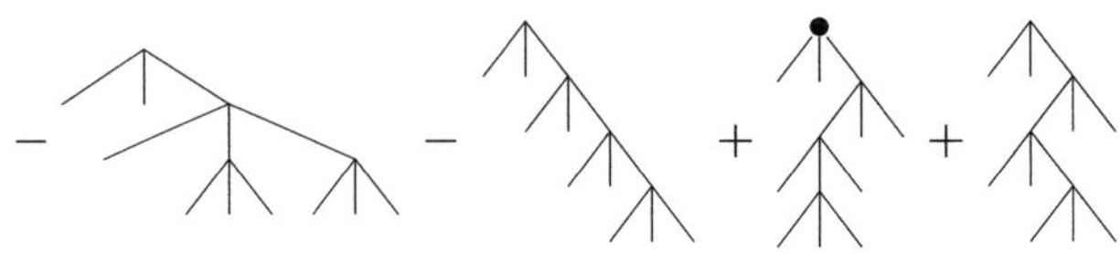

Terms 1, 2 cannot be reduced; terms 3, 4 can be reduced by α:

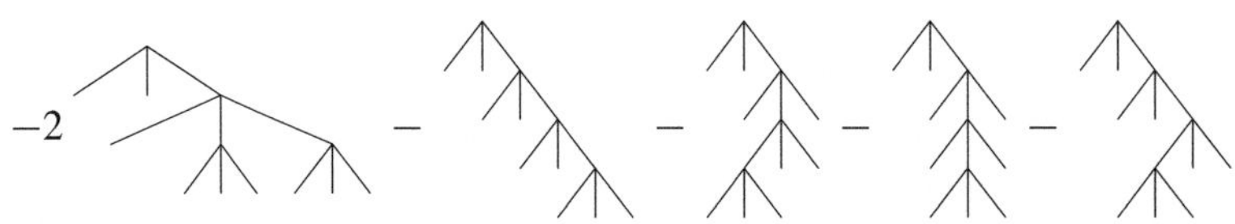

We reduce terms 3, 5 by α:

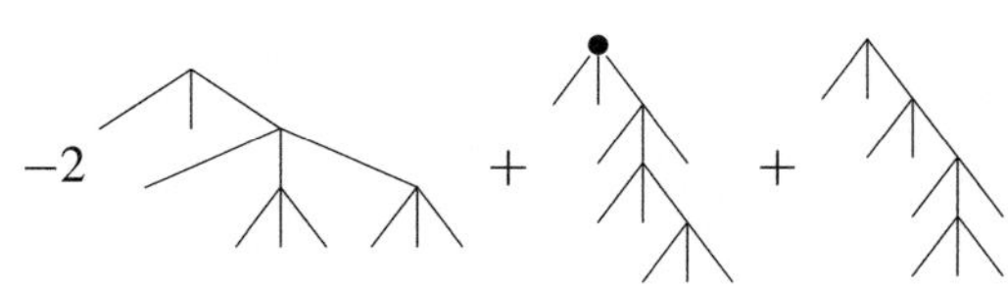

If we reduce term 2 using β, then two terms cancel and we obtain $-\gamma$. □

Lemma 6.9 *Identifying the first t of $\ell m(\alpha) = t \circ_1 t$ and the first t of $\ell m(\beta) = t \circ_2 (t \circ_3 t)$ we obtain the S-polynomial δ, and $\{\alpha, \beta, \delta\}$ is self-reduced:*

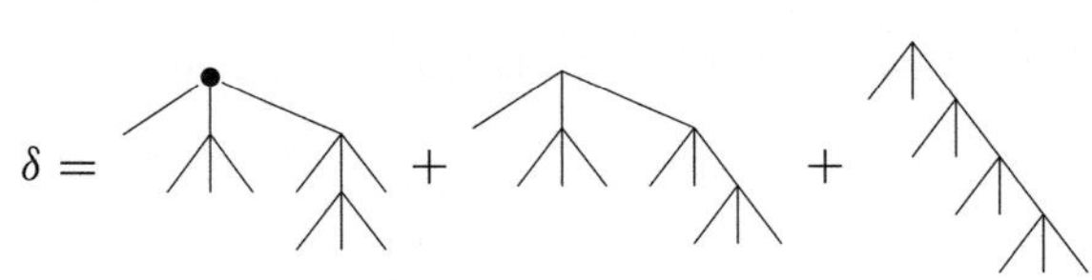

$$= (t \circ_3 (t \circ_2 t)) \circ_2 t) + (t \circ_3 (t \circ_3 t)) \circ_2 t) + (t \circ_3 (t \circ_3 (t \circ_3 t)).$$

Proof We have the equations

$$\ell m(\alpha) \circ_4 (t \circ_3 t) = (t \circ_1 t) \circ_4 (t \circ_3 t) =$$

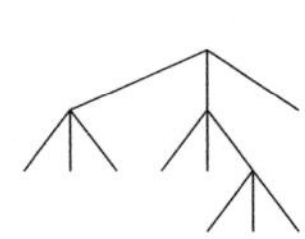

$$= (t \circ_2 (t \circ_3 t)) \circ_1 t = \ell m(\beta) \circ_1 t.$$

We apply the same partial compositions to α and β:

$$\alpha \circ_4 (t \circ_3 t) = (t \circ_1 t) \circ_4 (t \circ_3 t) + (t \circ_2 t) \circ_4 (t \circ_3 t) + (t \circ_3 t) \circ_4 (t \circ_3 t),$$
$$\beta \circ_1 t = (t \circ_2 (t \circ_3 t)) \circ_1 t + (t \circ_3 (t \circ_2 t)) \circ_1 t + (t \circ_3 (t \circ_3 t)) \circ_1 t.$$

The resulting S-polynomial is

$$(t \circ_1 t) \circ_4 (t \circ_3 t) + (t \circ_2 t) \circ_4 (t \circ_3 t) + (t \circ_3 t) \circ_4 (t \circ_3 t)$$
$$- (t \circ_2 (t \circ_3 t)) \circ_1 t - (t \circ_3 (t \circ_2 t)) \circ_1 t - (t \circ_3 (t \circ_3 t)) \circ_1 t.$$

Terms 1, 4 cancel, leaving

$$(t \circ_2 t) \circ_4 (t \circ_3 t) + (t \circ_3 t) \circ_4 (t \circ_3 t) - (t \circ_3 (t \circ_2 t)) \circ_1 t - (t \circ_3 (t \circ_3 t)) \circ_1 t$$

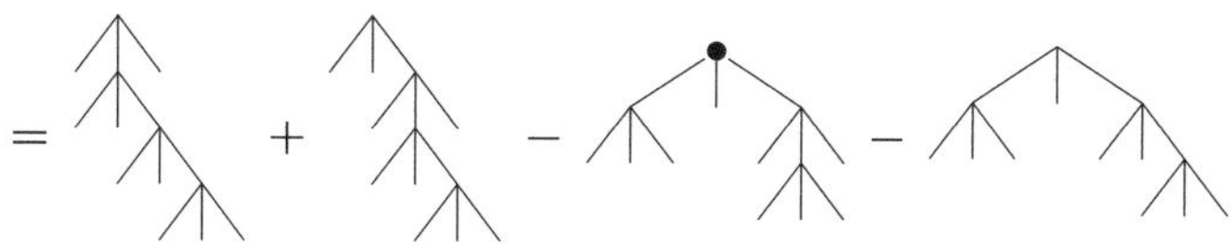

We reduce terms 3, 4 using α:

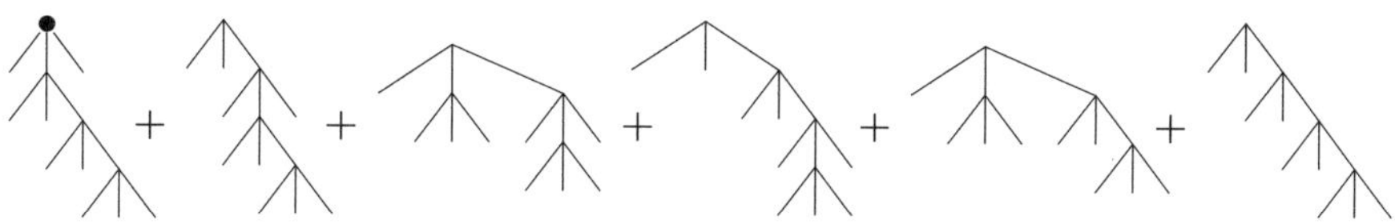

Reducing term 1 using β gives

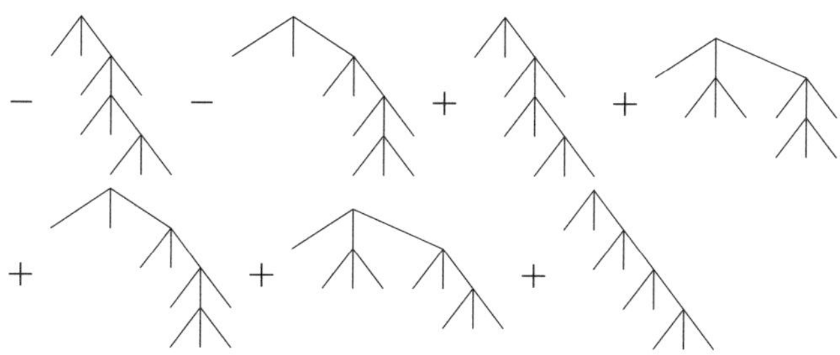

Terms 1, 3 and 2, 5 cancel; no further reduction is possible, producing δ. □

Lemma 6.10 *Identifying the first* t *of* $\ell m(\alpha) = t \circ_1 t$ *with the second* t *of* $\ell m(\beta) = t \circ_2 (t \circ_3 t)$ *we obtain the S-polynomial* ϵ *and* $\{\alpha, \beta, \epsilon\}$ *is self-reduced:*

$\epsilon =$ 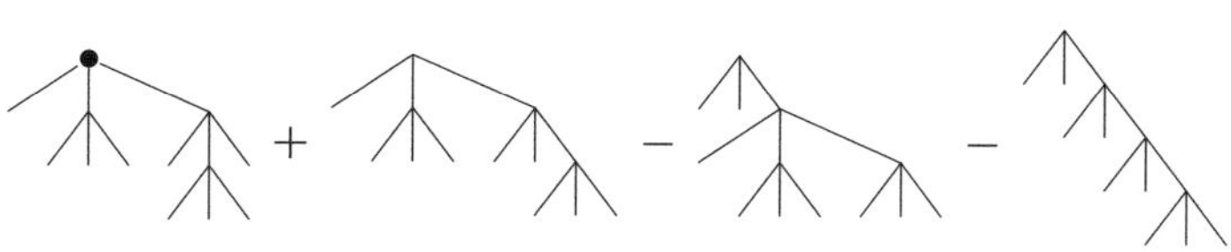

$$= ((t \circ_3 t) \circ_2 t) \circ_6 t + ((t \circ_3 t) \circ_2 t) \circ_7 t - (t \circ_3 (t \circ_3 t)) \circ_4 t - (t \circ_3 (t \circ_3 t)) \circ_7 t.$$

Proof We have the equations

$$t \circ_2 (\ell m(\alpha) \circ_5 t) = t \circ_2 ((t \circ_1 t) \circ_5 t) =$$

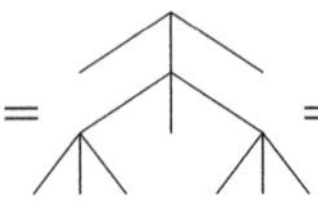

$$= (t \circ_2 (t \circ_3 t)) \circ_2 t = \ell m(\beta) \circ_2 t.$$

The resulting S-polynomial $t \circ_2 (\alpha \circ_5 t) - \beta \circ_2 t$ is

$$\begin{aligned} &t \circ_2 ((t \circ_1 t) \circ_5 t) + t \circ_2 ((t \circ_2 t) \circ_5 t) + t \circ_2 ((t \circ_3 t) \circ_5 t) \\ &- (t \circ_2 (t \circ_3 t)) \circ_2 t - (t \circ_3 (t \circ_2 t)) \circ_2 t - (t \circ_3 (t \circ_3 t)) \circ_2 t. \end{aligned}$$

Terms 1, 4 cancel, leaving

$$t \circ_2 ((t \circ_2 t) \circ_5 t) + t \circ_2 ((t \circ_3 t) \circ_5 t) - (t \circ_3 (t \circ_2 t)) \circ_2 t - (t \circ_3 (t \circ_3 t)) \circ_2 t$$

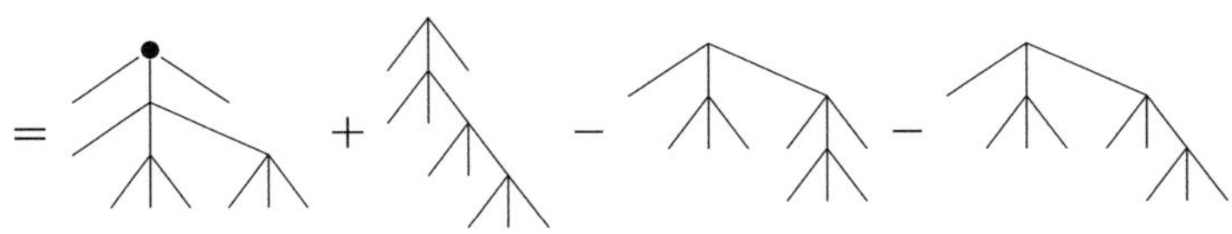

We reduce terms 1, 2 using β:

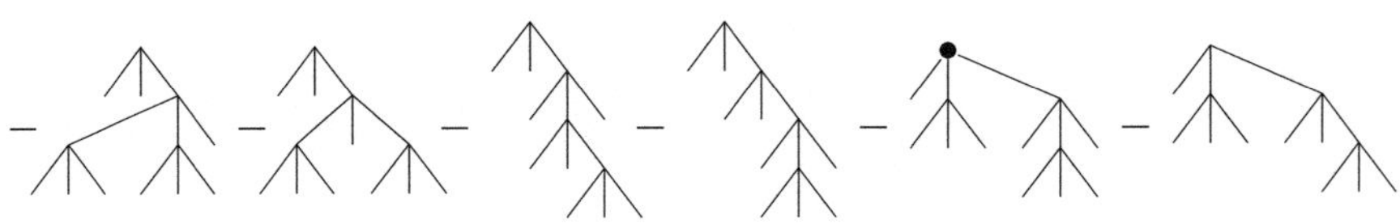

Reducing terms 1, 2 using α gives

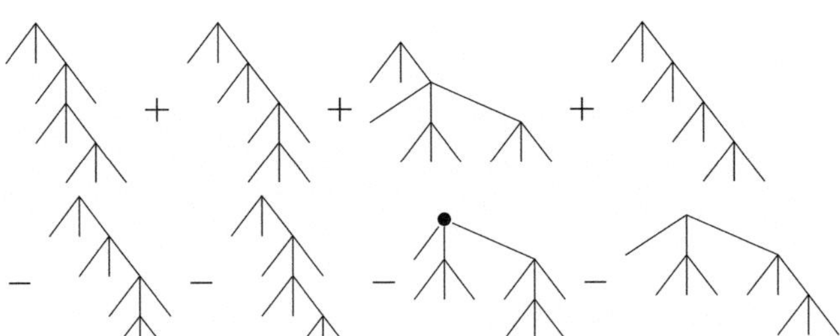

Terms 1, 6 and 2, 5 cancel. No further reduction is possible, giving $-\epsilon$. □

Lemma 6.11 *Identifying the first* t *of* $\ell m(\alpha) = t \circ_1 t$ *with the third* t *of* $\ell m(\beta) = t \circ_2 (t \circ_3 t)$ *we obtain new S-polynomial* ζ*, and* $\{\alpha, \beta, \zeta\}$ *is self-reduced:*

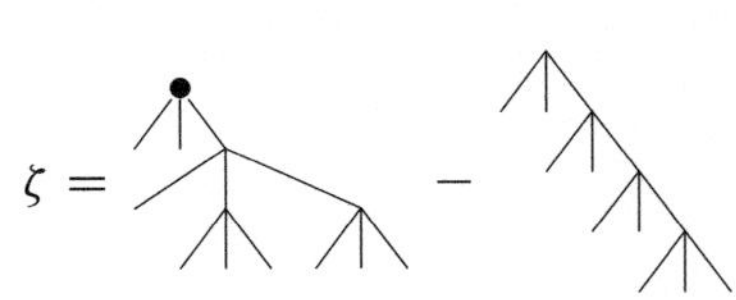

$$= t \circ_3 ((t \circ_2 t) \circ_5 t) - t \circ_3 (t \circ_3 (t \circ_3 t)).$$

Proof We have the equations

$$(t \circ_2 t) \circ_4 \ell m(\alpha) = (t \circ_2 t) \circ_4 (t \circ_1 t) =$$  $$= (t \circ_2 (t \circ_3 t)) \circ_4 t = \ell m(\beta) \circ_4 t.$$

The resulting S-polynomial $(t \circ_2 t) \circ_4 \alpha - \beta \circ_4 t$ is

$$\begin{aligned}&(t \circ_2 t) \circ_4 (t \circ_1 t) + (t \circ_2 t) \circ_4 (t \circ_2 t) + (t \circ_2 t) \circ_4 (t \circ_3 t)\\&- (t \circ_2 (t \circ_3 t)) \circ_4 t - (t \circ_3 (t \circ_2 t)) \circ_4 t - (t \circ_3 (t \circ_3 t)) \circ_4 t.\end{aligned}$$

Terms 1, 4 cancel, leaving

$$(t \circ_2 t) \circ_4 (t \circ_2 t) + (t \circ_2 t) \circ_4 (t \circ_3 t) - (t \circ_3 (t \circ_2 t)) \circ_4 t - (t \circ_3 (t \circ_3 t)) \circ_4 t$$

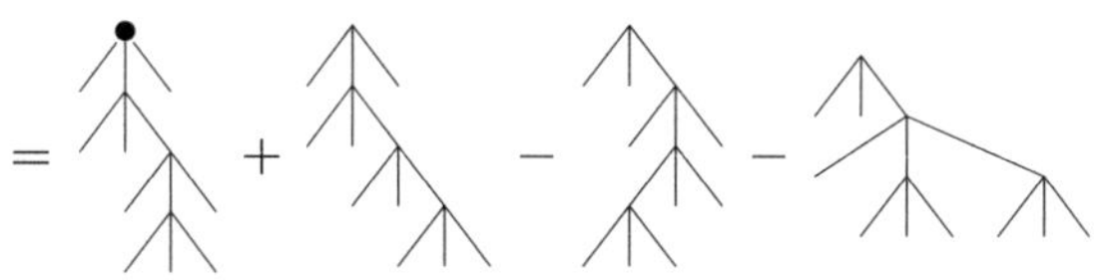

We use β to reduce terms 1, 2:

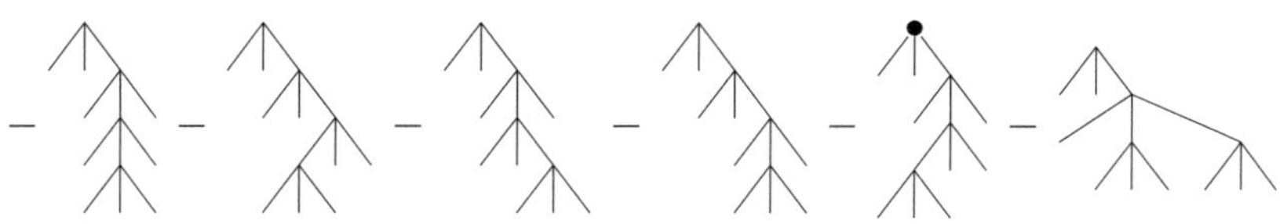

We use α to reduce terms 2, 5:

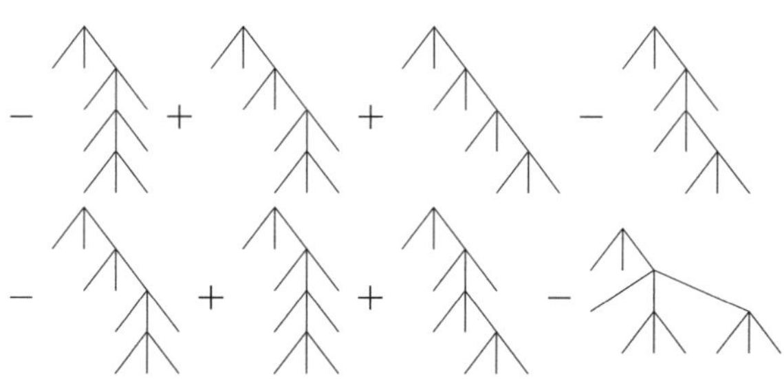

Terms 1, 6 and 2, 5 and 4, 7 cancel. No further reductions are possible, and the monic form of the last polynomial is ζ. □

Lemma 6.12 *The polynomials γ, δ, ϵ, ζ span a subspace with basis η, θ, ν.*

Proof It is easy to see that

$$\eta = \frac{1}{3}(\gamma + \delta + 2\epsilon), \qquad \theta = \frac{1}{3}(2\gamma - \delta + \epsilon), \qquad \nu = -\frac{1}{3}(\gamma - 2\delta + 2\epsilon),$$

and that these three polynomials form a basis of $\text{span}(\gamma, \delta, \epsilon, \zeta)$. □

Lemma 6.13 *Every S-polynomial obtained from α, β, η, θ, ν reduces to 0.*

Proof If either f or g is a monomial then clearly every S-polynomial obtained from f and g reduces to 0. We have already considered S-polynomials from α and β; the other cases are α, η and β, η and η, η with many subcases. We give details for the most difficult subcase and leave the others as exercises. These calculations can be simplified using the triangle lemma for nonsymmetric operads [2, Prop. 3.5.3.2].

We identify the second t of $\ell m(\alpha)$ with the first t of $\ell m(\eta)$ and obtain this SCM:

$$\ell m(\alpha) = t \circ_1 t, \quad \ell m(\eta) = (t \circ_2 t) \circ_5 (t \circ_2 t), \quad (\ell m(\alpha) \circ_2 t) \circ_5 (t \circ_2 t) = t \circ_1 \ell m(\eta).$$

To save space, we switch to nonassociative notation. We obtain the S-polynomial

$$\begin{aligned}
&(\alpha \circ_2 t) \circ_5 (t \circ_2 t) - t \circ_1 \eta = \\
&(*((***)(*(***)*)*)*) + (*(***)((*(***)*)**)) - ((*(***)(**(***))**).
\end{aligned}$$

Rewrite rules (6.2) and (6.3) have this form; the letters represent submonomials:

$$\begin{aligned}
((vwx)yz) &\longmapsto -(v(wxy)z) - (vw(xyz)), && (6.4)\\
(t(uv(wxy))z) &\longmapsto -(tu(v(wxy)z)) - (tu(vw(xyz))). && (6.5)
\end{aligned}$$

When we apply (6.4) or (6.5), we use bars to indicate the submonomials. To begin we reduce all three monomials in the S-polynomial using α and obtain

$$\begin{aligned}
&(*((\bar{*}\bar{*}\bar{*})\overline{(*(***)*)}\bar{*})*) + (*(***)((\bar{*}\overline{(***)}\bar{*})\bar{*}\bar{*})) - ((\bar{*}\overline{(***)}\,\overline{(**(***))}\bar{*}\bar{*}) = \\
&-(*(*(**(*(***)*))*)*) - (*(**(*(*(***)*)*))*) - (*(***)(*((\bar{*}\bar{*}\bar{*})\bar{*}\bar{*})*)) \\
&-(*(***)(*(***)(***))) + (*((\bar{*}\bar{*}\bar{*})\overline{(**(***))}\bar{*})*) + (*(***)((\bar{*}\bar{*}\overline{(***)})\bar{*}\bar{*})).
\end{aligned}$$

Terms 3, 5, 6 reduce using α as indicated; term 4 is $\theta \circ_2 t$ and reduces to 0:

$$\begin{aligned}
&-(*(\bar{*}(\bar{*}\bar{*}(\bar{*}\overline{(***)})\bar{*})\bar{*})*) - (\bar{*}(\bar{*}\bar{*}(\bar{*}\overline{(*(***)*)}\bar{*}))\bar{*}) + (*(***)(*(*(***)*)*)) \\
&+(*(***)(\bar{*}(\bar{*}\bar{*}(\bar{*}\bar{*}\bar{*}))\bar{*})) - (*(\bar{*}(\bar{*}\bar{*}(\bar{*}\bar{*}\overline{(***)}))\bar{*})*) - (\bar{*}(\bar{*}\bar{*}(\bar{*}\overline{(**(***))})\bar{*})\bar{*}) \\
&-(*(***)(*(*(***)*)*)) - (*(***)(**(***)*)*)).
\end{aligned}$$

Terms 3, 7 cancel, and terms 1, 2, 4, 5, 6 reduce using β as indicated:

$$
\begin{aligned}
&(\bar{*}(\bar{*}\bar{*}(\bar{*}(\overline{*(***)*}))\bar{*})\bar{*}) + (\bar{*}(\bar{*}\bar{*}(\bar{*}\bar{*}(\overline{(***)**})))\bar{*}) + (**(*(*(*(***)*)*)*)) \\
&+ (**(**((*(***)*)**)) - (*(***)(**(*(***)*))) - (*(***)(**(**(***)))) \\
&+ (\bar{*}(\bar{*}\bar{*}(\bar{*}(\overline{**(***)})\bar{*}))\bar{*}) + (\bar{*}(\bar{*}\bar{*}(\bar{*}\bar{*}(\overline{*(***)*}))\bar{*}) + (**(*(*(**(***))*))*) \\
&+ (**(**((**(***))**))) - (*(***)(**((***)**))).
\end{aligned}
$$

Terms $1, 2, 7, 8$ reduce using β as indicated; term 6 is $\nu \circ_2 t$ and reduces to 0; omitting terms which cancel, we obtain

$$
\begin{aligned}
&(**(*(**((\bar{*}\bar{*}\bar{*})\bar{*}\bar{*}))*)) + (**(**(*((\bar{*}\bar{*}\bar{*})\bar{*}\bar{*})*))) - (*(***)(**(*(***)*))) \\
&- (**(*(*(**(***))*)*)) - (**(*(**(*(***)*))*)) - (**(**(*(*(***))*)*)) \\
&+ (**(**((**(***))**))) - (*(***)(**((\bar{*}\bar{*}\bar{*})\bar{*}\bar{*}))).
\end{aligned}
$$

Terms $1, 2, 8$ reduce using α as indicated; omitting terms which cancel, we obtain

$$
\begin{aligned}
&- 2(**(\bar{*}(\bar{*}\bar{*}(\bar{*}(\overline{***})\bar{*}))\bar{*})) - 2(**(**(*(*(***)*)*))) - (**(\bar{*}(\bar{*}\bar{*}(\bar{*}\bar{*}(\overline{***})))\bar{*})) \\
&- (**(**(\bar{*}(\bar{*}\bar{*}(\bar{*}\bar{*}\bar{*}))\bar{*}))).
\end{aligned}
$$

Terms $1, 3, 4$ reduce using β as indicated. Some terms cancel, and others reduce to 0 using ν, leaving the single term $(**(**(\bar{*}(\bar{*}\bar{*}(\bar{*}\bar{*}\bar{*}))\bar{*})))$. We reduce using β and then both terms reduce to 0 using ν. □

We use Theorem 6.2 to calculate the dimensions of the homogeneous components of the ternary partially associative operad $\mathcal{TPA} = \mathcal{LT} / \langle \alpha \rangle$ with an operation of even (homological) degree. Theorem 6.3 below implies the conditional result of Goze and Remm [11, Theorem 15]; our proof using Gröbner bases is much simpler. For a more general conjecture, see [2, Conjecture 10.3.2.6, case 6].

Lemma 6.14 *For* $n = 1, 3, 5, 7$ *we have*

$$
\dim \mathcal{TPA}(1) = \dim \mathcal{TPA}(3) = 1, \qquad \dim \mathcal{TPA}(5) = 2, \qquad \dim \mathcal{TPA}(7) = 4.
$$

For $\mathcal{TPA}(5)$ *a monomial basis in increasing path-lex order is*

For $\mathcal{TPA}(7)$ *a monomial basis in increasing path-lex order is*

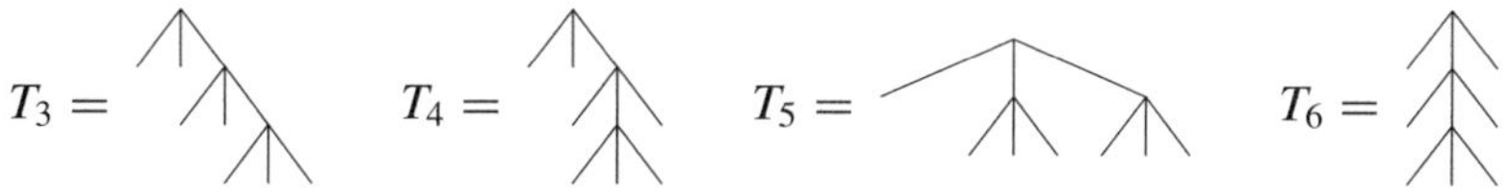

Proof The case $n = 1$ is trivial, and for $n = 3$ we have only the basic tree t. For $n = 5$, the monomial $t \circ_1 t$ reduces by α, leaving only $T_1 = t \circ_2 t$ and $T_2 = t \circ_3 t$.

For $n = 7$, we have (i) $T_1 \circ_i t$: if $i = 1, 3$ the result reduces by α, and if $i = 2, 4, 5$ we obtain T_5, T_4, T_3; (ii) $t \circ_i T_1$: if $i = 1, 2$ the result reduces by α, β, and if $i = 3$ we obtain T_3; (iii) $T_2 \circ_i t$: if $i = 1, 2$ the result reduces by α, if $i = 3$ we obtain T_6, if $i = 4$ the result reduces by β, and if $i = 5$ we obtain T_5; (iv) $t \circ_i T_2$: if $i = 1$ the result reduces by α, if $i = 2, 3$ we obtain T_6, T_4. Clearly $T_3, \ldots, T_6$ cannot be reduced using α or β, which proves linear independence. □

Theorem 6.3 *For weight $k \geq 3$ we have* $\dim \mathcal{TPA}(2k+1) = k+1$.

Proof Let M_0 be the tree with one vertex, set $M_1 = t$, and for $\ell \geq 2$ set

$$M_\ell = t \circ_2 (t \circ_2 (t \circ_2 \cdots (t \circ_2 t) \cdots)) \qquad (\ell \text{ copies of } t).$$

Consider the following $k+1$ monomials of weight k in increasing path-lex order; to save space we write $\boxed{\ell} = M_\ell$:

f_1	f_2	f_i $(3 \leq i \leq k-1)$	f_k	f_{k+1}
$(k, 3, k-2)$	$(k, 2, k-1)$	$(k, 3, k-2)$	$(k, 2, k-1)$	$(k, 1, k)$

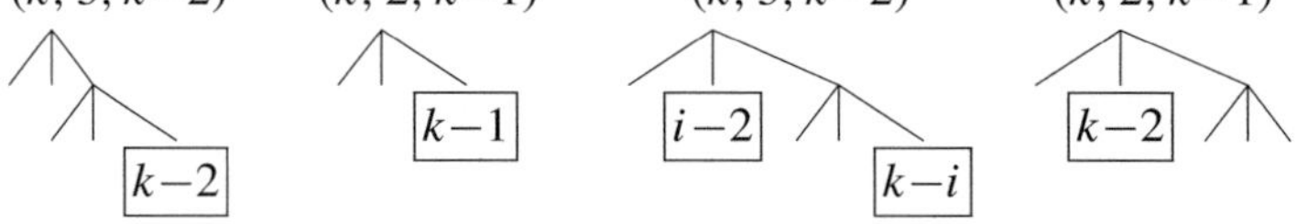

We say a leaf is left (middle, right) if it is the left (middle, right) child of its parent. The ordered triples above give the number of left (middle, right) leaves. We have $f_1 = t \circ_3 (t \circ_3 M_{k-2})$, $f_2 = t \circ_3 M_{k-1}$, and

$$f_i = (t \circ_3 (t \circ_3 M_{k-i})) \circ_2 M_{i-2} \quad (3 \leq i \leq k).$$

For $3 \leq i \leq k-1$, we obtain f_{i+1} from f_i by moving the bottom t of the right–right subtree to the middle subtree. We will show that $f_1, \ldots, f_{k+1}$ form a basis of $\mathcal{TPA}(2k+1)$. For linear independence, we simply observe that no f_i $(1 \leq i \leq k)$ can be reduced using any Gröbner basis element α, β, η, θ, ν.

To prove that $f_1, \ldots, f_{k+1}$ span $\mathcal{TPA}(2k+1)$ we use induction on $k \geq 3$. Basis: Lemma 6.14 gives $f_1 = T_3$, $f_2 = T_4$, $f_3 = T_5$, $f_4 = T_6$. Induction: Assume that $f_1, \ldots, f_{k+1}$ span $\mathcal{TPA}(2k+1)$ and write $f'_1, \ldots, f'_{k+2}$ for the monomials of weight $k+1$. For each f_i in $\mathcal{TPA}(2k+1)$ we obtain monomials of weight $k+1$ in two ways:

(1) $t \circ_j f_i$ for $j \in [3], i \in [k+1]$; (2) $f_i \circ_j t$ for $i \in [k+1], j \in [2k+1]$.

Case 1: If $j = 1$ then $t \circ_1 f_i$ reduces by α. If $j = 2$ then $t \circ_2 f_i$ reduces by β for $i \in [k]$, and $t \circ_2 f_{k+1} = M_{k+1} = f'_{k+2}$. If $j = 3$ then $t \circ_3 f_1$ reduces using ν, $t \circ_3 f_2 = f'_1$, $t \circ_3 f_i$ reduces using θ for $i \in [k]$, and $t \circ_3 f_{k+1} = f'_2$.

Case 2 has three subcases depending on where we attach t. If we attach to a left leaf of f_i then the result reduces by α. If we attach to a right leaf then for f_1 the result reduces by ν or β, for $f_2, \ldots, f_k$ the result reduces by β or θ, and for f_{k+1} either we obtain f'_{k+1} or the result reduces by β. If we attach to a middle leaf of f_1 then we obtain either f'_3 or f'_1 or the result reduces by θ. If we attach to a middle leaf of f_2 then we obtain either f'_3 or f'_2. If we attach to a middle leaf of f_i for $3 \leq i \leq k$ then we obtain f'_j for $3 \leq j \leq k{+}1$ or the result reduces by θ. If we attach to the middle leaf of f_{k+1} then we obtain f'_{k+2}. □

We now assume that the ternary operation t has odd (homological) degree. Thus every tree has even or odd parity depending the number of internal nodes. We write $|f| \in \{0, 1\}$ for the parity of f. We must include Koszul signs in the relations for partial compositions: transposing two odd elements introduces a minus sign.

Lemma 6.15 ([12, Def. 1.1]) *If* $p, q, r \in \mathcal{T}$ *then*

$$(p \circ_i q) \circ_j r = \begin{cases} p \circ_i (q \circ_{j-i+1} r) & i \leq j \leq i + \ell(q) - 1 \\ (-1)^{|q||r|} (p \circ_{j-\ell(q)+1} r) \circ_i q & i + \ell(q) \leq j \leq \ell(p) + \ell(q) - 1 \\ (-1)^{|q||r|} (p \circ_j r) \circ_{i+\ell(r)-1} q & 1 \leq j \leq i - 1 \end{cases}$$

Theorem 6.4 *The relation* α *is a Gröbner basis for* $\langle \alpha \rangle$ *in the free nonsymmetric operad with a ternary operation of odd homological degree.*

Proof The first few steps are identical to those for an even operation. The leading monomial $\ell m(\alpha) = t \circ_1 t$ overlaps with itself in one way to produce this SCM:

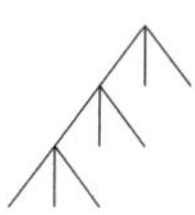

$$= \ell m(\alpha) \circ_1 t \;=\; t \circ_1 \ell m(\alpha).$$

We apply the same partial compositions to α instead of $\ell m(\alpha)$:

$$\alpha \circ_1 t = (t \circ_1 t) \circ_1 t + (t \circ_2 t) \circ_1 t + (t \circ_3 t) \circ_1 t$$

$=$ 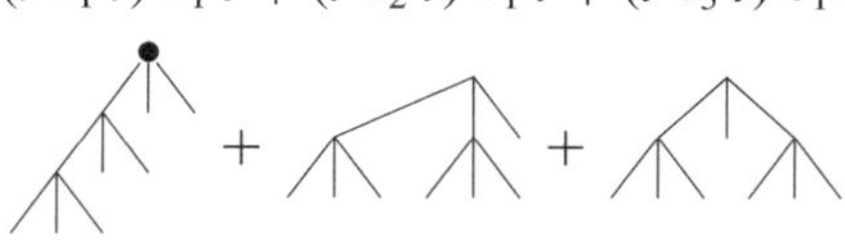

$$t \circ_1 \alpha = t \circ_1 (t \circ_1 t) + t \circ_1 (t \circ_2 t) + t \circ_1 (t \circ_3 t)$$

$=$ 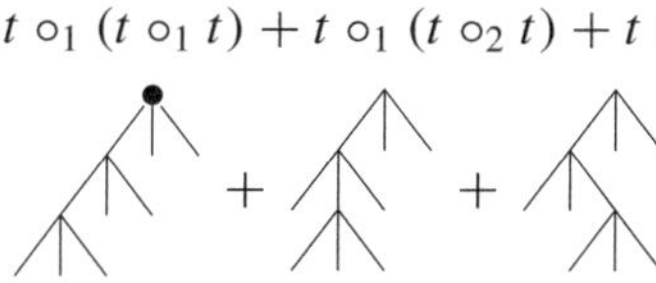

The difference is this (non-reduced) S-polynomial:

$$\alpha \circ_1 t - t \circ_1 \alpha = (t \circ_2 t) \circ_1 t + (t \circ_3 t) \circ_1 t - t \circ_1 (t \circ_2 t) - t \circ_1 (t \circ_3 t) =$$

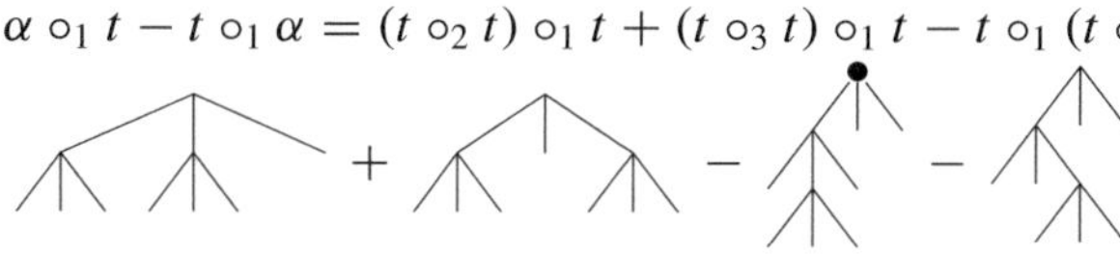 (6.6)

Lemma 6.15 (case 3), $p = q = r = t$, with $i, j = 2, 1$ and $i, j = 3, 1$ gives

$$(t \circ_2 t) \circ_1 t = -(t \circ_1 t) \circ_4 t, \qquad (t \circ_3 t) \circ_1 t = -(t \circ_1 t) \circ_5 t.$$

Lemma 6.15 (case 1), $p = q = r = t$, with $i, j = 2, 2$ and $i, j = 1, 3$ gives

$$-t \circ_1 (t \circ_2 t) = -(t \circ_1 t) \circ_2 t. \qquad -t \circ_1 (t \circ_3 t) = -(t \circ_1 t) \circ_3 t.$$

Therefore (6.6) equals

$$-(t \circ_1 t) \circ_4 t - (t \circ_1 t) \circ_5 t - (t \circ_1 t) \circ_2 t - (t \circ_1 t) \circ_3 t.$$

We reduce each monomial using α and obtain

$$\begin{aligned}&(t \circ_2 t) \circ_4 t + (t \circ_3 t) \circ_4 t + (t \circ_2 t) \circ_5 t + (t \circ_3 t) \circ_5 t \\ &+ (t \circ_2 t) \circ_2 t + (t \circ_3 t) \circ_2 t + (t \circ_2 t) \circ_3 t + (t \circ_3 t) \circ_3 t.\end{aligned}$$

Terms 3, 6 cancel by Lemma 6.15 (case 2), $(t \circ_2 t) \circ_5 t = -(t \circ_3 t) \circ_2 t$, leaving

$$\begin{aligned}&(t \circ_2 t) \circ_4 t + (t \circ_3 t) \circ_4 t + (t \circ_3 t) \circ_5 t \\ &\quad + (t \circ_2 t) \circ_2 t + (t \circ_2 t) \circ_3 t + (t \circ_3 t) \circ_3 t\end{aligned}$$

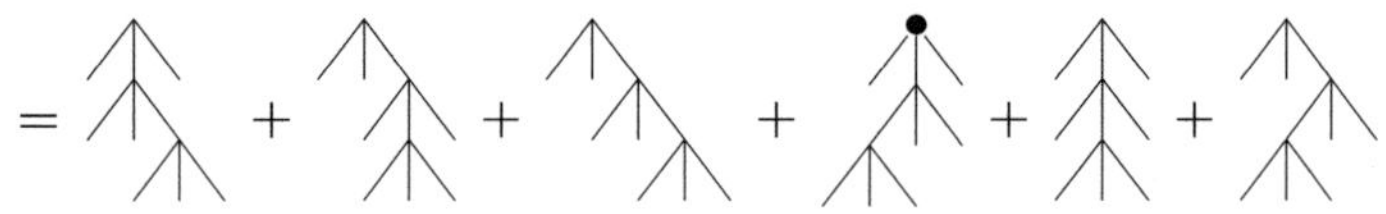

We reduce terms 4, 6 using α; this cancels terms 1, 5 and terms 2, 3. □

Theorem 6.5 *For an odd operation, the dimension of the ternary partially associative operad is the binary Catalan number (in the weight grading).*

Proof Relation α shows that any left subtree reduces, so the dimension for weight w is the number of binary trees of weight w. □

Acknowledgements This research was supported by the Discovery Grant *Algebraic Operads* (2016) from the Natural Sciences and Engineering Research Council of Canada (NSERC). The authors thank Vladimir Dotsenko for sending us his results without proofs on Gröbner bases and dimension formulas for ternary partially associative operads (emails of 16 and 26 February 2018) and for helpful comments on an earlier version of this paper.

References

1. M.R. Bremner, On free partially associative triple systems. Commun. Algebra **28**(4), 2131–2145 (2000)
2. M.R. Bremner, V. Dotsenko, *Algebraic Operads: An Algorithmic Companion* (Chapman and Hall/CRC, Boca Raton, 2016)

3. F. Chapoton, Sur une opérade ternaire liée aux treillis de Tamari. Ann. Fac. Sci. Toulouse Math. (6) **20**(4), 843–869 (2011)
4. C. De Concini, C. Procesi, Wonderful models of subspace arrangements. Sel. Math. (N.S.) **1**(3), 459–494 (1995)
5. V. Dotsenko, A. Khoroshkin, Gröbner bases for operads. Duke Math. J. **153**(2), 363–396 (2010)
6. V. Dotsenko, B. Vallette, Higher Koszul duality for associative algebras. Glasg. Math. J. **55**(A), 55–74 (2013)
7. V. Dotsenko, M. Markl, E. Remm, Non-Koszulness of operads and positivity of Poincaré series, arXiv:1604.08580 [math.KT] (submitted 28 April 2016)
8. V. Dotsenko, S. Shadrin, B. Vallette, Toric varieties of Loday's associahedra and noncommutative cohomological field theories. J. Topology **12**, 463–535 (2019)
9. A.V. Gnedbaye, Les algèbres k-aires et leurs opérades. C. R. Acad. Sci. Paris Sér. I Math. **321**(2), 147–152 (1995)
10. A.V. Gnedbaye, Opérades des algèbres $(k+1)$-aires, in *Operads: Proceedings of Renaissance Conferences*, ed. by J.-L. Loday, J.D. Stasheff, A.A. Voronov. Contemporary Mathematics, vol. 202 (American Mathematical Society, Providence, 1997), pp. 83–113
11. N. Goze, E. Remm, Dimension theorem for free ternary partially associative algebras and applications. J. Algebra **348**, 14–36 (2011)
12. M. Markl, Models for operads. Commun. Algebra **24**(4), 1471–1500 (1996)
13. M. Markl, E. Remm, Operads for n-ary algebras: calculations and conjectures. Arch. Math. (Brno) **47**(5), 377–387 (2011)
14. M. Markl, E. Remm, (Non-)Koszulness of operads for n-ary algebras, galgalim and other curiosities. J. Homotopy Relat. Struct. **10**(4), 939–969 (2015)
15. E. Remm, On the non-Koszulity of the ternary partially associative operad. Proc. Est. Acad. Sci. **59**(4), 355–363 (2010)

Chapter 7
A Survey on Koszul Algebras and Koszul Duality

Neeraj Kumar

7.1 Introduction

Let k be a field and A a standard graded k-algebra. We say that A is Koszul if the minimal graded free A-resolution of k has only linear maps. The study of Koszul algebras has greatly been accelerated in the past three decades, with an influx of new tools and ideas coming from diverse areas of mathematics. In this survey paper, we will discuss the appearance of Koszul algebras and Koszul duality phenomena in the literature in various fronts of mathematical topics.

There are many survey articles on this topic, namely on the theory of Koszul algebras due to Fröberg [22], on Koszul algebras and Gröbner basis of quadrics due to Conca [14], on Koszul algebras and regularity due to Conca, DeNegri and Rossi [15], Koszul algebras and their syzygies due to Conca [13], Koszul algebras in non-commutative settings due to Martínez-Villa [31], and a book containing several facts about Koszul algebras and Koszul duality due to Polishchuk [37].

Keeping in mind all these survey articles, we have tried to give an exposition on the topics that developed at a later stage. However, for the sake of completeness we have listed many old results and new results from these survey articles too for the optimal synchronization of results for an easy reading. Section 7.1 contains preliminaries, to set up the notation and basis terminologies of the paper, e.g. free resolution, Betti numbers, Hilbert series and Poincaré–Betti series, etc., with several examples. In Sect. 7.2, Koszul algebras are defined and several equivalent interpretation of it is discussed. We discuss Fröberg formula which gives a relation describing the Koszulness via Hilbert series and Poincaré–Betti series. Then we define quadratic dual algebra and discuss several classical examples. The notion of Koszul duality is discussed in Sect. 7.3. In Sect. 7.5, we reprove a theorem of Tate on the Poincaré–

N. Kumar (✉)
Department of Mathematics, Indian Institute of Technology Hyderabad, Sangareddy, Kandi 502285, India
e-mail: neeraj@iith.ac.in

A. A. Ambily et al. (eds.), *Leavitt Path Algebras and Classical K-Theory*,
Indian Statistical Institute Series,
https://doi.org/10.1007/978-981-15-1611-5_7

Betti series of quadratic complete intersection ring. Section 7.6 is devoted to highlight the study of Koszul algebras in combinatorics such as Stanley–Reisner rings, finite simple graphs, Hilbert series of algebras arising from finite directed graphs. The classical results on the Koszul algebras for combinatorial rational varieties and non-combinatorial non-rational varieties are discussed in Sect. 7.6. In this section, we discuss in detail the concept of diagonal subalgebra, recall classical results, construct defining equations for the diagonal subalgebras, and present recent published results. For the Koszul phenomena in the affine semigroup rings and for monomial projective curve, see Sect. 7.7. We end this survey article with an open problem due to Reiner and Welker: For which Koszul algebras, is the Hilbert series a Polya frequency sequence?

7.2 Preliminaries

Notations and Basic Facts

Let k be a field, and V be a finite-dimensional vector space over k with basis $\{x_1, \ldots, x_n\}$. The notation $V^{\otimes n}$ is short for $V \otimes_{\mathsf{k}} V \otimes_{\mathsf{k}} \cdots \otimes_{\mathsf{k}} V$ (n factors). Let $\mathrm{T(V)} = \oplus_{i \geq 0} V^{\otimes i}$ be the *tensor algebra* of V over k. The tensor algebra (or the non-commutative polynomial ring) is a graded k-algebra with $\dim_{\mathsf{k}} V^{\otimes i} = n^i$. We will use the notation $\mathsf{k}\langle x_1, \ldots, x_n \rangle$ for T(V).

Let $V^* = \mathrm{Hom}(V, \mathsf{k})$ be a dual k-vector space of V via $(af)(v) = af(v)$ for any $f \in V^*$, $v \in V$ and $a \in \mathsf{k}$. One may identify $V^* \otimes_{\mathsf{k}} V^*$ with $V \otimes_{\mathsf{k}} V$ by the rule $(f \otimes g)(v \otimes w) = f(v)g(w)$ for all $f, g \in V^*$ and $v, w \in V$. Inductively, one may identify $(V^*)^{\otimes j}$ with $(V^{\otimes j})^*$ for any $j \in \mathbb{N}$.

Let A be a finitely generated *graded* k*-algebra*, i.e., $A = \oplus_{i \geq 0} A_i$ is a graded ring with $A_0 = \mathsf{k}$. Let $A_+ = \oplus_{i \geq 1} A_i$ be the graded maximal ideal of A. We say that A is *standard graded* if A_+ is generated by A_1.

By a graded (non-commutative) k-algebra A, we mean an algebra of the form $A = \mathsf{k}\langle x_1, \ldots, x_n \rangle / I$ or simply $\mathrm{T(V)}\,/I$, where I is a two-sided ideal generated by homogeneous elements. By a graded (commutative) k-algebra A, we will mean an algebra of the form $A = \mathsf{k}[x_1, \ldots, x_n]/I$, where I is a homogeneous ideal in the polynomial ring $\mathsf{k}[x_1, \ldots, x_n]$. The graded algebra A is called quadratic if I is generated by elements of degree two.

Given a finitely generated graded A-module M and $j \in \mathbb{Z}$, the *shifted module* $M(j)$ is the graded A-module with ith graded component $M(j)_i = M_{i+j}$. In particular, $A(-j) = \oplus_{i \geq j} A_{i-j}$ is the graded free A-module of rank one, with generator in degree j.

Hilbert Series

The Hilbert series $\mathbb{F}_A(z)$ of a graded algebra $A = \oplus_{i \geq 0} A_i$ is a formal sum

$$F_A(z) = \sum_{i \geq 0} (\dim_{\mathsf{k}} A_i)\ z^i.$$

Since A is finitely generated, the dimension of A_i as a k- vector space is finite for each i, and so the Hilbert series $F_A(z)$ is well defined. Similarly, If $M = \oplus_{i\in\mathbb{Z}} M_i$ is finitely generated graded A-module, then it has a well-defined Hilbert series $F_M(z) = \sum_{i\in\mathbb{Z}} (\dim_{\mathsf{k}} M_i)\ z^i$. It is well known that if A is a commutative graded algebra, then $F_A(z)$ is a rational function.

Example 7.1 (i) If $A = \mathsf{k}\langle x_1, \ldots, x_n\rangle$, then $\dim_{\mathsf{k}} A_i = n^i$, so the Hilbert series $F_A(z) = 1 + nz + n^2z^2 + \cdots = \frac{1}{1-nz}$.

(ii) If $A = \mathsf{k}[x_1, \ldots, x_n]$, then $\dim_{\mathsf{k}} A_i = \binom{i+n-1}{n-1}$, and so $F_A(z) = \frac{1}{(1-z)^n}$.
(iii) If $A = \mathsf{k}[x_1, \ldots, x_n]/I$, where $I = (x_ix_j \mid i \neq j,\ i, j \in \{1, \ldots, n\})$, then $\dim_{\mathsf{k}} A_i = n$ and $F_A(z) = 1 + nz + nz^2 + \cdots = \frac{1+(n-1)z}{1-z}$.

Free Resolution and Betti Numbers

To describe certain structures of a module, Hilbert introduced the idea of associating a *free resolution* to a finitely generated module [28]. A minimal graded free resolution of a finitely generated A-module M is a complex of free A-modules

$$F_\bullet \quad \cdots \longrightarrow F_n \xrightarrow{\psi_n} \cdots\cdots \xrightarrow{\psi_2} F_1 \xrightarrow{\psi_1} F_0 \xrightarrow{\psi_0} 0,$$

such that $H_0(F_\bullet) = M$, $H_i(F_\bullet) = 0$ for all $i > 0$, and $\psi_{i+1}(F_{i+1}) \subseteq A_+F_i$ for every i. Such a minimal graded free resolution always exists and it is unique up to an isomorphism of complexes. The minimal graded free resolution of an A-module M in terms of shifted A-modules is

$$F_\bullet \quad \cdots \longrightarrow \oplus_j\ A(-j)^{\beta_{ij}} \longrightarrow \cdots \longrightarrow \oplus_j\ A(-j)^{\beta_{1,j}} \longrightarrow \oplus_j\ A(-j)^{\beta_{0,j}} \longrightarrow 0,$$

where $A(-j)$ is the graded free A-module $A(-j) = \oplus_{i\geq j} A_{i-j}$. Since the resolution $F_\bullet$ is *graded*, it means that the differential maps ψ_i of the complex are homogeneous homomorphisms of degree zero or equivalently non-zero entries of the matrices corresponding to differentials are homogeneous. The minimality condition $\psi_{i+1}(F_{i+1}) \subseteq A_+F_i$ for every i is equivalent to the fact that $F_\bullet \otimes_A \mathsf{k}$ has differential zero, that is, all non-zero entries corresponding to differentials have positive degrees. Hence by construction

$$\mathrm{Tor}_i^A(M, \mathsf{k}) = H_i(F_\bullet \otimes_A \mathsf{k}) = F_i \otimes_A \mathsf{k} = A^{\beta_i^A(M)},$$

where $\beta_i^A(M) = \dim_{\mathsf{k}} \mathrm{Tor}_i^A(M, \mathsf{k})$, and $\beta_{ij}^A(M) = \dim_{\mathsf{k}} \mathrm{Tor}_i^A(M, \mathsf{k})_j$ are called ith *Betti number* and (i, j)th *graded Betti number* of A-module M, respectively. For convenience, we simply write β_i and β_{ij} instead of $\beta_i^A(M)$ and $\beta_{i,j}^A(M)$.

Example 7.2 If $A = \mathsf{k}[x_1, \ldots, x_n]$, the Koszul complex on $x_1, \ldots, x_n$,

$$0 \longrightarrow A(-n)^{\beta_n} \xrightarrow{\partial_n} \cdots \longrightarrow A(-2)^{\beta_2} \xrightarrow{\partial_2} A(-1)^{\beta_1} \xrightarrow{\partial_1} A \to \mathsf{k} \to 0$$

is a minimal graded free resolution of $\mathsf{k} = A/A_+$ over A, where $\beta_i = \binom{n}{i}$.

Poincaré–Betti Series

Now we define the Poincaré–Betti series and bi-series.

(a) The *Poincaré–Betti series* $P_A(z)$ of A is the generating function for the k-dimensions of $\mathrm{Tor}_i^A(\mathsf{k}, \mathsf{k})$,

$$P_A(z) = \sum_{i \geq 0} \dim_{\mathsf{k}} \left(\mathrm{Tor}_i^A(\mathsf{k}, \mathsf{k})\right) z^i.$$

(b) The Poincaré–Betti bi-series $P_M^A(t, z)$ of a graded A-module M is a formal sum

$$P_M^A(t, z) = \sum_{i \geq 0} \left(\mathrm{Tor}_i^A(M, \mathsf{k})_j\right) t^j z^i = \sum_{i \geq 0} \beta_{ij}\, t^j z^i.$$

(c) The Poincaré–Betti series $P_M(z)$ of a graded A-module M is a formal sum

$$P_M^A(z) = \sum_{i \geq 0} \left(\mathrm{Tor}_i^A(M, \mathsf{k})\right) z^i = \sum_{i \geq 0} \beta_i\, z^i.$$

Notice that $P_M^A(z) = P_M^A(1, z)$.

Examples for Poincaré–Betti Series

(i) If $A = \mathsf{k}\langle x_1, \ldots, x_n\rangle$, a minimal free A-resolution of k is

$$0 \longrightarrow A^n \xrightarrow{\phi_1} A \to \mathsf{k}.$$

with $\phi_1 = \begin{bmatrix} x_1 \; x_2 \; \ldots \; x_n \end{bmatrix}$ and hence it is linear. We get that the Poincaré–Betti series of $\mathsf{k} = A/A_+$ over A is $1 + nz$, and the Poincaré–Betti bi-series is $1 + ntz$.

(ii) If $A = \mathsf{k}[x_1, \ldots, x_n]$, the Poincaré–Betti series of $\mathsf{k} = A/A_+$ over A is $1 + \binom{n}{1}z + \cdots + \binom{n}{n}z^n = (1 + z)^n$, and the Poincaré–Betti bi-series is given by $(1 + tz)^n$.

Definition 7.1 Let R be a standard graded k-algebra, and M be a finitely generated graded R-module with (i, j)th graded Betti number $\beta_{ij}^A(M) = \dim_{\mathsf{k}} \left(\mathrm{Tor}_i^A(M, \mathsf{k})_j\right)$.

(a) The *Castelnuovo–Mumford regularity* of M over A is

$$\operatorname{reg}_A(M) = \sup_{i \geq 0} \left\{ j - i \mid \beta_{ij}^A(M) \neq 0 \right\}.$$

(b) Let I be a homogeneous ideal in A generated in degree d. Then I has a linear resolution if $\operatorname{reg}_A(I) = d$, i.e., if for all i, $\beta_{ij}^A(I) = 0$ for $j \neq i + d$.
(c) We say that A is a *Koszul algebra* if $\operatorname{reg}_A(\mathsf{k}) = 0$, i.e., if for all i, we have $\beta_{ij}^A(\mathsf{k}) = 0$ for $j \neq i$ [41].

Remark 7.1 [18, Lemma 6.5] Let A be a standard graded k-algebra. Let I be a homogeneous ideal in A with $\operatorname{reg}_A(A/I) \leq 1$. If A is Koszul, then so is A/I.

7.3 Koszul Algebras

7.3.1 Poincaré Series and Hilbert Series Formula

Proposition 7.1 *Given a finitely generated graded A-module M, a formula relating the Hilbert series of A and M, and the Poincaré–Betti bi-series of M is given by*

$$\mathbb{F}_M(z) = \mathbb{F}_A(z) P_M^A(z, -1). \tag{7.1}$$

Proof The minimality of resolution $F_\bullet$ implies that $\min \beta_{i,j} \geq \min \beta_{i-1,j}$. Notice that for a given j, there exists only finitely many i such that $\beta_{i,j} \neq 0$. This implies that the series $P_M^A(z, -1)$ is well defined. By selecting the jth degree component from the minimal graded free resolution $F_\bullet$, we obtain a finite exact complex of finite-dimensional k-vector spaces. Now use the fact that Euler characteristic (alternating sum of the dimensions) of a finite exact complex of finite-dimensional k-vector space vanishes to get the relation (7.1). □

Definition 7.2 Let M be a finitely generated graded A-module. We say that M has a *linear resolution* if $\beta_{ij} = 0$ for all $i \neq j$. If M is generated in degree d, we say that M has a d-linear resolution if $\beta_{ij} = 0$ for all $j \neq i + d$.

A resolution is called linear if the non-zero entries of the matrices representing the differential maps between the free modules $A(-j)^{\beta_{ij}}$ are homogeneous of degree 1.

Remark 7.2 Notice that if the minimal graded free resolution $F_\bullet$ is linear, then the formula (7.1) becomes

$$F_M(z) = F_A(z) P_M^A(-z). \tag{7.2}$$

Now we focus on the minimal free graded A-resolution of k (we may consider $\mathsf{k} = A/A_+$ as an A-module).

Definition 7.3 If the minimal free graded A-resolution of k is linear, then we say that A is a *Koszul algebra*.

7.3.2 Fröberg Formula

Taking $M = \mathsf{k}$ in the formula (7.2), we get an equivalent definition of Koszul algebra, namely the following:

Definition 7.4 If A is a Koszul algebra then

$$F_A(z)P_{\mathsf{k}}^A(-z) = 1. \tag{7.3}$$

Let A be a Koszul algebra. We may also conclude (7.3) by using Euler characteristic in each degree of the following graded exact sequence:

$$\cdots \longrightarrow A(-m)^{\beta_m} \xrightarrow{\phi_m} \cdots \longrightarrow A(-2)^{\beta_2} \xrightarrow{\phi_2} A(-1)^{\beta_1} \xrightarrow{\phi_1} A \to \mathsf{k}.$$

We have from Fröberg formula (7.3) that $\frac{1}{F_A(-z)} = P_{\mathsf{k}}^A(z) \in \mathbb{N}[[t]]$.

Remark 7.3 In [30], Lofwall called (7.3) a Fröberg formula.

Theorem 7.1 ([30, Lofwall]) *A standard graded* k*-algebra* A *is Koszul if and only if* $F_A(z)P_{\mathsf{k}}^A(-z) = 1$.

One direction is clear from the previous discussion that Koszulness implies $F_A(z)P_{\mathsf{k}}^A(-z) = 1$. Hence we shall show that $F_A(z)P_{\mathsf{k}}^A(-z) = 1$ implies that A is Koszul.

Proof Let $F_\bullet = (F_i, \partial_i)$ be the minimal graded free resolution of k over A. Then we have

$$1 = F_{\mathsf{k}}(z) = \sum_{i \geq 0} F_{F_i}(z)(-1)^i.$$

On the other hand, since $F_i = \oplus A(-j)^{\beta_{i,j}}$, we see that $F_{F_i}(z) = \sum_{i \geq 0} \beta_{i,j} F_A(z) z^i$, and hence

$$1 = \sum_{i \geq 0} \sum_j \beta_{i,j} F_A(z) z^j (-1)^i = F_A(z) G_A(-1, z),$$

where $G_A(-1, z) = \sum_{i \geq 0} \beta_{i,j} z^j (-1)^i$. By uniqueness of inverse in the power series ring $\mathsf{k}[|z|]$, it is enough to show that A is Koszul if and only if $G_A(-1, z) \equiv P_{\mathsf{k}}^A(-z)$, that is,

$$\sum_{i \geq 0} \sum_j \beta_{i,j} z^j (-1)^i \equiv \sum_{i \geq 0} \beta_i z^i (-1)^i. \tag{7.4}$$

Note that if A is Koszul then $\beta_i = \beta_{i,i}$ and $\beta_{i,j} = 0$ for all $i \neq j$, hence (7.4) holds.

Now assume that A is not Koszul. Let s be the smallest index such that $\beta_{s,s} \neq \beta_s$. Also note that $\beta_{i,j} = 0$ for $j < i$. Therefore

$$\sum_{i\geq 0}\sum_{j\geq i}\beta_{i,j}z^j(-1)^i - \sum_{i\geq 0}\beta_i z^i(-1)^i$$

is still a formal series with $\beta_{s,s} - \beta_s(-1)^s z^s$ as lowest non-zero term. Then (7.4) does not hold which is a contradiction. □

Theorem 7.2 [30, Lofwall] *A standard graded* k*-algebra* $A = S/I$ *is Koszul then* I *is generated by quadrics.*

Proof For the proof see [22] and references therein. □

Serre–Kaplansky Problem

J.P. Serre and Kaplansky asked the following question independently: Let A be a local Artinian ring with maximal ideal $\mathfrak{m}_A$ and residue field $\mathsf{k} = A/\mathfrak{m}_A$. Is it true that the Poincaré–Betti series

$$P_A(z) = \sum_{i\geq 0} \dim_{\mathsf{k}}\left(\mathrm{Tor}_i^A(\mathsf{k},\mathsf{k})\right) z^i$$

is rational?

Remark 7.4 In [2], Anick gave the following example for which the Poincaré–Betti series $P_A(z)$ is irrational,

$$A = \frac{\mathbb{Q}[x_1, x_2, \ldots, x_5]}{(x_1^2, x_2^2, x_4^2, x_5^2, x_1x_2, x_4x_5, x_1x_3 + x_3x_4 + x_2x_5) + \mathfrak{m}_A^3}.$$

Remark 7.5 By formula (7.3), we see that if the Hilbert series $\mathbb{F}_A(z)$ is rational, so is the Poincaré–Betti series. Hence for Koszul algebras, Poincaré–Betti series is a rational function.

Example 7.3 Note that $\mathsf{k}[x_1, \ldots, x_n]$ and $\mathsf{k}\langle x_1, \ldots, x_n\rangle$ are examples of Koszul algebras, since k has a linear resolution over these algebras.

Example 7.4 Let $A = \mathsf{k}[x_1, \ldots, x_n]/I$, where $I = (x_ix_j \mid 1 \leq i, j \leq n)$. Then A is Koszul, see [24].

7.4 Koszul Duality

Notations

(a) For any subspace $Q \subseteq (V)^{\otimes j}$, define the perpendicular subspace $Q^{\perp} = \{\alpha \in (V^*)^{\otimes j} \mid \alpha(q) = 0 \text{ for all } q \in Q\}$. Denote by (Q) the two-sided ideal of T(V), and by $(Q^{\perp})$ the two-sided ideal of $\mathrm{T(V^*)}$.
(b) Let S(V) be the symmetric algebra of V. Then $\mathrm{S(V)} = \mathrm{T(V)}\,/(Q_S)$, where (Q_S) is an ideal in T(V) generated by $Q_S = \{v \otimes w - w \otimes v \mid v, w \in V\}$.
(c) In order to be consist with literature on the theory of Koszul duality, for a given algebra (), we shall denote by $()^!$ for the Koszul duality in stead of $()^*$.

Definition 7.5 Let $A = \mathrm{T(V)}\,/(Q)$ be a quadratic algebra, where (Q) is generated by the quadratic relation set $Q \subseteq V \oplus_{\mathsf{k}} V$. The *Quadratic dual algebra* of A is defined as $A^! = \mathrm{T(V^*)}/(Q^{\perp})$.

Notice that $A^!$ is a quadratic algebra. Thus, we can write $A^! = \oplus_{i \geq 0} A_i^!$ as a graded k-algebra with $A_0^! = \mathsf{k}$ and $A_1^! = V^*$.

Proposition 7.2 *Let A be a quadratic algebra, then we have* $(A^!)^! = A$.

The proof of $(A^!)^! = A$ follows immediately from the following two observations $\dim(Q^{\perp})^{\perp} = \dim Q$ and $Q = (Q^{\perp})^{\perp}$, whose proof is left to the reader as an exercise.

Example 7.5 By definition, S(V) is a quadratic algebra. We want to find dual algebra $\mathrm{S(V)}^!$.

To find dual algebra of S(V), It is enough to find $Q_S^{\perp}$. Define $Q_A = \{f \otimes f \mid f \in V^*\}$. Then for any $f \in V^*$ and any $v, w \in V$, we have $(f \otimes f)(v \otimes w - w \otimes v) = f(v)f(w) - f(w)f(v) = 0$. Hence $Q_A \subseteq Q_S^{\perp}$. Now regarding v, w as elements in V^{**}, we have $(v \otimes w - w \otimes v)(f \otimes f) = 0$. We get that $Q_S^{\perp} \subseteq Q_A$, and so $Q_S^{\perp} = Q_A$. Hence we conclude that

$$\mathrm{S(V)}^! = \mathrm{T(V^*)}/(\mathrm{Q_A}) = \wedge \mathrm{V}^*, \text{ and } (\wedge \mathrm{V}^*)^! = \mathrm{S(V)}^{!\,!} = \mathrm{S(V)}\,.$$

More Examples

(i) Note that the tensor algebra T(V) is naturally a quadratic algebra, since $\mathrm{T(V)} = \mathrm{T(V)}\,/(Q)$, where (Q) is the ideal generated by the quadratic relation set $Q = \{0\} \subseteq V \otimes_{\mathsf{k}} V$. Then, $Q^{\perp} = (V \otimes_{\mathsf{k}} V)^* = V^* \otimes_{\mathsf{k}} V^*$. Thus,

$$\mathrm{T(V)}^! = \mathrm{T(V^*)}/(\mathrm{Q}^{\perp}) = \mathrm{T(V^*)}/(\mathrm{V}^* \otimes_{\mathsf{k}} \mathrm{V}^*).$$

(ii) Let V be the k vector space generated by x. Then $\mathrm{T(V)} = \mathsf{k}[x]$. Thus,

$$(\mathsf{k}[x])^! = \mathsf{k}[x]/(x^2) \text{ and } (\mathsf{k}[x]/(x^2))^! = \mathsf{k}[x].$$

(iii) More generally, let V be the vector space with basis $\{x_1, \ldots, x_n\}$. Then the symmetric algebra S(V) is a polynomial ring $\mathsf{k}[x_1, \ldots, x_n]$. By definition, $\mathrm{S(V)} = \mathsf{k}[x_1, \ldots, x_n] = \mathsf{k}\langle x_1, \ldots, x_n\rangle/(Q_S)$, where $Q_S = \{v \otimes w - w \otimes v \mid v, w \in V\}$. Note that if $x_i \otimes x_j - x_j \otimes x_i = 0$ for all $1 \le i < j \le n$, then $v \otimes w - w \otimes v = 0$, where $v = a_1x_1 + \cdots + a_nx_n$ and $w = b_1x_1 + \cdots + b_nx_n$ for some $a_i, b_i \in \mathsf{k}$. Recall that we denote by $\mathsf{k}\langle x_1, \ldots, x_n\rangle$ the tensor algebra T(V). Then,

$$\mathrm{S(V)} = \mathsf{k}\langle x_1, \ldots, x_n\rangle/(x_i \otimes x_j - x_j \otimes x_i \mid 1 \le i < j \le n).$$

Let V^* be the dual space of V with the dual basis $\{\zeta_1, \ldots, \zeta_n\}$. Then,

$$\zeta_i \otimes \zeta_i,\ (\zeta_i + \zeta_j) \otimes (\zeta_i + \zeta_j) \in Q_A = \{f \otimes f \mid f \in V^*\}$$

for all $i, j \in \{1, \ldots, n\}$. The second relation implies that $\zeta_i \otimes \zeta_j + \zeta_j \otimes \zeta_i \in (Q_A)$ for all $i, j \in \{1, \ldots, n\}$.
Conversely, if $\zeta_i \otimes \zeta_i = 0$ and $\zeta_i \otimes \zeta_j + \zeta_j \otimes \zeta_i = 0$ for all $1 \le i < j \le n$, then $f \otimes f = 0$, where $f = a_1\zeta_1 + \cdots + a_n\zeta_n$ for some $a_i \in \mathsf{k}$. We get

$$\wedge V^* = \mathrm{T(V^*)}/(Q_A) = \mathrm{T(V^*)}/(\zeta_i \otimes \zeta_i, \zeta_i \otimes \zeta_j + \zeta_j \otimes \zeta_i \mid 1 \le i < j \le n).$$

By previous example, we have

$$\mathrm{S(V)}^! = \mathsf{k}\langle \zeta_1, \ldots, \zeta_n\rangle/(\zeta_i^2, \zeta_i\zeta_j + \zeta_j\zeta_i \mid 1 \le i < j \le n).$$

(iv) Let V be the vector space over the field k with basis $\{x, y\}$. Consider the quadratic algebra $A = \mathsf{k}[x_1, x_2]/(x_1^2)$. Then, $A = \mathrm{T(V)}/(Q)$, where $Q = \{x_1 \otimes x_1, x_1 \otimes x_2 - x_2 \otimes x_1\}$ is the quadratic relation set. Let V^* be the dual space of V with the basis $\{\zeta_1, \zeta_2\}$. Then $\mathrm{T(V^*)} = \mathsf{k}\langle \zeta_1, \zeta_2\rangle$ and $Q^{\perp}$ is generated by $\{\zeta_1\zeta_2 + \zeta_2\zeta_1, \zeta_2^2\}$. Thus

$$A^! = \mathrm{T(V^*)}/(Q^{\perp}) = \mathsf{k}\langle \zeta_1, \zeta_2\rangle/(\zeta_1\zeta_2 + \zeta_2\zeta_1, \zeta_2^2).$$

(v) If $A = \mathsf{k}\langle x_1, \ldots, x_n\rangle/I$, where I is generated by monomials of degree two, then $A^! = \mathsf{k}\langle \zeta_1, \ldots, \zeta_n\rangle/J$, where J is generated by those monomials $\zeta_i\zeta_j$ such that $x_ix_j \notin I$.

Definition 7.6 We say that A is *Koszul* if $P_A(z) = F_{A^!}(z)$, where $A^! = \mathrm{Hom}(A, \mathsf{k})$ is the Koszul dual of A.

$$F_{A^!}(-z)F_A(z) = 1.$$

Example 7.6 If $A = \mathsf{k}[x_1, \ldots, x_n]$, then $A^! = \mathsf{k}\langle \zeta_1, \ldots, \zeta_n\rangle/(\zeta_i^2, \zeta_i\zeta_j + \zeta_j\zeta_i \mid 1 \le i < j \le n)$. hence $F_A(z) = \frac{1}{(1-z)^n}$, $F_{A^!}(z) = (1+z)^n$. Therefore

$$F_{A^!}(-z)F_A(z) = (1-z)^n\frac{1}{(1-z)^n} = 1.$$

7.5 Koszul Phenomena in Literature

Poincaré–Betti Series of Complete Intersection

An (ideal theoretic) complete intersection ring is of the form S/I, where $S = \mathsf{k}[x_1, \ldots, x_n]$ is a polynomial ring and I an ideal of S generated by a regular sequence of homogeneous forms. We say that the homogeneous polynomials $g_1, g_2, \ldots, g_r$ form a regular sequence if g_i is a non-zero divisor on $S/(g_1, g_2, \ldots, g_{i-1})$ for every $i = 2, \ldots, r-1$. Now using Koszul algebra technique, we prove the following theorem due to Tate [43].

Theorem 7.3 ([43, Theorem 6]) *Let $A = \mathsf{k}[x_1, \ldots, x_n]/(g_1, g_2, \ldots, g_r)$ be a complete intersection ring with* $\deg g_i = 2$. *Then the Poincaré–Betti series of* k *over A is $\frac{(1+z)^n}{(1-z^2)^r}$.*

Proof Let $B_i = \mathsf{k}[x_1, \ldots, x_n]/(g_1, g_2, \ldots, g_{i-1})$. We have the following graded short exact sequence of B_i-modules

$$0 \longrightarrow B_i(-2) \xrightarrow{g_i} B_i \longrightarrow B_i/(g_i) \longrightarrow 0.$$

Using this short exact sequence repeatedly, we can compute the Hilbert series of A, namely $F_A(z) = \frac{(1-z^2)^r}{(1-z)^n}$. If we show that A is Koszul, then the Poincaré series formula follows immediately from the following relation $\mathbb{F}_A(z)P_{\mathsf{k}}^A(-z) = 1$. We know that $B_1 = \mathsf{k}[x_1, \ldots, x_n]$ is Koszul. Now using induction on i and applying regularity lemma (e.g., [20, Corollary 20.19]) we conclude that $\operatorname{reg}_{B_i} B_i/(g_i) \leq 1$ for all i. Using (7.1) we conclude that $B_{r+1} = A$ is Koszul. □

7.6 Combinatorics

There are interesting classes of quadratic monomial ideals coming from combinatorics.

Stanley–Reisner Rings

A simplicial complex $\triangle$ over a set of vertices $V = \{x_1, \ldots, x_n\}$ is a subset of the power set of V with the property that $F \in \triangle$ and $G \subseteq F$ imply $G \in \triangle$. If $F \in \triangle$, we define $x_F = \prod_{x_i \in F} x_i \in S = \mathsf{k}[x_1, \ldots, x_n]$ over the field k. The Stanley–Reisner ideal of $\triangle$, denoted by $I_\triangle$ is the ideal $(x_F \mid F \notin \triangle)$ in S, and the Stanley–Reisner ring of $\triangle$ is $\mathsf{k}[\triangle] = S/I_\triangle$. For more information on the theory of Stanley–Reisner ideals we refer the reader to [8]. Notice that $I_\triangle$ is a monomial ideal. We record the following theorem, see discussion [Section 3.1, [22]].

Theorem 7.4 *The Stanley–Reisner ring* $\mathsf{k}[\Delta] = S/I_\Delta$ *is Koszul if and only if* I_Δ *is quadratic if and only if* Δ *is a flag (clique) complex.*

Recall that a simplicial complex Δ is called flag (clique), if all minimal nonfaces consist of two elements, equivalently, I_Δ is generated by quadratic monomials.

Theorem 7.5 *Let* P *be a finite partially ordered set on* $\{x_1, \dots, x_n\}$. *The Stanley–Reisner ring associated to* P *is Koszul.*

Finite Simple Graphs

Let G be a graph on the vertex set $V = \{x_1, \dots, x_n\}$. Given a field k we define the edge ideal $I(G)$ of the polynomial ring $S = \mathsf{k}[x_1, \dots, x_n]$ generated by the set of monomials $x_i x_j$ such that x_i is adjacent to x_j. The following theorem follows from [24].

Theorem 7.6 *Let* $S = \mathsf{k}[x_1, \dots, x_n]$ *be a polynomial ring and* $I(G)$ *an edge ideal of* S *associated to a finite simple graph. Then* $S/I(G)$ *is Koszul.*

Matroids

Given $v_1, \dots, v_n$ spanning vector space V, define the Orlik–Solomon algebra

$$A = \bigwedge(x_1, \dots, x_n)/I,$$

where I is spanned by $\sum_{s=1}^{r}(-1)^s x_{i_1} \wedge \cdots \bar{x_{i_s}} \wedge \cdots \wedge x_{i_r}$ for all circuits, that is, minimal dependent subsets of $\{v_{i_1}, \dots, v_{i_r}\}$.

Question (Yuzvinskiĭ [45]) When is the Orlik–Solomon algebra A Koszul? □

Finite Directed Graphs

Let A_G be an algebras arising from walks in directed graphs G on $[n] = \{1, 2, \dots, n\}$. The following theorem is obtained by multiple authors, see Fröberg [24], Bruns, Herzog, and Vetter [9].

Theorem 7.7 *Let* G *be a directed graphs on* $[n] = \{1, 2, \dots, n\}$, *that is, a collection of ordered pairs* (i, j) *(with* $i = j$ *allowed). Then*

$$A_G = \mathsf{k}\langle x_1, \dots, x_n\rangle/(x_i x_j \mid (i, j) \notin G)$$

is Koszul.

The Koszul dual of A_G is $A_{\bar{G}}$, where $\bar{G}$ is the complementary directed graph of G. There is a nice description for the Hilbert series of A_G,

$$\mathbb{F}_{A_G} = \sum_{i \geq 0} |\{\text{walks of length } i \text{ in } G\}| \cdot z^i.$$

7.7 Geometry

Mumford's theorem [32] says that for any projective variety $X \subset \mathbb{P}^n$, its Veronese embedding $v_{d_0}(X) \subset \mathbb{P}^{N_0}$ is cut out by quadrics, for $d_0 \gg 0$. Backlin [4] proved that Veronese embedding $v_{d_1}(X) \subset \mathbb{P}^{N_1}$ for some $d_1 \gg 0$ is Koszul.

Eisenbud, Reeves, and Totaro in [21] proved that under sufficiently high Veronese embedding, $v_{d_2}(X) \subset \mathbb{P}^{N_2}$ for some $d_2 \gg 0$, is defined by Gröbner basis of quadrics. In general, we have $d_2 \geq d_1 \geq d_0$, and hence we have

$$\text{Gröbner basis of quadrics} \implies \text{Koszul} \implies \text{Quadratic.}$$

Example 7.7 An example of Quadratic algebra but not Koszul, let

$$R = K[x, y, z, t]/(x^2, y^2, z^2, t^2, xy + zt)$$

then one has $\beta^R_{34}(K) = 5$ and hence R is not Koszul. We may also see via Poincaré–Betti series

$$\mathbb{F}_R(z) = 1 + 4z + 5z^2 \text{ and } \frac{1}{\mathbb{F}_R(-z)} = 1 + 4z + \cdots + 44z^5 - 29z^6 \cdots$$

Example 7.8 An example of algebra with Gröbner basis of quadrics but not Koszul, let

$$R = K[x, y, z]/(x^2 + yz, y^2 + xz, z^2 + xy).$$

The Gröbner basis of ideal $I = (x^2 + yz, y^2 + xz, z^2 + xy)$ is $(x^2 + yz, y^2 + xz, z^2 + xy, 2yz^2, 2xz^2, z^4)$.

Kemph in [25] proved that any ring with a straightening law whose discrete algebra is defined by quadratic monomials is Koszul. He further gave an estimate for high enough bound to improve the Backlin result for the high Veronese subring of any graded k-algebra to be Koszul. There are several examples of coordinate rings of projective varieties that are Koszul.

Koszulness of Combinatorial Rational Varieties

Many combinatorial rational varieties, such as Grassmannians, Schubert varieties, determinantal varieties, etc., are cut out by quadrics. In fact, with respect to the system of coordinates and term orders, Grassmannians, Schubert varieties, determinantal varieties, etc., are defined by Gröbner basis of quadrics and hence are Koszul algebras.

Koszulness of Non-combinatorial, Non-rational Varieties:

For the non-combinatorial, non-rational varieties, the classical result is Petri's theorem [35] states that a smooth non-hyperelliptic curve of genus $g \geq 4$ in its canonical embedding is cut out by quadrics, with the exceptions of trigonal curves and plane quintics. Vishik and Finkelberg in [44] proved that the coordinate ring of general curve of genus $g \geq 5$ is Koszul. Pareschi and Purnaprajna [33] proved that the homogeneous coordinate ring of a canonical curve is Koszul unless the curve is hyperelliptic, trigonal, or isomorphic to a plane quintic. Butler in [19] proved that for a curve of genus g embedded by a complete linear system of degree $\geq 2g+2$, the homogeneous coordinate ring is Koszul. Polishchuk in [37] studied the Koszul property for the homogeneous coordinate ring of certain embeddings of degree $\geq g+3$. Conca, Rossi and Valla [16] proved that for a smooth, non-hyperelliptic, non-trigonal curve of genus $g \geq 5$ which is not a plane quintic, then its canonical ring is are defined by Gröbner basis of quadrics and hence are Koszul algebras.

Let X be a set of s distinct points in $\mathbb{P}^n_{\mathsf{k}}$ and let R_X be their coordinate ring. Kempf [26] proved that if $s \leq 2n$ and the points are in general linear position then R_X is Koszul (Kempf used the term Wonderful rings for Koszul). This result of Kempf was further extended by Conca, Trung, and Valla [17] in the following way: Let X be a set of generic points in $\mathbb{P}^n_{\mathsf{k}}$, then the coordinate ring R_X is Koszul if and only if $|X| \leq 1 + n + (n^2/4)$.

Diagonal Subalgebras

Let c, e positive integers, and $\Delta = \{(cs, es) \mid s \in \mathbb{Z}\}$ be the (c, e)-diagonal of $\mathbb{Z}^2$. Given a bigraded k-algebra $R = \oplus_{u,v\geq 0} R_{(u,v)}$, one can associate a graded k-algebra $R_\Delta = \oplus_{s\in\mathbb{Z}} R_{(cs,es)}$, the (c, e)-diagonal subalgebra of R. The diagonal subalgebra methods were introduced by Simis, Trung, and Valla, see [40]. The diagonal subalgebra R_Δ, being a graded k-algebra has a presentation of the form S/I, where S is some polynomial ring and I a homogeneous ideal of S, and hence R_Δ represents a homogeneous coordinate ring. See example 7.10 is discussed to illustrate this.

Example 7.9 Let $S = \mathsf{k}[x_1, \ldots, x_n, y_1, \ldots, y_m]$ be a bigraded polynomial ring with $\deg x_i = (1, 0)$ and $\deg t_j = (0, 1)$ for all i, j. Let $\Delta = (c, e)$ and set $A = \mathsf{k}[x_1, \ldots, x_n]$ and $B = \mathsf{k}[y_1, \ldots, y_m]$. Then the diagonal subalgebra S_Δ is the Segre product of the Veronese subrings $A^{(c)}$ and $B^{(e)}$, respectively, and is given by $\mathsf{k}[x_i y_j | 1 \leq i \leq n, 1 \leq j \leq m]$.

Example 7.10 Let $J = (x_1^3, x_2^3) \subset A = \mathsf{k}[x_1, x_2]$ and let $\Delta = (1, 1)$. The Rees algebra of an ideal J is $A[Jt] = \mathsf{k}[x_1, x_2, x_1^3 t, x_2^3 t]$. It is a well-known fact that the Rees algebra $A[Jt]$ is isomorphic to the quotient

$$\frac{\mathsf{k}[x_1, x_2, y_1, y_2]}{(x_1^3 y_2 - x_2^3 y_1)}.$$

Let $B = \mathsf{k}[x_1, x_2, y_1, y_2]$ and $K = (x_1^3 y_2 - x_2^3 y_1)$ an ideal in B. Set $A[Jt] = B/K$ and define $\deg x_i = (1, 0)$ and $\deg y_j = (0, 1)$, then $A[Jt]$ is a bigraded k-algebra. The $(1, 1)$-*diagonal subalgebra* of $A[Jt]$ is the graded k-algebra

$$A[Jt]_\Delta = \frac{B_\Delta}{K_\Delta} = \frac{\mathsf{k}[x_1 y_1, x_1 y_2, x_2 y_1, x_2 y_2]}{K_\Delta},$$

where

$$K_\Delta = \Big((x_1 y_1)^2 (x_1 y_2) - (x_2 y_1)^3, (x_2 y_1)(x_2 y_2)^2 - (x_1 y_2)^3, (x_1 y_1)(x_1 y_2)^2 - (x_2 y_1)^2 (x_2 y_2)\Big).$$

Using the isomorphism

$$\frac{\mathsf{k}[z_0, \ldots, z_3]}{(z_1 z_2 - z_0 z_3)} \longrightarrow \mathsf{k}[x_1 y_1, x_1 y_2, x_2 y_1, x_2 y_2]$$

where $z_0 \mapsto x_1 y_1,\ z_1 \mapsto x_1 y_2,\ z_2 \mapsto x_2 y_1,\ z_3 \mapsto x_2 y_2$, we can describe the $(1, 1)$-diagonal subalgebra of the Rees algebra $A[Jt]$ as follows:

$$A[Jt]_\Delta = \frac{\mathsf{k}[z_0, \ldots, z_3]}{(z_1 z_2 - z_0 z_3, z_1^3 - z_2 z_3^2,\ z_2^3 - z_0^2 z_1,\ z_0 z_1^2 - z_2^2 z_3)}.$$

A twisted quartic curve is defined from $\mathbb{P} \longrightarrow \mathbb{P}^3$ by the map $[x_0, x_1] \mapsto [x_0^4, x_0^3 x_1, x_0^2 x_1^2, x_0 x_1^3, x_1^4]$. The vanishing locus of the map above is the polynomials $z_1 z_2 - z_0 z_3, z_1^3 - z_2 z_3^2, z_2^3 - z_0^2 z_1, z_0 z_1^2 - z_2^2 z_3$. Therefore, the diagonal subalgebra $A[Jt]_\Delta$ is the homogeneous coordinate ring of the twisted quartic curve in the projective space $\mathbb{P}^3$.

Let $S = \mathsf{k}[x_1, \ldots, x_n]$ be a polynomial ring. Let $I = \langle f_1, \ldots, f_r \rangle$ be an ideal of S generated by a regular sequence of homogeneous forms of degree d. The *Rees algebra* of I is the subalgebra of $\mathsf{k}[x_1, \ldots, x_n, t]$ defined as $\mathcal{R}(I) = \mathsf{k}[x_1, \ldots, x_n, f_1 t, \ldots, f_r t]$. We list some important results in this direction of koszulness of diagonal subalgberas of bigraded algebras:

(a) Given any standard bigraded k-algebra R, one has R_Δ is Koszul for large Δ [18, Theorem 6.2].
(b) If standard bigraded k-algebra R (when viewed as a graded algebra) is Koszul, then R_Δ is Koszul for all Δ [6, Theorem 2.1].
(c) If $c \geq \frac{d(r-1)}{r}$ and $e > 0$, then, $\mathcal{R}(I)_\Delta$ is Koszul [18, Corollary 6.10].
(d) Let $n = 3$ and $f_1 = x^2$, $f_2 = y^2$, $f_3 = z^2$, and $\Delta = (1, 1)$, then $\mathcal{R}(I)_\Delta$ is the subring

$$R = \mathsf{k}[x^3, y^3, z^3, x^2 y, x y^2, x^2 z, x z^2, y^2 z, y z^2]$$

of $\mathsf{k}[x, y, z]$. R is Koszul [11].

(e) Let $r = 3$ and $\deg(f_i) = 2$. Then $\mathcal{R}(I)_\Delta$ is Koszul for $\Delta = (1, 1)$ [12, Theorem 3.2].
(f) Let $r = 3$ and $\deg(f_i) = d$. If $c \geq \frac{d}{2}$ and $e > 0$, then, $\mathcal{R}(I)_\Delta$ is Koszul [29].
(g) Suppose that S is replaced by a Koszul ring A and $= \langle f_1, \ldots, f_r \rangle$ be an ideal of A, then if $c \geq \frac{d(r-1)}{r}$ and $e > 0$, then, $\mathcal{R}(I)_\Delta$ is Koszul [29].
(h) Let $S = \mathsf{k}[x_1, \ldots, x_m, t_1, \ldots, t_n]$ be a polynomial ring bigraded by $\deg x_i = (1, 0)$ for $i = 1, \ldots, m$ and $\deg t_i = (0, 1)$ for $i = 1, \ldots, n$. Let I be an ideal of S generated by a regular sequence with elements all of bidegree $(d, 1)$ and $R = S/I$. Let $\frac{d}{2} \leq c < \frac{2d}{3}$ and $e > 0$. Then R_Δ is Koszul. Moreover $R(-a, -b)_\Delta$ have a linear resolution over R_Δ [29].

Question (Conca, Herzog, Trung, and Valla, [18]) Let $S = \mathsf{k}[x_1, \ldots, x_n]$ be a polynomial ring. Let $I = \langle f_1, \ldots, f_r \rangle$ be an ideal of S generated by a regular sequence of homogeneous forms of degree d. Let Δ be the (c, e)-diagonal of $\mathbb{Z}^2$, where c, e are positive integers. Is it true that $\mathcal{R}(I)_\Delta$ is Koszul for all $c \geq \frac{d}{2}$ and $e > 0$? □

This question is open for all $r \geq 4$.

Residual Intersections

Let R be a Noetherian local ring. Two ideals I and J are linked when $I = K : J$, $J = K : I$, and K is an ideal of R generated by a regular sequence. Residual intersection generalizes the notion of linkage. An ideal J of R is an m-residual intersection of I if there exists an m-generated ideal $K = \langle z_1, \ldots, z_m \rangle \subset I$ such that $J = K : I$ and $\mathrm{ht}(J) \geq m$, see [3]. An m-residual intersection J of I is a geometric m-residual intersection of I if $\mathrm{ht}(I + J) \geq m + 1$.

Given positive integers $m \geq n$, consider an ideal $J = \langle z_1, \ldots z_m \rangle + I_n(\phi)$ in the polynomial ring $S = \mathsf{k}[x_1, \ldots, x_n, y_1, \ldots, y_p]$, where ϕ is an $n \times m$ matrix, with entries linear in $y_1, \ldots, y_p$, such that $\begin{bmatrix} z_1 & z_2 & \ldots & z_m \end{bmatrix} = \begin{bmatrix} x_1 & x_2 & \ldots & x_n \end{bmatrix} \cdot \phi$. Such an ideal is shown to be a geometric residual intersection in [10, Lemma 4.7], when $\mathrm{ht}(J) \geq m$, and $\mathrm{ht}(I_n(\phi)) \geq m - n + 1$. The authors in [1] prove the following.

Theorem 7.8 *If $c \geq 1$ and $e \geq \frac{n}{2}$, then $(S/J)_\Delta$ is Koszul.*

7.8 Koszulness of Affine Semigroup Rings

A numerical semigroup is a set of non-negative integers $\mathbb{N}$ closed under addition, containing the zero element, and with finite complement in the set of non-negative integers. By an affine semigroup we mean a finitely generated sub-semigroup S of the additive monoid $\mathbb{N}^d$, where d is some positive integer. Let $\mathsf{k}[\mathsf{S}]$ denote the semigroup ring of S over a field k. Then one can identify $\mathsf{k}[\mathsf{S}]$ with the subring of a polynomial ring $\mathsf{k}[y_1, \ldots, y_d]$ generated by the monomials $y^\mu = y_1^{\mu_1} \cdots y_d^{\mu_d}$, where $\mu = (\mu_1, \ldots, \mu_d) \in \mathbb{N}^d$. Define the map $\phi : \mathsf{k}[x_1, \ldots, x_n] \longmapsto \mathsf{k}[y_1, \ldots, y_d]$ by sending x_i to the generating monomials $y^{\lambda_i} = y_1^{\lambda_{1,i}} \cdots y_d^{\lambda_{d,i}}$ of the affine semigroup S. The toric ideal I_{S} is the kernel of the map ϕ. Hence we have

$$\mathsf{k}[x_1, \ldots, x_n]/(I_{\mathsf{S}}) \cong \mathsf{k}[\mathsf{S}] \subset \mathsf{k}[y_1, \ldots, y_d].$$

Let J be a semigroup ideal generated by semigroup elements in $\mathsf{k}[\mathsf{S}]$. The following theorem is due to Herzog, Reiner, and Welker [27].

Theorem 7.9 *Let S be an affine semigroup and let J be a semigroup ideal in the affine semigroup ring $\mathsf{k}[\mathsf{S}]$. Set $A = \mathsf{k}[\mathsf{S}]/J$. Then A is Koszul in the following case:*

(i) $\mathsf{S} = \mathbb{N}^d$ *(so $A = \mathsf{k}[x_1, \ldots, x_d]$) and J is a quadratic monomial ideal, or*
(ii) $J = 0$ *and S has minimal generators $\lambda_1, \ldots, \lambda_d$ for which the toric ideal I_{S} has a quadratic Gröbner basis.*

The following theorem is due to Peeva, Reiner, and Sturmfels [34].

Theorem 7.10 $\mathsf{k}[\mathsf{S}]$ *is Koszul if and only if S is a Cohen–Macaulay poset over* k *when ordered by divisibility.*

Monomial Projective Curves

It is a standard fact that Gröbner basis of quadrics implies Koszulness, and Koszulness implies quadratic defining ideal, however the converse may not be true in general. In the context of monomial projective curves (let $d = 2$ in the definition of affine semigroup rings), the following two questions have been extensively studied in the literature.

Question Whether all monomial projective curves with quadratic defining ideal are Koszul? □

Question Whether all Koszul monomial projective curves have quadratics Gröbner basis? □

The following Table 7.1 due to Roos and Sturmfels [39, Sect. 2] discusses the computational outcome for the monomial projective curves (toric curves) in $\mathbb{P}^n$ for $n \leq 8$.

Example 7.11 We refer to Table 7.1.

(i) For $n = 3$, we have precisely two toric curves which are the twisted cubic [0, 1, 2, 3] and the complete intersection [0, 1, 2, 4], see [42].
(ii) For $n \leq 6$, positive answer for 7.8 that all Koszul algebras have quadratic Gröbner bases. This is not true if $n = 7$, see [42].

Table 7.1 Census analysis of toric curves defined by quadrics in $\mathbb{P}^n$

n	Candidates	Quadratics	Koszul	Quadratics Gröbner basis
2	1	1	1	1
3	2	2	2	2
4	9	8	8	8
5	67	47	46	46
6	752	384	358	358
7	11320	3794	≥ 3321	3320
8	≥ 60000	–	–	–

(iii) For $n = 5$, there is precisely one case, for which quadratically defined ideal does not imply koszulness. Let $\mathsf{S} = [0, 3, 5, 7, 11]$, then the corresponding toric ideal I_{S} is defined by quadratic relations

$$(x_2^2 - x_1x_4, x_3^2 - x_2x_5, x_4^2 - x_3x_5, x_5^2 - x_2x_6, x_3x_4 - x_1x_6).$$

The semigroup ring $\mathsf{k}[\mathsf{S}] = \mathsf{k}[x_1, \ldots, x_6]/(I_{\mathsf{S}})$ is quadratic but not Koszul, see [42].

(iv) In $\mathbb{P}^8$, let S be the monoid generated by

$$(36, 0), (33, 3), (30, 6), (28, 8), (26, 10), (25, 11), (24, 12), (18, 18), (0, 36).$$

Then the Poincaré–Betti series of $\mathsf{k}[\mathsf{S}]$ is irrational [39, Theorem 1].

(v) Following [39, Theorem 1], it was asked whether affine numerical semigroup ring has a irrational Poincaré–Betti series. Fröberg and Roos [23] modified the example given by Roos and Sturmfels [39] and proved that an affine numerical semigroup ring can have irrational Poincaré–Betti series.

The following theorem is due to Bermejo, García-Llorente and García-Marco [5].

Theorem 7.11 *Let $a_1 < \cdots < a_n$ be a generalized arithmetic sequence of relatively prime integers. Consider the projective monomial curve $C \subset \mathbb{P}^n$ over k parametrically defined by*

$$x_1 = s^{a_1}t^{a_n - a_1}, \ldots, x_{n-1} = s^{a_{n-1}}t^{a_n - a_{n-1}}, x_n = s^{a_n}, x_{n+1} = t^{a_n}.$$

Then, the homogeneous coordinate ring $\mathsf{k}[C]$ is Koszul if and only if $a_1 < \cdots < a_n$ are consecutive numbers and $n > a_1$.

7.9 Koszulness and Pólya Frequency Sequences

Let A be a finitely generated graded k-algebra. Note that $F_A(z)$ and $P_A(z)$ are only power series in z, and not rational functions of z in general. If we assume that A is commutative, then $F_A(z)$ is a rational and can be written in the form

$$F_A(z) = \sum_{i\geq 0} (\dim_{\mathsf{k}} A_i)\ z^i = \frac{h(A, z)}{(1-z)^d},$$

where $h(A, z) = h_0 + h_1 z + \cdots + h_{\alpha(A)} z^{\alpha(A)}$, where $\alpha(A) \neq 0$ is the degree of the Hilbert polynomial $h(A, z)$ and d is the Krull dimension of A. If A is Cohen–Macaulay, then $h(A, z) \in \mathbb{N}[z]$. Moreover if A is commutative and Gorenstein then A is Cohen–Macaulay and $h(A, z) = h_0 + h_1 z + \cdots + h_{\alpha(A)} z^{\alpha(A)}$ with $h_{\alpha(A)-i} = h_i$ for $i \in \{0, 1, \ldots, \alpha(A)\}$, see [8].

Pólya Frequency Sequences

Let $H(z) = \sum_{i\geq 0} s_i z^i \in \mathbb{R}[[t]]$ be a formal sum. We say that a sequence of real numbers $(s_0, s_1, s_2, \ldots)$ is a Pólya frequency (in short, PF) sequence if the (infinite) Toeplitz matrix $\left(a_{j-i}\right)_{i,j=0,1,2,\ldots}$ has all minor determinants non-negative, see [7]. We say that a formal power series $H(z) = \sum_{i\geq 0} s_i z^i \in \mathbb{R}[[t]]$ generates a PF-sequence if the sequence $(s_0, s_1, s_2, \ldots)$ is a PF-sequence.

Question (Reiner and Welker [38]) For which Koszul algebras, is the Hilbert function a Pólya frequency sequence? □

References

1. H. Ananthnarayan, N. Kumar, V. Mukundan, Diagonal subalgebras of residual intersections. Proc. Am. Math. Soc. **148**(1), 41–52 (2020)
2. D.J. Anick, A counterexample to a conjecture of Serre. Ann. Math. (2). **115**(1), 1–33 (1982)
3. M. Artin, M. Nagata, Residual intersections in Cohen-Macaulay rings. J. Math. Kyoto Univ. **12**, 307–323 (1972)
4. J. Backelin, On the rates of growth of the homologies of Veronese subrings. in *Algebra, Algebraic Topology and Their Interactions (Stockholm, 1983)*. Lecture Notes in Mathematics, vol. 1183 (Springer, Berlin, 1986), pp. 79–100
5. I. Bermejo, G. García-Llorente, I. García-Marco, Algebraic invariants of projective monomial curves associated to generalized arithmetic sequences. J. Symbol. Comput. **81**, 1–19 (2017)
6. S. Blum, Subalgebras of bigraded Koszul algebras. J. Algebra **242**(2), 795–809 (2001)
7. F. Brenti, Unimodal, log-concave and Pólya frequency sequences in combinatorics. Mem. Am. Math. Soc. **81**(413), viii+106 (1989)
8. W. Bruns, J. Herzog, *Cohen–Macaulay Rings*. Cambridge Studies in Advanced Mathematics, vol. 39 (Cambridge University Press, Cambridge, 1993), pp. xii+403

9. W. Bruns, J. Herzog, U. Vetter, *Syzygies and walks, in Commutative Algebra (Trieste, 1992)* (World Scientific Publishing, River Edge, NJ, 1994), pp. 36–57
10. W. Bruns, A.R. Kustin, M. Miller, The resolution of the generic residual intersection of a complete intersection. J. Algebra **128**(1), 214–239 (1990)
11. G. Caviglia, The pinched Veronese is Koszul. J. Algebr. Combin. **30**(4), 539–548 (2009)
12. G. Caviglia, A. Conca, Koszul property of projections of the Veronese cubic surface. Adv. Math. **234**, 404–413 (2013)
13. A. Conca, Koszul algebras and their syzygies, in *Combinatorial Algebraic Geometry*. Lecture Notes in Mathematics, vol. 2108, (Springer, Cham, 2014), pp. 1–31
14. A. Conca, Koszul algebras. Boll. Unione Mat. Ital. (9). **1**(2), 429–437 (2008)
15. A. Conca, E. De Negri, M.E. Rossi, *Koszul Algebras and Regularity, Commutative Algebra*, vol. 1 (Springer, New York, 2013), pp. 285–315
16. A. Conca, Maria Evelina Rossi and Giuseppe Valla, Gröbner flags and Gorenstein algebras. Compos. Math. **129**(1), 95–121 (2001)
17. A. Conca, Ngô Viêt Trung and Giuseppe Valla, Koszul property for points in projective spaces. Math. Scand. **89**(2), 201–216 (2001)
18. A. Conca, J. Herzog, Ngô Viêt Trung and Giuseppe Valla, diagonal subalgebras of bigraded algebras and embeddings of blow-ups of projective spaces. Am. J. Math. **119**(4), 859–901 (1997)
19. C. David, Butler, normal generation of vector bundles over a curve. J. Differ. Geom. **39**(1), 1–34 (1994)
20. D. Eisenbud, *Commutative Algebra*. Graduate Texts in Mathematics, With a View Toward Algebraic Geometry, vol. 150 (Springer, New York, 1995), pp. xvi+785
21. D. Eisenbud, A. Reeves, B. Totaro, Initial ideals, Veronese subrings, and rates of algebras. Adv. Math. **109**(2), 168–187 (1994)
22. R. Fröberg, *Koszul algebras, in Advances in Commutative Ring Theory (Fez, 1997)*, vol. 205 (Taylor & Francis Limited, London, 1999), pp. 337–350
23. R. Fröberg, J.-E. Roos, An affine monomial curve with irrational Poincaré-Betti series. J. Pure Appl. Algebra **152**(1–3), 89–92 (2000)
24. R. Fröberg, Determination of a class of Poincaré series. Math. Scand. **37**(1), 29–39 (1975)
25. R. George, Kempf, Some wonderful rings in algebraic geometry. J. Algebra **134**(1), 222–224 (1990)
26. R. George, Kempf, Syzygies for points in projective space. J. Algebra **145**(1), 219–223 (1992)
27. J. Herzog, V. Reiner, V. Welker, The Koszul property in affine semigroup rings. Pac. J. Math. **186**(1), 39–65 (1998)
28. D. Hilbert, Ueber die Theorie der algebraischen Formen. Math. Ann. **36**(4), 473–534 (1890)
29. N. Kumar, Koszul property of diagonal subalgebras. J. Commut. Algebra **6**(3), 385–406 (2014)
30. C. Löfwall, On the subalgebra generated by the one-dimensional elements in the Yoneda Ext-algebra, in *Algebra, Algebraic Topology and Their Interactions (Stockholm, 1983)*, vol. 1183 (Springer, Berlin)291–338
31. R. Martínez-Villa, Introduction to Koszul algebras. Rev. Un. Mat. Argent. **48**, 67–95 (2008)
32. D. Mumford, Varieties defined by quadratic equations, in *Questions on Algebraic Varieties (C.I.M.E., III Ciclo, Varenna, 1969)* (Edizioni Cremonese, Rome, 1970), pp. 29–100
33. G. Pareschi, B.P. Purnaprajna, Canonical ring of a curve is Koszul: a simple proof. Ill. J. Math. **41**(2), 266–271 (1997)
34. I. Peeva, V. Reiner, B. Sturmfels, How to shell a monoid. Math. Ann. **310**(2), 379–393 (1998)
35. K. Petri, Über die invariante Darstellung algebraischer Funktioneneiner Veränderlichen. Math. Ann. **88**(3–4), 242–289 (1923)
36. A. Polishchuk, L. Positselski, *Quadratic algebras*. University Lecture Series, vol. 37 (American Mathematical Society, Providence, RI, 2005), pp. xii+159
37. A. Polishchuk, On the Koszul property of the homogeneous coordinate ring of a curve. J. Algebra **178**(1), 122–135 (1995)
38. V. Reiner, V. Welker, On the Charney-Davis and Neggers-Stanley conjectures. J. Combin. Theory Ser. A **109**(2), 247–280 (2005)

39. J.-E. Roos, B. Sturmfels, A toric ring with irrational Poincaré-Betti series, C. R. Acad. Sci. Paris Sér. I Math. **326**(2), pp. 141–146 (1998)
40. A. Simis, N.V. Trung, G. Valla, The diagonal subalgebra of a blow-up algebra. J. Pure Appl. Algebra **125**(1–3), 305–328 (1998)
41. B. Stewart, Priddy, Koszul resolutions. Trans. Am. Math. Soc. **152**, 39–60 (1970)
42. B. Sturmfels, *Gröbner bases and convex polytopes*. University Lecture Series, vol. 8 (American Mathematical Society, Providence, RI, 1996), pp. xii+162
43. J. Tate, Homology of Noetherian rings and local rings. Ill. J. Math. **1**, 14–27 (1957)
44. A. Vishik, M. Finkelberg, The coordinate ring of general curve of genus $g \geq 5$ is Koszul. J. Algebra **162**(2), 535–539 (1993)
45. S. Yuzvinskiĭ, Orlik-Solomon algebras in algebra and topology. Uspekhi Mat. Nauk **56**(338), 87–166 (2001)

Part II
Classical K-Theory

The theory of Leavitt path algebras has created an astonishingly large amount of recent activities in ring theory. Besides being a beautiful subject in its own right, it is closely related to several other areas in mathematics, which might explain the burst of activity in the subject. The first part of this volume exclusively deals with Leavitt path algebras and the related areas.

With initial impetus from the theory of graph C*-algebras, the K-theory of Leavitt Path Algebras has also begun to be developed. For instance, the deep Kirchberg-Philips theorem asserting that a surprisingly small amount of similarity between the K-theoretic data of two such C*-algebras is sufficient to yield an isomorphism between them, has been applied to LPAs corresponding to certain Cayley graphs by Abrams and others. In order to introduce K-theory to the whole community of researchers, the second part of the volume carries some articles in K-theory and on other aspects of K-theory not necessarily connected with LPAs directly also appear here.

Chapter 8
Symplectic Linearization of an Alternating Polynomial Matrix

Ravi A. Rao and Ram Shila

8.1 Introduction

Let R be a commutative ring with 1. Let $M_r(R[X])$ denote the set of all polynomial matrices over R. Let $\alpha(X) \in M_r(R[X])$ be a polynomial matrix of degree $d \geq 1$, where degree of a polynomial matrix is defined as the maximum degree of its entries.

One can ask if $\alpha(X)$ can 'stably' be made a linear polynomial matrix. Specifically, one asks if one can find a *linear* matrix $(\alpha(0) \perp I_s) + nX$, for some s, and for some $n \in M_{r+s}(R)$, by making elementary row and column operations on $(\alpha(X) \perp I_s)$, for some s; i.e. is there an $\varepsilon(X) \in E_{r+s}(R[X])$ such that

$$(\alpha(X) \perp I_s)\varepsilon(X) = (\alpha(0) \perp I_s) + nX.$$

The process of linearization of a polynomial matrix was started by the well-known 'Higman trick'; which solved the linearization problem above.

This idea can also be applied when we deal with alternating matrices (of Pfaffian 1) with respect to stable equivalence under congruence with respect to the subgroup of elementary matrices. The reader will find a treatment in [7, p. 945] of how to linearize an alternating polynomial matrix (of Pfaffian 1) in this way; viz. given an alternating

R. A. Rao (✉)
School of Mathematics, Tata Institute of Fundamental Research, 1, Dr. Homi Bhabha Road, Mumbai 400005, India
e-mail: ravi@math.tifr.res.in

R. Shila
School of Physical Sciences, JNU, Delhi 110067, India
e-mail: ramshila01@gmail.com

Department of Science, Forbesganj College, Forbesganj 854318, Bihar, India

A. A. Ambily et al. (eds.), *Leavitt Path Algebras and Classical K-Theory*, Indian Statistical Institute Series,
https://doi.org/10.1007/978-981-15-1611-5_8

matrix $\varphi(X) \in SL_{2r}(R[X])$, there is a $s \geq 0$, and an $\varepsilon(X) \in E_{2(r+s)}(R[X])$ such that

$$\varepsilon(X)(\varphi(X) \perp \psi_s)\varepsilon(X)^t = (\varphi(0) \perp \psi_s) + nX,$$

for some $n \in M_{2(r+s)}(R)$.

We prove an 'elementary symplectic analogue' of Karoubi's linearization process. This asserts that one can stably linearize an alternating polynomial matrix by conjugating it by an *elementary symplectic matrix*.

8.2 Preliminaries

For matrices $\alpha \in M_r(A)$, $\beta \in M_s(A)$, we will denote the operations $\perp$, $\top$ as follows:

$$\alpha \perp \beta = \begin{pmatrix} \alpha & 0 \\ 0 & \beta \end{pmatrix} \in M_{r+s}(A),$$
$$\alpha \top \beta = \begin{pmatrix} 0 & \alpha \\ \beta & 0 \end{pmatrix} \in M_{r+s}(A).$$

$\perp$, $\top$ are associative operations.

The standard hyperbolic matrix $\{1\}\top\{-1\}$ will be denoted by ψ_1, and ψ_r will denote $\psi_1 \perp \cdots \perp \psi_1$, the sum being over r terms.

In [4], we found it useful to work with a cousin of ψ_r, which we denote by $\boxslash_r \in SL_{2r}(\mathbb{Z})$. The notation is suggestive, $\boxslash_r$ is the alternating matrix of size $2r$ whose anti-diagonal entries are 1 to the right, and -1 to the left, and all other entries are zero. Thus

$$\begin{aligned} \boxslash_r &= (\{1\}\top \cdots \top\{1\})\top(\{-1\}\top \cdots \top\{-1\}) \text{ (each } r \text{ times)} \\ &= \boxslash_r^+ \top \boxslash_r^-, \text{ where} \\ \boxslash_r^+ &= (\{1\}\top \cdots \top\{1\}) \ (r \text{ times}), \\ \boxslash_r^- &= (\{-1\}\top \cdots \top\{-1\}) \ (r \text{ times}). \end{aligned}$$

We will also, for the sake of simplicity, let $\boxslash$ denote $\boxslash_r$ for some r, etc., when we do not wish to specify the size.

Note that $\boxslash_r$ is of Pfaffian 1: It is easy to show directly that $\boxslash_r \sim_E \psi_r$: Let $\tau_{2r} = I_{2r-1}\top\{-1\} \in E_{2r}(\mathbb{Z})$, then $\tau_{2r}(\boxslash_{r-1} \perp \psi_1)\tau_{2r}^t = \boxslash_r$. Now induct on r and conclude that $\boxslash_r \sim_E \psi_r$.

8.3 Elementary Symplectic Linearization

We will use the following notation below:

$$\begin{aligned}\varphi(0)^* &= \varphi(0) \perp \varphi(0) \perp \varphi(0),\\ \varphi^{**} &= \varphi \perp (\varphi(0) \top \varphi(0)).\end{aligned}$$

Let $\varphi(X) = \varphi_0 + \varphi_1 X + \ldots + \varphi_k X^k$. Linearization is usually achieved by spreading a degree k term $a_k X^k$ as two terms $a_k X^{k-1}$ and X in space. In the usual linearization process, and one just 'eliminates' φ_k first, by conjugating by a suitable elementary symplectic matrix—the 'elimination' being done by the intersection at the center. This is illustrated in the key Proposition 8.1 next.

The main aim of Proposition 8.1 is to show that stably one can reduce the degree of an invertible alternating polynomial matrix by conjugating it with an elementary symplectic matrix (w.r.t. some alternating form) equivalent to the standard hyperbolic form ψ_t, for $t >> 0$.

In the sequel, we use the notation $E_{ij}(Z)$ to denote the block matrix with the block Z on the (ij)th entry, the Identity block matrix of the same size as that of Z in the diagonal entries, and the 0 block matrices elsewhere.

Proposition 8.1 *Let $\varphi(X) \in GL_{2r}(R[X])$ be an alternating polynomial matrix of degree $k > 1$. Then there is an elementary symplectic matrix*

$$\varepsilon \in ESp(\boxslash_{6r}^{-} \top \boxslash_{3r} \top \boxslash_{6r}^{+})$$

such that $\varepsilon \Phi \varepsilon^t$ has degree $(k-1)$, where $\Phi = \varphi(0)^* \top \varphi^{**} \top \varphi(0)^*$

Proof We shall think of Φ as a 9×9 block matrix with blocks of size $2r \times 2r$ below. Similarly, we think of

$$\boxslash_{6r}^{-} \top \boxslash_{3r} \top \boxslash_{6r}^{+} = \boxslash_{2r}^{-} \top \boxslash_{2r}^{-} \top \boxslash_{2r}^{-} \top \boxslash_{2r}^{-} \top \boxslash_{r} \top \boxslash_{2r}^{+} \top \boxslash_{2r}^{+} \top \boxslash_{2r}^{+} \top \boxslash_{2r}^{+},$$

as a 9×9 block matrix.

Let $Y, Z \in M_{2r}(R[X])$, $Z_1 = Z\varphi(0)^{-1}\boxslash_{2r}^{+}$, $Z_2^t = Z_1^t \boxslash_{2r}^{+} \varphi(0)$, $Y_1 = Y\varphi(0)$, $Y_2 = -Y\boxslash_{2r}^{+}$, $Y_3 = -Y_2^t \boxslash_{2r}^{+} \varphi(0)$. Let

$$\varepsilon = E_{36}(-Y_2^t \boxslash_{2r}^{+}) E_{47}(Y) E_{96}(-Z_1^t \boxslash_{2r}^{+}) E_{41}(Z\varphi(0)^{-1}).$$

Then a direct check shows that ε is symplectic. Moreover, ε is the image of a symplectic matrix over $\mathbb{Z}[t_1, \ldots, t_N]$, for some N. Using the Quillen–Suslin theory principles established in [3, 6], it is proved in [2] that $Sp(\mathbb{Z}[t_1, \ldots, t_N]) = ESp(\mathbb{Z}[t_1, \ldots, t_N])$. Hence, ε is elementary symplectic.

Let $\varphi(X) = \varphi_0 + \varphi_1 X + \cdots + \varphi_k X^k$, for some $\varphi \in M_{2r}(R)$, $0 \leq i \leq k$, and $\varphi_k = \nu - \nu^t$, for some $\nu \in M_{2r}(R)$. Take $Z = -\nu X^{k-1}$, $Y = X I_{2r}$, above. Then

$\varphi' = \varphi - YZ^t + ZY^t$ has degree $(k-1)$. Therefore if one calculates $\varepsilon\Phi\varepsilon^t$ we find that it has degree $(k-1)$. □

Theorem 8.1 *An invertible alternating polynomial matrix $\varphi(X)$ can be stably reduced to a linear form $\Phi(X) = \Phi(0) + NX$, by conjugating it by an elementary symplectic matrix $\varepsilon(X) \in ESp(\theta)$, with θ an alternating matrix over $\mathbb{Z}$ which is a conjugate of ψ_t, for some $t >> 0$.*

Proof We have shown in the above Proposition 8.1 that degree reduction can be achieved by conjugating by an elementary symplectic matrix. Apply the same process successively to reduce to the linear case. □

Remark 8.1 We shall refer to $\Phi(0)$ as the stable constant form, and θ as the stable form of the symplectic matrix, in the sequel.

Remark 8.2 Elementary Symplectic Higman Linearization: By using a similar argument as in the above proposition, one can show that a polynomial matrix can be stably linearized by means of elementary symplectic matrices.

Remark 8.3 In the above process, the reader will notice that the 'stable constant form' i.e. $\Phi(0)$ of the given alternating matrix $\Phi(X)$ and the 'stable form' of the symplectic matrix θ do not commute. It is a moot point if one can do the linearization process so that these two forms commute. This is work in progress; and we hope to write up our results in [5] shortly.

Acknowledgements The second named author gratefully acknowledges the support from the NBHM for his Ph.D. fellowship. He also acknowledges Dr. Amala Bhave and Prof. Riddhi Shah for their support.

References

1. M. Karoubi, *Periodicite de la K-theorie hermitienne. Algebraic K-theory, III: Hermitian K-theory and geometric applications (Proc. Conf., Battelle Inst., Seattle, Wash., 1972)*. Lecture Notes in Mathematics, vol. 343 (Springer, Berlin, 1973), pp. 301–411
2. V.I. Kopeiko, *The Stabilization of Symplectic Groups over a Polynomial Ring*. Math. USSR Sbornik 34 (1978), pp. 655–669
3. D. Quillen, Projective modules over polynomial rings. Inventionnes Math. **36**, 166–172 (1976)
4. R.A. Rao, R.G. Swan, *Excerpts from: A Regenerative Property of a Fibre of Invertible Alternating Matrices*, http://math.uchicago.edu/~swan/
5. R.A. Rao, R. Shila, *Symplectic Linearization*, in progress
6. A.A. Suslin, On the structure of the special linear group over polynomial rings. Math. USSR. Izv. **11**, 221–238 (1977)
7. A.A. Suslin, L.N. Vaserstein, Serre's problem on projective modules over polynomial rings and algebraic K-theory. Math. USSR Izv. **10**, 937–1001 (1976)

Chapter 9
Actions on Alternating Matrices and Compound Matrices

Bhatoa Joginder Singh and Selby Jose

9.1 Introduction

In this note, we consider the action of $\mathrm{SL}_n(R)$ on $\mathrm{Alt}_n(R)$, the space of alternating matrices of order n over R, by conjugation: $V \mapsto \sigma V \sigma^t$, for $\sigma \in \mathrm{SL}_n(R)$, $V \in \mathrm{Alt}_n(R)$. We prove (See Theorem 9.2) that the matrix of the above linear transformation (associated to σ) is $\wedge^2\sigma$.

These results are well known to experts when R is a field, but we worked it, as we will need it, in a sequel, over any commutative ring R. (The book [5] gives some details.)

In the last section, we restrict to the case when $n = 4$. We show that by taking a suitable basis of $\mathrm{Alt}_4(R)$ we can get a map from $\mathrm{SL}_4(R)$ to $\mathrm{SO}_6(R)$. Moreover, this map induces an injection from $\mathrm{SL}_4(R)/\mathrm{E}_4(R)$ to $\mathrm{SO}_6(R)/\mathrm{EO}_6(R)$ (See Theorem 9.3). The case when $R = \mathbb{C}$ is proved in [1].

In some sense, this result is reminiscent to the Jose–Rao Theorem in [3, Theorem 4.14], when $n = 2$, where it was shown that

$$\mathrm{SUm}_r(R)/\mathrm{EUm}_r(R) \to \mathrm{SO}_{2(r+1)}(R)/\mathrm{EO}_{2(r+1)}(R)$$

is injective. (We refer the reader to [3] for details.)

In recent article [4], Jose–Rao have shown that for $v, w \in R^{r+1}$, $\sigma \in \mathrm{SL}_{r+1}(R)$, the Suslin matrix

$$S_r(v\sigma, w\sigma^{t^{-1}}) = AS_r(v, w)B,$$

B. J. Singh
Department of Mathematics, Government College, Nani-Daman 396210, India
e-mail: jogiojogi@gmail.com

S. Jose (✉)
Department of Mathematics, The Institute of Science,
Madam Cama Road, Mumbai 400032, India
e-mail: selbyjose@gmail.com

A. A. Ambily et al. (eds.), *Leavitt Path Algebras and Classical K-Theory*,
Indian Statistical Institute Series,
https://doi.org/10.1007/978-981-15-1611-5_9

for some $A, B \in \mathrm{SL}_{2^r}(R)$, with AB, the Euler characteristic of σ. We may regard Theorem 9.2 as a prelude to this result; it signifies that the Suslin form brings out the Euler characteristic, whereas the alternating form only displays the initial $\wedge^1$ and $\wedge^2$.

9.2 Preliminaries

In this section, we recall a few definitions, state some results and fix some notations which will be used throughout this paper.

Let R be a commutative ring with 1. Let $\mathrm{M}_r(R)$ denote the set of all $r \times r$ matrices with entries in R.

Definition 9.1 The General Linear group $\mathrm{GL}_r(R)$ is defined as the group of $r \times r$ invertible matrices with entries in R.

Definition 9.2 The Special Linear group is denoted by $\mathrm{SL}_r(R)$ and is defined as $\mathrm{SL}_r(R) = \{\alpha \in \mathrm{GL}_r(R) : \det(\alpha) = 1\}$. It is a normal subgroup of $\mathrm{GL}_r(R)$.

Definition 9.3 The group of elementary matrices $\mathrm{E}_r(R)$ is a subgroup of $\mathrm{GL}_r(R)$ generated by matrices of the form $E_{ij}(\lambda) = I_r + \lambda e_{ij}$, where $\lambda \in R$, $i \neq j$ and $e_{ij} \in M_r(R)$ with ijth entry is 1 and all other entries are zero.

Note that $e_{ij}e_{rs} = \begin{cases} e_{is} & \text{if } j = r \\ 0 & \text{if } j \neq r \end{cases}$.

Following are some well-known properties of the elementary generators:

Lemma 9.1 *For $\lambda, \mu \in R$,*

(1) (*Splitting Property*) $E_{ij}(\lambda + \mu) = E_{ij}(\lambda)E_{ij}(\mu)$, $1 \leq i, j \leq r$, $i \neq j$.
(2) (*Steinberg relation*) $[E_{ij}(\lambda), E_{jk}(\mu)] = E_{ik}(\lambda\mu)$, $1 \leq i, j, k \leq r$, $i \neq j$, $i \neq k$, $j \neq k$.

Remark 9.1 In view of the Steinberg relation, $\mathrm{E}_r(R)$ is generated by

$$\{E_{1i}(\lambda), E_{i1}(\mu) : 2 \leq i \leq r, \lambda, \mu \in R\}.$$

Note that $E_{ij}(\lambda), i \neq j, \lambda \in R$, is invertible with inverse $E_{ij}(-\lambda)$. In fact, $E_{ij}(\lambda)$ belongs to $\mathrm{SL}_r(R)$. Hence, $\mathrm{E}_r(R) \subseteq \mathrm{SL}_r(R) \subseteq \mathrm{GL}_r(R)$.

We now recall the notion of the compound matrix:

Definition 9.4 (*Minors of a matrix*) Given an $n \times m$ matrix $A = (a_{ij})$ over R, a minor of A is the determinant of a smaller matrix formed from its entries by selecting only some of the rows and columns. Let $K = \{k_1, k_2, \ldots, k_p\}$ and $L = \{l_1, l_2, \ldots, l_p\}$ be subsets of $\{1, 2, \ldots, n\}$ and $\{1, 2, \ldots, m\}$, respectively. The indices are chosen such that $k_1 < k_2 < \cdots < k_p$ and $l_1 < l_2 < \cdots < l_p$. The pth-order minor defined by K and L is the determinant of the submatrix of A obtained by considering the rows $k_1, k_2, \ldots, k_p$ and columns $l_1, l_2, \ldots, l_p$ of A. We denote this submatrix as $A\begin{pmatrix} k_1 & k_2 & \cdots & k_p \\ l_1 & l_2 & \cdots & l_p \end{pmatrix}$ or $A(K \mid L)$.

Theorem 9.1 (The Cauchy–Binet formula) *Let A be an $m \times n$ matrix and B an $n \times m$ matrix over R, where $m \leq n$. Then the determinant of their product $C = AB$ can be written as a sum of products of minors of A and B, i.e.,*

$$|C| = \sum_{1 \leq k_1 < k_2 < \cdots < k_m \leq n} \det A\begin{pmatrix} 1 & 2 & \cdots & m \\ k_1 & k_2 & \cdots & k_m \end{pmatrix} \det B\begin{pmatrix} k_1 & k_2 & \cdots & k_m \\ 1 & 2 & \cdots & m \end{pmatrix}.$$

The sum is over the maximal (mth order) minors of A and the corresponding minor of B. In particular, $\det(AB) = \det(A)\det(B)$, *if A, B are $n \times n$ matrices.*

Definition 9.5 Suppose that A is an $m \times n$ matrix with entries from a ring R and $1 \leq r \leq \min\{m, n\}$. The rth compound matrix $C_r(A)$ or rth adjugate of A is the $\binom{m}{r} \times \binom{n}{r}$ matrix whose entries are the minors of order r, arranged in lexicographic order, i.e.

$$C_r(A) = \left(\det A\begin{pmatrix} i_1 & i_2 & \ldots & i_r \\ j_1 & j_2 & \ldots & j_r \end{pmatrix}\right).$$

Lemma 9.2 (Properties, See [2, 5]) *Let A and B be $n \times n$ matrices over R and $r \leq n$. Then*

(i) $C_1(A) = A$.
(ii) $C_n(A) = \det(A)$.
(iii) $C_r(AB) = C_r(A)C_r(B)$.
(iv) $C_r(A^t) = (C_r(A))^t$.

9.3 Associated Linear Transformations

We shall always work over a commutative ring R with 1. In this section, we find the linear transformation of the action of $\mathrm{SL}_n(R)$ on the space of alternating matrices.

Definition 9.6 A matrix $A \in M_n(R)$ is said to be alternating if $a_{ij} = -a_{ji}$ and $a_{ii} = 0$, for $1 \leq i, j \leq n$.

Notation The space of all alternating $n \times n$ matrices over a commutative ring R will be denoted by $\mathrm{Alt}_n(R)$. It is clearly a free R-module of rank $1 + 2 + \cdots + (n-1) = \binom{n}{2}$ with basis $B_{ij} = e_{ij} - e_{ji}$, $1 \leq i < j \leq n$. □

One has the action of $\mathrm{SL}_n(R)$ on $\mathrm{Alt}_n(R)$ by

$$\begin{aligned} \mathrm{SL}_n(R) \times \mathrm{Alt}_n(R) &\to \mathrm{Alt}_n(R) \\ (\sigma, A) &\mapsto \sigma A \sigma^t. \end{aligned}$$

This action enables one to associate a linear transformation $T_\sigma : \mathrm{Alt}_n(R) \to \mathrm{Alt}_n(R)$ for $\sigma \in \mathrm{SL}_n(R)$, via $T_\sigma(A) = \sigma A \sigma^t$.

We input the next observation for completeness; which can be found in [5, pp. 399–400].

Lemma 9.3 *Let $\sigma : R^n \longrightarrow R^m$ be a R-linear map. Then the matrix of the linear transformation $\wedge^r \sigma : \wedge^r R^n \longrightarrow \wedge^r R^m$ is $C_r(M(\sigma))$, where $M(\sigma)$ is the matrix of σ and $r \leq \min\{n, m\}$.*

Proof This is well-known to experts when R is a field. We compute it as follows:

Let $e_1, \ldots, e_n$ be a basis of R^n and $f_1, \ldots, f_m$ be a basis of R^m. Let us compute the matrix of $\wedge^r \sigma$ w.r.t. the standard basis $e_{i_1} \wedge \cdots \wedge e_{i_r}$ of $\wedge^r R^n$ and $f_{i_1} \wedge \cdots \wedge f_{i_r}$ of $\wedge^r R^m$ ordered lexicographically. Suppose $1 \leq i_1 < \cdots < i_r \leq n$ as usual. Then

$$\begin{aligned} \wedge^r(\sigma)(e_{i_1} \wedge \cdots \wedge e_{i_r}) &= \sigma(e_{i_1}) \wedge \cdots \wedge \sigma(e_{i_r}) \\ &= \sum_{j_1=1}^{m} d_{j_1 i_1} f_j \wedge \cdots \wedge \sum_{j_r=1}^{m} d_{j_r i_r} f_j \\ &= \sum_{1 \leq j_1 < \cdots < j_r \leq n} \det A \begin{pmatrix} j_1 \, j_2 \, \ldots \, j_r \\ i_1 \, i_2 \, \ldots \, i_r \end{pmatrix} \left(f_{j_1} \wedge \cdots \wedge f_{j_r}\right), \end{aligned}$$

where A^t denotes the matrix of the linear transformation σ. □

Since $\wedge^r(\sigma \circ \tau) = \wedge^r(\sigma) \circ \wedge^r(\tau)$, it is clear from Lemma 9.3 that the multiplicative property of compound matrices hold, i.e.

$$C_r(AB) = C_r(A) C_r(B),$$

where A is an $m \times n$ matrix, B is an $n \times m$ matrix and $r \leq \min\{m, n\}$.

Let us compute the matrix associated to T_σ for $\sigma \in \mathrm{SL}_n(R)$. We prove that it is the matrix $\wedge^2 \sigma$.

Theorem 9.2 *Let $\sigma \in \mathrm{SL}_n(R)$. Then the matrix of the linear transformation T_σ w.r.t. the basis $\{B_{ij} : 1 \leq i < j \leq n\}$ is the same as the matrix of the linear transformation $\wedge^2 \sigma : \wedge^2 R^n \longrightarrow \wedge^2 R^n$; which is the compound matrix of order 2 associated to σ.*

Proof Let $\sigma = (a_{ij})$. For $1 \leq i < j \leq n$, by definition,

$$\begin{aligned} T_\sigma(B_{ij}) &= \sigma B_{ij}\sigma^t = \sigma(e_{ij} - e_{ji})\sigma^t = \sigma e_{ij}\sigma^t - \sigma e_{ji}\sigma^t \\ &= \sum_{r=1}^{n}\sum_{s=r+1}^{n} a_{ri}a_{sj}B_{rs} - \sum_{r=1}^{n}\sum_{s=r+1}^{n} a_{rj}a_{si}B_{rs} \\ &= \sum_{r=1}^{n}\sum_{s=r+1}^{n} (a_{ri}a_{sj} - a_{rj}a_{si})B_{rs} \\ &= \sum_{r=1}^{n}\sum_{s=r+1}^{n} \det\sigma\begin{pmatrix} r\ s \\ i\ j\end{pmatrix} B_{rs}. \end{aligned}$$

Thus $[T_\sigma] = \left(\det\sigma\begin{pmatrix} i\ j \\ r\ s\end{pmatrix}\right) = C_2(\sigma)$. The rest follows via Lemma 9.3. □

The following Corollary gives the explicit form of $[T_{E_{1i}(\lambda)}]$, where $E_{1i}(\lambda) \in E_n(R)$. Since $E_{i1}(\lambda) = E_{1i}(\lambda)^t$, by Lemma 9.2(iv) one has, $[T_{E_{i1}(\lambda)}] = [T_{E_{1i}(\lambda)}]^t$.

Corollary 9.1 *Let $A = E_{1i}(\lambda) \in E_n(R)$, $\lambda \in R$. Let $\alpha = \{i_1, i_2\}$, $\beta = \{j_1, j_2\}$, where $1 \le i_1 < i_2 \le n$ and $1 \le j_1 < j_2 \le n$. Then the $(\alpha\beta)$th entry $\det A(\alpha|\beta)$ of $\wedge^2 A$ is given by*

$$\det A(\alpha|\beta) = \begin{cases} 1 & \text{if } \alpha = \beta \\ (-1)^r\lambda & \text{if } |\alpha\cap\beta| = 1, 1 \in \alpha, i \in \beta \text{ and } 1, i \notin \alpha\cap\beta \\ 0 & \text{otherwise,} \end{cases}$$

where r is the number of integers in $\alpha\cap\beta$ between 1 and i.

Proof Clearly if $\alpha = \beta$, then $\det A(\alpha|\beta) = 1$ as the submatrix $A(\alpha|\beta) = A\begin{pmatrix} i_1\ i_2 \\ j_1\ j_2\end{pmatrix}$ is either I_2 or an upper triangular matrix $\begin{pmatrix} 1 & \lambda \\ 0 & 1\end{pmatrix}$.

If $1 \in \alpha, i \in \beta$ and $1, i \notin \alpha\cap\beta$, then for $r \in \alpha\cap\beta$, $A(\alpha|\beta)$ is of the form $A\begin{pmatrix} 1\ r \\ r\ i\end{pmatrix}$ if $1 < r < i$ and is of the form $A\begin{pmatrix} 1\ r \\ i\ r\end{pmatrix}$ if $i < r \le n$. Note that if $A = (a_{ij})$, then

$$a_{jk} = \begin{cases} 1 & \text{if } j = k \\ \lambda & \text{if } j = 1, k = i. \\ 0 & \text{otherwise .} \end{cases}$$

Thus if $1 < r < i$, $A\begin{pmatrix} 1\ r \\ r\ i\end{pmatrix} = \begin{pmatrix} a_{1r} & a_{1i} \\ a_{rr} & a_{ri}\end{pmatrix} = \begin{pmatrix} 0 & \lambda \\ 1 & 0\end{pmatrix}$ and hence $\det A(\alpha|\beta) = -\lambda$. Also if $i < r \le n$, $A\begin{pmatrix} 1\ r \\ i\ r\end{pmatrix} = \begin{pmatrix} a_{1i} & a_{1r} \\ a_{ri} & a_{rr}\end{pmatrix} = \begin{pmatrix} \lambda & 0 \\ 0 & 1\end{pmatrix}$ and hence $\det A(\alpha|\beta) = \lambda$. All other entries of $\wedge^2 A$ contains either a zero row or a zero column. □

9.4 The 4 × 4 Case

L. N. Vaserstein studied the case when $n = 4$ in [6]. We consider the Vaserstein space $V = \text{Alt}_4(R)$ of dimension 6.

Definition 9.7 Let π denote the permutation $(1\ r+1)\cdots(r\ 2r)$ corresponding to the form $\begin{pmatrix} 0 & I_r \\ I_r & 0 \end{pmatrix}$. The elementary orthogonal matrices over R are defined by

$$oe_{ij}(\lambda) = I_{2r} + \lambda e_{ij} - \lambda e_{\pi(j)\pi(i)}, \text{ if } i \neq \pi(j),$$

where $1 \leq i, j \leq 2r$ and $\lambda \in R$.

Definition 9.8 The elementary orthogonal group $\text{EO}_{2r}(R)$ is the subgroup of $\text{SO}_{2r}(R)$ generated by the matrices $oe_{ij}(\lambda)$, where $1 \leq i < j \leq 2r$, $i \neq \pi(j)$ and $\lambda \in R$.

It is observed that the matrix $[T_\sigma]$ w.r.t. the basis $\{B_{12}, B_{13}, B_{14}, B_{23}, B_{24}, B_{34}\}$, where $\sigma = E_{1i}$ or E_{i1}, $2 \leq i \leq 4$ are not orthogonal w.r.t. the standard form $\begin{pmatrix} 0 & I_3 \\ I_3 & 0 \end{pmatrix}$. However, we have the following lemma.

Lemma 9.4 *With respect to the ordered basis* $\{B_{12}, B_{13}, B_{14}, B_{34}, -B_{24}, B_{23}\}$, *the matrix* $[T_{E_{1i}(\lambda)}]$ *and* $[T_{E_{i1}(\lambda)}]$, $2 \leq i \leq 4$ *are elementary orthogonal w.r.t. the standard form.*

Proof By Lemma 9.3, w.r.t. the basis $B_1 = \{B_{12}, B_{13}, B_{14}, B_{23}, B_{24}, B_{34}\}$, the matrix of $T_{E_{12}(\lambda)}$ is the compound matrix of order 2 associated to $A = E_{12}(\lambda)$. By Corollary 9.1, $\det A(\{1,3\},\{2,3\}) = \lambda$ and $\det A(\{1,4\},\{2,4\}) = \lambda$ and all other $\det A(\alpha|\beta) = 0$ if $\alpha \neq \beta$. If $\alpha = \beta$, then $\det A(\alpha|\beta) = 1$. Thus (24)th and (35)th entry of $[T_{E_{12}(\lambda)}]_{B_1}$ are λ. Hence we have $[T_{E_{12}(\lambda)}]_{B_1} = E_{24}(\lambda)E_{35}(\lambda)$. Then w.r.t. the basis $B_2 = \{B_{12}, B_{13}, B_{14}, B_{34}, -B_{24}, B_{23}\}$, the matrix $[T_{E_{12}(\lambda)}]_{B_2} = E_{26}(\lambda)E_{35}(-\lambda)$ which is by definition $oe_{26}(\lambda)$ w.r.t. the permutation $\pi = (14)(25)(36)$. Similarly w.r.t. the basis B_2 one has

$$\begin{aligned}
\left[T_{E_{13}(\lambda)}\right]_{B_2} &= E_{34}(\lambda)E_{16}(-\lambda) = oe_{34}(\lambda). \\
\left[T_{E_{14}(\lambda)}\right]_{B_2} &= E_{15}(\lambda)E_{24}(-\lambda) = oe_{15}(\lambda). \\
\left[T_{E_{21}(\lambda)}\right]_{B_2} &= E_{62}(\lambda)E_{53}(-\lambda) = oe_{62}(\lambda). \\
\left[T_{E_{31}(\lambda)}\right]_{B_2} &= E_{43}(\lambda)E_{61}(-\lambda) = oe_{43}(\lambda). \\
\left[T_{E_{41}(\lambda)}\right]_{B_2} &= E_{51}(\lambda)E_{42}(-\lambda) = oe_{51}(\lambda).
\end{aligned}$$

Hence the result. □

In general one has the following.

Proposition 9.1 *Let $\sigma \in \mathrm{SL}_4(R)$. Then the matrix of the linear transformation T_σ on the Vaserstein space V w.r.t. the ordered basis $\{B_{12}, B_{13}, B_{14}, B_{34}, -B_{24}, B_{23}\}$ is an orthogonal matrix w.r.t. the standard form.*

Proof Let $\widetilde{\psi}_3 = \begin{pmatrix} 0 & I_3 \\ I_3 & 0 \end{pmatrix}$. Let β be the matrix of T_σ. We show that β is in the orthogonal group of $\widetilde{\psi}_3$.

Let $\mathfrak{p}$ be a prime ideal of R. It suffices to show that $\beta_{\mathfrak{p}}$ is in the orthogonal group of $\widetilde{\psi}_3$, for all prime ideals $\mathfrak{p}$ of R. (Note that of $T_{\sigma_{\mathfrak{p}}}$ is the same as the matrix of $(T_\sigma)_{\mathfrak{p}}$.)

As $R_{\mathfrak{p}}$ is a local ring, $\mathrm{SL}_r(R_{\mathfrak{p}}) = \mathrm{E}_r(R_{\mathfrak{p}})$, for all $r \geq 2$. Hence, $\sigma_{\mathfrak{p}}$ is an elementary matrix, i.e. it is a product of elementary generators $\varepsilon_1, \ldots, \varepsilon_k$, for some k. We may assume that ε_i is of type $E_{1i}(x)$ or $E_{i1}(x)$, for some i, and arbitrary $x \in R$.

Now, $T_{\sigma_{\mathfrak{p}}} = \prod T_{\varepsilon_k}$. By Lemma 9.4, the matrix of each T_{ε_j} is an elementary orthogonal matrix w.r.t. the ordered basis $\{B_{12}, B_{13}, B_{14}, B_{34}, -B_{24}, B_{23}\}$. Hence, so is $T_{\sigma_{\mathfrak{p}}}$, for all prime ideals $\mathfrak{p}$ of R. □

But one has the following:

Remark 9.2 Let $\sigma \in \mathrm{SL}_4(R)$. Then the matrix of the linear transformation T_σ on the Vaserstein space V w.r.t. the ordered basis $\{B_{12}, B_{13}, B_{14}, B_{23}, B_{24}, B_{34}\}$ is an orthogonal matrix with respect to the form $\begin{pmatrix} 0 & \alpha \\ \alpha & 0 \end{pmatrix}$, where $\alpha = \begin{pmatrix} 0 & 0 & 1 \\ 0 & -1 & 0 \\ 1 & 0 & 0 \end{pmatrix}$.

Proof Let A and B denote the matrices of T_σ w.r.t. the bases

$$B_1 = \{B_{12}, B_{13}, B_{14}, B_{34}, -B_{24}, B_{23}\} \text{ and } B_2 = \{B_{12}, B_{13}, B_{14}, B_{23}, B_{24}, B_{34}\},$$

respectively. Let P denote the transition matrix from B_1 to B_2. Then clearly $P = I_3 \perp \alpha$ and $P^{-1}AP = B$. Note that $P^{-1} = P^T = P$. Hence $P^{-1}A^tP = (P^{-1}AP)^t = B^t$. By Proposition 9.1, A is orthogonal w.r.t. the standard form $\widetilde{\psi}_3 = \begin{pmatrix} 0 & I_3 \\ I_3 & 0 \end{pmatrix}$. Thus we have

$$A\widetilde{\psi}_3A^t = \widetilde{\psi}_3 \Rightarrow P^{-1}(A\widetilde{\psi}_3A^t)P = P^{-1}\widetilde{\psi}_3P \Rightarrow B(P^{-1}\widetilde{\psi}_3P)B^t = P^{-1}\widetilde{\psi}_3P,$$

which means B is orthogonal w.r.t. the form $P^{-1}\widetilde{\psi}_3P = \begin{pmatrix} 0 & \alpha \\ \alpha & 0 \end{pmatrix}$. □

9.5 Injectivity

In this section, we show that we can obtain a map from $\mathrm{SL}_4(R) \to \mathrm{SO}_6(R)$ and this map induces an injection $\dfrac{\mathrm{SL}_4(R)}{\mathrm{E}_4(R)} \hookrightarrow \dfrac{\mathrm{SO}_6(R)}{\mathrm{EO}_6(R)}$.

Proposition 9.2 *The map $\varphi : \mathrm{E}_4(R) \to \mathrm{EO}_6(R)$ is defined as $\varphi(\sigma) = [T_\sigma]$ is surjective.*

Proof Note that $\mathrm{EO}_6(R)$ is generated by the elementary orthogonal matrices $oe_{12}(\lambda)$, $oe_{21}(\lambda)$, $oe_{13}(\lambda)$, $oe_{31}(\lambda)$, $oe_{23}(\lambda)$, $oe_{32}(\lambda)$, $oe_{24}(\lambda)$, $oe_{42}(\lambda)$, $oe_{34}(\lambda)$, $oe_{43}(\lambda)$, $oe_{35}(\lambda)$ and $oe_{53}(\lambda)$. By the same argument as that of Lemma 9.4, one has

$$\begin{aligned}
&\left[T_{E_{23}(\lambda)}\right] = oe_{12}(\lambda), \left[T_{E_{32}(\lambda)}\right] = oe_{21}(\lambda), \left[T_{E_{24}(\lambda)}\right] = oe_{13}(\lambda),\\
&\left[T_{E_{42}(\lambda)}\right] = oe_{31}(\lambda), \left[T_{E_{34}(\lambda)}\right] = oe_{23}(\lambda), \left[T_{E_{43}(\lambda)}\right] = oe_{32}(\lambda),\\
&\left[T_{E_{14}(-\lambda)}\right] = oe_{24}(\lambda), \left[T_{E_{41}(-\lambda)}\right] = oe_{42}(\lambda), \left[T_{E_{13}(\lambda)}\right] = oe_{34}(\lambda),\\
&\left[T_{E_{31}(\lambda)}\right] = oe_{43}(\lambda), \left[T_{E_{12}(-\lambda)}\right] = oe_{35}(\lambda), \left[T_{E_{21}(-\lambda)}\right] = oe_{53}(\lambda).
\end{aligned}$$

Hence φ is surjective. □

Lemma 9.5 *Let u be a unit in R with $u^2 = 1$. Then $uI_4 \in \mathrm{E}_4(R)$.*

Proof This follows from Whitehead's lemma. Explicitly, if

$$\alpha_1 = \begin{pmatrix} I_2 & (1-u)I_2 \\ 0 & I_2 \end{pmatrix}, \alpha_2 = \begin{pmatrix} I_2 & 0 \\ -I_2 & I_2 \end{pmatrix}, \alpha_3 = \begin{pmatrix} I_2 & 0 \\ uI_2 & I_2 \end{pmatrix},$$

then clearly $\alpha_1, \alpha_2, \alpha_3 \in \mathrm{E}_4(R)$ and the direct computation shows $uI_4 = \alpha_1\alpha_2\alpha_1\alpha_3$. Hence the result. □

Proposition 9.3 *Let $\alpha \in \mathrm{M}_4(R)$ such that $\alpha A \alpha^t = A$ for all $A \in \mathrm{Alt}_4(R)$. Then $\alpha = uI_4$, where $u^2 = 1$.*

Proof Let $\alpha = (\alpha_{ij})_{4\times 4}$. Consider the generators $\{B_{ij} : 1 \le i < j \le 4\}$ of $\mathrm{Alt}_4(R)$. From $\alpha B_{1i}\alpha^t = B_{1i}$, $2 \le i \le 3$, one has

$$\alpha_{11}\alpha_{ki} - \alpha_{1i}\alpha_{k1} = 0, \quad i+1 \le k \le 4, \tag{9.1}$$
$$\alpha_{i1}\alpha_{ki} - \alpha_{ii}\alpha_{k1} = 0, \quad i+1 \le k \le 4, \tag{9.2}$$
$$\alpha_{11}\alpha_{ii} - \alpha_{1i}\alpha_{i1} = 1. \tag{9.3}$$

Now (9.1) $\times\alpha_{ii}-$ (9.2) $\times\alpha_{1i} \Rightarrow (\alpha_{11}\alpha_{ii} - \alpha_{1i}\alpha_{i1})\alpha_{ki} = 0$. Thus by (9.3),

$$\alpha_{ki} = 0, \quad i+1 \le k \le 4.$$

Also (9.1) $\times\alpha_{i1}-$ (9.2) $\times\alpha_{11} \Rightarrow (\alpha_{11}\alpha_{ii} - \alpha_{1i}\alpha_{i1})\alpha_{k1} = 0$. Again by (9.3), $\alpha_{k1} = 0$ for $k = 3, 4$.

Now we show that $\alpha_{21} = 0$. Consider $\alpha B_{13}\alpha^t = B_{13}$, we get

$$\alpha_{11}\alpha_{23} - \alpha_{13}\alpha_{21} = 0, \tag{9.4}$$
$$\alpha_{21}\alpha_{33} - \alpha_{23}\alpha_{31} = 0, \tag{9.5}$$
$$\alpha_{11}\alpha_{33} - \alpha_{13}\alpha_{31} = 1. \tag{9.6}$$

Now (9.4) $\times\alpha_{31}-$ (9.5) $\times\alpha_{11} \Rightarrow (\alpha_{11}\alpha_{33} - \alpha_{13}\alpha_{31})\alpha_{21} = 0$. Thus by (9.6), $\alpha_{21} = 0$. Hence

$$\alpha_{ij} = 0 \text{ for } 1 \le j < i \le 4. \tag{9.7}$$

Similarly using $\alpha B_{i4}\alpha^t = B_{i4}$, $1 \le i \le 3$, one can show that

$$\alpha_{ij} = 0 \text{ for } 1 \le i < j \le 4. \tag{9.8}$$

From (9.7) and (9.8), $\alpha_{ij} = 0, \ \forall \ i \neq j$.

Now from (9.3) and the relations obtained from $\alpha B_{i4}\alpha^t = B_{i4}$, $1 \le i \le 3$ one get, $\alpha_{11}\alpha_{22} = \alpha_{11}\alpha_{33} = \alpha_{11}\alpha_{44} = \alpha_{22}\alpha_{44} = 1$ and hence $\alpha_{11} = \alpha_{22} = \alpha_{33} = \alpha_{44} = u$, where $u \in R$ with $u^2 = 1$. Hence the result. □

Theorem 9.3 *One has an injective homomorphism*

$$\overline{\varphi} : \frac{\mathrm{SL}_4(R)}{\mathrm{E}_4(R)} \hookrightarrow \frac{\mathrm{SO}_6(R)}{\mathrm{EO}_6(R)}$$

($\overline{\varphi}$ is induced by the homomorphism $\varphi : \mathrm{SL}_4(R) \to \mathrm{SO}_6(R)$).

Proof Let $\alpha \in \mathrm{SL}_4(R)$ with $[T_\alpha] = I_6$. Then $\alpha V \alpha^t = V$, for all $V \in \mathrm{Alt}_4(R)$. Thus by Proposition 9.3, $\alpha = uI_4$ with $u^2 = 1$. By Lemma 9.5, $\alpha \in \mathrm{E}_4(R)$. Hence $\frac{\mathrm{SL}_4(R)}{\mathrm{E}_4(R)} \hookrightarrow \frac{\mathrm{SO}_6(R)}{\mathrm{EO}_6(R)}$. □

Acknowledgements The second author thanks the Science and Engineering Research Board (SERB), Department of Science and Technology, Government of India, for the funding of project MTR/2017/000875 under Mathematical Research Impact Centric Support (MATRICS).

References

1. P. Garrett, Sporadic isogenies to orthogonal groups, http://www.math.umn.edu/~garrett/m/v/sporadic_isogenies.pdf
2. R.A. Horn, C.R. Johnson, *Matrix Analysis* (Cambridge University Press, Cambridge)
3. S. Jose, R.A. Rao, A fundamental property of Suslin matrices. J. K-Theory: K-Theory Appl. Algebra Geom. Topol. **5**, 407–436 (2010)
4. S. Jose, R.A. Rao, Suslin forms and the Euler characteristics, in preparation
5. B.R. McDonald, *Linear Algebra over Commutative Rings*. Pure and Applied Mathematics, vol. 87 (Marcel Dekker Inc., New York, 1984)
6. A.A. Suslin, L.N. Vaserstein, Serre's problem on projective modules over polynomial rings and algebraic K-theory. Math. USSR Izv. **10**, 937–1001 (1976)

Chapter 10
A Survey on the Non-injectivity of the Vaserstein Symbol in Dimension Three

Neena Gupta, Dhvanita R. Rao and Sagar Kolte

10.1 Introduction

L. N. Vaserstein in [20] proved that the orbit space of unimodular rows of length three modulo elementary action have a Witt group structure over two-dimensional rings.

R. A. Rao–W. van der Kallen in [14] showed that the Vaserstein symbol V_A is not injective for $A = \Gamma(S^3_{\mathbb{R}})$, the coordinate ring of the real 3-sphere but is injective over three-dimensional smooth affine algebras over a field of cohomological dimension one whose characteristic $\neq 2, 3$.

R. G. Swan–R. A. Rao–J. Fasel in [17] gave another example of a real affine algebra of dimension three for which the Vaserstein symbol is not injective.

It was shown by N. Gupta–D. Rao in [12] that there is an uncountable family of affine algebras of dimension three over the real field for which the Vaserstein symbol is not injective.

It was also shown by S. Kolte–D. Rao in [13] that there is a countable family of smooth affine threefold for which the Vaserstein symbol is not injective.

We recall these results in this article.

N. Gupta (✉)
Statistics and Mathematics Unit, Indian Statistical Institute,
203 B.T. Road, Kolkata 700108, India
e-mail: neenag@isical.ac.in

D. R. Rao
Bhavan's College, Andheri (W), Munshi Nagar, Mumbai 400058, India
e-mail: dhvanita18@gmail.com

S. Kolte
Credit Suisse, Powai, Mumbai 400076, India
e-mail: sagar.kolte@credit-suisse.com

Department of Mathematics, Indian Institute of Technology, Mumbai 400076, India

A. A. Ambily et al. (eds.), *Leavitt Path Algebras and Classical* K*-Theory*,
Indian Statistical Institute Series,
https://doi.org/10.1007/978-981-15-1611-5_10

We finally recall the conjecture of J. Fasel about when one can expect the Vaserstein symbol to be injective or not in the case of a real threefold. And end with a brief description of his solution to it in [5].

10.2 The Witt Group $W_G(A)$

Let A be a commutative ring with 1. We shall also assume that A is a noetherian ring.

A matrix from $M_r(A)$ is said to be alternating if it has the form $\nu - \nu^t$, where $\nu \in M_r(A)$ and the superscript 't' denotes the transpose, i.e. it is skew-symmetric and its diagonal elements are equal to zero.

For α from $M_r(A)$ and β from $M_s(A)$ we denote by $\alpha \perp \beta$ the matrix in $M_{r+s}(A)$ given by

$$\begin{pmatrix} \alpha & 0 \\ 0 & \beta \end{pmatrix}$$

The operation $\perp$ is obviously associative.

We define inductively an alternating matrix ψ_r in $E_{2r}(A)$, setting

$$\psi_1 = \begin{pmatrix} 0 & 1 \\ -1 & 0 \end{pmatrix}$$
$$\psi_r = \psi_{r-1} \perp \psi_1$$

It is well known that there exists a Pfaffian—a polynomial pf in the matrix elements with coefficients ± 1 such that $\det(\varphi) = (pf(\varphi))^2$, for all alternating matrices φ.

On matrices of odd order the Pfaffian is identically equal to 0, and on matrices of even order it is defined up to sign, to fix which we insist that $pf(\psi_r) = 1$, for all r.

For any α from $M_r(A)$ and any alternating φ from $M_r(A)$ we have $pf(\alpha^t \varphi \alpha) = pf(\varphi).det(\alpha)$. For any alternating matrices φ_1, φ_2, it is easy to check that $pf(\varphi_1 \perp \varphi_2) = pf(\varphi_1) pf(\varphi_2)$.

As usual, $GL_r(A)$ is the group of all invertible matrices over A, and $SL_r(A)$ is the subgroup of $GL_r(A)$ consisting of matrices of determinant one.

Let $SL(A)$ denote the infinite linear group $\cup_r SL_r(A)$, where $SL_r(A)$ is thought of as a subgroup of $SL_{r+1}(A)$ under the usual identification $\alpha \mapsto (1) \perp \alpha$.

Let $E(A)$ denote the infinite elementary subgroup of $SL(A)$ consisting of $\cup_r E_r(A)$, where $E_r(A)$ denotes the usual subgroup of $SL_r(A)$ generated by the elementary generators $E_{ij}(a)$, $i \neq j$, $a \in A$. (Of course, here $E_r(A)$ is regarded as a subgroup of $SL_r(A)$, and so sits inside $E_{r+1}(A)$ by the previous identification.)

Note that by Whitehead's Lemma $[SL(A), SL(A)] = E(A)$.

In particular, $E(A)$ is a normal subgroup of $SL(A)$ (and even $GL(A)$). In fact, Suslin showed in [16, Corollary 1.4] that $E_r(A)$ is a normal subgroup of $GL_r(A)$, for $r \geq 3$.

We fix some subgroup G of $SL(A)$, containing $E(A)$. This G is automatically normal in $GL(A)$ as $GL(A)/E(A)$ is an abelian group. (In view of Whitehead's Lemma above.)

Two alternating matrices α from $M_{2r}(A)$ and β from $M_{2s}(A)$ are said to be equivalent relative to G (written $\alpha \sim \beta$) if

$$\alpha \perp \psi_{s+p} = \gamma^t(\beta \perp \psi_{r+p})\gamma,$$

for some natural number p and some matrix γ from $G \cap SL_{2(r+s+p)}(A)$.

This relation is reflexive, symmetric and transitive, i.e. it is an equivalence relation on the set of all alternating matrices. Two equivalent alternating matrices have the same Pfaffian, and it follows that this relation is also an equivalence relation on the set of alternating matrices of Pfaffian one.

Note: One can see that $\alpha \perp \beta \sim \beta \perp \alpha$ as the matrix

$$\begin{pmatrix} 0 & I_s \\ I_r & 0 \end{pmatrix} \in E_{r+s}(A),$$

when r, s is even.

Vaserstein showed (cf. [20, Sect. 3]) that the operation $\perp$ induces the structure of an abelian group on the set of equivalence classes relative to G of alternating matrices with Pfaffian 1; this group is denoted by $W_G(A)$.

10.3 The Vaserstein Symbol $V : Um_3(A)/E_3(A) \longrightarrow W_E(A)$

A row $v = (v_0, \ldots, v_r)$ is called unimodular of length $(r+1)$ if there is a row $w = (w_0, \ldots, w_r)$ such that $\langle v, w\rangle = v \cdot w^t = \sum_i v_i w_i = 1$.

(This is the case when the ideal generated by the coordinates of v is the unit ideal. Hence a row can be checked to be unimodular if it is a non-zero vector over the field A/m, for every maximal ideal m of A.)

The set of all unimodular rows of length $(r+1)$ over a ring A is denoted by $\mathrm{Um}_{r+1}(A)$.

There is a very natural association of a unimodular row of length 3 with an alternating matrix, which was pointed out by L. N. Vaserstein in [20]:

Given a pair of unimodular rows $v = (a, b, c)$, $w = (a', b', c')$, with a relation $\langle v, w\rangle = aa' + bb' + cc' = 1$, one can associate an alternating matrix $V(v, w)$ as follows:

$$V(v, w) = \begin{pmatrix} 0 & a & b & c \\ -a & 0 & c' & -b' \\ -b & -c' & 0 & a' \\ -c & b' & -a' & 0 \end{pmatrix} \in SL_4(A)$$

It is easily checked that $V(v, w)$ has Pfaffian $(aa' + bb' + cc') = 1$.

Vaserstein considered the map from $Um_3(A) \longrightarrow W_E(A)$ given by $v \mapsto [V(v, w)] \in W_E(A)$.

He showed that it did not depend on the choice of w. Moreover, if v was replaced by an elementary transformation $v\varepsilon$ of v (and w replaced by the corresponding row $w\varepsilon^{t^{-1}}$) then $[V(v, w)] = [V(v\varepsilon, w\varepsilon^{t^{-1}})] \in W_E(A)$.

Vaserstein studied the map $Um_3(A)/E_3(A) \longrightarrow W_E(A)$ given by $[v] \mapsto [V(v, w)]$. (This map is known as Vaserstein symbol.)

Theorem 10.1 (Vaserstein [20]) *The (Vaserstein) symbol*

$$V(= V_A) : Um_3(A)/E_3(A) \longrightarrow W_E(A)$$

is an isomorphism when A is of Krull dimension 2.

The four most important ingredients needed to prove the above theorem which were used by L. N. Vaserstein are as follows:

1. *Unimodular rows of odd length* ≥ 5 *can be completed to an elementary matrix:* $Um_{2r-1}(A) = e_1 E_{2r-1}(A)$, for $r \geq 3$.
2. *Equality of elementary and symplectic orbits for unimodular rows of even length:*

$$e_1 E_{2r}(A) = e_1\{Sp_\varphi(A) \cap E_{2r}(A)\},$$

   for $r \geq 3$, where $Sp_\varphi(A) = \{\alpha \in SL_{2r}(A) \mid \alpha^t \varphi \alpha = \varphi\}$ is the isotropy group of the invertible alternating matrix φ.
3. *Elementary completion of the first row of an odd sized* 1*-stably elementary matrix:* If $\rho \in SL_{2r-1}(A) \cap E_{2r}(A)$, $r \geq 2$, then $e_1\rho = e_1\varepsilon$, for some $\varepsilon \in E_{2r-1}(A)$. (The case when $r = 2$ is the one needed to prove Vaserstein's theorem.)
4. *Injective stability starts from size* 4: By the Bass–Milnor–Serre theorem in [3], and Vaserstein theorem in [18] one knows that if A is a commutative ring of dimension d then

$$SL_r(A) \cap E(A) = E_r(A),$$

   for $r \geq \max\{3, d + 2\}$. In particular, if A is 2 dimensional then $SL_4(A) \cap E(A) = E_4(A)$.

Note that (1)–(3) hold for three-dimensional rings; whereas (4) does not hold for three-dimensional rings in general; but only for certain special kinds of three-dimensional rings.

For instance, if A is a smooth affine algebra over a field k of cohomological dimension ≤ 1, with $\mathrm{char}(k) \neq 2$, then it was shown in [14, Theorem 1] that (4) holds. Consequently, the Vaserstein symbol V_A is injective for such smooth threefolds A.

On the other hand for the coordinate ring $A = \Gamma(S^3_{\mathbb{R}})$ of the real 3-sphere it was shown in [14, Proposition 4.2] that the Vaserstein symbol V_A is not injective.

An Euler Class Approach

In these Proceedings, Anjan Gupta, Raja Sridharan and Sunil K. Yadav have given an Euler class approach to establish Vaserstein's results: Given two unimodular rows over a two-dimensional ring, one adds their Euler classes in the Euler class group. The unimodular row corresponding to this Euler class is the sum of the two unimodular row. We refer the reader to the paper [9] for details.

10.4 The Vaserstein Symbol in Dimension Three and Four

Ravi A. Rao–Wilberd van der Kallen Examples

When A is a non-singular affine algebra of dimension d over a nice field (say an algebraically closed field, say the complex numbers $\mathbb{C}$, or a field like $\mathbb{C}(t)$ which is a function field in one variable over $\mathbb{C}$; more generally, a field of cohomological dimension at most one) then Ravi A. Rao and Wilberd van der Kallen showed that the injective stability estimate for $K_1(A)$ improves by 1, i.e. $SL_{d+1}(A) \cap E(A) = E_{d+1}(A)$, As a consequence they could prove:

Theorem 10.2 (Ravi Rao–Wilberd van der Kallen, see [14]) *Let A be a non-singular affine threefold over a field of cohomological dimension at most one and of characteristic $\neq 2, 3$. Then the Vaserstein symbol V_A is an isomorphism.*

R.G. Swan's Topological Example

In dimension 3 it is known that the Vaserstein symbol V_A is always **surjective** due to results of L. N. Vaserstein in [20]. Its kernel was computed by Anuradha Garge and Ravi A. Rao in [8] when A is an affine algebra of dimension three over a field of cohomological dimension at most 1, and of characteristic $\neq 2, 3$, and shown to be $\{[e_1\rho] \mid \rho \in SL_3(A) \cap E_5(A)\}$. (The latter set is always contained in the kernel.)

In the homepage of R. G. Swan (see [17]) another example has been given of a real algebra A of dimension three for which V_A is not injective.

Theorem 10.3 (Swan–Rao–Fasel, see [17]) *If $2n - 1 \equiv 3$ mod 8, there is an affine domain A over $\mathbb{R}$ of dimension $2n - 1$ and a 2-stably elementary element $\rho \in SL_{2n-1}(A)$ such that $(1 \perp \rho) \notin E_{2n}(A)Sp_{2n}(A)$. Moreover, if $n = 2$, we can choose ρ in such a way that its first row is not completable to an elementary matrix or, equivalently, $e_1\rho$ is not elementarily equivalent to e_1.*

The last sentence will give the desired example as it shows that the kernel has a non-trivial element.

Four Dimensional Examples

Recently, there is a theorem of Jean Fasel, Ravi A. Rao and Richard G. Swan (FRS theorem) on a problem of A. Suslin which has as a consequence the following theorem:

Theorem 10.4 (Fasel–Rao–Swan, see [6]) *Let A be a non-singular affine algebra of dimension* 4 *over an algebraically closed field of characteristic* $\neq 2, 3$. *Then the Vaserstein symbol* V_A *is injective.*

10.5 An Uncountable Family of Singular Counterexamples

The second named author and Neena Gupta began the search of more examples of affine algebras A over the real field for which the Vaserstein symbol V_A is not injective. In [12] they found an uncountable family of examples, viz. For $\lambda \in \mathbb{R}$, let

$$R(\lambda) = \mathbb{R}[X_1, X_2, \ldots, X_6]/({X_1}^2 + {X_2}^2 - 1, {X_3}^2 + \cdots + {X_6}^2 - 1, X_3 + \lambda {X_4}^2),$$

that the Vaserstein symbol $V_{R(\lambda)}$ is not injective is shown by an argument going back to an example of W. van der Kallen in [10]. We refer to [12] for more details.

The hard part was to show that these threefolds were not isomorphism as $\mathbb{R}$-algebras. This was shown by computing the units in these algebras. This was achieved via the following observation, which depended on an observation of Daigle in [4].

Theorem 10.5 *For any integer* $\lambda \in \mathbb{R}$, *let*

$$A(\lambda) := \mathbb{C}[X_1, X_2, \ldots, X_6]/({X_1}^2 + {X_2}^2 - 1, {X_3}^2 + \cdots + {X_6}^2 - 1, X_3 + \lambda {X_4}^2).$$

Then $A(\lambda) \cong A(\lambda')$ *if and only if* $\lambda = \pm\lambda'$. *Thus, if*

$$R(\lambda) = \mathbb{R}[X_1, X_2, \ldots, X_6]/({X_1}^2 + {X_2}^2 - 1, {X_3}^2 + \cdots + {X_6}^2 - 1, X_3 + \lambda {X_4}^2),$$

then $R(\lambda) \cong R(\lambda)$ *if and only if* $\lambda = \pm\lambda'$.

10.6 Smooth Counterexamples

The initial counterexample $A = \Gamma(S^3_{\mathbb{R}})$, the coordinate ring of the real 3-sphere, was a smooth variety. The Rao–van der Kallen example was shown to be related to the existence of an orthogonal 3×3 matrix $\rho \in SL_3(A)$, for which $[e_1\rho^2] \neq [1]$ in $W_E(A)$. (See [14] or [13] for details of the construction of ρ.)

By the usual arguments in the Witt group one can show that $[V_A(e_1\rho)]^2 = [V_A(\chi_2(e_1\rho))]$, where $\chi_2 : Um_3(A)/E_3(A) \to Um_3(A)/E_3(A)$ is defined by

$\chi_2((a_1, a_2, a_3)E_3(A)) = ((a_1{}^2, a_2, a_3)E_3(A))$ (cf. [19]). Since ρ is orthogonal it is easy to show that $\chi_2[e_1\rho] = 1$. Hence $[V_A(e_1\rho)]^2 = 1$.

But by Whitehead's Lemma, $[V_A(e_1\rho^2)]=[V_A(e_1\rho)]\perp[V_A(e_1\rho)]=[V_A(e_1\rho)]^2=1$. So if V_A is injective then $[e_1\rho^2] = 1$. A contradiction.

So it is natural to take the most general example of a 3×3 orthogonal matrix and proceed.

Let $\overline{X} = (X_1, X_2, X_3), \overline{Y} = (Y_1, Y_2, Y_3), \overline{Z} = (Z_1, Z_2, Z_3)$. Let

$$\sigma = \begin{pmatrix} X_1 & X_2 & X_3 \\ Y_1 & Y_2 & Y_3 \\ Z_1 & Z_2 & Z_3 \end{pmatrix}$$

be an orthogonal matrix, i.e. $\sigma\sigma^T = I_3$. Then we may regard $\sigma \in SO_3(A_{\mathbb{Z}})$, where $A_{\mathbb{Z}}$ is the three-dimensional algebra over $\mathbb{Z}$, viz. the quotient of the polynomial ring in nine variables over the integers $\mathbb{Z}$: $\mathbb{Z}[X_1, X_2, X_3, Y_1, Y_2, Y_3, Z_1, Z_2, Z_3]$ modulo the ideal generated by $(\langle\overline{X}, \overline{X}\rangle - 1, \langle\overline{Y}, \overline{Y}\rangle - 1, \langle\overline{Z}, \overline{Z}\rangle - 1, \langle\overline{X}, \overline{Y}\rangle, \langle\overline{Y}, \overline{Z}\rangle, \langle\overline{Z}, \overline{X}\rangle)$.

Let K be a field of characteristic zero, and let A_K denote the threefold over K: $A_{\mathbb{Z}} \otimes K$. Clearly, there is a natural evaluation homomorphism $A_{\mathbb{R}} \longrightarrow \Gamma(S^3_{\mathbb{R}})$.

In view of the existence of ρ it follows that

Lemma 10.1 *With the above notation,* $[e_1\sigma] \neq 1$, $[e_1\sigma^2] \neq 1$ *in* $Um_3(A_{\mathbb{R}})/E_3(A_{\mathbb{R}})$.

Note that if $f \in A[\mathbb{R}]$ and $f \notin \mathfrak{m}_r$, for every real point $\mathfrak{m}_r$ of $A[\mathbb{R}]$, then its image $\varphi(f)$ under the natural homomorphism $\varphi : A[\mathbb{R}] \longrightarrow \Gamma(S^3_{\mathbb{R}})$ does not vanish on $S^3_{\mathbb{R}}$. Consequently, the induced map $SO_3(A[\mathbb{R}]) \longrightarrow SO_3(\Gamma(S^3_{\mathbb{R}}))$ will factor through $SO_3(A[\mathbb{R}]_f)$.

Corollary 10.1 *If* $D(f)$ *is a principal open set which contains all the real points of* $A_{\mathbb{R}}$ *then the Vaserstein symbol* $(V =)\ V_{(A_{\mathbb{R}})_f} : \mathrm{Um}_3(A_{\mathbb{R}f})/E_3(A_{\mathbb{R}f}) \longrightarrow W_E((A_f)_{\mathbb{R}})$ *is not injective.*

Corollary 10.1 will be the moot reason for the existence of a countable family of smooth non-isomorphic affine threefolds over $\mathbb{R}$ which are birationally equivalent but for which the Vaserstein symbol is not injective. This will follow from the following observation:

Given any affine algebra A over $\mathbb{R}$ *(or* $\mathbb{C}$*), there is a countable collection of* $f \in A \otimes \mathbb{C}$ *such that the coordinate rings of the open sets D(f) are not isomorphic as* $\mathbb{C}$*-algebras. Moreover, in case of the orthogonal group of the algebraic 3-sphere* $SO_3(\mathbb{R})$ *we can find such a collection of basic open sets* $D(f), f \in A = \Gamma(SO_3(\mathbb{R}))$*, not passing through the real points of Spec(A).*

Again, the main idea is to look at the following well-known property of the quotient of the group of units of A, viz.

Lemma 10.2 *Let A be an affine domain over an algebraically closed field k. Then* A^*/k^* *is a finitely generated abelian group.*

We later found that in [7, Lemma 1.1] another proof is given in the case when A is a normal affine variety. T. J. Ford attributes the result to P. Samuel in [15, Lemma 1].

From this one can deduce

Theorem 10.6 *Let k be a field of characteristic zero. Let $X = Spec(A)$ be a smooth affine algebra of dimension ≥ 2 over k. Then there exists an infinite family of principal basic open subsets $D(f_n)$ of X which are not isomorphic to each other.*

Finally, one can deduce the existence of a countable collection of birational algebras A_n to $\Gamma(SO_3(\mathbb{R}))$ which are not isomorphic, and for which the Vaserstein symbol V_{A_n} is not injective.

Theorem 10.7 *Let A be the coordinate ring of the algebraic $SO_3(\mathbb{R})$ then there is a sequence $f_1, f_2, \ldots$ of elements of A such that the Vaserstein symbol*

$$V_{A_{f_1 \ldots f_n}} : Um_3(A_{f_1 \ldots f_n})/E_3(A_{f_1 \ldots f_n}) \longrightarrow W_E(A_{f_1 \ldots f_n})$$

is not injective.

10.7 Jean Fasel's Conjecture

After the articles [12, 13], J. Fasel shared his views about the question of when V_A is not injective for affine threefolds over $\mathbb{R}$: Here is his very precise conjecture, which indicates it will be non-injective generally for 'almost all' smooth real threefolds:

Conjecture: (Jean Fasel)
The Vaserstein symbol V_A is injective on $X = \text{Spec}(A)$, for A a smooth affine algebra of dimension three over the reals $\mathbb{R}$ if and only if $X(\mathbb{R})$, endowed with the Euclidean topology, has no compact connected components.

The smooth counterexamples we give in Sect. 10.6 are birational threefolds which have connected compact components in the Euclidean topology.

Later, Jean Fasel himself settled his conjecture in [5]. He proved

Theorem 10.8 *Let $X = Spec(R)$ be a smooth affine real threefold. Let C be the set of compact connected components of $X(\mathbb{R})$ (in the Euclidean topology). Then, the Vaserstein symbol*

$$Um_3(R)/E_3(R) \to W_E(R)$$

is injective if and only if $C = \emptyset$.

We refer the reader to [5] for the detailed proof. We give a gist next taken essentially verbatim from the introduction of [5]: The method is as follows.

In [2, proof of Theorem 4.3.1], it was observed that the Vaserstein symbol has an interpretation in the realm of $\mathbb{A}^1$-homotopy theory. More precisely, let k be a perfect field. The smooth affine quadric Q_5 with $k[Q_5]=k[x_1, x_2, x_3, y_1, y_2, y_3]/\langle \sum x_i y_i - 1\rangle$ is isomorphic to the quotient of algebraic varieties SL_4/Sp_4.

The latter is in turn isomorphic to the affine scheme A'_4 representing the functor assigning to a ring R the set of invertible skew-symmetric matrices of size 4 with trivial Pfaffian.

The composite $Q_5 \to SL_4/Sp_4 \to A'_4$ associates to a 6-tuple $(a_1, a_2, a_3, b_1, b_2, b_3)$ such that $\sum a_i b_i = 1$ the matrix $V(a_1, a_2, a_3)$ described by

$$V(a_1, a_2, a_3) = \begin{pmatrix} 0 & -a_1 & -a_2 & -a_3 \\ a_1 & 0 & -b_3 & b_2 \\ a_2 & b_3 & 0 & -b_1 \\ a_3 & -b_2 & b_1 & 0 \end{pmatrix}$$

where b_1, b_2, b_3 are such that $\sum_{i=1}^{3} a_i b_i = 1$.

Now, there is a stabilization map $SL_4/Sp_4 \to SL_6/Sp_6$ and it turns out that this map actually determines the injectivity of the Vaserstein symbol.

Indeed, let $\mathscr{H}_{\mathbb{A}^1}(k)$ be the $\mathbb{A}^1$-homotopy category defined by Morel and Voevodsky in [11]. We then have $\mathrm{Hom}_{\mathscr{H}_{\mathbb{A}^1}(k)}(X, Q_5) = Um_3(R)/E_3(R)$ for any smooth affine threefold $X = SpecR$ and $\mathrm{Hom}_{\mathscr{H}_{\mathbb{A}^1}(k)}(X, SL_6/Sp_6) = W_E(R)$, while the stabilization map $Q_5 \to SL_6/Sp_6$ precisely induces the Vaserstein symbol.

In this context, Fasel used the computation of the homotopy sheaves of $Q_5 \simeq \mathbb{A}^3 \setminus 0$ obtained in [1] to prove that the symbol V is injective if C is empty.

To prove that this condition is also necessary, Fasel produced explicitly a morphism $\mathbb{A}^4 \setminus 0 \to Q_5$ whose composite with $Q_5 \to SL_6/Sp_6$ is homotopy trivial and showed that its real realization is the Hopf map $S^3_{\mathbb{R}} \to S^2_{\mathbb{R}}$. If $C \neq \emptyset$, this allows to produce non-trivial elements in $Um_3(R)/E_3(R)$ whose image under V is trivial.

References

1. A. Asok, J. Fasel, Splitting vector bundles outside the stable range and $\mathbb{A}^1$-homotopy sheaves of punctured affine spaces. J. Am. Math. Soc. **28**(4), 1031–1062 (2015)
2. A. Asok, B. Doran, J. Fasel, Smooth models of motivic spheres and the clutching construction. Int. Math. Res. Not. IMRN **6**, 1890–1925 (2017)
3. H. Bass, J. Milnor, J.-P. Serre, Solution of the congruence subgroup problem for SL_n $(n \geq 3)$ and Sp_{2n} $(n \geq 2)$. Inst. Hautes Études Sci. Publ. Math. **33**, 59–137 (1967)
4. D. Daigle, Locally nilpotent derivations and Danielewski surfaces. Osaka J. Math. **41**, 37–80 (2004)
5. J. Fasel, The Vaserstein symbol on real smooth affine threefolds, to appear in the *Proceedings of the International Colloquium on K-theory, T.I.F.R*
6. J. Fasel, R.A. Rao, R.G. Swan, On stably free modules over affine algebras. Publ. Math. Inst. Hautes Études Sci. **116**, 223–243 (2012)
7. T.J. Ford, The group of units on an affine variety. J. Algebra Appl. **13**(8), 1450065 (2014), 27 pp
8. A.S. Garge, R.A. Rao, A nice group structure on the orbit space of unimodular rows. *K*-Theory **38**(2), 113–133 (2008)

9. A. Gupta, R. Sridharan, S.K. Yadav, On a group structure on unimodular rows of length three over a two dimensional ring, to appear in *Leavitt Path Algebras and Classical K-Theory*. Indian Statistical Institute Series (Springer, Berlin, 2019)
10. W. van der Kallen, A module structure on certain orbit sets of unimodular rows. J. Pure Appl. Algebra **57**, 657–663 (1975)
11. F. Morel, V. Voevodsky, A^1-homotopy theory of schemes. Inst. Hautes Études Sci. Publ. Math. **90** (1999), 45–143 (2001)
12. D.R. Rao, N. Gupta, On the non-injectivity of the Vaserstein symbol in dimension three. J. Algebra **399**, 378–388 (2014)
13. D.R. Rao, S. Kolte, Odd orthogonal matrices and the non-injectivity of the Vaserstein symbol. J. Algebra **510**, 458–468 (2018)
14. R.A. Rao, W. van der Kallen, Improved stability for K_1 and WMS_d of a non-singular affine algebra. K-Theory (Strasbourg, 1992). Asterisqué no. 226(11), 411–420 (1994)
15. P. Samuel, Á propos du théorème des unités. (French) Bull. Sci. Math. (2) **90**, 89–96 (1966)
16. A.A. Suslin, The structure of the special linear group over rings of polynomials. Izv. Akad. Nauk SSSR Ser. Mat. **41**, 235–252 (1977)
17. R.G. Swan, On some actions of stably elementary matrices on alternating matrices (with Ravi A. Rao and Jean Fasel), see homepage of R.G. Swan at http://www.math.uchicago.edu/~swan/
18. L.N. Vaserstein, On the stabilization of the general linear group over a ring. Mat. Sb. (N.S.) **79**(121), 405–424 (Russian); translated in Math. USSR-Sb. **8**, 383–400 (1969)
19. L.N. Vaserstein, Operation on orbits of unimodular vectors. J. Algebra **100**, 456–461 (1986)
20. L.N. Vaserstein, A.A. Suslin, Serre's problem on projective modules over polynomial rings, and algebraic K-theory. (Russian) Funkcional. Anal. i Prilozen. **8**(2), 65–66 (1974). English translation in: Izv. Akad. Nauk SSSR Ser. Mat. **40**(5), 993–1054 (1976)

Chapter 11
Two Approaches to the Bass–Suslin Conjecture

Ravi A. Rao and Selby Jose

11.1 Introduction

In this short note, we give a glimpse of two ongoing attempts to resolve the well-known Bass–Suslin conjecture regarding completing unimodular polynomial rows over a local ring. We call the first approach the Suslin–Vaserstein symbol approach and the second approach the Unhampered Descent approach.

Let us begin with the known results about the Bass–Suslin conjecture (in small dimensions and rank); as our attempt evolved from these methods.

The primordial idea of L. N. Vaserstein to study unimodular rows of length 3 is by studying the alternating matrix which one can associate with them.

In [13], L. N. Vaserstein observed that given a pair $v = (a_1, a_2, a_3)$, $w = (b_1, b_2, b_3) \in R^3$ with $\langle v, w\rangle = a_1b_1 + a_2b_2 + a_3b_3 = 1$ one can associate an alternating matrix $V(v, w) \in SL_4(R)$ of pfaffian $\langle v, w\rangle = 1$ with it.

$$V_R(v) := V(v, w) = \begin{pmatrix} 0 & a_1 & a_2 & a_3 \\ -a_1 & 0 & b_3 & -b_2 \\ -a_2 & -b_3 & 0 & b_1 \\ -a_3 & b_2 & -b_1 & 0 \end{pmatrix}$$

R. A. Rao (✉)
School of Mathematics, Tata Institute of Fundamental Research,
1, Dr. Homi Bhabha Road, Mumbai 400005, India
e-mail: ravi@math.tifr.res.in

S. Jose
Department of Mathematics, Institute of Science,
Madam Cama Road, Mumbai 400032, India
e-mail: selbyjose@gmail.com

A. A. Ambily et al. (eds.), *Leavitt Path Algebras and Classical K-Theory*,
Indian Statistical Institute Series,
https://doi.org/10.1007/978-981-15-1611-5_11

L. N. Vaserstein then defined the elementary symplectic Witt groups $W_E(R)$ over any commutative ring with 1. His idea was to take the set of alternating matrices of pfaffian one, define an equivalence relation of stable equivalence under the action of the (infinite) elementary group on it by conjugation, and 'add' by 'placing diagonally' (the $\perp$ process). This enabled him to define a map $V_R : Um_3(R)/E_3(R) \to W_E(R)$; which we now call the Vaserstein symbol.

In [13, Theorem 5.2] L. N. Vaserstein proved that over a two dimensional ring R the orbit space of unimodular rows $Um_3(R)$ of length 3 modulo the action of the elementary group $E_3(R)$ is bijective with this Witt group $W_E(R)$. In this way, he gave a Witt group structure to the orbit space over two-dimensional rings.

But more importantly, since the symbol V_R exists in any dimension; the Bass–Suslin conjecture raises a basic question: can two unimodular polynomial rows v_1, v_2 of length 3 over $A := R[X]$, be equivalent 'stably', i.e., are the associated elements in the Witt group $[V_A(v_1)], [V_A(v_2)]$ equivalent in $W_E(A)$.

The theorem of M. Karoubi (see [13]) states that $W_{SL}(R[X]) = 0$ if R is local and $1/2 \in R$. So one gets that, over a polynomial ring $R[X]$ over a local ring R, $V_A(v_1)$, $V_A(v_2)$ are stably SL-equivalent in $W_{SL}(R[X])$. If dimension $R \leq 3$ one can then see, by the descent methods of L. N. Vaserstein, that there is $\sigma \in SL_4(R[X])$ such that $\sigma^t V_A(v_1)\sigma = V_A(v_2)$. In particular, if one takes $v_2 = e_1$ then one can say that v_1 is 'stably completable'.

The question raised by Hyman Bass in the early 70s in the Bateille conference (see [1]), and reiterated with a rider by A. Suslin in mid 70s (see ([14], Problem 4, page 491), was whether polynomial unimodular rows over a local ring R, of length $(r + 1)$ are always completable if $1/r! \in R$.

We call this the **Bass–Suslin conjecture**.

When R is a regular local ring it is referred to as the **Bass–Quillen conjecture**; here Suslin's rider $1/r! \in R$ is dropped. **The question of Quillen** in [9] was the following: Let $(R, \mathfrak{m})$ be a regular local ring. Let $\pi \in \mathfrak{m} \setminus \mathfrak{m}^2$ be a regular parameter of R. Are projective R_π-modules free? The affirmative solution of the Bass–Suslin conjecture follows as a consequence of the affirmative solution of the question of Quillen.

This conjecture is known when R contains a field due to results of H. Lindel in [7], from which Popescu [8] structure theorem of such regular local rings could derive the general case. The Quillen question has also been answered when R is a regular local ring containing a field in [6].

Much less is known of the Bass–Suslin conjecture $BS_r(R[X])$: every unimodular polynomial row of length $(r + 1)$ over a local ring R is completable to an invertible matrix, if $1/r! \in R$. More generally, I think one expects that the stronger statement that a 'factorial unimodular row' of the type $(s_0, s_1, s_2^2, s_3^3, \ldots, s_r^r)$ lies in the elementary orbit of any unimodular row of length $(r + 1)$ is true.

Basically, when dimension$(R) = 2$, then $BS_2(R[X])$ was completely solved by a remark of M. P. Murthy that Karoubi's theorem $W_{SL}(R[X]) = 0$, and Vaserstein's theorem that $V_{R[X]}$ is an isomorphism, put together solves it. In fact, due to arguments of M. Roitman in [12], the stronger $BS_2(R[X])$ statement is also true. As a consequence, $BS_d(R[X])$ is solved if dimension $R = d$.

When dimension$(R) = 3$ then $BS_2(R[X])$ is completely solved in dimension 3 by the first named author in [10, 11]. However, the stronger version is not clear for $Um_3(R[X])$, when dimension $R = 3$.

In a sense, the missing link to settle $BS_{d-1}(R[X])$ is whether the Vaserstein symbol $V_{R[X]}$ is injective when R is a local ring of dimension 3, with $1/2 \in R$.

We have not been able to answer that. But we expect it to be true.

So we have taken the approach that for unimodular rows of length 3 the Vaserstein symbol is a way to analyse the problem. For unimodular rows of length ≥ 3 we felt that a different symbol on the orbit space was desirable. Here is what we started to do.

11.2 The Suslin–Vaserstein Symbol

In this section we give a quick preview on some recent development in the study of the Suslin–Vaserstein symbols S_R^r from the orbit space of unimodular rows of length $r+1$

$$S_R^r : Um_{r+1}(R)/E_{r+1}(R) \longrightarrow W^r_{EUm^{\geq r}}(R),$$

to the elementary unimodular vector Witt group; which mimics L. N. Vaserstein's construction of the elementary symplectic Witt group symbol in [13] from

$$V_R : Um_3(R)/E_3(R) \longrightarrow W_E(R),$$

the elementary symplectic Witt group. Note that in the sequel we will be replacing the Witt group on the right-hand side by a variant $W^r_{EUm^{s\equiv r \bmod (4)}}(R)$; and asserting similar results with those symbols. Why have we taken $r \equiv s$ modulo (4): this is because the Suslin matrices corresponding to unimodular rows of the same length r satisfy similar properties according to the *Suslin identities* in ([14], Lemma 5.3).

We remark that our calculations seem to show that we may also take $r \equiv s$ modulo (2) and work with the union of those elementary unimodular vector groups $\cup EUm_{s\equiv r \bmod (2)}$. We find it works just as fine, when r is even; with the usual action.

We refer the reader to [2, 3] for the terminology below.

In a nutshell, the elementary symplectic Witt group is got by stable equivalence, under the action by conjugation of the infinite elementary linear group on the set of alternating matrices of pfaffian one. Whereas the elementary unimodular vector group is got by stable equivalence, under the action by conjugation of the infinite elementary unimodular vector group, under the action of the natural involution, on the set of all special Suslin matrices $S_r(v, w)$, $\langle v, w\rangle = 1$. (Since the involution on $SUm_r(R)$ is only defined up to a unit when r is odd (see [2]), one needs to take a bit of care here.)

Let us now discuss the 'positioning' of the Suslin matrices, which leads to the addition in the new Witt group similar to the effect of $\perp$ in the Witt group $W_E(R)$.

If we regard each $S_r(v, w)$ as a 2×2 block matrix, say

$$S_r(v_i, w_i) = \begin{pmatrix} a_1 I_{2^{r-1}} & S_i \\ T_i & b_i I_{2^{r-1}} \end{pmatrix}$$

for $i = 1, 2$, then the placement of the 'sum' $S_r(v_1, w_1) \odot S_r(v_2, w_2)$ is given by

$$S_r(v_1, w_1) \odot S_r(v_2, w_2) = \begin{pmatrix} a_i I_{2^{r-1}} & 0 & 0 & S_1 \\ 0 & a_2 I_{2^{r-1}} & S_2 & 0 \\ 0 & T_2 & b_2 I_{2^{r-1}} & 0 \\ T_2 & 0 & 0 & b_1 I_{2^{r-1}} \end{pmatrix}.$$

We call this operation $\odot$ as 'circle placement' and read it as $S_r(v_1, w_1)$ circles $S_r(v_2, w_2)$. This placement will be used for the 'addition' operation in the elementary unimodular vector Witt group.

There is nothing sacrosanct about the circle placement; one can try several other natural choices too. We tried a few, and were convinced that all of them gave isomorphic groups.

11.3 The Suslin–Vaserstein Symbol

11.3.1 The Elementary Unimodular Vector Witt Groups

We need to construct **a variant of the elementary symplectic Witt group** which we christen the elementary unimodular vector group $W_{EUm^r_{\geq r}}(R)$. This is done for each size r in [4].

The basic idea of this construction is to replace the alternating matrices by special Suslin matrices corresponding to unimodular rows of length $(r + 1)$. However, even in the case $r = 2$ the groups $W_{EUm^2_{\geq 2}}(R)$ (or even $W^2_{EUm_{r\equiv 2\bmod(4)}}(R)$) and $W_E(R)$ need not be isomorphic. One can show that they are isomorphic if the Vaserstein symbol V_R is injective.

11.3.2 An Analogue of Karoubi's Linearization Process

This part is work in progress and is yet to be finalised.

Following Murthy's remark, we next hope to show that the analogous Karoubi theorem also holds in this context; i.e., if R is a local ring of dimension d with $1/2 \in R$, then the group $W^r_{SUm_{\geq r}}(R[X])$ (defined in a similar manner) is trivial.

In particular, one can deduce that $W^r_{EUm_{\geq r}}(R)$ is k-divisible when $1/k \in R$. (We are in the process of doing this step; which is again basically establishing a linearization process using the group $EUm_r^{\infty}(R[X]) := \cup_{t\geq r} EUm_r(R[X])$ or the groups $W^r_{EUm_{s\equiv r \bmod i}}(R)$, for $i = 2, 4$).

Then one needs the analogue of Vaserstein symbol that the Vaserstein symbol is bijective.

11.3.3 Analogue of Vaserstein's Theorem

Our Main Theorem so far is that there is a natural map from

$$S_R^r : Um_{r+1}(R)/E_{r+1}(R) \longrightarrow W^r_{EUm_{s\equiv r \bmod 4}}(R),$$

for all r, which is **injective**. Moreover, if $d = \dim(R)$, and $d \leq 2(r+1) - 3 = 2r - 1$ then this map is also surjective.

The surjectivity is due to the fact that the **Mennicke–Newmann lemma** become available in this range, i.e., given any two unimodular rows $v_1, v_2 \in Um_{r+1}(R)$, there exists $\varepsilon_1, \varepsilon_2 \in E_{r+1}(R)$ such that

$$\begin{aligned} v_1\varepsilon_1 &= (x, a_1, \ldots, a_r), \\ v_2\varepsilon_2 &= (y, a_1, \ldots, a_r). \end{aligned}$$

One may even arrange that $x + y = 1$, i.e., after elementary transformations one may arrange that the rows have all but one coordinate the same.

As a consequence, after completing the symplectic linearization process, we can deduce that Bass–Suslin conjecture holds *in the metastable range*, i.e., unimodular polynomial rows of length $r + 1$ over a local ring R of dimension d, with $1/r! \in R$, are completable provided $d \leq 2(r+1) - 3 = 2r - 1$.

Remark 11.1 Since the groups $W^r_{EUm_{s\equiv r \bmod i}}(R)$ are defined by an action of $\cup_s EUm_{s\equiv r \bmod i}$, when $i = 4$ (or even $\cup_s EUm_{s\equiv r \bmod i}$, when $i = 2$), and when r is even; and not otherwise, so the Witt group is only well defined in these situations. For the other situations, when the vectors are of even length, a slightly different argument needs to be done.

In particular, the estimates of the existence of the group structure on the orbit space $Um_{r+1}(R)/E_{r+1}(R)$ is when $d \leq 2(r+1) - 4 = 2r - 2$; as in van der Kallen's theorem in [5].

Note 11.1 Since many parts of the above approach are available for the general dimension and size cases; one may hope to also be able to extend the results by above method for all r, i.e., to encompass the Bass–Suslin theorem.

But this is far from clear at the moment.

In any case, the above method shows that we need to *concentrate at a certain rank r*, fixed once and for all. In that sense, one should first stick to the rank 2 case. Here one needs to play in $W_E(R[X])$. The Vaserstein approach leads via Karoubi's linearization to resolve the issue stably. The Vaserstein descent process causes difficulties beyond dimension three, due to the injective and surjective stability issues.

11.4 A Descent Approach

Let us concentrate on the case $r = 2$ of the Bass–Suslin conjecture. Since we are studying unimodular rows of length 3, we intend to play with the associate alternating matrices. These can be linearized, as shown by L. N. Vaserstein in [13]. Then L. N. Vaserstein has developed a method of descent. However, this method has some severe limitations, and seems to work up to dimension of $R \leq 5$ at most.

So we have to develop another method of *descent*. Here is a fleeting glimpse of a method which is being developed (with J. Fasel, R. G. Swan) in [16]:
Descent Lemma: Let R be a commutative ring with 1. Let φ, φ^* be invertible alternating matrices of the same size. Assume that one has a relation of the form

$$\Theta(\psi_r \perp \varphi \perp \psi_r)\Theta^t = \psi_r \perp \varphi^* \perp \psi_r,$$

for some $r \geq 0$, and for some invertible alternating matrix Θ. If pf $(\varphi^{-1} + \varphi^*)$ is a non-zero-divisor in R, then there is an invertible alternating matrix Θ^* such that $\Theta^*\varphi\Theta^{*^t} = \varphi^*$. Moreover, if one writes Θ as a 3×3 block matrix $\Theta = (\Theta_{ij})_{1\leq i,j\leq 3}$, of appropriate sizes, then

$$\begin{aligned} 0 = \Theta_{12} = \Theta_{21} = \Theta_{13} = \Theta_{31} = \Theta_{23} = \Theta_{32}, \\ \Theta_{22} = \Theta^*. \end{aligned}$$

How does one reach a situation where the above lemma can be applicable?

Remark 11.2 Our feeling is that a more symmetric placement approach should remove the injective stability issue. This is because we feel that it is only *a symplectic matrix* which is hindering the descent. If one could 'position' properly when we linearize then one could bring out this symplectic matrix which is jamming things. We hope to achieve this by doing a 'symplectic linearization' process (instead of the usual linearization process used so far).

Our attempt is to do a type of Karoubi linearization—only to do it via symplectic matrix. (A bit more than just symplectic linearization is actually needed.) We are attempting to complete this process.

Finally, we suspect that if the above process can be accomplished in the case when $r = 2$ then we hope to imitate that process to cover $BS_r(R[X])$, $r > 2$, via a similar 'stable positioning play' with the Suslin matrices.

Acknowledgements The second author thank the Science and Engineering Research Board (SERB), Department of Science and Technology, Government of India, for the funding of project MTR/2017/000875 under Mathematical Research Impact Centric Support (MATRICS).

References

1. H. Bass, *Some Problems in "Classical" Algebraic K-Theory. Algebraic K-Theory, II: "Classical" Algebraic K-Theory and Connections with Arithmetic (Proc. Conf., Battelle Memorial Inst., Seattle, Wash., 1972)*. Lecture Notes in Mathematics, vol. 342 (Springer, Berlin, 1973), pp. 3–73
2. Selby Jose, Ravi A. Rao, A structure theorem for the elementary unimodular vector group. Trans. Am. Math. Soc. **358**(7), 3097–3112 (2006)
3. S. Jose, R.A. Rao, A fundamental property of Suslin matrices. J. *K*-Theory **5**(3), 407–436 (2010)
4. S. Jose, R.A. Rao, *A Witt group structure on Suslin matrices*, in preparation
5. W. van der Kallen, A module structure on certain orbit sets of unimodular rows. J. Pure Appl. Algebr. **57**(3), 281–316 (1989)
6. S.M. Bhatwadekar, R.A. Rao, On a question of Quillen. Trans. Am. Math. Soc. **279**(2), 801–810 (1983)
7. H. Lindel, On the Bass–Quillen conjecture concerning projective modules over polynomial rings. Invent. Math. **65**(2), 319–323 (1981/82)
8. D. Popescu, General Néron desingularization. Nagoya Math. J. **100**, 97–126 (1985)
9. D. Quillen, Projective modules over polynomial rings. Invent. Math. **36**, 167–171 (1976)
10. R.A. Rao, The Bass–Quillen conjecture in dimension three but characteristic $\neq 2,3$ via a question of A. Suslin. Invent. Math. **93**(3), 609–618 (1988)
11. R.A. Rao, On completing unimodular polynomial vectors of length three. Trans. Am. Math. Soc. **325**(1), 231–239 (1991)
12. M. Roitman, On stably extended projective modules over polynomial rings. Proc. Am. Math. Soc. **97**(4), 585–589 (1986)
13. A.A. Suslin, L.N. Vaserstein, Serre's problem on projective modules over polynomial rings and algebraic K-theory. Math. USSR Izvestija **10**, 937–1001 (1976)
14. A.A. Suslin, Stably free modules. (Russian) Mat. Sb. (N.S.) **102** (144)(4), 537–550 (1977)
15. A.A. Suslin, The structure of the special linear group over rings of polynomials. Izv. Akad. Nauk SSSR Ser. Mat. **41**, 235–252 (1977)
16. R.G. Swan, R.A. Rao, J. Fasel, A regenerative property of a fibre of invertible alternating matrices. see Excerpts on home page of R.G. Swan at http://math.uchicago.edu/~swan/

Chapter 12
The Pillars of Relative Quillen–Suslin Theory

Rabeya Basu, Reema Khanna and Ravi A. Rao

12.1 Introduction

The main pillars of the Horrocks–Quillen–Suslin theory were developed in the papers [11, 19, 26]. In [11] the *Monic Inversion Principle*, in [19] the *Local-Global Principle*, and in [26] the *Normality of the Elementary subgroup* $\mathrm{E}_n(R)$, were established. In [26], the K_1 analogues of both the Monic Inversion Principle and the Local-Global Principle were developed. In addition, Suslin established the *Normality of the Elementary Linear subgroup* $\mathrm{E}_n(R)$ in the general linear group $\mathrm{GL}_n(R)$ over a module finite ring A, when $n \geq 3$. This was appeared in [27].

In [7] the authors had established, for classical linear groups, viz. the linear, symplectic and orthogonal groups, that the Quillen–Suslin's Local-Global Principle for the pair $(\mathrm{GL}_n(R[X]), \mathrm{E}_n(R[X]))$ and Suslin's Normality Principle were *equivalent* in the sense that if one holds then so does the other. Recently, in [20] a further unification of these three principles was achieved.

In this article, we develop the equivalence of a *relative Quillen's Local-Global Principle* and a *normality of the relative elementary subgroup*; cf. Theorem 18.1 for the precise equivalent statements.

R. Basu
Indian Institute of Science Education and Research,
Dr. Homi Bhabha Road, Pune 400008, India
e-mail: rabeya.basu@gmail.com; rbasu@iiserpune.ac.in

R. Khanna
Somaiya College, Vidyavihar, Mumbai 400077, India
e-mail: reemag16@gmail.com

R. A. Rao (✉)
Tata Institute of Fundamental Research,
1, Dr. Homi Bhabha Road, Mumbai 400005, India
e-mail: ravi@math.tifr.res.in

A. A. Ambily et al. (eds.), *Leavitt Path Algebras and Classical* K*-Theory*,
Indian Statistical Institute Series,
https://doi.org/10.1007/978-981-15-1611-5_12

We refer the reader to the Introduction of [7] where recent developments of the Quillen–Suslin theory are discussed in detail. The study of the relative Local-Global Principle with respect to an extended ideal began in [1]; and was developed in [2] for the Chevalley groups.

The proofs of the equivalent statements in this paper are done in an analogous manner to that done in [7]. This was possible due to a recent argument, which is detailed in [13], and which first appeared in the thesis of Anjan Gupta [8]. This argument works with the Noetherian excision ring $R \oplus I$ rather than the use of the (non-Noetherian) Excision ring $\mathbb{Z} \oplus I$, and the Excision theorem of W. van der Kallen in [28], as is commonly used. We refer [9] to see other interesting applications of the Noetherian Excision rings.

For the sake of being self contained we have detailed the arguments of the various equivalences. However, we note that we could have alternatively *deduced* the implications from the corresponding implications done in [7] via this Noetherian Excision ring argument.

12.2 Definitions and Notations

Let R be a commutative ring with 1, and $I \subset R$ an ideal. We refer [7] for the standard definitions and facts of the general linear, symplectic and orthogonal groups, and their elementary subgroups. Let σ denote the permutation of the natural numbers given by $\sigma(2i) = 2i-1$ and $\sigma(2i-1) = 2i$. With respect to this permutation, we define following classical groups.

For an integer $m > 0$, the symplectic group of size $2m \times 2m$ is defined with respect to the alternating matrix ψ_m corresponding to the standard symplectic form

$$\psi_m = \sum_{i=1}^{m} e_{2i-1,2i} - \sum_{i=1}^{m} e_{2i,2i-1}.$$

For the orthogonal group we have considered symmetric matrix $\widetilde{\psi}_m$ corresponding to the standard hyperbolic form

$$\widetilde{\psi}_m = \sum_{i=1}^{m} e_{2i-1,2i} + \sum_{i=1}^{m} e_{2i,2i-1}.$$

Definition 12.1 (*Symplectic Group* $\mathrm{Sp}_{2m}(R)$) The group of all non-singular $2m \times 2m$ matrices $\{\alpha \in \mathrm{GL}_{2m}(R) \mid \alpha^t \psi_m \alpha = \psi_m\}$.

Definition 12.2 (*Orthogonal Group* $\mathrm{O}_{2m}(R)$) The group of all non-singular $2m \times 2m$ matrices $\{\alpha \in \mathrm{GL}_{2m}(R) \mid \alpha^t \widetilde{\psi}_m \alpha = \widetilde{\psi}_m\}$.

Definition 12.3 (*Elementary Symplectic Group* $\mathrm{ESp}_{2m}(R)$) For $1 \leq i \neq j \leq 2m$ we define,

$$se_{ij}(z) = \mathrm{I}_{2m} + ze_{ij} \text{ if } i = \sigma(j)$$
$$= \mathrm{I}_{2n} + ze_{ij} - (-1)^{i+j} ze_{\sigma(j)\sigma(i)} \text{ if } i \neq \sigma(j) \text{ and } i < j.$$

It is clear that when $z \in R$ all these matrices belong to $\mathrm{Sp}_{2m}(R)$. We call them the elementary symplectic matrices over R and the group generated by them is called elementary symplectic group.

Definition 12.4 (*Elementary Orthogonal Group* $\mathrm{ESp}_{2m}(R)$) For $1 \leq i \neq j \leq 2m$ we define,

$$oe_{ij}(z) = \mathrm{I}_{2n} + ze_{ij} - ze_{\sigma(j)\sigma(i)} \text{ if } i \neq \sigma(j) \text{ and } i < j.$$

It is clear that when $z \in R$ all these matrices belong to $\mathrm{O}_{2m}(R)$. We call them the elementary orthogonal matrices over R and the group generated by them is called elementary orthogonal group.

Notation In the sequel $\mathrm{M}(n, R)$ will denote the set of all $n \times n$ matrices, $\mathrm{G}(n, R)$ will denote either the linear group $\mathrm{GL}_n(R)$, the symplectic group $\mathrm{Sp}_{2m}(R)$, or the orthogonal group $\mathrm{O}_{2m}(R)$, where $2m = n$. $\mathrm{S}(n, R)$ will denote either the special linear group $\mathrm{SL}_n(R)$, the symplectic group $\mathrm{Sp}_{2m}(R)$, or the special orthogonal group $\mathrm{SO}_{2m}(R)$, when R is a commutative ring. Similarly, $\mathrm{E}(n, R)$ will denote the corresponding elementary subgroups $\mathrm{E}_n(R)$, $\mathrm{ESp}_{2m}(R)$, $\mathrm{EO}_{2m}(R)$ respectively. To denote the generators of $\mathrm{E}(n, R)$ we shall use the symbol $ge_{ij}(x)$, $x \in R$. □

Definition 12.6 The elementary subgroup $\mathrm{E}(n, I)$ with respect to the ideal I is the subgroup of $\mathrm{E}(n, R)$ generated as a group by the elements $ge_{ij}(x)$, for $x \in I$. The *relative elementary group* $\mathrm{E}(n, R, I)$ is the normal closure of $\mathrm{E}(n, I)$ in $\mathrm{E}(n, R)$. Thus $\mathrm{E}(n, R, I)$ is generated by elements of the form $ge_{ij}(a) ge_{kl}(x) ge_{ij}(-a)$ where $a \in R$ and $x \in I$.

Notation The relative subgroups of $\mathrm{G}(n, R)$ and $\mathrm{S}(n, R)$ will be denoted by $\mathrm{G}(n, R, I)$ and $\mathrm{S}(n, R, I)$ respectively, i.e.,

$$\mathrm{G}(n, R, I) = \{\alpha \in \mathrm{G}(n, R) \mid \alpha \equiv \mathrm{I}_n \text{ modulo } I\},$$

$$\mathrm{S}(n, R, I) = \{\alpha \in \mathrm{S}(n, R) \mid \alpha \equiv \mathrm{I}_n \text{ modulo } I\}.$$

For an ideal I in R, its extension in the ring $R[X]$, i.e., $I \otimes_R R[X]$ will be denoted by $I[X]$.

Similarly, $\mathrm{Um}_n(R, I)$ will denote the set of all unimodular rows of length n which are congruent to $e_1 = (1, 0, \ldots, 0)$ modulo I.

We will mostly use localizations with respect to two types of multiplicatively closed subsets of R. viz. $S = \{1, s, s^2, \ldots\}$, where $s \in R$ is a non-nilpotent, non-zero divisor, and $S = R \setminus \mathfrak{m}$ for $\mathfrak{m} \in \mathrm{Max(R)}$. By $I_s[X]$ and $I_{\mathfrak{m}}[X]$ we shall mean the extension of $I[X]$ in $R_s[X]$ and $R_{\mathfrak{m}}[X]$ respectively.

Blanket Assumption: We assume that $n \geq 3$, when dealing with the linear case and $n = 2m$, with $m \geq 2$, when considering the symplectic and orthogonal cases. While dealing with the orthogonal groups we shall consider only isotropic vectors; i.e., all such non-zero vectors which are orthogonal to themselves with respect to the given non-degenerate bilinear form. Throughout the article we shall assume 2 is invertible in the ring R.

Notation For any column vector $v \in R^n$ we denote by $\widetilde{v} = v^t.\psi_n$ in the symplectic case and $\widetilde{v} = v^t.\widetilde{\psi}_n$ in the orthogonal case. □

Definition 12.9 We define the map $M : R^n \times R^n \to M(n, R)$ and the inner product $\langle\ ,\ \rangle$ as follows: Let v, w be column vectors in R^n. Then,

$$
\begin{aligned}
M(v, w) &= v.w^t, \quad \text{when dealing with the case } \mathrm{G}(n, R) = \mathrm{GL}_n(R).\\
&= v.\widetilde{w} + w.\widetilde{v}, \quad \text{when } \mathrm{G}(n, R) = \mathrm{Sp}_{2m}(R).\\
&= v.\widetilde{w} - w.\widetilde{v}, \quad \text{when } \mathrm{G}(n, R) = \mathrm{O}_{2m}(R).\\
\langle v, w\rangle &= v^t.w, \quad \text{when } \mathrm{G}(n, R) = \mathrm{GL}_n(R).\\
&= \widetilde{v}.w, \quad \text{when } \mathrm{G}(n, R) = \mathrm{Sp}_{2m}(R) \text{ or } O_{2m}(R).
\end{aligned}
$$

Notation For any $\alpha \in \mathrm{G}(n, R)$, as usual $\alpha \perp \mathrm{I}_r$ denotes its embedding in $\mathrm{G}(n + r, R)$, where r is even for non-linear cases. □

To deduce the relative case from the absolute case we consider the 'Noetherian Excision ring'.

Definition 12.11 (*The ring $R \oplus I$*) Let I be an ideal in the ring R. We construct the new ring $R \oplus I$ by defining addition and coordinate wise multiplication as follows:

$$(r \oplus j)(s \oplus i) = rs \oplus (sj + ri + ij) \text{ for } r, s \in R \text{ and } i, j \in I.$$

There is a natural homomorphism $\phi : R \oplus I \longrightarrow R$ given by $(r \oplus i) \to r + i \in R$.

Note that when R is a Noetherian ring then the ring $R \oplus I$ is also a Noetherian ring; whereas, the Excision ring $\mathbb{Z} \oplus I$ need not be a Noetherian ring.

Notation Let $\mathrm{E}(n, I) = \{\alpha \in \mathrm{S}(n, R) \mid \alpha \equiv \mathrm{I}_n \text{ modulo } I\}$. In general, $\mathrm{E}(n, I)$ is not normal in $\mathrm{G}(n, R)$. By $\mathrm{E}(n, R, I)$ we mean the the normalisation of $\mathrm{E}(n, I)$ in $\mathrm{G}(n, R)$, i.e., the relative elementary group generated by elements of the type $ge_{ij}(f)ge_{ji}(h)(ge_{ij}(f))^{-1}$, where $f \in R$ and $h \in I$. While working on the polynomial ring $R[X]$, by writing $\alpha(X) \in \mathrm{E}(n, R[X], I[X])$ we mean $\alpha(X)$ is I_n modulo I, and of the form $ge_{ij}(f(X))ge_{ji}(h(X))(ge_{ij}(f(X)))^{-1}$, where $f(X) \in R[X]$ and $h(X) \in I[X]$, as $\mathrm{E}(n, R[X], I[X])$ is the normalisation of $\mathrm{E}(n, I[X])$ in $\mathrm{G}(n, R[X])$. □

Lemma 12.1 *If $\epsilon \in \mathrm{E}(n, R, I)$, then there exists $\epsilon' \in \mathrm{E}(n, R \oplus I)$ such that $\phi(\epsilon')=\epsilon$. (In fact, the converse is also true).*

Proof Let $\epsilon = (\epsilon_{ij})$ be a generator of the type $ge_{ij}(a)ge_{kl}(x)ge_{ij}(-a)$, where $a \in R$ and $x \in I$. Let

$$\epsilon' = ge_{ij}((a,0))ge_{kl}((0,x))ge_{ij}(-(a,0)) \in \mathrm{E}(n, R \oplus I).$$

Then by applying the homomorphism ϕ to it we obtain $\phi(\epsilon') = \epsilon$.

Any $\gamma \in \mathrm{E}(n, R)$ can be written as $\prod_{r=1}^{s} ge_{i_r j_r}(\lambda_r)$, where $\lambda_r \in R$, and for $x \in I$, $\gamma ge_{kl}(x)\gamma^{-1}$ corresponds to $\prod_{r=1}^{s} ge_{i_r j_r}(\lambda_r, 0)ge_{kl}(0, x)(\prod_{r=1}^{s} ge_{i_r j_r}(\lambda_r, 0))^{-1} \in \mathrm{E}(n, R \oplus I)$. □

Lemma 12.2 *Let* $\alpha \in \mathrm{G}(n, R, I)$. *Then there exists* $\alpha' \in \mathrm{G}(n, R \oplus I)$ *such that* $\phi(\alpha') = \alpha$.

Proof Let $\alpha = (\alpha_{ij}) \in \mathrm{G}(n, R, I)$. Then $\alpha_{ii} = 1 + a_{ii}$ and $\alpha_{ij} = a_{ij}$ for $i \neq j$ where $a_{ij} \in I$ for all i, j. We get a new matrix $\alpha' = \alpha'_{ij}$, where $\alpha'_{ii} = (u_i, a_{ii})$ and $\alpha'_{ij} = (0, a_{ij})$ for $i \neq j$. The entries in α' are in the ring $R \oplus I$. Using the definition of multiplication in the ring $R \oplus I$, we can see that $\alpha' \in \mathrm{G}(n, R \oplus I)$ and applying the homomorphism ϕ we obtain $\phi(\alpha') = \alpha$. □

Now we state the main theorem of this article. For the absolute case; i.e., for $I = R$ we refer to [7].

12.3 Equivalence: Relative L-G Principle and Normality

Theorem 12.1 *Let R be a commutative ring with identity, and* $I \subsetneq R$ *an ideal of the ring R. Let v, w be column vectors in* R^n *with* $w \in I^n$. *Then the followings are equivalent:*

(1) *(***Normality***):* $\mathrm{E}(n, R, I)$ *is a normal subgroup of* $\mathrm{G}(n, R)$.
(2) $\mathrm{I}_n + M(v, w) \in E_n(R, I)$ *if* $v \in \mathrm{Um}_n(R, I)$ *and* $\langle v, w\rangle = 0$ *and* $w \in I^n$.
(3) *(***Local-Global Principle***):*
If $\alpha(X) \in \mathrm{G}(n, R[X], I[X])$; $\alpha(0) = \mathrm{I}_n$ *and* $\alpha_{\mathfrak{m}}(X) \in \mathrm{E}(n, R_{\mathfrak{m}}[X], \mathrm{I}_{\mathfrak{m}}[X])$ *for all* $\mathfrak{m} \in \mathrm{Max(R)}$ *then* $\alpha(X) \in \mathrm{E}(n, R[X], I[X])$.
(4) *(***Dilation Principle***):*
If $\alpha(X) \in \mathrm{G}(n, R[X], I[X])$; $\alpha(0) = \mathrm{I}_n$ *and* $\alpha_s(X) \in \mathrm{E}(n, R_s[X], I_s[X])$ *for some non-nilpotent element* $s \in R$, *then* $\alpha(bX) \in \mathrm{E}(n, R[X], I[X])$ *for* $b \in (s^l)$, $l \gg 0$. *(Actually, we mean there exists some* $\beta(X) \in \mathrm{E}(n, R[X], I[X])$ *such that* $\beta(0) = \mathrm{I}_n$ *and* $\beta_s(X) = \alpha_s(bX)$. *But, since there is no ambiguity, for simplicity we are using the notation* $\alpha(bX)$ *instead of* $\beta_s(X)$*).*
(5) *Let* $\alpha(X) = \mathrm{I}_n + X^d M(v, w)$ *for some integer* $d \gg 0$, $v \in \mathrm{E}(n, R, I)e_1$, $w \in I^n$ *with* $\langle v, w\rangle = 0$. *Then one gets* $\alpha(X) \in \mathrm{E}(n, R[X], I[X])$. *Moreover,* $\alpha(X)$ *can be expressed as a product decomposition of the form* $\Pi ge_{ij}(Xh(X))$ *for* $d \gg 0$ *and* $h(X) \in I[X]$.
(6) $\mathrm{I}_n + M(v, w) \in \mathrm{E}(n, R, I)$ *if* $v \in \mathrm{E}(n, R, I)e_1$, $w \in I^n$ *and* $\langle v, w\rangle = 0$.

(7) $\mathrm{I}_n + M(v, w) \in \mathrm{E}(n, R, I)$ *if* $v \in \mathrm{G}(n, R, I)e_1$, $w \in I^n$ *and* $\langle v, w\rangle = 0$.

Remark 12.1 Since (6) will be established in Lemma 12.8, it follows that all the above statement (1)–(7) of Theorem 18.1 hold for commutative (In fact, for almost commutative) rings.

Before proving the theorem we first collect a few lemmas.

Lemma 12.3 *The group* $\mathrm{E}(n, R, I)$ *satisfies the property:*

$$[\mathrm{E}(n, R, I), \mathrm{E}(n, R)] = \mathrm{E}(n, R, I).$$

Proof Cf. [4] for the general linear groups, ([15], Theorem 1.1) for the symplectic groups and ([24], §2) for the orthogonal groups. □

Below we state a few useful well-known lemmas. For the proofs cf. [4] for the linear groups, [15] for the symplectic groups, [24] for the orthogonal groups. For a uniform proof cf. [6, 7]. The analogous results for the relative cases follow from the proofs of the absolute cases.

Lemma 12.4 (Splitting Property) $ge_{ij}(x + y) = ge_{ij}(x)ge_{ij}(y)$, $\forall\, x, y \in R$.

Lemma 12.5 *Let* G *be a group, and* $a_i, b_i \in G$, *for* $i = 1, \ldots, n$. *Then* $\prod_{i=1}^{n} a_i b_i = \prod_{i=1}^{n} r_i b_i r_i^{-1} \prod_{i=1}^{n} a_i$, *where* $r_i = \prod_{j=1}^{i} a_j$.

Lemma 12.6 *The group* $\mathrm{G}(n, R[X], (X)) \cap \mathrm{E}(n, R[X], I[X])$ *is generated by the elements of the type* $\epsilon ge_{ij}(Xh(X))\epsilon^{-1}$, *where* $\epsilon \in \mathrm{E}(n, R[X])$, $h(X) \in I[X]$.

Lemma 12.7 *For* $m > 0$, *and* $h(Y) \in I[Y]$, *there are* $h_t(X, Y, Z) \in I[X, Y, Z]$ *such that*

$$ge_{pq}(Z)ge_{ij}(X^{2m}h(Y))ge_{pq}(-Z) = \prod_{t=1}^{k} ge_{p_tq_t}(X^m h_t(X, Y, Z)).$$

Corollary 12.1 *If* $\epsilon = \epsilon_1\epsilon_2\cdots\epsilon_r$, *where each* ϵ_j *is an elementary generator, and* $h(Y) \in I[Y]$, *then there are* $h_t(X, Y) \in I[X, Y]$ *such that*

$$\epsilon ge_{pq}(X^{2^r m}h(Y))\epsilon^{-1} = \prod_{t=1}^{k} ge_{p_tq_t}(X^m h_t(X, Y)).$$

Proof Follows by induction on r and using Lemma 12.7. □

We show that statement (6) of Theorem 18.1 is true over an arbitrary associative ring R with 1.

Lemma 12.8 *Let* R *be a ring and* $v \in \mathrm{E}(n, R, I)e_1$. *Let* $w \in I^n$ *be a column vector such that* $\langle v, w\rangle = 0$. *Then* $\mathrm{I}_n + M(v, w) \in \mathrm{E}(n, R, I)$.

Proof Let $v = \epsilon e_1$, where $\epsilon = (\epsilon_{ij}) \in \mathrm{E}(n, R, I)$. Hence $\epsilon_{ii} = 1 + a_{ii}$ and $\epsilon_{ij} = a_{ij}$ for $i \neq j$, where $a_{ij} \in I$ for all i, j. Let $\epsilon' = (\epsilon'_{ij})$, where $\epsilon'_{ii} = (1, a_{ii})$, and $\epsilon'_{ij} = (0, a_{ij})$ for $i \neq j$. Let $e'_1 = ((1, 0), (0, 0), \ldots, (0, 0))$, and

$$v' = ((1, v_1), (0, v_2), \ldots, (0, v_n)) \in (R \oplus I)^n,$$

$$w' = ((0, w_1), (0, w_2), \ldots, (0, w_n)) \in (0 \oplus I)^n.$$

Then it follows that

$$\mathrm{I}_n + M(v', w') = \epsilon'(\mathrm{I}_n + M(e'_1, w'_1))(\epsilon')^{-1},$$

$$\text{and } w'_1 = \begin{cases} (\epsilon')^t w' & \text{for linear case} \\ (\epsilon')^{-1} w' & \text{otherwise.} \end{cases}$$

Since $\langle (e'_1, w'_1) \rangle = \langle v', w' \rangle = 0$, we get

$$(w'_1)^t = \begin{cases} ((0, 0), (0, w_{12}), (0, w_{13}), \ldots, (0, w_{1n})) & \text{for linear case} \\ ((0, w_{11}), (0, 0), (0, w_{13}), \ldots, (0, w_{1n})) & \text{otherwise.} \end{cases}$$

Therefore,

$$\mathrm{I}_n + M(v', w') = \begin{cases} \prod_{j=2}^{n} \epsilon' \, ge_{1j}(0, w_{1j}) \, (\epsilon')^{-1} & \text{for linear case} \\ \prod_{\substack{j=1 \\ j \neq 2}}^{n} \epsilon' \, ge_{1j}(0, w_{1j}) \, (\epsilon')^{-1} & \text{otherwise .} \end{cases}$$

Hence $\mathrm{I}_n + M(v', w') \in \mathrm{E}(n, R \oplus I, 0 \oplus I)$. Now applying the homomorphism ϕ it follows that $\mathrm{I}_n + M(v, w) \in \mathrm{E}(n, R, I)$; as desired. □

Note that the above implication is true for any associative ring with identity.

Remark 12.2 It is well known that every ring is a direct limit of Noetherian rings. Hence we may consider R to be Noetherian.

We shall use following lemma frequently and sometime in a subtle way; e.g., for the implication (4) ⇒ (3).

Lemma 12.9 ([10], Lemma 5.1) *Let R be a Noetherian ring and $s \in R$. Then there exists a natural number k such that the canonical homomorphism $\mathrm{G}(n, s^k R) \to \mathrm{G}(n, R_s)$ (induced by localization homomorphism $R \to R_s$) is injective. Moreover, it follows that the map $\mathrm{E}(n, R, s^k R) \to \mathrm{E}(n, R_s)$ for $k \in \mathbb{N}$ is injective.*

Proof of Theorem 18.1 We shall assume the result for the absolute case; i.e., when $I = R$. The implication (7) ⇒ (6): Obvious. We prove, (6) ⇒ (5):

Note that we have assumed that (6) holds for any commutative ring, in particular for the ring $R[X]$, and the matrix $I_n + XM(v, w)$. Replacing R by $R[X]$ in (6) we get that $\mathrm{I}_n + XM(v, w) \in \mathrm{E}(n, R[X], I[X])$. Let $v = \epsilon e_1$, where $\epsilon \in \mathrm{E}(n, R, I)$. As before, let $v' = ((1, v_1), (0, v_2), \ldots, (0, v_n)) \in (R \oplus I)^n$, and $w' = ((0, w_1), (0, w_2), \ldots, (0, w_n)) \in (0 \oplus I)^n$. Hence as in the proof of Lemma 12.8, we can write

$$\mathrm{I}_n + XM(v', w') = \begin{cases} \prod_{j=2}^{n} \epsilon' \, ge_{1j}((0, Xw_{1j})) \, (\epsilon')^{-1} & \text{for linear case} \\ \prod_{\substack{j=1 \\ j \neq 2}}^{n} \epsilon' \, ge_{1j}((0, Xw_{1j})) \, (\epsilon')^{-1} & \text{otherwise} . \end{cases} \qquad (\star)$$

Now we split the proof into following two cases:

Case I: ϵ is an elementary generator of the type $ge_{pq}(x)$, $x \in R$. First applying the homomorphism $X \mapsto X^2$ and then applying Lemma 12.7 over $R[X]$ we get

$$\mathrm{I}_n + X^2 M(v', w') = \prod_j \left(\prod_{t=1}^{k} ge_{p_{j(t)} q_{j(t)}}(Xh'_{j(t)}(X)) \right),$$

where $h'_{j(t)}(X) \in ((0 \oplus I)[X])$. Again, as before applying the homomorphism ϕ it follows that

$$\mathrm{I}_n + X^2 M(v, w) = \prod_j \left(\prod_{t=1}^{k} ge_{p_{j(t)} q_{j(t)}}(Xh_{j(t)}(X)) \right),$$

where $h_{j(t)}(X) \in I[X]$; as desired. Hence the result also follows for $d \gg 0$.

Case II: ϵ is a product of elementary generators of the type $ge_{pq}(x)$. Let $\mu(\epsilon) = r$. Arguing as before, the result follows by applying the homomorphism $X \mapsto X^{2^r}$ using the Corollary 12.1.

(5) $\Rightarrow$ (4): Given that $\alpha_s(X) \in \mathrm{E}(n, R_s[X], I_s[X])$, where s is non-nilpotent element in the ring R, and $\alpha(0) = \mathrm{I_n}$. By Lemma 12.1, there exists $\alpha'_{(s,0)}(X) \in \mathrm{E}(n, R_s[X] \oplus I_s[X])$, where the element $(s, 0)$ will remain non-nilpotent in the ring $R \oplus I$, and $\phi(\alpha'_{(s,0)}(X)) = \alpha_s(X)$.

Also, by Lemma 12.6, $\alpha_s(X)$ can be written as a product of the matrices of the form $\epsilon_s ge_{ij}(Xh(X))\epsilon_s^{-1}$, with $h(X) \in I_s[X]$ and $\epsilon_s \in \mathrm{E}(n, R_s)$. Hence using the proof of Lemma 12.1 it follows that $\alpha'_{(s,0)}(X)$ can be written as a product of the matrices of the form $\epsilon'_{(s,0)} ge_{ji}((0, Xh(X)))(\epsilon'_{(s,0)})^{-1}$, where $\phi(\epsilon'_{(s,0)}) = \epsilon_s$ and $(0, Xh(X)) \in ((R \oplus I)_{(s,0)}[X])$.

Applying the homomorphism $X \mapsto XT^d$, where $d \gg 0$, from the polynomial ring $R[X]$ to the polynomial ring $R[X, T]$, we consider $\alpha'_{(s,0)}(XT^d)$. Note that $R_s[X, T] \cong (R_s[X])[T]$. Now, using the Equation ($\star$) as in the proof of (6) $\Rightarrow$ (5), we can rewrite $\alpha'_{(s,0)}(XT^d)$ as the form $\mathrm{I}_n + XT^d M(v, w)$; for some suitable v, w over the ring $(R_s[X] \oplus I_s[X])[T]$. Hence by (5) we can write $\alpha'_{(s,0)}(XT^d)$ as a product of elementary generators of general linear (symplectic/orthogonal resp.) group such that each of those elementary generator is congruent to identity modulo the ideal

(T) over the ring $((R_s \oplus I_s)[X])[T]$. Let l be the maximum of the powers occurring in the denominators of those elementary generators. Again, as R assumed to be Noetherian, by applying the homomorphism $T \mapsto (s, 0)^m T$, for $m \geq l$, it follows from Lemma 12.9 that by (uniquely) identifying it's lift over the ring $(R \oplus I)[X, T]$ we can write $\alpha'_{(s,0)}((s, 0)^m XT^d)$ as a product of elementary generators of the general linear (symplectic/orthogonal resp.) group such that each of those elementary generator is congruent to identity modulo (T). i.e., there exists some $\beta'(X, T) \in \mathrm{E}(n, (R \oplus I)[X, T])$ such that $\beta'(0, 0) = \mathrm{I}_n$ and $\beta'_{(s,0)}(X, T) = \alpha'_{(s,0)}((b, 0)XT^d)$ for some $(b, 0) \in (s, 0)^m(R \oplus I)$. Finally, by substituting $T = (1, 0)$ and using Lemma 12.9, we get $\alpha'((b, 0)X) \in \mathrm{E}(n, (R \oplus I)[X])$. Hence the result follows applying ϕ as before.

(4) $\Rightarrow$ (3): Since $\alpha_{\mathfrak{m}}(X) \in \mathrm{E}(n, R_{\mathfrak{m}}[X], I_{\mathfrak{m}}[X])$, for all $\mathfrak{m} \in \mathrm{Max}(R)$, for each $\mathfrak{m}$ there exists $s \in R \setminus \mathfrak{m}$ such that $\alpha_s(X) \in \mathrm{E}(n, R_s[X], I_s[X])$. Observe that

$$R_s[X] \oplus I_s[X] \cong (R_s \oplus I_s)[X] \cong (R \oplus I)_s[X].$$

Hence by Lemma 12.1, applied to the base ring $R_s[X]$, there exists $\alpha'_{(s,0)}(X) \in \mathrm{E}(n, (R \oplus I)_{(s,0)}[X])$ such that $\phi_s(\alpha'_{(s,0)}) = \alpha_s$. Let

$$\theta'(X, T) = \alpha'_{(s,0)}(X + T)\alpha'_{(s,0)}(T)^{-1}.$$

Then $\theta'(X, T) \in \mathrm{E}(n, (R \oplus I)_{(s,0)}[X, T])$ and $\theta'(0, T) = \mathrm{I}_n$. By the condition (4) of the Theorem, applied to the base ring $(R \oplus I)[T]$, there exists $\beta'(X) \in \mathrm{E}(n, (R \oplus I)[X, T])$ such that

$$\beta'_{(s,0)}(X) = \theta'((b, 0)X, T). \tag{12.1}$$

with $(b, 0) \in (s, 0)^l(R \oplus I)$ for some $l \gg 0$.

Now, using the Noetherian property of $R \oplus I$, as mentioned in the Remark 12.2, we may consider a finite cover of $R \oplus I$, say $(s_1, 0) + \cdots + (s_r, 0) = (1, 0)$. Since for $l \gg 0$, the ideal $\langle (s_1, 0)^l, \ldots, (s_r, 0)^l \rangle = R \oplus I$, we choose $(b_1, 0), \ldots, (b_r, 0) \in R \oplus I$, with $(b_i, 0) \in (s_i, 0)^l(R \oplus I)$, $l \gg 0$ such that (12.1) holds and $(b_1, 0) + \cdots + (b_r, 0) = (1, 0)$. Hence for each $i = 1, \ldots, r$, there exists $(\beta')^i(X) \in \mathrm{E}(n, (R \oplus I)[X, T])$ such that $(\beta')^i_{(s_i,0)}(X) = \theta'((b_i, 0)X, T)$. Now,

$$\prod_{i=1}^{r} (\beta')^i(X) \in \mathrm{E}(n, (R \oplus I)[X, T]).$$

But,

$$\alpha'_{s'_1\cdots s'_r}(X) = \left(\prod_{i=1}^{r-1} \theta'_{s'_1\cdots \widehat{s'_i}\cdots s'_r}(b'_iX, T)|_{T=b'_{i+1}X+\cdots+b'_rX}\right) \theta'_{s'_1\cdots\cdots s'_{r-1}}(b'_rX, 0),$$

where $s'_i = (s_i, 0)$ and $b'_i = (b_i, 0)$ for each $i = 1, \ldots, r$. Now $\alpha'(0) = \mathrm{I}_n$. Also, as a consequence of the Lemma 12.9 it follows that the map

$$\mathrm{E}(n, R, (s, 0)^k (R \oplus I)[X]) \to \mathrm{E}(n, (R \oplus I)_{(s,0)}[X])$$

for $k \in \mathbb{N}$ is injective for each $s = s_i$. Hence by (uniquely) identifying $\alpha'_{s'_1 \cdots s'_r}(X)$ with its lift, we conclude $\alpha'(X) \in \mathrm{E}(n, R[X] \oplus I[X])$. Finally, applying the map ϕ we get $\alpha(X) \in \mathrm{E}(n, R[X], I[X])$; as desired.

(3) $\Rightarrow$ (2): This is the implication where we use the commutative property of the base ring R. Let $\alpha(X) = \mathrm{I}_n + XM(v, w)$, where $v \in \mathrm{Um}_n(R, I)$ and $\langle v, w\rangle = 0$ and $w \in I^n$. Then $\alpha(0) = \mathrm{I}_n$. Let $v = (1 + v_1, v_2, \ldots, v_n)$ and $w = (w_1, w_2, \ldots, w_n) \in I^n$, with $v_i, w_i \in I$ for $i = 1, \ldots, n$. Then by Lemma 12.8, $\alpha_{\mathfrak{m}}(X)$ is elementary for every maximal ideal $\mathfrak{m}$ in R. Hence $\alpha(X)$ is elementary by (3).

(2) $\Rightarrow$ (1): Let $\epsilon \in \mathrm{E}(n, R, I)$ and $\gamma = (\gamma_{ij}) \in \mathrm{G}(n, R)$. There exist $\epsilon' \in \mathrm{E}(n, R \oplus I)$ and $\gamma' = ((\gamma_{ij}, 0)) \in \mathrm{G}(n, R \oplus I)$ respectively such that $\phi(\epsilon') = \epsilon$ and $\phi(\gamma') = \gamma$. Using (2) $\Rightarrow$ (1) of the absolute case we get $\gamma' \epsilon' (\gamma')^{-1} \in \mathrm{E}(n, R \oplus I)$ as $\mathrm{E}(n, R \oplus I) \lhd \mathrm{G}(n, R \oplus I)$ and applying the homomorphism ϕ it follows $\gamma \epsilon \gamma^{-1} \in \mathrm{E}(n, R, I)$; as required.

(1) $\Rightarrow$ (7): Let $v = \gamma e_1$ where $\gamma \in \mathrm{G}(n, R)$. Then there exists $\gamma' \in \mathrm{G}(n, R \oplus I)$ such that $\phi(\gamma') = \gamma$. Let $v' = \gamma' e_1$ and $w' = ((0, w_1), (0, w_2), \ldots, (0, w_n)) \in (0 \oplus I)^n$. We have $\langle v', w'\rangle = 0$. Hence, using (1) $\Rightarrow$ (7) of the absolute case it follows that $\mathrm{I}_n + M(v', w') \in E_n(R \oplus I)$. Now applying the homomorphism ϕ we get $\mathrm{I}_n + M(v, w) \in \mathrm{E}(n, R, I)$; as required.

The above implications prove the equivalence of the statements. □

Remark 12.3 Assuming the result for the absolute case treated in [7] one can give simpler proofs of the steps (5) $\Rightarrow$ (4), and (4) $\Rightarrow$ (3). But, there is a gap in the proof of the absolute case in [7], as mentioned in [5]. The gap was filled in [5] by proving results for Bak's unitary groups, which cover linear, symplectic and orthogonal groups, and some more classical type groups. To make this note self contained, we have given the detailed proofs of those steps.

12.4 Relative L-G Principle for Transvection Subgroups

In this section, we shall state auxiliary results without detailed proofs. For definitions of symplectic and orthogonal modules and their transvection subgroups, we refer to [3].

In [3], the first and third authors together with Anthony Bak generalised Quillen–Suslin's local-global principle for the transvection subgroups of the projective, symplectic and orthogonal modules. As before, all three cases were treated uniformly. We observe below how to obtain relative versions of that local-global principle. To state the results we need to recall a few notations.

Notation In the sequel P will denote either a finitely generated projective R-module of rank n, a symplectic R-module or an orthogonal R-module of even rank $n = 2m$

with a fixed form $\langle\, ,\rangle$. And Q will denote $P \oplus R$ in the linear case, and $P \perp R^2$, otherwise. We will use the notation $Q[X]$ to denote $(P \oplus R)[X]$ in the linear case and $(P \perp R^2)[X]$, otherwise. We assume that the rank of the projective module is $n \geq 2$, when dealing with the linear case, and $n \geq 6$, when considering the symplectic and the orthogonal cases. For a finitely generated projective R-module M we use the notation $\mathrm{G}(M)$ to denote $\mathrm{Aut}(M)$, $\mathrm{Sp}(M, \langle\, ,\rangle)$ and $\mathrm{O}(M, \langle\, ,\rangle)$ respectively; denote $\mathrm{SL}(M)$, $\mathrm{Sp}(M, \langle\, ,\rangle)$ and $\mathrm{Trans}(M)$, $\mathrm{Trans}_{\mathrm{Sp}}(M)$ and $\mathrm{Trans}_{\mathrm{O}}(M)$ respectively; and $\mathrm{ET}(M)$ to denote $\mathrm{ETrans}(M)$, $\mathrm{ETrans}_{\mathrm{Sp}}(M)$ and $\mathrm{ETrans}_{\mathrm{O}}(M)$ respectively.

We shall also assume the following hypotheses:
(H1) for every maximal ideal $\mathfrak{m}$ of R, the symplectic (orthogonal) module $Q_{\mathfrak{m}}$ is isomorphic to $R_{\mathfrak{m}}^{2m+2}$ for the standard bilinear form $\mathbb{H}(R_{\mathfrak{m}}^{m+1})$.

(H2) for every non-nilpotent $s \in R$, if the projective module Q_s is free R_s-module, then the symplectic (orthogonal) module Q_s is isomorphic to R_s^{2m+2} for the standard bilinear form $\mathbb{H}(R_s^{m+1})$.

We recall the following fact just to remind the reader that in the free case the transvection subgroups coincide with the elementary subgroups. Here the maps φ, φ_p, σ and τ are as defined in [3].

Lemma 12.10 *If the projective module P is free of finite rank n (in the symplectic and the orthogonal cases we assume that the projective module is free for the standard bilinear form), then* $\mathrm{Trans}(P) = \mathrm{E}_n(R)$, $\mathrm{Trans}_{\mathrm{Sp}}(P) = \mathrm{ESp}_n(R)$ *and* $\mathrm{Trans}_{\mathrm{O}}(P) = \mathrm{EO}_n(R)$ *for $n \geq 3$ in the linear case and for $n \geq 6$ otherwise.*

Proof In the linear case, for $p \in P$ and $\varphi \in P^*$ if $P = R^n$ then $\varphi_p : R^n \to R \to R^n$. Hence $1 + \varphi_p = \mathrm{I}_n + v.w^t$ for some column vectors v and w in R^n. Since $\varphi(p) = 0$, it follows that $\langle v, w\rangle = 0$. Since either v or w is unimodular, it follows that $1 + \varphi_p = \mathrm{I}_n + v.w^t \in \mathrm{E}_n(R)$. Similarly, in the non-linear cases we have $\sigma_{(u,v)}(p) = \mathrm{I}_n + v.\widetilde{w} + w.\widetilde{v}$, and $\tau_{(u,v)}(p) = \mathrm{I}_n + v.\widetilde{w} - w.\widetilde{v}$, where either v or w is unimodular and $\langle v, w\rangle = 0$. (Here $\sigma_{(u,v)}$ and $\tau_{(u,v)}$ are as in the definition of symplectic and orthogonal transvections.) Classically, these are known to be elementary matrices—for details see [25] for the linear case, [15] for the symplectic case, and [24] for the orthogonal case. □

Remark 12.4 Lemma 12.10 holds for $n = 4$ in the symplectic case. This will follow from Remark 12.5.

Remark 12.5 $\mathrm{ESp}_4(A)$ is a normal subgroup of $\mathrm{Sp}_4(A)$ by ([15], Corollary 1.11). Also $\mathrm{ESp}_4(A[X])$ satisfies the Dilation Principle and the Local-Global Principle by ([15], Theorem 3.6). Since we were intent on a uniform proof, these cases have not been covered above by us.

Proposition 12.1 (Relative Dilation Principle) *Let R be a commutative ring with identity, and $I \subsetneq R$ an ideal in R. Let P and Q be as in 12.13. Assume that (H2) holds. Let s be a non-nilpotent in R such that P_s is free, and let $\sigma(X) \in \mathrm{G}(Q[X], I[X])$ with $\sigma(0) = Id$. Suppose*

$$\sigma_s(X) \in \begin{cases} \mathrm{E}(n+1, R_s[X], I_s[X]) & \textit{in the linear case,} \\ \mathrm{E}(n+2, R_s[X], I_s[X]) & \textit{otherwise.} \end{cases}$$

Then there exists $\widehat{\sigma}(X) \in \mathrm{ET}(Q[X], I[X])$ *and* $l > 0$ *such that* $\widehat{\sigma}(X)$ *localises to* $\sigma(bX)$ *for some* $b \in (s^l)$ *and* $\widehat{\sigma}(0) = Id$.

Proof Follows by imitating the technique explained in [3], and following steps mentioned in Sect. 12.2. □

Theorem 12.2 (Relative Local-Global Principle) *Let R be a commutative ring with identity, and* $I \subsetneq R$ *an ideal in R. Let P and Q be as in 12.13. Assume that (H1) holds. Suppose* $\sigma(X) \in \mathrm{G}(Q[X], I[X])$ *with* $\sigma(0) = Id$. *If*

$$\sigma_{\mathfrak{p}}(X) \in \begin{cases} \mathrm{E}(n+1, R_{\mathfrak{p}}[X], I_{\mathfrak{p}}[X]) & \textit{in the linear case,} \\ \mathrm{E}(n+2, R_{\mathfrak{p}}[X], I_{\mathfrak{p}}[X]) & \textit{otherwise} \end{cases}$$

for all $\mathfrak{p} \in \mathrm{Spec}(R)$, *then* $\sigma(X) \in \mathrm{ET}(Q[X], I[X])$.

Proof Follows by using similar technique as in (4) ⇒ (3) in Theorem 3.1 of [7], and and arguing as in Sect. 12.2. □

Remark 12.6 The authors believe that the above method using the 'Noetherian Excision ring', makes it possible to deduce the relative versions of almost all the results mentioned in [3, 5, 7], and the results mentioned in [6] between pages 35–40.

Acknowledgements Research by the first author was supported by SERB-MATRICS grant (File No. MTR/2017/000886) for the financial year 2018–2019.

References

1. H. Apte, P. Chattopadhyay, R.A. Rao, *A local global theorem for extended ideals*. J.Ramanujan Math. Soc. **27**(1), 1–20 (2012)
2. H. Apte, A. Stepanov, Local-global principle for congruence subgroups of Chevalley groups. Cent. Eur. J. Math. **12**(6), 801–812 (2014)
3. A. Bak, R. Basu, R.A. Rao, Local-Global Principle for Transvection groups, in *Proceedings of the American Mathematical Society*, vol. 138 (2010), pp. 1191–1204
4. H. Bass, *Algebraic K -Theory* (W. A. Benjamin Inc, New York, 1968)
5. R. Basu, *Local-global principle for the general quadratic and the general Hermitian groups and the nilpotence of* KH_1, Zap. Nauchn. Sem. S.-Peterburg. Otdel. Mat. Inst. Steklov, (POMI) 452 (2016), Voprosy Teorii Predstavlenii Algebr i Grupp. 30, 5–31. (English version; J. Math. Sci. **232**(5) (2018))
6. R. Basu, *Results in classical algebraic K-theory*. Ph.D. thesis, Tata Institute of Fundamental Research (2007), p. 70
7. R. Basu, R.A. Rao, R. Khanna, On Quillen's local global principle, in *Commutative Algebra and Algebraic Geometry*. Contemporary Mathematics, vol. 390 (American Mathematical Society, Providence, RI, 2005), pp. 17–30

8. A. Gupta, *Structures over commutative rings*. Ph.D. Thesis, Tata Institute of Fundamental Research (2014)
9. A. Gupta, A. Garge, R.A. Rao; *A nice group structure on the orbit space of unimodular rows–II*. J. Algebr. **407**, 201–223 (2014)
10. R. Hazrat, N. Vavilov, K_1 of Chevalley groups are nilpotent. J. Pure Appl. Algebr. **179**(1–2), 99–116 (2003)
11. G. Horrocks, Projective modules over an extension of a local ring. Proc. Lond. Math. Soc. **14**(3), 714–718 (1964)
12. F. Ischebek, R.A. Rao, *Ideals and Reality*. Springer Monographs in Mathematics (Springer, Berlin, 2005), ISBN 3-540-23032-7
13. S. Jose, R. Khanna, R. Rao, S. Sharma, *The quotient Unimodular Vector group is nilpotent*, to appear in these Proceedings
14. S. Jose, R.A. Rao, A Local global principle for the elementary unimodular vector group, in *Commutative Algebra and Algebraic Geometry (Bangalore, India, 2003), 119–125*. Contemporary Mathematics, vol. 390 (American Mathematical Society, Providence, RI, 2005)
15. V.I. Kopeiko, The stabilization of Symplectic groups over a polynomial ring. Math. USSR. Sbornik **34**, 655–669 (1978)
16. T.Y. Lam, *Serre's Conjecture*. Lecture Notes in Mathematics, vol. 635 (Springer, Berlin, 1978)
17. J. Milnor, *Introduction to Algebraic K-Theory*. Annals of Mathematics Studies, vol. 72 (Princeton University Press, Princeton)
18. M.P. Murthy, S.K. Gupta, *Suslin's Work on Linear Groups Over Polynomial Rings and Serre's Problem*. ISI Lecture Notes, vol. 8 (Macmillan Co. of India, Ltd., New Delhi, 1980)
19. D. Quillen, Projective modules over polynomial rings. Invent. Math. **36**, 167–171 (1976)
20. R.A. Rao, S.K. Yadav, *Relating the principles of Quillen–Suslin theory*, to appear in these Proceedings
21. R.A. Rao, Stably elementary homotopy. Proc. Am. Math. Soc. **137**(11), 3637–3645 (2009)
22. J.-P. Serre, *Faisceaux algébriques cohérents*. (French) Ann. Math. **61**(2), 197–278 (1955)
23. A.A. Suslin, Stably free modules. (Russian) Mat. Sb. (N.S.) **102**(144)(4), 537–550 (1977)
24. A.A. Suslin, V.I. Kopeiko, *Quadratic modules and Orthogonal groups over polynomial rings*. Nauchn. Sem., LOMI **71**, 216–250 (1978)
25. A.A. Suslin, On the structure of special linear group over polynomial rings. Math. USSR. Izv. **11**, 221–238 (1977)
26. A.A. Suslin, The structure of the special linear group over rings of polynomials. Izv. Akad. Nauk SSSR Ser. Mat. **41**, 235–252 (1977)
27. M.S. Tulenbaev, *Schur multiplier of a group of elementary matrices of finite order*. Zapiski Nauchnykh Seminarov Leningradskogo Otdeleniya Matematicheskogo Instuta im V.A. Steklova Akad. Nauk SSSR, vol. 86 (1979), pp. 162–169
28. W. van der Kallen, A group structure on certain orbit sets of unimodular rows. J. Algebr. **82**(2), 363–397 (1983)
29. L.N. Vaserstein, On the normal subgroups of GL_n over a ring, in *Algebraic K-Theory, Evanston (1980) (Proc. Conf., Northwestern Univ., Evanston Ill., (1980))*. Lecture Notes in Mathematics, vol. 854 (Springer, Berlin, 1981), pp. 456–465
30. L.N. Vaserstein, A.A. Suslin, *Serre's problem on projective modules over polynomial rings, and algebraic K-theory*. (Russian) Izv. Akad. Nauk SSSR Ser. Mat. **40**(5), 993–1054 (1976)

Chapter 13
The Quotient Unimodular Vector Group is Nilpotent

Reema Khanna, Selby Jose, Sampat Sharma and Ravi A. Rao

13.1 Introduction

R will be a commutative[1] ring with 1, in which 2 is invertible. $\mathrm{Um}_{r+1}(\mathrm{R})$ will denote the set of unimodular vectors $v \in R^{r+1}$, i.e. those vectors v for which there is a vector $w \in R^{r+1}$, with $\langle v, w\rangle = v \cdot w^T = 1$.

Suslin introduced the Suslin matrix in ([17], Sect. 5), and indicated its properties as well as how he felt they will be useful.

In [10], we initiated the study of the special unimodular vector group $\mathrm{SUm}_r(\mathrm{R})$, which is a subgroup of $\mathrm{GL}_{2^r}(R)$ related to $\mathrm{Um}_{r+1}(\mathrm{R})$. We also introduced the elementary unimodular vector subgroup $\mathrm{EUm}_r(\mathrm{R})$ of $\mathrm{SUm}_r(\mathrm{R})$, which is related to the $(r+1)$-unimodular vectors which have a completion to an elementary matrix. We developed the calculus for $\mathrm{EUm}_r(\mathrm{R})$ in [10], and got a nice set of generators for it. In [9], we showed that $\mathrm{EUm}_r(\mathrm{R})$ is a normal subgroup of $\mathrm{SUm}_r(\mathrm{R})$, for $r \geq 2$.

[1] Section 13.4 is part of the doctoral thesis of the first named author under the second named author; Section 13.5 is part of the doctoral thesis of the third named author under the fourth named author.

R. Khanna (✉)
K.J. Somaiya College, Vidyavihar, Mumbai 400077, India
e-mail: reemag16@gmail.com

S. Jose
Department of Mathematics, Institute of Science, Madam Cama Road, Mumbai 400032, India
e-mail: selbyjose@gmail.com

S. Sharma · R. A. Rao
School of Mathematics, Tata Institute of Fundamental Research, 1, Dr. Homi Bhabha Road, Mumbai 400005, India
e-mail: sampan@math.tifr.res.in

R. A. Rao
e-mail: ravi@math.tifr.res.in

A. A. Ambily et al. (eds.), *Leavitt Path Algebras and Classical* K-*Theory*, Indian Statistical Institute Series,
https://doi.org/10.1007/978-981-15-1611-5_13

In [18] Suslin, inspired by Quillen's methods in [12], applied them to the study of unstable K_1-theory of polynomial rings. He proved the K_1-analogue of the Local-Global Principle and the Monic Inversion Principle. The theory built up in [12, 18] is known as the Quillen–Suslin theory.

Using Quillen–Suslin Local–Global principle, Bak established in [2], that the linear quotient $SL_n(R)/E_n(R)$, for $n \geq 3$, is nilpotent. This theme has been revisited several times for different classical groups, see [7, 16], and ([3], Sect. 3.3) for instance.

Now we apply Bak's approach to the pair ($SUm_r(R)$, $EUm_r(R)$), for $r \geq 2$, when R is a Noetherian ring of Krull dimension d. We give a direct approach to reprove the result in [8] that the unimodular vector quotient $SUm_r(R)/EUm_r(R)$ is a nilpotent group of class d. (The latter had been established in [8] via the Jose–Rao theorem that the unimodular vector quotient group was a subgroup of the special orthogonal quotient group; which was nilpotent in view of [7].)

We also deduce a relative version of this result from the absolute case. This argument does not depend on the Excision ring argument of W. van der Kallen, which is normally used to deduce 'relative' results; and is much more flexible. (This approach evolved from the work [14] according to Anjan Gupta; who used it in his thesis ([6], Sect. 2.2) to reprove a theorem of Chattopadhyay–Rao in [5].)

Finally, we consider $SUm_r(R)/EUm_r(R)$, the unimodular vector quotient group, when $R = A[X]$ is a polynomial extension of a local ring A. In this case we show, arguing as in [15] that the unimodular quotient group is an abelian group. A relative version for extended ideals is also deduced.

13.2 Recap About the Suslin Matrix $S_r(v, w)$

Given two row vectors $v, w \in R^{r+1}$, A. Suslin constructed in ([17], Sect. 5), a matrix $S_r(v, w)$, which is of determinant one if $\langle v, w\rangle = v \cdot w^T = 1$. He defined this inductively, as follows: Let $v = (a_0, a_1, \ldots, a_r) = (a_0, v_1)$, with $v_1 = (a_1, \ldots, a_r)$, $w = (b_0, b_1, \ldots, b_r) = (b_0, w_1)$, with $w_1 = (b_1, \ldots, b_r)$. Set $S_0(v, w) = a_0$, and set

$$S_r(v, w) = \begin{pmatrix} a_0 I_{2^{r-1}} & S_{r-1}(v_1, w_1) \\ -S_{r-1}(w_1, v_1)^T & b_0 I_{2^{r-1}} \end{pmatrix}.$$

The reader will find more details about these matrices in this amazing Sect. 13.5; with several unresolved questions.

These matrices have been studied by Jose–Rao in [8, 9]. The survey article [13] gives a quick glimpse at the known results today.

We shall denote by $SUm_r(R)$ the subgroup of $GL_{2^r}(R)$ generated by the set $\{S_r(v, w) | v, w \in R^{r+1}, \langle v, w\rangle = 1\}$, and $EUm_r(R)$ its subgroup generated by the set $\{S_r(v, w) | v, w \in R^{r+1}, \langle v, w\rangle = 1, v = e_1\varepsilon$, for some $\varepsilon \in E_{r+1}(R)\}$. It was shown in [9], that $EUm_r(R)$ is a normal subgroup of $SUm_r(R)$, for $r \geq 2$.

For a matrix $\alpha \in \mathrm{M}_k(R)$, we define α^{top} as the matrix whose entries are the same as that of α above the diagonal, and on the diagonal, and is zero below the diagonal. Similarly, we define α^{bot}. Moreover, we use α^{tb} for α^{top} or α^{bot}.

In [10], a structure theorem for $\mathrm{EUm_r(R)}$ was proved. The following nice set of generators of $\mathrm{EUm_r(R)}$ was established:

For $2 \leq i \leq r+1$, $\lambda \in R$, let

$$\begin{aligned} E(e_i)(\lambda) &= S_r(e_1 + \lambda e_i, e_1),\ E(e_i^*)(\lambda) = S_r(e_1, e_1 + \lambda e_i), \\ E(e_{i1})(\lambda) &= S_r(e_i + \lambda e_1, e_i),\ E(e_{i1}^*)(\lambda) = S_r(e_i, e_i + \lambda e_1). \end{aligned}$$

It was shown that the group $\mathrm{EUm_r(R)}$ can be generated by either

(a) $E(c)(x)$, $E(d)(x)S_r(e_i, e_i)^{-1}$, if 2 is invertible in R, or by
(b) $E(c)(x)^{top}$, $E(c)(x)^{bot}$,

where $c = e_i$ or e_i^*, $d = e_{i1}$ or e_{i1}^*, $2 \leq i \leq r+1$, $x \in R$.

In [8, 10], Jose–Rao noted a fundamental property which is satisfied by the Suslin matrices. Let $v, w, s, t \in \mathrm{M}_{1,r+1}(R)$. Then

$$\begin{aligned} S_r(s,t)S_r(v,w)S_r(s,t) &= S_r(v',w') \\ S_r(t,s)S_r(w,v)S_r(t,s) &= S_r(w',v'), \end{aligned}$$

for some $v', w' \in \mathrm{M}_{1,r+1}(R)$, which depend linearly on v, w and quadratically on s, t. Consequently, $v' \cdot w'^T = (s \cdot t^T)^2 (v \cdot w^T)$.

This fundamental property enables one to define an involution $\star$ on the group $\mathrm{SUm_r(R)}$, details of which can be found in [8]. This involution is then used to give an action of $\mathrm{SUm_r(R)}$ on the Suslin space, viz. the free R-module of rank $2(r+1)$

$$S = \{S_r(v,w) | v, w \in \mathrm{M}_{1,r+1}(R)\}.$$

(For a basis one can take $se_1, \ldots, se_{r+1}, se_1^*, \ldots, se_{r+1}^*$, where $se_i = S_r(e_i, 0), se_i^* = S_r(0, e_i)$, for $1 \leq i \leq r$.)

In [8] they associated a linear transformation T_g of the Suslin space with a Suslin matrix g, via

$$T_g(x,y) = (x', y'),$$

where $gS_r(x,y)g^* = S_r(x', y')$. Moreover, if g is a product of Suslin matrices $S_r(v_i, w_i)$, with $\langle v_i, w_i \rangle = 1$, for all i, then $T_g \in \mathrm{SO}_{2(r+1)}(R)$, i.e.

$$\langle T_g(v,w), T_g(s,t) \rangle = \langle (v,w), (s,t) \rangle = v \cdot w^T + s \cdot t^T.$$

13.3 Computation of the Matrix of the Linear Transformation

In ([8], Sect. 4), via the fundamental property, Jose–Rao observed that the above action induces a canonical homomorphism

$$\varphi : \mathrm{SUm}_r(R) \to \mathrm{SO}_{2(r+1)}(R),$$
$$\varphi(S_r(v, w)) = T_{S_r(v,w)} = \tau_{(v,w)} \circ \tau_{(e_1 e_1)},$$

where $\tau_{(v,w)}$ is the standard reflection with respect to the vector $(v, w) \in R^{2(r+1)}$ (of length one) given by the formula

$$\tau_{(v,w)}(s, t) = \langle v, w\rangle(s, t) - (\langle v, t\rangle + \langle s, w\rangle)(v, w).$$

The following simple computation gives an alternate way to prove this:

Lemma 13.1 *Let R be a commutative ring with 1. Let $v, w \in \mathrm{Um}_{r+1}(R)$, then the matrix of the linear transformation $T_{S_r(v,w)}$ with respect to the (ordered) basis*

$$\{S_r(e_1, 0), S_r(e_2, 0), \ldots, S_r(e_{r+1}, 0), S_r(0, e_1), S_r(0, e_2), \ldots, S_r(0, e_{r+1})\}$$

is

$$\left(I - \begin{pmatrix} v^T \\ w^T \end{pmatrix} \begin{pmatrix} w & v \end{pmatrix}\right)\left(I - \begin{pmatrix} e_1^T \\ e_1^T \end{pmatrix} \begin{pmatrix} e_1 & e_1 \end{pmatrix}\right).$$

Proof Let $v = (a_0, a_1, \ldots, a_r)$, $w = (b_0, b_1, \ldots, b_r)$. By the definition of $T_{S_r(v,w)}$,

$$\begin{aligned}
T_{S_r(v,w)}(e_1, 0) &= \tau_{(v,w)} \circ \tau_{(e_1,e_1)}(e_1, 0) \\
&= \tau_{(v,w)}(0, -e_1) = (0, -e_1) + a_0(v, w) = (a_0 v, a_0 w - e_1). \\
T_{S_r(v,w)}(e_j, 0) &= \tau_{(v,w)} \circ \tau_{(e_1,e_1)}(e_j, 0) \\
&= \tau_{(v,w)}(e_j, 0) = (e_j, 0) - b_{j-1}(v, w) = (e_j - b_{j-1} v, -b_{j-1} w). \\
T_{S_r(v,w)}(0, e_1) &= \tau_{(v,w)} \circ \tau_{(e_1,e_1)}(0, e_1) \\
&= \tau_{(v,w)}(-e_1, 0) = (-e_1, 0) + b_0(v, w) = (b_0 v - e_1, b_0 w). \\
T_{S_r(v,w)}(0, e_j) &= \tau_{(v,w)} \circ \tau_{(e_1,e_1)}(0, e_j) \\
&= \tau_{(v,w)}(0, e_j) = (0, e_j) - a_{j-1}(v, w) = (-a_{j-1} v, e_j - a_{j-1} w).
\end{aligned}$$

Thus the matrix of $T_{S_r(v,w)}$ is

$$\begin{pmatrix} a_0 v & e_2 - b_1 v \cdots e_{r+1} - b_r v & b_0 v - e_1 & -a_1 v & \cdots & -a_r v \\ a_0 w - e_1 & -b_1 w \quad \cdots \quad -b_r w & b_0 w & e_2 - a_1 w & \cdots & e_{r+1} - a_r w \end{pmatrix}.$$

Right multiply the above matrix by the matrix $\left(I - \begin{pmatrix} e_1^T \\ \\ e_1^T \end{pmatrix} \begin{pmatrix} e_1 & e_1 \end{pmatrix}\right)$ will interchange the 1-st and $(r+2)$-th columns with sign changed. Hence, the matrix of $T_{S_r(v,w)}$ is

$$\left(I - \begin{pmatrix} v^T \\ \\ w^T \end{pmatrix} \begin{pmatrix} w & v \end{pmatrix}\right)\left(I - \begin{pmatrix} e_1^T \\ \\ e_1^T \end{pmatrix} \begin{pmatrix} e_1 & e_1 \end{pmatrix}\right)$$

as required. □

Notation We denote the matrix of the linear transformation $T_{S_r(v,w)}$ by $[T_{S_r(v,w)}]$. □

Let us recollect the matrix of the linear transformations corresponding to the generators of $\mathrm{EUm_r(R)}$, $r \geq 2$, computed in [8].

For the sake of completeness, we give a slightly simpler argument than the one given in [8] below. However, in this approach, unlike in [8], we need that 2 is invertible in R.

Lemma 13.2 *For $2 \leq i, j \leq r+1$, one has the following relations in* $\mathrm{EUm}_r(R)$*:*

$$\begin{aligned} E(e_i^*)(-2\lambda)^{bot} &= S_r(e_1 - e_j, e_1 - e_i)S_r((1+\lambda)e_1 + e_j, e_1 - \lambda e_j) \\ &\quad S_r(e_1 - e_j, e_1 + e_i)S_r((1-\lambda)e_1 + e_j, e_1 + \lambda e_j) \\ &\quad \left[E(e_j^*)(\lambda), E(e_i^*)(1)\right]. \\ E(e_i)(-2\lambda)^{bot} &= \left[E(e_i)(-1), E(e_j)(-\lambda)\right] \\ &\quad S_r(e_1 + \lambda e_j, (1-\lambda)e_1 + e_j)S_r(e_1 + e_i, e_1 + e_j) \\ &\quad S_r(e_1 - \lambda e_j, (1+\lambda)e_1 + e_j)S_r(e_1 - e_i, e_1 - e_j). \end{aligned}$$

(Note that by reversing the elements in the product in the above relation we can obtain the formulae for $E(e_i^*)(-2\lambda)^{top}$ and $E(e_i)(-2\lambda)^{top}$.)

Proof We prove the first relation; the others are verified similarly. Put $x = 1$, $y = \lambda$, and $z = 1$ in the proof of ([10], Proposition 5.6), to get
$E(e_i^*)(-2\lambda)^{bot}$

$$\begin{aligned} &= \{E(e_j)(1)^{-1}\}\{E(e_j)(1/2)E(e_i^*)(1/2)^{-1}E(e_i^*)(1/2)^{-1}E(e_j)(1/2)\} \\ &\quad \{E(e_j)(1)^{-1}\}\{S_r((1+\lambda)e_1 + e_j, e_1 - \lambda e_j)\}\{E(e_j)(1)^{-1}\} \\ &\quad \{E(e_j)(1/2)E(e_i^*)(1/2)E(e_i^*)(1/2)E(e_j)(1/2)\}\{E(e_j)(1)^{-1}\} \\ &\quad \{S_r((1-\lambda)e_1 + e_j, e_1 + \lambda e_j)\}\left[E(e_i^*)(1), E(e_j^*)(\lambda)\right]^{-1}. \end{aligned}$$

Now by ([10], Lemma 5.2),

$$\begin{aligned} E(e_i^*)(-2\lambda)^{bot} &= S_r(e_1 - e_j, e_1 - e_i) S_r((1+\lambda)e_1 + e_j, e_1 - \lambda e_j) \\ &\quad S_r(e_1 - e_j, e_1 + e_i) S_r((1-\lambda)e_1 + e_j, e_1 + \lambda e_j) \\ &\quad \left[E(e_j^*)(\lambda), E(e_i^*)(1)\right] \end{aligned}$$

as required. □

Corollary 13.1 ([8], Lemma 4.9, Proposition 4.10) *Let R be a commutative ring with 1 in which 2 is invertible. For $2 \le i \le r+1$,*

$$\text{the matrix of } T_X = \begin{cases} oe_{\pi(1)i}(\lambda) & \text{if } X = E(e_i^*)^{bot}(-\lambda) \\ oe_{i\pi(1)}(-\lambda) & \text{if } X = E(e_i)^{top}(-\lambda) \\ oe_{1i}(\lambda) & \text{if } X = E(e_i^*)^{top}(-\lambda) \\ oe_{i1}(-\lambda) & \text{if } X = E(e_i)^{bot}(-\lambda). \end{cases}$$

Proof By Lemma 13.1, the matrix A of $T_{S_r(e_1-e_j, e_1-e_i)}$ is given by

$$\begin{aligned} A &= \left(I - \begin{pmatrix} (e_1 - e_j)^T \\ (e_1 - e_i)^T \end{pmatrix} \begin{pmatrix} e_1 - e_i & e_1 - e_j \end{pmatrix}\right) \left(I - \begin{pmatrix} e_1^T \\ e_1^T \end{pmatrix} \begin{pmatrix} e_1 & e_1 \end{pmatrix}\right) \\ &= \begin{pmatrix} I + e_{1i} - e_{j1} - e_{ji} & e_{1j} - e_{j1} - e_{jj} \\ e_{1i} - e_{i1} - e_{ii} & I + e_{1j} - e_{i1} - e_{ij} \end{pmatrix}. \end{aligned}$$

Similarly, the matrix B of $T_{S_r((1+\lambda)e_1+e_j, e_1-\lambda e_j)}$ is $B = \begin{pmatrix} B_{11} & B_{12} \\ B_{21} & B_{22} \end{pmatrix}$, where

$$\begin{aligned} B_{11} &= I + \lambda(\lambda+2)e_{11} + \lambda(1+\lambda)e_{1j} + (1+\lambda)e_{j1} + \lambda e_{jj}, \\ B_{12} &= \lambda e_{11} - (1+\lambda)e_{1j} + e_{j1} - e_{jj}, \\ B_{21} &= \lambda e_{11} + \lambda e_{1j} - \lambda(1+\lambda)e_{j1} - \lambda^2 e_{jj} \\ B_{22} &= I - e_{1j} - \lambda e_{j1} + \lambda e_{jj}, \end{aligned}$$

the matrix C of $T_{S_r(e_1-e_j, e_1+e_i)}$ is

$$C = \begin{pmatrix} I - e_{1i} - e_{j1} + e_{ji} & e_{1j} - e_{j1} - e_{jj} \\ -e_{1i} + e_{i1} - e_{ii} & I + e_{1j} + e_{i1} + e_{ij} \end{pmatrix}$$

and the matrix D of $T_{S_r((1-\lambda)e_1+e_j, e_1+\lambda e_j)}$ is $D = \begin{pmatrix} D_{11} & D_{12} \\ D_{21} & D_{22} \end{pmatrix}$, where

$$\begin{aligned} D_{11} &= I + \lambda(\lambda-2)e_{11} + \lambda(\lambda-1)e_{1j} + (1-\lambda)e_{j1} - \lambda e_{jj}, \\ D_{12} &= -\lambda e_{11} + (\lambda-1)e_{1j} + e_{j1} - e_{jj}, \\ D_{21} &= -\lambda e_{11} - \lambda e_{1j} + \lambda(1-\lambda)e_{j1} - \lambda^2 e_{jj}, \\ D_{22} &= I - e_{1j} + \lambda e_{j1} - \lambda e_{jj}. \end{aligned}$$

Now $AB = \begin{pmatrix} \alpha_{11} & \alpha_{12} \\ \alpha_{21} & \alpha_{22} \end{pmatrix}$, where

$$\begin{aligned}
\alpha_{11} &= I + \lambda e_{11} + \lambda e_{1j} - \lambda e_{j1} - \lambda e_{jj} + e_{1i} - e_{ji} \\
\alpha_{12} &= 0 \\
\alpha_{21} &= e_{1i} - (1+2\lambda)e_{i1} - 2\lambda e_{ij} - e_{ii} - \lambda^2 e_{11} + \lambda(1-\lambda)e_{1j} - \lambda(1+\lambda)e_{j1} - \lambda^2 e_{jj} \\
\alpha_{22} &= I - e_{i1} + e_{ij} - \lambda e_{j1} + \lambda e_{jj} - \lambda e_{11} + \lambda e_{1j}.
\end{aligned}$$

Also $CD = \begin{pmatrix} \beta_{11} & \beta_{12} \\ \beta_{21} & \beta_{22} \end{pmatrix}$, where

$$\begin{aligned}
\beta_{11} &= I - \lambda e_{11} - \lambda e_{1j} + \lambda e_{j1} + \lambda e_{jj} - e_{1i} + e_{ji} \\
\beta_{12} &= 0 \\
\beta_{21} &= -e_{1i} - \lambda^2 e_{11} - \lambda(1+\lambda)e_{1j}(1-2\lambda)e_{i1} - 2\lambda e_{ij} - e_{ii} + \lambda(1-\lambda)e_{j1} - \lambda^2 e_{jj} \\
\beta_{22} &= I - e_{ij} - e_{1j} + \lambda e_{j1} - \lambda e_{jj} + \lambda e_{11} + e_{i1}.
\end{aligned}$$

Thus

$$ABCD = \begin{pmatrix} I & 0 \\ 2\lambda e_{1i} - 2\lambda e_{i1} - 2\lambda e_{ij} + 2\lambda e_{ji} & I \end{pmatrix}.$$

Also by Lemma 13.1, the matrix P of $T_{E(e_j^*)(\lambda)}$ is given by

$$\begin{aligned}
P &= \left(I - \begin{pmatrix} e_1^T \\ (e_1 + \lambda e_j)^T \end{pmatrix} (e_1 + \lambda e_j \; e_1)\right)\left(I - \begin{pmatrix} e_1^T \\ e_1^T \end{pmatrix} (e_1 \; e_1)\right) \\
&= \begin{pmatrix} I - \lambda e_{1j} & 0 \\ \lambda e_{j1} - \lambda e_{1j} - \lambda^2 e_{jj} & I + \lambda e_{j1} \end{pmatrix}.
\end{aligned}$$

Clearly $P^{-1} = \begin{pmatrix} I + \lambda e_{1j} & 0 \\ -\lambda e_{j1} + \lambda e_{1j} - \lambda^2 e_{jj} & I - \lambda e_{j1} \end{pmatrix}$, which is the matrix of $T_{E(e_j^*)(-\lambda)}$. Similarly, the matrix Q of $T_{E(e_i^*)(1)}$ and its inverse Q^{-1} of $T_{E(e_i^*)(-1)}$ are

$$Q = \begin{pmatrix} I - e_{1i} & 0 \\ -e_{1i} + e_{i1} - e_{ii} & I + e_{i1} \end{pmatrix}, \; Q^{-1} = \begin{pmatrix} I + e_{1i} & 0 \\ e_{1i} - e_{i1} - e_{ii} & I - e_{i1} \end{pmatrix}.$$

Thus the matrix

$$[P, Q] = \begin{pmatrix} I & 0 \\ 2\lambda e_{ij} - 2\lambda e_{ji} & I \end{pmatrix}.$$

Hence the product of the matrices $ABCD$ and $[P, Q]$ is

$$\begin{pmatrix} I & 0 \\ 2\lambda e_{1i} - 2\lambda e_{i1} & I \end{pmatrix} = I + 2\lambda e_{\pi(1)i} - 2\lambda e_{\pi(i)1} = oe_{\pi(1)i}(2\lambda).$$

Since φ is a homomorphism, the matrix of $T_{E(e_i^*)(-2\lambda)^{bot}}$ is $oe_{\pi(1)i}(2\lambda)$. This proves the first relation. The second relation is its transpose-inverse. Similarly, one can prove the third and fourth relations. □

Corollary 13.2 *Let R be a commutative ring with* 1 *in which* 2 *is invertible. For* $2 \le i \ne j \ne \pi(i) \le r+1$,

$$\text{the matrix of } T_X = \begin{cases} oe_{i\pi(1)}(\lambda)oe_{i1}(\lambda) & \text{if } X = E(e_i)(\lambda) \\ oe_{\pi(1)i}(-\lambda)oe_{1i}(-\lambda) & \text{if } X = E(e_i^*)(\lambda) \\ oe_{1i}(\lambda)oe_{1\pi(i)}(\lambda)\pi_{1i}(-1) & \text{if } X = E(e_{1i})(\lambda) \\ \pi_{1i}(-1)oe_{1i}(\lambda)oe_{1\pi(i)}(\lambda) & \text{if } X = E(e_{1i}^*)(\lambda) \\ oe_{ij}(\lambda) & \text{if } X = [E(e_j^*)(\lambda)^{top}, E(e_i)(1)^{bot}] \\ oe_{i\pi(j)}(\lambda) & \text{if } X = [E(e_j)(\lambda)^{top}, E(e_i)(1)^{bot}] \\ oe_{\pi(i)j}(\lambda) & \text{if } X = [E(e_j^*)(\lambda)^{top}, E(e_i^*)(1)^{bot}]. \end{cases}$$

(Here $\pi_{1i}(-1)$ denote the matrix of $T_{S_r(e_i,e_i)}$.)

Proof Follows immediately from Corollary 13.1. □

Proposition 13.1 *Let R be a commutative ring with* 1 *in which* 2 *is invertible. Then the map* $\varphi : \mathrm{EUm}_r(R) \to \mathrm{EO}_{2(r+1)}(R)$ *given by* $\varphi(S_r(v,w)) = T_{S_r(v,w)}$ *is surjective.*

Proof Follows from Corollary 13.1. □

13.4 $\mathbf{SUm_r(R)/EUm_r(R)}$ is Nilpotent

Notation Let s be a non-zero divisor, $\mathrm{SUm_r(R, s^nR)}$ denote the subgroup of $\mathrm{SUm_r(R)}$ consisting of matrices which are identity modulo (s^n), and $\mathrm{EUm_r(R, s^nR)}$ denote the corresponding elementary subgroup.

Lemma 13.3 *Let R be a commutative ring with* 1. *Let s be a non-zero divisor in Jacobson radical $J(R)$ of R and* $\beta \in \mathrm{SUm_r(R, s^nR)}$ *for* $n \ge 0$. *Then the matrix of the linear transformation* T_β *is in* $\mathrm{SO}_{2(r+1)}(R, s^nR)$.

Proof Since $\beta \in \mathrm{SUm}_r(R, s^nR)$, $\beta = S_r(v,w)$ where $v \equiv e_1 \mathrm{mod}(s^n)$ and $w \equiv e_1 \mathrm{mod}(s^n)$. Let $v = (a_0, a_1, \ldots, a_r)$ and $w = (b_0, b_1, \ldots, b_r)$, where a_0 and b_0 are $\equiv 1 \bmod (s^n)$, a_i and b_i are $\equiv 0 \bmod (s^n)$. By definition, the matrix of T_β, $[T_\beta] \in \mathrm{SO}_{2(r+1)}(R)$ and by Lemma 13.1,

$$\begin{aligned} [T_\beta] &= \left(I_{2(r+1)} - \begin{pmatrix} v^T \\ w^T \end{pmatrix}(w\ v)\right)\left(I_{2(r+1)} - \begin{pmatrix} e_1^T \\ e_1^T \end{pmatrix}(e_1\ e_1)\right) \\ &= I_{2(r+1)} - \begin{pmatrix} v^T \\ w^T \end{pmatrix}(w\ v) - \begin{pmatrix} e_1^T \\ e_1^T \end{pmatrix}(e_1\ e_1) + (a_0+b_0)\begin{pmatrix} v^T \\ w^T \end{pmatrix}(e_1\ e_1) \\ &= \begin{pmatrix} I_{r+1} - v^Tw - e_1^Te_1 + (a_0+b_0)v^Te_1 & -v^Tv - e_1^Te_1 + (a_0+b_0)v^Te_1 \\ -w^Tw - e_1^Te_1 + (a_0+b_0)w^Te_1 & I_{r+1} - w^Tv - e_1^Te_1 + (a_0+b_0)w^Te_1 \end{pmatrix}. \end{aligned}$$

Therefore,

$$[T_\beta]\text{mod}(s^n) = \begin{pmatrix} I_{r+1} - e_1^T e_1 - e_1^T e_1 + 2e_1^T e_1 & -e_1^T e_1 - e_1^T e_1 + 2e_1^T e_1 \\ -e_1^T e_1 - e_1^T e_1 + 2e_1^T e_1 & I_{r+1} - e_1^T e_1 - e_1^T e_1 + 2e_1^T e_1 \end{pmatrix}$$
$$= I_{2(r+1)}.$$

Hence $[T_\beta] \in SO_{2(r+1)}(R, s^n R)$. □

Lemma 13.4 *Let R be a commutative ring with* 1. *In* $\mathrm{EUm_r(R[X, Y, Z])}$, $E(c)(Z)^{tb}$ $E(d)(X^3Y)^{tb}E(c)(-Z)^{tb}$, *where* $c = e_i$ *or* e_i^* *and* $d = e_j$ *or* e_j^* *is a product of elementary generators in* $\mathrm{EUm_r(R[X, Y, Z])}$ *each of which is* $\equiv I_{2^r}$ *modulo* (X).

Proof If necessary, the reader can consult ([9], Lemma 3.1) for details. □

Lemma 13.5 *Let R be a commutative ring with* 1. *Let s be a non-zero divisor in Jacobson radical* $J(R)$ *of R. Then we can write* $E(c)(1)^{bot}E(d)(s^3x)^{top}E(c)(-1)^{bot}$, *where* $c = e_i$ *or* e_i^*, $d = e_j$ *or* e_j^* *and* $x \in R$, *as a product of elementary generators in* $\mathrm{EUm}_r(R)$ *which are* $\equiv I_{2^r}$ *modulo* (s).

Proof Put $Z = 1$, $X = s$ and $Y = x$ in Lemma 13.4. □

Lemma 13.6 *Let R be a commutative ring with* 1. *Let s be a non-zero divisor in Jacobson radical* $J(R)$ *of R. If* $u \equiv 1 \bmod (s^9)$ *where* $u \in R$ *with* $u^2 = 1$, *then* $[u] \perp [u^{-1}]$ *is a product of elementary generators in* $\mathrm{EUm}_r(R)$ *each of which is* $\equiv I_{2^r}$ *modulo* (s).

Proof Note that

$$\begin{aligned}[u] \perp [u^{-1}] = &\{E(e_2)(1-u^{-1})^{bot}E(e_2^*)(1-u^{-1})^{bot}\}\{E(e_2^*)(-1)^{bot}E(e_2)(-1)^{bot}\}\\ &\{E(e_2)(1-u)^{top}E(e_2^*)(1-u)^{top}\}\{E(e_2)(1)^{bot}E(e_2^*)(1)^{bot}\}\\ &\{E(e_2)(1-u^{-1})^{top}E(e_2^*)(1-u^{-1})^{top}\}.\end{aligned}$$

Let $u^{-1} = u = 1 + s^9x$ for some $x \in R$. Then

$$\begin{aligned}[u] \perp [u^{-1}] &= \{E(e_2)(-s^9x)^{bot}E(e_2^*)(-s^9x)^{bot}\}\{E(e_2^*)(-1)^{bot}E(e_2)(-1)^{bot}\}\\ &\quad\{E(e_2)(-s^9x)^{top}E(e_2^*)(-s^9x)^{top}\}\{E(e_2)(1)^{bot}E(e_2^*)(1)^{bot}\}\\ &\quad\{E(e_2)(-s^9x)^{top}E(e_2^*)(-s^9x)^{top}\}\\ &= \{E(e_2)(-s^9x)^{bot}E(e_2^*)(-s^9x)^{bot}\}\alpha\{E(e_2)(-s^9x)^{top}E(e_2^*)(-s^9x)^{top}\},\end{aligned}$$

where

$$\begin{aligned}\alpha &= \{E(e_2^*)(-1)^{bot}E(e_2)(-1)^{bot}\}\{E(e_2)(-s^9x)^{top}E(e_2^*)(-s^9x)^{top}\}\\ &\quad\{E(e_2)(1)^{bot}E(e_2^*)(1)^{bot}\}\\ &= E(e_2^*)(-1)^{bot}\{E(e_2)(-1)^{bot}E(e_2)(-s^9x)^{top}E(e_2)(1)^{bot}\}\\ &\quad\{E(e_2)(-1)^{bot}E(e_2^*)(-s^9x)^{top}E(e_2)(1)^{bot}\}E(e_2^*)(1)^{bot}.\end{aligned}$$

By Lemma 13.5, each element in the bracket is a product of elementary generators in $\mathrm{EUm_r(R)}$ which are $\equiv I_{2^r}$ modulo (s^3). Thus

$$\alpha = E(e_2^*)(-1)^{bot}\left(\prod \alpha_i \prod \beta_i\right) E(e_2^*)(1)^{bot},$$

where each $\alpha_i, \beta_i \in \mathrm{EUm_r(R)}$ with each one $\equiv I_{2^r}$ mod (s^3). Also we can write

$$\alpha = \prod \left(E(e_2^*)(-1)^{bot}\alpha_i E(e_2^*)(1)^{bot}\right) \prod \left(E(e_2^*)(-1)^{bot}\beta_i E(e_2^*)(1)^{bot}\right).$$

Again by Lemma 13.5, each element in the product of α is a product of elementary generators in $\mathrm{EUm_r(R)}$ which are $\equiv I_{2^r}$ modulo (s). Thus $[u] \perp [u^{-1}]$ is a product of elementary generators in $\mathrm{EUm_r(R)}$ each of which are $\equiv I_{2^r}$ modulo (s). □

Lemma 13.7 *Let R be a commutative ring with 1. Let s be a non-zero divisor in Jacobson radical $J(R)$ of R and $\beta \in \mathrm{SUm}_r(R, s^n R)$ for $n \gg 9$. Then β can be written as a product of elementary generators in $\mathrm{EUm_r(R)}$ where each is $\equiv I_{2^r}$ mod (s).*

Proof By Lemma 13.3, $[T_\beta] \in \mathrm{SO}_{2(r+1)}(R, s^n R)$. Thus by ([7], Lemma 2.2), $\varphi(\beta) = [T_\beta] = \varepsilon_1 \dots \varepsilon_k$ where each $\varepsilon_i \in \mathrm{EO}_{2(r+1)}(R)$ which is $\equiv I_{2^r}$ mod (s). For sufficiently large n, we may assume that each $\varepsilon_i \equiv I_{2^r}$ mod (s^p) where $n > p \geq 9$. By Proposition 13.1, $\varepsilon_i = \varphi(\varepsilon_i')$ where each $\varepsilon_i' \in \mathrm{EUm}_r(R, s^p R)$. Thus $\varphi(\beta) = \varphi(\varepsilon_1' \dots \varepsilon_k')$. Hence $\beta(\varepsilon_1' \dots \varepsilon_k')^{-1} \in \ker \varphi = Z(\mathrm{SUm}_r(R)) \subseteq \mathrm{EUm}_r(R)$. By ([8], Corollary 3.5), $\beta(\varepsilon_1' \dots \varepsilon_k')^{-1} = u I_{2^r}$ where u is a unit with $u^2 = 1$. Since β and ε_i' are $\equiv I_{2^r}$ mod (s^p) $(n > p \geq 9)$, $u \equiv 1$ mod (s^p). Therefore, by Lemma 13.6, $\beta = u\varepsilon_1' \dots \varepsilon_k'$ is a product of elementary generators each of which is $\equiv I_{2^r}$ mod (s). □

Lemma 13.8 *Let R be a commutative ring with 1 in which 2 is invertible, $s \in R$ a non-zero-divisor and $a \in R$. Then for $n \gg 0$ and $c = e_i$, or e_i^*,*

$$\left[E(c)\left(\frac{a}{s}X\right)^{tb}, SUm_r(R, s^n R)\right] \subseteq \mathrm{EUm}_r(R[X]).$$

More generally, given $p > 0$, for $n \gg 0$,

$$\left[\mathrm{EUm}_r(R_s[X]), \mathrm{SUm}_r(R, s^n R)\right] \subseteq \mathrm{EUm}_r(R[X], s^p R[X])$$

Proof Let $\alpha(X) = \left[E(c)\left(\frac{a}{s}X\right), \beta\right]$ where $\beta \in \mathrm{SUm}_r(R, s^n R)$. Then $\varphi(\beta) \in \mathrm{SO}_{2(r+1)}(R, s^n R)$, where $\varphi : \mathrm{SUm}_r(R, s^n R) \to \mathrm{SO}_{2(r+1)}(R, s^n R)$ is the canonical homomorphism. By Corollary 13.2,

$$\varphi\left(E(c)\left(\frac{a}{s}X\right)\right) \in \mathrm{EO}_{2(r+1)}(R_s[X]).$$

Thus by [[7], Lemma 2.4],

$$\varphi(\alpha(X)) \in [\mathrm{EO}_{2(r+1)}(R_s[X]), \mathrm{SO}_{2(r+1)}(R, s^n R)] \subseteq \mathrm{EO}_{2(r+1)}(R[X]),$$

and hence by Proposition 13.1, there exists $\varepsilon \in \mathrm{EUm}_r(R[X])$ such that $\varphi(\alpha(X)) = \varphi(\varepsilon)$. This implies, $\varphi(X)\varepsilon^{-1} \in \ker\varphi \subseteq Z(\mathrm{SUm}_r(R[X])) \subseteq \mathrm{EUm}_r(R[X])$. Hence $\alpha(X) \in \mathrm{EUm}_r(R[X])$. □

Lemma 13.9 *Let R be a commutative ring with 1 in which 2 is invertible, $s \in R$ a non-zero divisor and $a \in R$. Then for $n \gg 0$ and $c = e_i$, or e_i^*,*

$$\left[E(c)\left(\frac{a}{s}\right), \mathrm{SUm}_r(R, s^n R)\right] \subseteq \mathrm{EUm}_r(R).$$

More generally, $[\mathrm{EUm}_r(R_s), \mathrm{SUm}_r(R, s^n R)] \subseteq \mathrm{EUm}_r(R)$ *for $n \gg 0$.*

Proof Put $X = 1$ in Lemma 13.8. □

In ([8], Corollary 4.15) the quotient group $\mathrm{SUm_r(R)/EUm_r(R)}$, $r \geq 2$ was shown to be nilpotent. This was obtained as a consequence of the Jose–Rao Theorem in ([8], Theorem 4.14) which asserts that this quotient unimodular vector group is a subgroup of the orthogonal quotient group $SO_{2(r+1)}(R)/EO_{2(r+1)}(R)$; which has been shown to be nilpotent in [7]. (Also see [16] for another proof.) We give a direct proof of the result following Bak's methods in [2].

Theorem 13.1 *Let R be a commutative Noetherian ring with 1 in which 2 is invertible and let* $\dim R = d$. *Then the group $SUm_r(R)/\mathrm{EUm}_r(R)$ is nilpotent of class d for $r \geq 2$.*

Proof Let $G = \mathrm{SUm}_r(R)/\mathrm{EUm}_r(R)$. We prove that $Z^d = \{1\}$. We prove by induction on $d = \dim R$. When $d = 0$, the ring R is Artinian, so is semilocal. Hence $\mathrm{Um}_{r+1}(R) = e_1 E_{r+1}(R)$ and so any generator $S_r(v, w)$, $\langle v, w\rangle = 1$ is in $\mathrm{EUm_r(R)}$.

Suppose $d > 0$, Let $\alpha \in Z^d$, then $\alpha = [\beta, \gamma]$, where $\beta \in G$ and $\gamma \in Z^{d-1}$. Let β' be the preimage of β in $\mathrm{SUm_r(R)}$.

Choose a non-zero-divisor s in R such that $\beta'_s \in \mathrm{EUm}_r(R_s)$ (such s exists as $d > 0$). Consider $\overline{G} = \frac{\mathrm{SUm}_r(R/s^n R)}{\mathrm{EUm}_r(R/s^n R)}$ for some $n \gg 0$. By induction, $\overline{\gamma} = \{1\}$ in $\overline{G}$. Since $\mathrm{EUm_r(R)}$ is normal in $\mathrm{SUm_r(R)}$, by modifying γ we may assume that $\gamma' \in SUm_r(R, s^n R)$ where γ' is the preimage of γ in $\mathrm{SUm_r(R, s^nR)}$. Thus by Lemma 13.8, $[\beta', \gamma'] \in \mathrm{EUm}_r(R)$. Hence $\alpha = \{1\}$ in G. □

The Relative Case

In this section, we deduce the relative case of Theorem 13.1 from the absolute case. We use the 'Excision ring' $R \oplus I$ below instead of the usual non-Noetherian Excision ring $\mathbb{Z} \oplus I$ as is usually done due to the work of van der Kallen in [19].

Notation By ([10, Proposition 5.6]), the elementary generators,

$$E(c)(x)^{top}, E(c)(x)^{bot},$$

where $c = e_i$ or e_j^*, and for $2 \leq i \leq r+1$, and with $x \in R$, generate the Elementary Unimodular vector group $\mathrm{EUm_r(R)}$. For simplicity, we shall denote these by $ge_i(x)$ below. □

Theorem 13.2 *Let R be a commutative Noetherian ring with* 1 *in which* 2 *is invertible and with* $\dim R = d$. *Let I be an ideal of R. Then the group* $\mathrm{SUm_r(R, I)}/\mathrm{EUm_r(R, I)}$ *is nilpotent of class d for* $r \geq 2$.

Proof Let $G = \mathrm{SUm}_r(R, I)/\mathrm{EUm}_r(R, I)$. We prove that $Z^d = \{1\}$. We prove by induction on $d = \dim R$. When $d = 0$, the ring R is Artinian, so is semilocal. Hence $Um_{r+1}(R, I) = e_1 E_{r+1}(R, I)$ and so any generator $S_r(v, w)$, $\langle v, w\rangle = 1$ is in $\mathrm{EUm_r(R, I)}$.

Suppose $d > 0$, Let $\alpha \in Z^d$, then $\alpha = [\beta, \gamma]$, where $\beta \in G$ and $\gamma \in Z^{d-1}$. We can write $\beta = Id + \beta'$, $\gamma = Id + \gamma'$ for some $\beta', \gamma' \in M_{2^r}(I)$. Let $\alpha = Id + \alpha'$ for some $\alpha' \in M_{2^r}(I)$. Let $\tilde{\alpha} = (Id, \alpha') \in \mathrm{SUm_r(R \oplus I, 0 \oplus I)}$. In view of ([1, Lemma 3.3]),

$$\tilde{\alpha} \in \mathrm{EUm_r(R \oplus I)} \cap \mathrm{SUm_r(R \oplus I, 0 \oplus I)} = \mathrm{EUm_r(R \oplus I, 0 \oplus I)}$$

as $\frac{R \oplus I}{0 \oplus I} \simeq R$ is a retract of $R \oplus I$. Thus,

$$\tilde{\alpha} = \prod_{k=1}^{m} \varepsilon_k ge_{i_k}(0, a_k)\varepsilon_k^{-1}, \quad \varepsilon_k \in \mathrm{EUm_r(R \oplus I)},\ \mathrm{a_k} \in \mathrm{I}.$$

Now, consider the homomorphism

$$\begin{aligned} f : R \oplus I &\longrightarrow R \\ (r, i) &\longmapsto r + i. \end{aligned}$$

This f induces a map

$$\tilde{f} : \mathrm{EUm_r(R \oplus I, 0 \oplus I)} \longrightarrow \mathrm{EUm_r(R)}.$$

Clearly,

$$\begin{aligned} \alpha &= \tilde{f}(\tilde{\alpha}) \\ &= \prod_{k=1}^{m} \gamma_k ge_{i_k}(0 + a_k)\gamma_k^{-1} \\ &= \prod_{k=1}^{m} \gamma_k ge_{i_k}(a_k)\gamma_k^{-1} \in \mathrm{EUm_r(R, I)}; \quad \text{since } \mathrm{a_k} \in \mathrm{I}, \end{aligned}$$

where, $\gamma_k = \tilde{f}(\varepsilon_k)$. □

13.5 Abelian Quotients over Polynomial Extensions of a Local Ring

In this section we use the Quillen–Suslin Local–Global Principle, following the ideas of Bak in [2], to prove that if $R = A[X]$, with A a local ring, then the quotient Unimodular Vector group is abelian. The method is similar to the one in [15] where we had used it to analyse the quotients of the linear, symplectic, and orthogonal groups.

We begin with a few simple observations.

The following observation is well known, we record it here for future use:

Lemma 13.10 *Let R be a commutative ring and $v, w \in Um_n(R)$ be such that $v \cdot w^t = 1$. If $v = e_1\sigma$ for some $\sigma \in E_n(R)$ then there exists $\varepsilon \in E_n(R)$ such that $v = e_1\varepsilon$ and $w = e_1(\varepsilon^{-1})^t$.*

Proof In view of ([17, Corollary 2.8]), $w\zeta = e_1(\sigma^{-1})^t$, where $\zeta = I_n + v^t(e_1(\sigma^{-1})^t - w) \in E_n(R)$. We see that $v\zeta^t = v$. Thus $e_1\sigma\zeta^t = e_1\sigma = v$. Upon taking $\varepsilon = e_1\sigma\zeta^t$, we have $v = e_1\varepsilon$ and $w = e_1(\varepsilon^{-1})^t$. □

Corollary 13.3 *Let R be a local ring. For $r \geq 1$,* $\mathrm{SUm_r(R)} = \mathrm{EUm_r(R)}$.

Proof Let $\alpha = S_r(v, w) \in \mathrm{SUm_r(R)}$. Since R is a local ring, therefore $v = e_1\sigma$ for some $\sigma \in E_{r+1}(R)$. Since $v \cdot w^t = 1$, by Lemma 13.10, there exists $\varepsilon \in E_{r+1}(R)$ such that $v = e_1\varepsilon$ and $w = e_1(\varepsilon^{-1})^t$. Thus $\alpha = S_r(v, w) = S_r(e_1\varepsilon, e_1(\varepsilon^{-1})^t) \in \mathrm{EUm_r(R)}$. □

Lemma 13.11 *Let R be a local ring and $\alpha(X)$, $\beta(X) \in$* $\mathrm{SUm_r(R[X])}$. *Then, for $r \geq 2$, the commutator,*

$$[\alpha(X),\ \beta(X)] \in [\alpha(X)\alpha(0)^{-1},\ \beta(X)\beta(0)^{-1}]\mathrm{EUm_r(R[X])}.$$

Proof Since R is a local ring, in view of Corollary 13.3, $\mathrm{SUm_r(R)} = \mathrm{EUm_r(R)}$ for all $r \geq 1$. Thus $\alpha(0), \beta(0) \in \mathrm{EUm_r(R)}$.

Let $\eta = \alpha(X)\alpha(0)^{-1}$, $\tau = \beta(X)\beta(0)^{-1}$. Then,

$$\begin{aligned}[\alpha(X),\ \beta(X)] &= [\alpha(X)\alpha(0)^{-1}\alpha(0),\ \beta(X)\beta(0)^{-1}\beta(0)]\\ &= \eta\alpha(0)\tau\beta(0)(\eta\alpha(0))^{-1}(\tau\beta(0))^{-1}\\ &= \eta\tau\eta^{-1}\tau^{-1}(\tau\eta\tau^{-1}\alpha(0)\tau\eta^{-1}\tau^{-1})(\tau\eta\beta(0)\alpha(0)^{-1}\eta^{-1}\tau^{-1})(\tau\beta(0)^{-1}\tau^{-1}).\end{aligned}$$

By ([8, Corollary 4.12]), $\mathrm{EUm_r(R[X])}$ is a normal subgroup of $\mathrm{SUm_r(R[X])}$ for $r \geq 2$, hence

$$\begin{aligned}(\tau\eta\tau^{-1}\alpha(0)\tau\eta^{-1}\tau^{-1}) &\in \mathrm{EUm_r(R[X])},\\ (\tau\eta\beta(0)\alpha(0)^{-1}\eta^{-1}\tau^{-1}) &\in \mathrm{EUm_r(R[X])},\\ (\tau\beta(0)^{-1}\tau^{-1}) &\in \mathrm{EUm_r(R[X])}.\end{aligned}$$

Hence the result. □

Theorem 13.3 *Let R be a local ring. Then the group* $\frac{\mathrm{SUm_r(R[X])}}{\mathrm{EUm_r(R[X])}}$ *is an abelian group for* $r \geq 2$.

Proof Let $\alpha(X), \beta(X) \in \mathrm{SUm_r(R[X])}$, we need to prove $[\alpha(X), \beta(X)] \in \mathrm{EUm_r(R[X])}$. In view of Lemma 13.11, we may assume that $\alpha(0) = \beta(0) = Id$. Define

$$\gamma(X, T) = [\alpha(XT), \beta(X)].$$

Then for every maximal ideal $\mathfrak{m}$ of $R[X]$,

$$\gamma(X, T)_{\mathfrak{m}} = [\alpha(XT)_{\mathfrak{m}}, \beta(X)_{\mathfrak{m}}].$$

Since $\beta(X)_{\mathfrak{m}} \in \mathrm{SUm_r(R[X]_{\mathfrak{m}})} = \mathrm{EUm_r(R[X]_{\mathfrak{m}})}$, and in view of the normality of $\mathrm{EUm_r(R[X]_{\mathfrak{m}}}[T]) \trianglelefteq \mathrm{SUm_r(R[X]_{\mathfrak{m}}[T])}$, for $r \geq 2$, one has $\gamma(X, T)_{\mathfrak{m}} \in EUm_r (R[X]_{\mathfrak{m}}[T])$ and $\gamma(X, 0) = Id$. Thus by the Local–Global Principle, ([8, Corollary 4.11]), $\gamma(X, T) \in \mathrm{EUm_r(R[X, T])}$, by putting $T = 1$, one gets, $\gamma(X, 1) = [\alpha(X), \beta(X)] \in \mathrm{EUm_r(R[X])}$. □

Theorem 13.4 *Let R be a local ring and I be an ideal of R. Then the group* $\frac{\mathrm{SUm_r(R[X],I[X])}}{\mathrm{EUm_r(R[X],I[X])}}$ *is an abelian group for* $r \geq 2$.

Proof Let $\alpha, \beta \in \mathrm{SUm_r(R[X], I[X])}$. We can write $\alpha = Id + \alpha^{'}, \beta = Id + \beta^{'}$ for some $\alpha^{'}, \beta^{'} \in M_{2^r}(I[X])$. Let $\sigma = [\alpha, \beta] = Id + \sigma^{'}$ for some $\sigma^{'} \in M_{2^r}(I[X])$. Let $\tilde{\sigma} = (Id, \sigma^{'}) \in \mathrm{SUm_r(R[X] \oplus I[X], 0 \oplus I[X])}$. In view of ([1, Lemma 3.3]) and Theorem 13.3, $\tilde{\sigma} \in \mathrm{EUm_r(R[X] \oplus I[X])} \cap \mathrm{SUm_r(R[X] \oplus I[X], 0 \oplus I[X])} = \mathrm{EUm_r}$ $\mathrm{(R[X] \oplus I[X], 0 \oplus I[X])}$ as $\frac{R[X] \oplus I[X]}{0 \oplus I[X]} \simeq R[X]$ is a retract of $R[X] \oplus I[X]$. Thus,

$$\tilde{\sigma} = \prod_{k=1}^{m} \varepsilon_k g e_{i_k}(0, a_k)\varepsilon_k^{-1}, \quad \varepsilon_k \in \mathrm{EUm_r(R[X] \oplus I[X])}, \mathrm{a_k} \in \mathrm{I[X]}.$$

Now, consider the homomorphism

$$f : R[X] \oplus I[X] \longrightarrow R[X]$$
$$(r, i) \longmapsto r + i.$$

This f induces a map

$$\tilde{f} : \mathrm{EUm_r(R[X] \oplus I[X], 0 \oplus I[X])} \longrightarrow \mathrm{EUm_r(R[X])}$$

Clearly,

$$\sigma = \tilde{f}(\tilde{\sigma}) = \prod_{k=1}^{m} \gamma_k g e_{i_k}(0 + a_k)\gamma_k^{-1}$$
$$= \prod_{k=1}^{m} \gamma_k g e_{i_k}(a_k)\gamma_k^{-1} \in E(n, R, I); \quad \text{since } a_k \in I,$$

where, $\gamma_k = \tilde{f}(\varepsilon_k)$. □

Acknowledgements The second author thanks the Science and Engineering Research Board (SERB), Department of Science and Technology, Government of India, for the funding of project MTR/2017/000875 under Mathematical Research Impact Centric Support (MATRICS).

References

1. H. Apte, P. Chattopadhyay, R.A. Rao, A local-global theorem for extended ideals. Ramanujan Math. Soc. **27**(1), 1–20 (2012)
2. A. Bak, Nonabelian K-theory: the nilpotent class of K_1 and general stability. K-Theory **4**(4), 363–397 (1991)
3. R. Basu, Topics in classical algebraic K-theory. Ph.D. thesis, Tata Institute of Fundamental Research, 2006
4. R. Basu, R. Khanna, R.A. Rao, On Quillen's local global principle, in *Commutative Algebra and Algebraic Geometry (Bangalore, India, 2003), Contemporary Mathematics*, vol. 390 (American Mathematical Society, Providence, RI, 2005), pp. 17–30
5. P. Chattopadhyay, R.A. Rao, Elementary symplectic orbits and improved K_1-stability. J. K-Theory **7**(2), 389–403 (2011)
6. A. Gupta, Structures over commutative rings. Ph.D. thesis, Tata Institute of Fundamental Research, 2014
7. R. Hazrat, N. Vavilov, K_1 of Chevalley groups are nilpotent. J. Pure Appl. Algebra **179**(1–2), 99–116 (2003)
8. S. Jose, R.A. Rao, A fundamental property of Suslin matrices. J. K-Theory **5**(3), 407–436 (2010)
9. S. Jose, R.A. Rao, A local global principle for the elementary unimodular vector group, in *Commutative Algebra and Algebraic Geometry (Banglore, India, Contemporary Mathematics, 2005*, vol. 390 (American Mathematical Society, Providence, RI, 2003), pp. 119–125
10. S. Jose, R.A. Rao, A structure theorem for the elementary unimodular vector group. Trans. Am. Math. Soc. **358**(7), 3097–3112 (2006)
11. V.I. Kopeĭko, A.A. Suslin, Quadratic modules over polynomial rings (Russian), algebraic numbers and finite groups. Zap. Nauchn. Sem. Leningrad. Otdel. Mat. Inst. Steklov. (LOMI) **86**(114–124), 190–191 (1979)
12. D. Quillen, Projective modules over polynomial rings. Invent. Math. **36**, 167–171 (1976)
13. R.A. Rao, S. Jose, A study of Suslin matrices: their properties and uses, in *Algebra and its Applications, Springer Proceedings in Mathematics Statistics*, vol. 174 (Springer, Singapore, 2016), pp. 89–121
14. R.A. Rao, A stably elementary homotopy. Proc. Am. Math. Soc. **137**(11), 3637–3645 (2009)
15. R.A. Rao, S. Sharma, Homotopy and commutativity principle. J. Algebra **484**, 23–46 (2017)

16. A.V. Stepanov, *Nonabelian K-theory of Chevalley groups over rings (Russian)*, Zap. Nauchn. Sem. S.-Peterburg. Otdel. Mat. Inst. Steklov. (POMI) 423 (2014), Voprosy Teorii Predstavleni Algebr i Grupp. 26, 244–263; translation in J. Math. Sci. (N.Y.) 209 (2015), no. 4, 645–656
17. A.A. Suslin, On stably free modules. Math. USSR Sbornik **31**, 479–491 (1977)
18. A.A. Suslin, The structure of the special linear group over rings of polynomials. Izv. Akad. Nauk SSSR Ser. Mat. **41**, 235–252 (1977)
19. W. van der Kallen, A group structure on certain orbit sets of unimodular rows. J. Algebra **82**(2), 363–397 (1983)

Chapter 14
On a Theorem of Suslin

Raja Sridharan and Sunil K. Yadav

14.1 Introduction

In this paper, we give a new proof of Suslin's $n!$ theorem on unimodular rows. Suslin's theorem (see [27]) states the following:

Theorem 14.1 *Let A be a ring and $[u, a_1, a_2, \ldots, a_n]$ be a unimodular row in A. Then $[u^{n!}, a_1, a_2, \ldots, a_n]$ can be completed to a matrix in $\mathrm{SL}_{n+1}(A)$.*

To prove this, we give a new proof of the following Proposition of Suslin [12, 27] which was used by Suslin to prove Theorem 14.1.

Proposition 14.1 *Let $[u, a_1, \ldots, a_n]$ be a unimodular row in A. Suppose $[\overline{a}_1, \ldots, \overline{a}_n]$ is completable to a matrix belonging to $\mathrm{SL}_n(A/Au)$. Then the row $[u^n, a_1, \ldots, a_n]$ is completable to a matrix belonging to $\mathrm{SL}_{n+1}(A)$.*

Different proofs of this proposition have also been given by Mohan Kumar [16, Lemma 3] and Nori (unpublished). Our proof uses a result of Bhatwadekar–Lindel–Rao [6, 3.10].

We briefly sketch our proof of the Proposition of Suslin.

Let $P = \frac{A^{n+1}}{A[u^n, a_1, \ldots, a_n]}$ be the projective module associated to the unimodular row $[u^n, a_1, \ldots, a_n]$. We choose $a, b \in A$ in a suitable manner such that the ideal $aA + bA = A$ and P_a and P_b are free A_a and A_b-modules, respectively. In such a situation

R. Sridharan (✉)
Tata Institute of Fundamental Research, 1, Dr. Homi Bhabha Road,
Mumbai 400005, India
e-mail: sraja@math.tifr.res.in

S. K. Yadav
Department of Mathematics, Indian Institute of Technology Bombay,
Powai, Mumbai 400076, India
e-mail: skyadav@math.iitb.ac.in; skymath.bhu@gmail.com

A. A. Ambily et al. (eds.), *Leavitt Path Algebras and Classical K-Theory*,
Indian Statistical Institute Series,
https://doi.org/10.1007/978-981-15-1611-5_14

one can associate to P a one cocycle. We show, using Quillen splitting and a result of Bhatwadekar–Lindel–Rao [6] mentioned above, that this cocycle splits. Hence we conclude that P is free and the unimodular row is completable.

The arrangement of this paper is as follows:

In Sect. 14.2, we recall some preliminary results which are needed for the proof of Suslin's theorem. In Sect. 14.3, we collect together some known results on one cocycles. In Sect. 14.4, we give a proof of Suslin's Theorem and also a theorem of Murthy–Swan [22].

14.2 Some Preliminaries

In this section, we record some known results.

Definition 14.1 (i) Let A be a ring. A row $[a_1, a_2, \dots, a_n] \in A^n$ is said to be **unimodular** (of length n) if the ideal $(a_1, a_2, \dots, a_n) = A$. The set of unimodular rows of length n is denoted by $\mathrm{Um_n(A)}$.

(ii) A unimodular row $[a_1, a_2, \dots, a_n]$ is said to be **completable** if there is a matrix in $\mathrm{SL}_n(A)$ whose first row is $[a_1, a_2, \dots, a_n]$.

(iii) We define $E_n(A)$ to be the subgroup of $\mathrm{GL_n(A)}$ generated by all matrices of the form $e_{ij}(\lambda) = I_n + \lambda E_{ij}, \lambda \in A, i \neq j$, where E_{ij} is a matrix whose (i, j)-th entry is 1 and all other entries are zero. The matrices $e_{ij}(\lambda)$ will be referred to as elementary matrices.

Since $M_n(A)$ acts on A^n via matrix multiplication, the group $E_n(A)$ which is a subset of $M_n(A)$ also acts on A^n. This induces an action of $E_n(A)$ on $\mathrm{Um}_n(A)$. The equivalence relation on $\mathrm{Um}_n(A)$ given by this action is denoted by $\overset{E_n(A)}{\sim}$. Similarly one can define $\overset{\mathrm{GL_n(A)}}{\sim}$ and $\overset{\mathrm{SL}_n(A)}{\sim}$.

It is not hard to see that a unimodular row $v \in A^n$ is completable if and only if $v \overset{\mathrm{GL_n(A)}}{\sim} (1, 0, \dots, 0)$.

Theorem 14.2 (see [3])

(i) *Let A be a ring and $[a_1, a_2, \dots, a_n] \in A^n$ be a unimodular row of length n which contains a unimodular row of shorter length. Then the row $[a_1, a_2, \dots, a_n]$ is completable. In fact*

$$(a_1, a_2, \dots, a_n) \overset{E_n(A)}{\sim} (1, 0, \dots, 0).$$

(ii) *Let A be a semilocal ring. Then any unimodular row $[a_1, a_2, \dots, a_n]$ of length $n \geq 2$ is completable. In fact*

$$(a_1, a_2, \dots, a_n) \overset{E_n(A)}{\sim} (1, 0, \dots, 0).$$

Definition 14.2 Two matrices α and β in $\mathrm{SL}_n(A)$ are said to be **connected** if there exists $\sigma(X) \in \mathrm{SL}_n(A[X])$ such that $\sigma(0) = \alpha$ and $\sigma(1) = \beta$. By considering the matrix $\sigma(1 - X)$, it follows that if α is connected to β then β is connected to α.

Lemma 14.1 *Any matrix in $E_n(A)$ can be connected to the identity matrix.*

Proof Every matrix $\alpha \in E_n(A)$ can be written as a product of elementary matrices of the form $e_{ij}(\lambda) = I_n + \lambda E_{ij}$ for $i \neq j$, that is, $\alpha = \prod_{i=1}^{r} e_{ij}(\lambda)$. We define $\sigma(X) = \prod_{i=1}^{r} e_{ij}(\lambda X)$. Then $\sigma(X) \in \mathrm{SL}_n(A[X])$, $\sigma(0) = I_n$ and $\sigma(1) = \alpha$. This proves the lemma. □

Lemma 14.2 *Let A be a ring and I be an ideal of A. Then the map $E_n(A) \to E_n(A/I)$ is surjective.*

Proof The proof follows from the fact that the generators $e_{ij}(\bar{\lambda})$ of $E_n(A/I)$ for $\lambda \in A$ can be lifted to generators $e_{ij}(\lambda)$ of $E_n(A)$. □

Let us recall Quillen's Splitting Lemma [23] with the proof following the exposition of [3]. In what follows, $(\psi_1(X))_t$ denotes the image of $\psi_1(X)$ in $\mathrm{GL_n(A_{st}[X])}$ and $(\psi_2(X))_s$ denotes the image of $\psi_2(X)$ in $\mathrm{GL_n(A_{st}[X])}$.)

Lemma 14.3 (see [23]) *Let A be a domain and $s, t \in A$ be such that $sA + tA = A$. Suppose there exists $\sigma(X) \in \mathrm{GL_n(A_{st}[X])}$ with the property that $\sigma(0) = I_n$. Then there exist $\psi_1(X) \in \mathrm{GL_n(A_s[X])}$ with $\psi_1(0) = I_n$ and $\psi_2(X) \in \mathrm{GL_n(A_t[X])}$ with $\psi_2(0) = I_n$ such that $\sigma(X) = (\psi_1(X))_t(\psi_2(X))_s$.*

Proof Since $\sigma(0) = I_n$, $\sigma(X) = I_n + X\tau(X)$, where $\tau(X) \in M_n(A_{st}[X])$, we choose a large integer N_1 such that $\sigma(\lambda s^k X) \in \mathrm{GL_n(A_t[X])}$ for all $\lambda \in A$ and for all $k \geq N_1$. Define $\beta(X, Y, Z) \in \mathrm{GL_n(A_{st}[X, Y, Z])}$ as follows:

$$\beta(X, Y, Z) = \sigma((Y + Z)X)\sigma(YX)^{-1}. \tag{14.1}$$

Then $\beta(X, Y, 0) = I_n$, and hence there exists a large integer N_2 such that for all $k \geq N_2$ and for all $\mu \in A$ we have $\beta(X, Y, \mu t^k Z) \in \mathrm{GL_n(A_s[X, Y, Z])}$. This means

$$\beta(X, Y, \mu t^k Z) = (\sigma_1(X, Y, Z))_t, \tag{14.2}$$

where $\sigma_1(X, Y, Z) \in \mathrm{GL_n(A_s[X, Y, Z])}$ with $\sigma_1(X, Y, 0) = I_n$.

Taking $N = \max(N_1, N_2)$, it follows by the comaximality of sA and tA that $s^N A + t^N A = A$. Pick $\lambda, \mu \in A$ such that $\lambda s^N + \mu t^N = 1$. Setting $Y = \lambda s^N$, $Z = \mu t^N$ in (14.1) and $Z = 1$, $Y = \lambda s^N$ in (14.2) we get

$$\beta(X, \lambda s^N, \mu t^N) = \sigma(X)\sigma(\lambda s^N X)^{-1}$$

and

$$\beta(X, \lambda s^N, \mu t^N) = (\sigma_1(X, \lambda s^N, \mu t^N))_t = (\psi_1(X))_t, \text{ where } \psi_1(X) \in \mathrm{GL_n(A_s[X])}.$$

Hence, we conclude $\sigma(X)\sigma(\lambda s^N X)^{-1} = (\psi_1(X))_t$. Let $\sigma(\lambda s^N X) = (\psi_2(X))_s$, where $(\psi_2(X))_s \in \mathrm{GL_n}(\mathrm{A_t}[\mathrm{X}])$. Since $\sigma(0) = I_n$, $\psi_1(0) = \psi_2(0) = I_n$, the result follows by using the identity $\sigma(X) = \sigma(X)\sigma(\lambda s^N X)^{-1}\sigma(\lambda s^N X)$. □

Remark 14.1 In the above lemma by interchanging the roles of s and t we can write $\sigma(X) = (\tau_1(X))_s(\tau_2(X))_t$, where $\tau_1(X) \in \mathrm{GL_n}(\mathrm{A_t}[\mathrm{X}])$ with $\tau_1(0) = I_n$ and $\tau_2(X) \in \mathrm{GL_n}(\mathrm{A_s}[\mathrm{X}])$ with $\tau_2(0) = I_n$.

Lemma 14.4 *Let A be a domain and $s, t \in A$ be such that $sA + tA = A$. If $\sigma_1 \in \mathrm{SL}_n(A_s)$, $\sigma_2 \in E_n(A_t)$, then $\sigma_1\sigma_2 = \beta_1\beta_2$, where $\beta_1 \in \mathrm{SL}_n(A_t)$ and $\beta_2 \in \mathrm{SL}_n(A_s)$.*

Proof We can write $\sigma_1\sigma_2 = \sigma_1\sigma_2\sigma_1^{-1}\sigma_1$. Therefore, it suffices to show that $\sigma_1\sigma_2\sigma_1^{-1} = \gamma_1\gamma_2$, where $\gamma_1 \in \mathrm{SL}_n(A_t)$ and $\gamma_2 \in \mathrm{SL}_n(A_s)$. Then the result follows by setting $\beta_1 = \gamma_1$ and $\beta_2 = \gamma_2\sigma_1$. Since any elementary matrix can be connected to the identity matrix, we can find $\alpha(X) \in \mathrm{SL}_n(A_t[X])$ such that $\alpha(0) = I_n$ and $\alpha(1) = \sigma_2$. Let $\delta(X) = \sigma_1\alpha(X)\sigma_1^{-1}$. Then $\delta(1) = \sigma_1\sigma_2\sigma_1^{-1}$. Since $\delta(X) \in \mathrm{SL}_n(A_{st}[X])$ and $\delta(0) = I_n$, by Remark 14.1, $\delta(X) = \delta_1(X)\delta_2(X)$, where $\delta_1(X) \in \mathrm{SL}_n(A_t[X])$ and $\delta_2(X) \in \mathrm{SL}_n(A_s[X])$. Setting $\gamma_1 = \delta_1(1)$ and $\gamma_2 = \delta_2(1)$, the lemma follows. □

Lemma 14.5 (see [6]) *Let A be a domain and $s, t \in A$ be such that $sA + tA = A$. Let $\sigma \in \mathrm{SL}_n(A_{st})$ and $\varepsilon \in E_n(A_{st})$. Then $\sigma\varepsilon = \tau_1\sigma\tau_2$, where $\tau_1 \in \mathrm{SL}_n(A_s)$ and $\tau_2 \in \mathrm{SL}_n(A_t)$.*

Proof Let $\varepsilon = \varepsilon_1\varepsilon_2$, where $\varepsilon_1 \in \mathrm{SL}_n(A_s)$ is chosen such that $\varepsilon_1 = I_n \bmod (t^N)$ for sufficiently large N and $\varepsilon_2 \in \mathrm{SL}_n(A_t)$. So, we have $\sigma\varepsilon = \sigma\varepsilon_1\varepsilon_2 = \sigma\varepsilon_1\sigma^{-1}\sigma\varepsilon_2$. Now, since $\varepsilon_1 = I_n \bmod (t^N)$ for sufficiently large N, therefore $\sigma\varepsilon_1\sigma^{-1} \in \mathrm{SL}_n(A_s)$. Now by taking $\tau_1 = \sigma\varepsilon_1\sigma^{-1}$ and $\tau_2 = \varepsilon_2$, we have $\sigma\varepsilon = \tau_1\sigma\tau_2$. □

Lemma 14.6 *Let A be a domain and I be an ideal of A. Let $a, c \in A$ be such that $aA + cA = A$. Then*

$$\begin{array}{ccc} I & \longrightarrow & I_a \\ \downarrow & & \downarrow \\ I_c & \longrightarrow & I_{ac} \end{array}$$

is a pullback diagram. This means that if two elements $x \in I_a$, $y \in I_c$ are equal in I_{ac}, then there exists a unique $z \in I$ such that $\frac{z}{1} = x$ in I_a and $\frac{z}{1} = y$ in I_b.

Proof Let $x = \frac{b}{a^r}$ and $y = \frac{d}{c^s}$ be such that $\frac{b}{a^r} = \frac{d}{c^s}$ in I_{ac}, where $b, c \in A$. Hence $bc^s = da^r$ in A. Since $aA + cA = A$, $a^rA + c^sA = A$. We choose $\lambda, \mu \in A$ such that $\lambda a^r + \mu c^s = 1$. Let $z = \lambda b + \mu d$. Then $a^r z = a^r\lambda b + a^r\mu d = a^r\lambda b + c^s\mu b = b(a^r\lambda + c^s\mu) = b$ and $c^s z = c^s\lambda b + c^s\mu d = a^r\lambda d + c^s\mu d = d(a^r\lambda + c^s\mu) = d$. Hence we have $\frac{z}{1} = \frac{b}{a^r}$ in I_a and $\frac{z}{1} = \frac{d}{c^s}$ in I_c. The uniqueness is proved in a similar manner. □

14.3 On Some Results on Cocycles

In this section, we recall some known results on one cocycles. We give many details to make the paper readable. The reader who is familiar with one cocycles could skip this section. For rest of the paper we assume that A is a domain.

Let A be a domain and $[v_0, v_1, \ldots, v_n]$ be a unimodular row in A. We describe the cocycle associated to the unimodular row $[v_0, v_1, \ldots, v_n]$ and as well as the cocycle associated to the projective module $\frac{A^{n+1}}{A[v_0, v_1, \ldots, v_n]}$ and explain how they are related to each other.

For notational simplicity, we do this for unimodular rows of length 3 and modifications needed for the general case are cosmetic.

Suppose $s, t \in A$ are such that $sA + tA = A$ and assume that the row $[v_0, v_1, v_2]$ is completable[1] to matrices $\alpha \in \mathrm{SL}_3(A_s)$ and $\beta \in \mathrm{SL}_3(A_t)$. Let

$$\alpha = \begin{pmatrix} v_0 & & \\ v_1 & p_1 & p_2 \\ v_2 & & \end{pmatrix} \text{ and } \beta = \begin{pmatrix} v_0 & & \\ v_1 & q_1 & q_2 \\ v_2 & & \end{pmatrix}. \tag{14.3}$$

It is clear that the first column of the matrix $\alpha^{-1}\beta$ is $\begin{pmatrix} 1 \\ 0 \\ 0 \end{pmatrix}$, whereby

$$\alpha^{-1}\beta = \begin{pmatrix} 1 & \mu_1 & \mu_2 \\ 0 & \lambda_{22} & \lambda_{23} \\ 0 & \lambda_{32} & \lambda_{33} \end{pmatrix} \in \mathrm{SL}_2(A_{st}). \tag{14.4}$$

Definition 14.3 The matrix $\sigma \in \mathrm{SL}_2(A_{st})$ given by $\sigma = \begin{pmatrix} \lambda_{22} & \lambda_{23} \\ \lambda_{32} & \lambda_{33} \end{pmatrix}$ is called a "**cocycle**" associated to the unimodular row $[v_0, v_1, v_2]$.

Rewriting (14.4) as

$$\begin{pmatrix} v_0 & & \\ v_1 & q_1 & q_2 \\ v_2 & & \end{pmatrix} = \begin{pmatrix} v_0 & & \\ v_1 & p_1 & p_2 \\ v_2 & & \end{pmatrix} \begin{pmatrix} 1 & \mu_1 & \mu_2 \\ 0 & \lambda_{22} & \lambda_{23} \\ 0 & \lambda_{32} & \lambda_{33} \end{pmatrix} \tag{14.5}$$

we get

$$q_1 = \mu_1 \begin{pmatrix} v_0 \\ v_1 \\ v_2 \end{pmatrix} + \lambda_{22} p_1 + \lambda_{32} p_2; \; q_2 = \mu_2 \begin{pmatrix} v_0 \\ v_1 \\ v_2 \end{pmatrix} + \lambda_{23} p_1 + \lambda_{33} p_2. \tag{14.6}$$

Quotienting modulo $[v_0, v_1, v_2]$, we get

[1]This assumption is needed to define cocycles.

$$\overline{q_1} = \lambda_{22}\overline{p_1} + \lambda_{32}\overline{p_2};\ \overline{q_2} = \lambda_{23}\overline{p_1} + \lambda_{33}\overline{p_2}. \tag{14.7}$$

In matrix form we can write this as follows:

$$\begin{pmatrix} \overline{q_1} \\ \overline{q_2} \end{pmatrix} = \begin{pmatrix} \lambda_{22} & \lambda_{32} \\ \lambda_{23} & \lambda_{33} \end{pmatrix} \begin{pmatrix} \overline{p_1} \\ \overline{p_2} \end{pmatrix} \tag{14.8}$$

in $\frac{A_{st}^3}{A_{st}[v_0, v_1, v_2]}$.

Definition 14.4 The matrix $\begin{pmatrix} \lambda_{22} & \lambda_{32} \\ \lambda_{23} & \lambda_{33} \end{pmatrix} = \tau$, which is obviously the transpose of σ, is called a "**cocycle**" associated to the projective module $\frac{A^3}{A[v_0, v_1, v_2]}$.

Therefore, there are two ways of associating a cocycle to a projective module given by a unimodular row. Depending on the context, both methods are useful.

Now, we see how the cocycle associated to a unimodular row depends on the choice of completions. Let us consider completions $\alpha \in \mathrm{SL}_3(A_s)$ and $\beta \in \mathrm{SL}_3(A_t)$ of the unimodular row $[v_0, v_1, v_2]$. Then we get (14.4).

Now, we choose different completions α', β' of $[v_0, v_1, v_2]$ in A_s and A_t as follows: Let $\delta \in \mathrm{SL}_3(A_s)$ be such that δ has first column $(1, 0, 0)^t$ then $\alpha' = \alpha\delta \in \mathrm{SL}_3(A_s)$ and has first column $(v_0, v_1, v_2)^t$, and any other completion of the unimodular row $[v_0, v_1, v_2]$ to a matrix in $\mathrm{SL}_3(A_s)$ can be obtained in this way. Similarly, the matrix $\beta' = \beta\gamma$, where $\gamma \in \mathrm{SL}_3(A_t)$ has first column $(1, 0, 0)^t$.

Again it is clear that the first column of $\alpha'^{-1}\beta'$ is $\begin{pmatrix} 1 \\ 0 \\ 0 \end{pmatrix}$. We can write

$$\alpha'^{-1}\beta' = \begin{pmatrix} 1 & \lambda'_{st} & \mu'_{st} \\ 0 & & \\ 0 & \multicolumn{2}{c}{\sigma'} \end{pmatrix},$$

where $\lambda'_{st}, \mu'_{st} \in A_{st}$ and $\sigma' \in \mathrm{SL}_2(A_{st})$. This σ' is the cocycle associated to the unimodular row $[v_0, v_1, v_2]$ with respect to the new completions of the row $[v_0, v_1, v_2]$ in A_s and A_t. Now putting the values of α' and β' and rearranging them we get

$$\delta^{-1} \begin{pmatrix} 1 & \lambda_{st} & \mu_{st} \\ 0 & & \\ 0 & \multicolumn{2}{c}{\sigma} \end{pmatrix} \gamma = \begin{pmatrix} 1 & \lambda'_{st} & \mu'_{st} \\ 0 & & \\ 0 & \multicolumn{2}{c}{\sigma'} \end{pmatrix}.$$

The matrices δ^{-1}, γ have the form $\begin{pmatrix} 1 & \lambda_s & \mu_s \\ 0 & & \\ 0 & \multicolumn{2}{c}{\tau_1^{-1}} \end{pmatrix}, \begin{pmatrix} 1 & \lambda_t & \mu_t \\ 0 & & \\ 0 & \multicolumn{2}{c}{\tau_2} \end{pmatrix}$, respectively, where $\lambda_s, \mu_s \in A_s$, $\tau_1 \in \mathrm{SL}_2(A_s)$ and $\lambda_t, \mu_t \in A_t$, $\tau_2 \in \mathrm{SL}_2(A_t)$. Therefore, we can write

$$\begin{pmatrix} 1 & \lambda_s \ \mu_s \\ 0 & \\ 0 & \tau_1^{-1} \end{pmatrix} \begin{pmatrix} 1 & \lambda_{st} \ \mu_{st} \\ 0 & \\ 0 & \sigma \end{pmatrix} \begin{pmatrix} 1 & \lambda_t \ \mu_t \\ 0 & \\ 0 & \tau_2 \end{pmatrix} = \begin{pmatrix} 1 & \lambda'_{st} \ \mu'_{st} \\ 0 & \\ 0 & \sigma' \end{pmatrix}.$$

This implies that the new cocycle σ' associated to the unimodular row $[v_0, v_1, v_2]$ is obtained from the old cocycle σ after multiplying by $\tau_1^{-1} \in \mathrm{SL}_2(A_s)$ from the left and multiplying by $\tau_2 \in \mathrm{SL}_2(A_t)$ from the right.

The above discussion leads us to the following definition:

Definition 14.5 Let A be a domain. Suppose $s, t \in A$ are such that $sA + tA = A$ and $\sigma, \sigma' \in \mathrm{SL}_2(A_{st})$. We say that the cocycle σ is **equivalent** to the cocycle σ' if there exist $\tau_1 \in \mathrm{SL}_2(A_s)$ and $\tau_2 \in \mathrm{SL}_2(A_t)$ such that $\sigma' = \tau_1^{-1}\sigma\tau_2$.

We see that the cocycles arising out of different completions of a unimodular row are equivalent.

Let $P = \frac{A^3}{A[v_0, v_1, v_2]}$ be the projective module associated to the unimodular row $[v_0, v_1, v_2]$. Consider the diagram

$$\begin{array}{ccc} P & \longrightarrow & P_s \\ \downarrow & & \downarrow \\ P_t & \longrightarrow & P_{st}. \end{array}$$

Let $\sigma_1 \in \mathrm{SL}_2(A_{st})$ be the cocycle associated to the projective module P corresponding to the basis $\{p_1, p_2\}$ of P_s and the basis $\{q_1, q_2\}$ of P_t which give completions of the unimodular row $[v_0, v_1, v_2]$ in $\mathrm{SL}_2(A_s)$ and $\mathrm{SL}_2(A_t)$. Then we have

$$\begin{pmatrix} q_1 \\ q_2 \end{pmatrix} = \sigma_1 \begin{pmatrix} p_1 \\ p_2 \end{pmatrix}. \tag{14.9}$$

Now, we choose new bases for P_s and P_t in the following manner: Let $\tau_1' \in \mathrm{SL}_2(A_s)$ and $\tau_2' \in \mathrm{SL}_2(A_t)$ such that

$$\tau_1' \begin{pmatrix} p_1 \\ p_2 \end{pmatrix} = \begin{pmatrix} p_1' \\ p_2' \end{pmatrix}; \ \tau_2' \begin{pmatrix} q_1 \\ q_2 \end{pmatrix} = \begin{pmatrix} q_1' \\ q_2' \end{pmatrix}. \tag{14.10}$$

So, we have new bases $\{p_1', p_2'\}$ of P_s and $\{q_1', q_2'\}$ of P_t. Suppose that $\sigma_2 \in \mathrm{SL}_2(A_{st})$ is the cocycle associated to the projective module P corresponding to the new bases. Thus we have

$$\begin{pmatrix} q_1' \\ q_2' \end{pmatrix} = \sigma_2 \begin{pmatrix} p_1' \\ p_2' \end{pmatrix}. \tag{14.11}$$

Now, combining (14.9)–(14.11) we get

$$\sigma_1 \begin{pmatrix} p_1 \\ p_2 \end{pmatrix} = \tau_2'^{-1}\sigma_2\tau_1' \begin{pmatrix} p_1 \\ p_2 \end{pmatrix}.$$

This implies that $\sigma_1 = \tau_2'^{-1}\sigma_2\tau_1'$ or $\tau_2'\sigma_1\tau_1'^{-1} = \sigma_2$. Hence, the new cocycle σ_2 associated to the projective module P is obtained from the old cocycle σ_1 after multiplying by $\tau_2' \in \mathrm{SL}_2(A_t)$ from the left and multiplying by $\tau_1'^{-1} \in \mathrm{SL}_2(A_s)$ from the right.

The above discussion leads us to the following definition:

Definition 14.6 Let A be a domain. Suppose $s, t \in A$ are such that $sA + tA = A$ and $\sigma_1, \sigma_2 \in \mathrm{SL}_2(A_{st})$. Then we say that σ_1 is **equivalent** to σ_2 if there exist $\tau_1' \in \mathrm{SL}_2(A_s)$ and $\tau_2' \in \mathrm{SL}_2(A_t)$ such that $\sigma_2 = \tau_2'\sigma_1\tau_1'^{-1}$.

Let A be a domain and $[v_0, v_1, v_2]$ be a unimodular row in A. Suppose $s, t \in A$ are such that $sA + tA = A$ and the unimodular row $[v_0, v_1, v_2]$ is completable to a matrix $\alpha \in \mathrm{SL}_3(A_s)$ and $\beta \in \mathrm{SL}_3(A_t)$ as in (14.3).

Now, we want to give conditions under which the unimodular row $[v_0, v_1, v_2]$ is completable to a matrix $\gamma \in \mathrm{SL}_3(A)$. In order to find whether the completion of the unimodular row $[v_0, v_1, v_2]$ in A exists we use Lemma 14.6. According to that we can get the required matrix $\gamma \in \mathrm{SL}_3(A)$ by patching together $\alpha \in \mathrm{SL}_3(A_s)$ and $\beta \in \mathrm{SL}_3(A_t)$ which satisfy $\alpha_t = \beta_s$ in A_{st}.

Now, we see what the sufficient condition is under which we can patch together $\alpha \in \mathrm{SL}_3(A_s)$ and $\beta \in \mathrm{SL}_3(A_t)$ in A_{st} which give completions of the column $[v_0, v_1, v_2]^t$. Referring to (14.4), that is,

$$\alpha^{-1}\beta = \begin{pmatrix} 1 & \lambda_{11} & \lambda_{12} \\ 0 & \multicolumn{2}{c}{\multirow{2}{*}{σ}} \\ 0 & & \end{pmatrix},$$

for some $\lambda_{11}, \lambda_{12} \in A_{st}$ and $\sigma \in \mathrm{SL}_2(A_{st})$.

Now, suppose σ splits as $\sigma = \sigma_1\sigma_2$, where $\sigma_1 \in \mathrm{SL}_2(A_s)$ and $\sigma_2 \in \mathrm{SL}_2(A_t)$. Then we can write

$$\alpha^{-1}\beta = \begin{pmatrix} 1 & \lambda_{11} & \lambda_{12} \\ 0 & \multicolumn{2}{c}{\multirow{2}{*}{$\sigma_1\sigma_2$}} \\ 0 & & \end{pmatrix}.$$

Let $\alpha' = \alpha \begin{pmatrix} 1 & 0 & 0 \\ 0 & \multicolumn{2}{c}{\multirow{2}{*}{σ_1}} \\ 0 & & \end{pmatrix}$ and $\beta' = \beta \begin{pmatrix} 1 & 0 & 0 \\ 0 & \multicolumn{2}{c}{\multirow{2}{*}{σ_2^{-1}}} \\ 0 & & \end{pmatrix}$. Thus we have

$$\alpha'^{-1}\beta' = \begin{pmatrix} 1 & 0 & 0 \\ 0 & \multicolumn{2}{c}{\multirow{2}{*}{σ_1^{-1}}} \\ 0 & & \end{pmatrix} \alpha^{-1}\beta \begin{pmatrix} 1 & 0 & 0 \\ 0 & \multicolumn{2}{c}{\multirow{2}{*}{σ_2^{-1}}} \\ 0 & & \end{pmatrix} = \begin{pmatrix} 1 & \lambda_{11}' & \lambda_{12}' \\ 0 & 1 & 0 \\ 0 & 0 & 1 \end{pmatrix}.$$

Since $sA + tA = A$, we have $s^rA + t^rA = A$ and there exist $\lambda, \mu \in A$ such that $\lambda s^r + \mu t^r = 1$. By using this relation for sufficiently large r, we can write $\lambda_{11}' = \lambda_1 + \mu_1, \lambda_{12}' = \lambda_2 + \mu_2$, where $\lambda_1, \lambda_2 \in A_s$ and $\mu_1, \mu_2 \in A_t$. Therefore,

$$\begin{pmatrix} 1 & \lambda'_{11} & \lambda'_{12} \\ 0 & 1 & 0 \\ 0 & 0 & 1 \end{pmatrix} = \begin{pmatrix} 1 & \lambda_1 & \lambda_2 \\ 0 & 1 & 0 \\ 0 & 0 & 1 \end{pmatrix} \begin{pmatrix} 1 & \mu_1 & \mu_2 \\ 0 & 1 & 0 \\ 0 & 0 & 1 \end{pmatrix} = \varepsilon_1 \varepsilon_2,$$

where $\varepsilon_1 \in E_3(A_s)$ and $\varepsilon_2 \in E_3(A_t)$.

Therefore, we change α' to $\alpha'' = \alpha'\varepsilon_1$ and β' to $\beta'' = \beta'\varepsilon_2^{-1}$. Clearly $\alpha''^{-1}\beta'' = I_3$. That is $\alpha'' = \beta''$ in A_{st}. Therefore by using Lemma 14.6, we get the required completion of the unimodular row $[v_0, v_1, v_2]$ in A. Therefore, any unimodular row $[v_0, v_1, v_2]$ is completable in A if the cocycle associated to the unimodular row $[v_0, v_1, v_2]$ splits.

We have the following theorem:

Theorem 14.3 *Let A be a domain and $[v_0, v_1, v_2] \in A^3$ be a unimodular row. Suppose there exist $s, t \in A$ such that $sA + tA = 1$ and $[v_0, v_1, v_2]$ is completable in A_s and A_t. Then $[v_0, v_1, v_2]$ is completable in A if and only if the cocycle associated to the unimodular row $[v_0, v_1, v_2]$ splits.*

Proof We have already shown one part we show the other part. Suppose $[v_0, v_1, v_2]$ is completable to a matrix $\delta \in \mathrm{SL}_3(A)$. Then we have to show that the cocycle associated to the unimodular row $[v_0, v_1, v_2]$ splits.

Choose completions of the unimodular row $[v_0, v_1, v_2]$ given by $\alpha \in \mathrm{SL}_3(A_s)$ and $\beta \in \mathrm{SL}_3(A_t)$. Then from (14.4), we have

$$\alpha^{-1}\beta = \begin{pmatrix} 1 & * & * \\ 0 & \multicolumn{2}{c}{\multirow{2}{*}{σ}} \\ 0 & & \end{pmatrix}.$$

Since $\delta \in \mathrm{SL}_3(A)$ is a matrix whose first column is $(v_0, v_1, v_2)^t$, using (14.4), we can write

$$\delta^{-1}\alpha = \begin{pmatrix} 1 & * & * \\ 0 & & \\ 0 & \sigma_1 & \end{pmatrix}, \text{ where } \sigma_1 \in \mathrm{SL}_2(A_s)$$

and

$$\delta^{-1}\beta = \begin{pmatrix} 1 & * & * \\ 0 & & \\ 0 & \sigma_2 & \end{pmatrix}, \text{ where } \sigma_2 \in \mathrm{SL}_2(A_t).$$

Then we have $(\delta^{-1}\alpha)^{-1}\delta^{-1}\beta = \alpha^{-1}\delta\delta^{-1}\beta = \alpha^{-1}\beta$. Hence

$$\begin{pmatrix} 1 & * & * \\ 0 & & \\ 0 & \sigma_1^{-1} & \end{pmatrix} \begin{pmatrix} 1 & * & * \\ 0 & & \\ 0 & \sigma_2 & \end{pmatrix} = \begin{pmatrix} 1 & * & * \\ 0 & & \\ 0 & \sigma & \end{pmatrix}.$$

This implies $\sigma = \sigma_1^{-1}\sigma_2$, where $\sigma_1 \in \mathrm{SL}_2(A_s)$ and $\sigma_2 \in \mathrm{SL}_2(A_t)$. That is, the cocycle associated to the unimodular row $[v_0, v_1, v_2]$ splits. □

Theorem 14.4 *Let A be a domain and $[v_0, v_1, v_2] \in A^3$ be a unimodular row. Suppose there exist $s, t \in A$ such that $sA + tA = A$ and $[v_0, v_1, v_2]$ is completable in A_s and A_t. Then $[v_0, v_1, v_2]$ is completable if and only if the cocycle associated to the projective module $\frac{A^3}{A[v_0,v_1,v_2]}$ splits.*

Proof Suppose the cocycle associated to the projective module $\frac{A^3}{A[v_0,v_1,v_2]}$ splits. Then we have to show that the unimodular row $[v_0, v_1, v_2]$ is completable. Let $\alpha \in \mathrm{SL}_3(A_s)$ and $\beta \in \mathrm{SL}_3(A_t)$ be as given in (14.3), two completions of the unimodular row $[v_0, v_1, v_2]$.

Therefore, $\alpha^{-1}\beta$ as in (14.4). Rewriting (14.4) and using (14.5)–(14.7), we get the following relation:

$$\begin{pmatrix} \overline{q_1} \\ \overline{q_2} \end{pmatrix} = \begin{pmatrix} \lambda_{22} & \lambda_{32} \\ \lambda_{23} & \lambda_{33} \end{pmatrix} \begin{pmatrix} \overline{p_1} \\ \overline{p_2} \end{pmatrix}$$

in $\frac{A_{st}^3}{A_{st}[v_0,v_1,v_2]}$.

Therefore, we have a cocycle $\tau = \begin{pmatrix} \lambda_{21} & \lambda_{22} \\ \lambda_{31} & \lambda_{32} \end{pmatrix}$ associated to the projective module $\frac{A_{st}^3}{A_{st}[v_0,v_1,v_2]}$. If τ splits, that is, $\tau = \tau_1\tau_2$, where $\tau_1 \in \mathrm{SL}_2(A_t)$ and $\tau_2 \in \mathrm{SL}_2(A_s)$, then

$$\begin{pmatrix} \overline{q_1} \\ \overline{q_2} \end{pmatrix} = \tau_1\tau_2 \begin{pmatrix} \overline{p_1} \\ \overline{p_2} \end{pmatrix}.$$

That is,

$$\tau_1^{-1} \begin{pmatrix} \overline{q_1} \\ \overline{q_2} \end{pmatrix} = \tau_2 \begin{pmatrix} \overline{p_1} \\ \overline{p_2} \end{pmatrix}.$$

Suppose that $\tau_1^{-1} \begin{pmatrix} \overline{q_1} \\ \overline{q_2} \end{pmatrix} = \begin{pmatrix} \overline{q_1'} \\ \overline{q_2'} \end{pmatrix}$ and $\tau_2 \begin{pmatrix} \overline{p_1} \\ \overline{p_2} \end{pmatrix} = \begin{pmatrix} \overline{p_1'} \\ \overline{p_2'} \end{pmatrix}$. Thus

$$\begin{pmatrix} \overline{q_1'} \\ \overline{q_2'} \end{pmatrix} = \begin{pmatrix} \overline{p_1'} \\ \overline{p_2'} \end{pmatrix} \bmod [v_0, v_1, v_2]. \tag{14.12}$$

Then we have $q_1' - p_1' = \lambda_{st}[v_0, v_1, v_2]$ and $q_2' - p_2' = \mu_{st}[v_0, v_1, v_2]$, where $\lambda_{st}, \mu_{st} \in A_{st}$. Since $sA + tA = A$, (14.12) implies following:

$$q_1' - p_1' = (\lambda_s + \lambda_t)[v_0, v_1, v_2];\ q_2' - p_2' = (\mu_s + \mu_t)[v_0, v_1, v_2],$$

where $\lambda_s, \mu_s \in A_s$, and $\lambda_t, \mu_t \in A_t$. This relation can be written as follows:

$$q_1' - \lambda_t[v_0, v_1, v_2] = p_1' + \lambda_s[v_0, v_1, v_2];\ q_2' - \mu_t[v_0, v_1, v_2] = p_2' + \mu_s[v_0, v_1, v_2].$$

Let the column $(v_0, v_1, v_2)^t$ be denoted by C_1. Then we transform

$$\begin{pmatrix} v_0 & & \\ v_1 & q_1' & q_2' \\ v_2 & & \end{pmatrix}$$

to

$$\begin{pmatrix} v_0 & & \\ v_1 & q_1' - \lambda_t C_1 & q_2' - \mu_t C_1 \\ v_2 & & \end{pmatrix} \in \mathrm{SL}_3(A_t). \tag{14.13}$$

and

$$\begin{pmatrix} v_0 & & \\ v_1 & p_1' & p_2' \\ v_2 & & \end{pmatrix},$$

to

$$\begin{pmatrix} v_0 & & \\ v_1 & p_1' + \lambda_s C_1 & p_2' + \mu_s C_1 \\ v_2 & & \end{pmatrix} \in \mathrm{SL}_3(A_s). \tag{14.14}$$

Now, patching together (14.13) and (14.14) in A_{st} and using Lemma 14.6, we get a completion of the row $[v_0, v_1, v_2]$ in A.

Conversely, let $[v_0, v_1, v_2]$ be completable in A. Then we have to show that the cocycle $\sigma \in \mathrm{SL}_2(A_{st})$ associated to the projective module $P = \frac{A^3}{A[v_0, v_1, v_2]}$ splits. Let $\{r_1, r_2\}$ be a basis for the projective module P which gives a completion of $[v_0, v_1, v_2]$ in $\mathrm{SL}_3(A)$. We have completions of the unimodular row $[v_0, v_1, v_2]$ in $\mathrm{SL}_3(A_s)$ and $\mathrm{SL}_3(A_t)$. Now, suppose $\{\widetilde{p_1}, \widetilde{p_2}\}$ and $\{\widetilde{q_1}, \widetilde{q_2}\}$ are the bases for P_s and P_t, respectively, and τ is the connecting matrix for these two bases. Thus using (14.8), we have

$$\begin{pmatrix} \widetilde{q_1} \\ \widetilde{q_2} \end{pmatrix} = \tau \begin{pmatrix} \widetilde{p_1} \\ \widetilde{p_2} \end{pmatrix}. \tag{14.15}$$

Let $\delta_1 \in \mathrm{SL}_2(A_s)$, $\delta_2 \in \mathrm{SL}_2(A_t)$ be the respective connecting matrices for the basis $\{r_1, r_2\}$ of P and the bases $\{\widetilde{p_1}, \widetilde{p_2}\}$ of P_s and $\{\widetilde{q_1}, \widetilde{q_2}\}$ of P_t. That is,

$$\begin{pmatrix} r_1 \\ r_2 \end{pmatrix} = \delta_1 \begin{pmatrix} \widetilde{p_1} \\ \widetilde{p_2} \end{pmatrix} \text{ and } \begin{pmatrix} r_1 \\ r_2 \end{pmatrix} = \delta_2 \begin{pmatrix} \widetilde{q_1} \\ \widetilde{q_2} \end{pmatrix}. \tag{14.16}$$

Combining (14.15) and (14.16), we get

$$\delta_2^{-1} \begin{pmatrix} r_1 \\ r_2 \end{pmatrix} = \tau \delta_1^{-1} \begin{pmatrix} r_1 \\ r_2 \end{pmatrix}.$$

That is $\delta_2^{-1} = \tau \delta_1^{-1}$ or $\delta_2^{-1}\delta_1 = \tau$. This implies that τ splits. □

Corollary 14.1 *Let A be a domain and $[a_1, a_2, a_3]$, $[b_1, b_2, b_3]$ be unimodular rows in A. Suppose $s, t \in A$ are such that $sA + tA = A$. Let $P_1 = \frac{A^3}{A[a_1, a_2, a_3]}$ and*

$P_2 = \frac{A^3}{A[b_1,b_2,b_3]}$ *be projective modules such that* $(P_1)_s$, $(P_1)_t$, $(P_2)_s$, $(P_2)_t$ *are free of rank* 2. *Let* σ_1 *be the cocycle associated to the projective module* P_1 *and* σ_2 *be the cocycle associated to the projective module* P_2. *Suppose that* σ_1 *is equivalent to* σ_2. *Then we can choose bases of* $(P_1)_s$ *and* $(P_1)_t$ *such that the cocycle associated to* P_1 *corresponding to these bases is equal to* σ_2.

Proof Suppose $\sigma_1 \in \mathrm{SL}_2(A_{st})$ is the cocycle associated to the projective module P_1 corresponding to a basis $\{\overline{p_1}, \overline{p_2}\}$ of $(P_1)_s$ and a basis $\{\overline{q_1}, \overline{q_2}\}$ of $(P_1)_t$. Then we have

$$\begin{pmatrix} \overline{q_1} \\ \overline{q_2} \end{pmatrix} = \sigma_1 \begin{pmatrix} \overline{p_1} \\ \overline{p_2} \end{pmatrix}. \tag{14.17}$$

Similarly, let $\sigma_2 \in \mathrm{SL}_2(A_{st})$ be the cocycle associated to the projective module P_2 corresponding to a basis $\{\overline{p_1'}, \overline{p_2'}\}$ of $(P_2)_s$ and a basis $\{\overline{q_1'}, \overline{q_2'}\}$ of $(P_2)_t$. Then we have

$$\begin{pmatrix} \overline{q_1'} \\ \overline{q_2'} \end{pmatrix} = \sigma_2 \begin{pmatrix} \overline{p_1'} \\ \overline{p_2'} \end{pmatrix}. \tag{14.18}$$

Since σ_1 is equivalent to σ_2, there exist $\tau_1 \in \mathrm{SL}_2(A_s)$ and $\tau_2 \in \mathrm{SL}_2(A_t)$ such that $\sigma_2 = \tau_2\sigma_1\tau_1^{-1}$. Now, we find another basis of $(P_1)_s$ and $(P_1)_t$ by using $\tau_1 \in \mathrm{SL}_2(A_s)$ and $\tau_2 \in \mathrm{SL}_2(A_t)$ as follows: Let

$$\tau_1 \begin{pmatrix} \overline{p_1} \\ \overline{p_2} \end{pmatrix} = \begin{pmatrix} \overline{p_1''} \\ \overline{p_2''} \end{pmatrix} \text{ and } \tau_2 \begin{pmatrix} \overline{q_1} \\ \overline{q_2} \end{pmatrix} = \begin{pmatrix} \overline{q_1''} \\ \overline{q_2''} \end{pmatrix}. \tag{14.19}$$

Now, suppose σ_3 is the cocycle associated to the projective module P_1 corresponding to the basis $\{\overline{p_1''}, \overline{p_2''}\}$ of $(P_1)_s$ and the basis $\{\overline{q_1''}, \overline{q_2''}\}$ of $(P_1)_t$. Then we have

$$\begin{pmatrix} \overline{q_1''} \\ \overline{q_2''} \end{pmatrix} = \sigma_3 \begin{pmatrix} \overline{p_1''} \\ \overline{p_2''} \end{pmatrix}. \tag{14.20}$$

Now, combining (14.17)–(14.20) we get

$$\sigma_1 \begin{pmatrix} \overline{p_1} \\ \overline{p_2} \end{pmatrix} = \tau_2^{-1}\sigma_3\tau_1 \begin{pmatrix} \overline{p_1} \\ \overline{p_2} \end{pmatrix}.$$

This implies that $\sigma_1 = \tau_2^{-1}\sigma_3\tau_1$. Since we have $\sigma_2 = \tau_2\sigma_1\tau_1^{-1}$ or $\tau_2^{-1}\sigma_2\tau_1 = \sigma_1$, this implies $\tau_2^{-1}\sigma_3\tau_1 = \tau_2^{-1}\sigma_2\tau_1$. Hence $\sigma_2 = \sigma_3$. □

Corollary 14.2 *Let* A *be a domain and* $[a_1, a_2, a_3]$, $[b_1, b_2, b_3]$ *be two unimodular rows in* A. *Suppose* $s, t \in A$ *are such that* $sA + tA = A$. *Suppose the unimodular rows are completable in* A_s *and* A_t. *Let* $\sigma_1 \in \mathrm{SL}_2(A_{st})$ *be the cocycle associated to the projective module* $P_1 = \frac{A^3}{A[a_1,a_2,a_3]}$ *corresponding to the basis* $\{p_1, p_2\}$ *of* $(P_1)_s$ *and* $\{q_1, q_2\}$ *of* $(P_1)_t$ *which give* SL_3 *completions and* $\sigma_2 \in \mathrm{SL}_2(A_{st})$ *is the cocycle associated to the projective module* $P_2 = \frac{A^3}{A[b_1,b_2,b_3]}$ *corresponding to the* *basis* $\{p_1', p_2'\}$ *of* $(P_2)_s$ *and* $\{q_1', q_2'\}$ *of* $(P_2)_t$ *which give* SL_3 *completions. Suppose* $\sigma_1 = \sigma_2$. *Let* $T_1 : A_s^3 \to A_s^3$ *given by* $T_1([a_1, a_2, a_3]) = [b_1, b_2, b_3]$, $T_1(p_1) = p_1'$, $T_1(p_2) = p_2'$ *and* $T_2 : A_t^3 \to A_t^3$ *given by* $T_2([a_1, a_2, a_3]) = [b_1, b_2, b_3]$, $T_2(q_1) = q_1'$, $T_2(q_2) = q_2'$. *Then* $T_1 = T_2$.

Proof Since $\sigma_1 \in \mathrm{SL}_2(A_{st})$ is the cocycle associated to the projective module P_1 corresponding to the basis $\{p_1, p_2\}$ of $(P_1)_s$ and the basis $\{q_1, q_2\}$ of $(P_1)_t$, we have

$$\begin{pmatrix} q_1 \\ q_2 \end{pmatrix} = \sigma_1 \begin{pmatrix} p_1 \\ p_2 \end{pmatrix}. \tag{14.21}$$

Similarly, $\sigma_2 \in \mathrm{SL}_2(A_{st})$ is the cocycle associated to the projective module P_2 corresponding to the basis $\{p_1', p_2'\}$ of $(P_2)_s$ and the basis $\{q_1', q_2'\}$ of $(P_2)_t$, we have

$$\begin{pmatrix} q_1' \\ q_2' \end{pmatrix} = \sigma_2 \begin{pmatrix} p_1' \\ p_2' \end{pmatrix}. \tag{14.22}$$

Suppose $\sigma_1 = \begin{pmatrix} \lambda_{11} & \lambda_{12} \\ \lambda_{21} & \lambda_{22} \end{pmatrix} = \sigma_2$, then we have $q_1 = \lambda_{11}p_1 + \lambda_{12}p_2$ and $q_2 = \lambda_{21}p_1 + \lambda_{22}p_2$. Applying the transformation T_1, we get

$$T_1(q_1) = T_1(\lambda_{11}p_1 + \lambda_{12}p_2) = \lambda_{11}p_1' + \lambda_{12}p_2'$$

and

$$T(q_2) = T_1(\lambda_{21}p_1 + \lambda_{22}p_2) = \lambda_{21}p_1' + \lambda_{22}p_2'.$$

These equations can be written in matrix form as follows:

$$\begin{pmatrix} T_1(q_1) \\ T_1(q_2) \end{pmatrix} = \begin{pmatrix} \lambda_{11} & \lambda_{12} \\ \lambda_{21} & \lambda_{22} \end{pmatrix} \begin{pmatrix} p_1' \\ p_2' \end{pmatrix}. \tag{14.23}$$

Combining (14.22) and (14.23), we get

$$\begin{pmatrix} T_1(q_1) \\ T_1(q_2) \end{pmatrix} = \begin{pmatrix} \lambda_{11} & \lambda_{12} \\ \lambda_{21} & \lambda_{22} \end{pmatrix} \begin{pmatrix} p_1' \\ p_2' \end{pmatrix} = \begin{pmatrix} q_1' \\ q_2' \end{pmatrix}.$$

Since $T_2(q_1) = q_1'$ and $T_2(q_2) = q_2'$, we have

$$\begin{pmatrix} T_1(q_1) \\ T_1(q_2) \end{pmatrix} = \begin{pmatrix} \lambda_{11} & \lambda_{12} \\ \lambda_{21} & \lambda_{22} \end{pmatrix} \begin{pmatrix} p_1' \\ p_2' \end{pmatrix} = \begin{pmatrix} q_1' \\ q_2' \end{pmatrix} = \begin{pmatrix} T_2(q_1) \\ T_2(q_2) \end{pmatrix}.$$

Therefore, $T_1(q_1) = T_2(q_1)$ and $T_1(q_2) = T_2(q_2)$. □

Remark 14.2 The transformation obtained by patching T_1 and T_2 in Corollary 14.2 belongs to $\mathrm{SL}_3(A)$. Hence, from Corollaries 14.1 and 14.2, it follows that if the cocycles corresponding projective modules given by two unimodular rows are equivalent then the unimodular rows are in the the same $\mathrm{SL}_3(A)$ orbit.

Corollary 14.3 *Let A be a domain, and $a, b \in A$ be such that $sA + tA = A$. Let $[a_1, a_2, a_3], [b_1, b_2, b_3] \in A^3$ be unimodular rows which are completable in A_s and A_t. Suppose the cocycle associated to $[a_1, a_2, a_3]$ is the element $\sigma \in \mathrm{SL}_2(A_{st})$. Suppose the cocycle associated to $[b_1, b_2, b_3]$ is $\sigma\varepsilon$ (or $\varepsilon\sigma$), where $\varepsilon \in E_2(A_{st})$. Then $[a_1, a_2, a_3]$ and $[b_1, b_2, b_3]$ are in the same $\mathrm{SL}_3(A)$ orbit.*

14.4 On a Proof of Suslin's Theorem

In this section, we derive Suslin's theorem by giving a new proof of the following Proposition also due to Suslin [27] (see also [12]).

Proposition 14.2 *Let $[u, a_1, \ldots, a_n]$ be a unimodular row in A. Suppose $[\overline{a_1}, \ldots, \overline{a_n}]$ is completable to a matrix belonging to $\mathrm{SL}_n(A/Au)$. Then the row $[u^n, a_1, \ldots, a_n]$ is completable to a matrix belonging to $\mathrm{SL}_{n+1}(A)$.*

Proof We prove the proposition by using the method of cocycles. Let $J = (a_1, \ldots, a_n)$. Since $[\overline{a_1}, \ldots, \overline{a_n}]$ is completable in A/Au, there exists a matrix $\overline{\alpha} \in \mathrm{SL}_n(A/Au)$ such that

$$\overline{\alpha} = \begin{pmatrix} \overline{a_1} & \cdots & \overline{a_n} \\ & & \\ & * & \\ & & \end{pmatrix}.$$

So $\det(\overline{\alpha}) = \overline{1} \in A/Au$. Therefore $\det(\alpha) = \det \begin{pmatrix} a_1 & \cdots & a_n \\ & & \\ & * & \\ & & \end{pmatrix} = 1 + uv \in A$ for some $v \in A$. Further expanding the determinant along the first row we see that $\det(\alpha) \in J$. Suppose $1 + uv = j \in J$. Then we have $u\frac{v}{1-j} = -1$. Let $v' = \frac{v}{1-j}$ then we have $uv' = -1$.

We first handle the case where $n = 2$. We give two completions of unimodular row $v = [u^2, a_1, a_2]$ in $\mathrm{SL}_3(A_{1-j})$ and $\mathrm{SL}_3(A_j)$ as follows:

$$\begin{pmatrix} u^2 & 0 & 0 \\ a_1 & v' & 0 \\ a_2 & 0 & v' \end{pmatrix} \in \mathrm{SL}_3(A_{1-j}) \text{ and } \begin{pmatrix} u^2 & 0 & 1/j \\ a_1 & b_1 & 0 \\ a_2 & b_2 & 0 \end{pmatrix} \in \mathrm{SL}_3(A_j).$$

We now compute the cocycle associated to the projective module $\frac{A^3}{[u^2, a_1, a_2]}$. We can write

$$\begin{pmatrix} 0 \\ b_1 \\ b_2 \end{pmatrix} = \lambda_1 \begin{pmatrix} u^2 \\ a_1 \\ a_2 \end{pmatrix} + \lambda_2 \begin{pmatrix} 0 \\ v' \\ 0 \end{pmatrix} + \lambda_3 \begin{pmatrix} 0 \\ 0 \\ v' \end{pmatrix}$$

and

$$\begin{pmatrix} 1/j \\ 0 \\ 0 \end{pmatrix} = \mu_1 \begin{pmatrix} u^2 \\ a_1 \\ a_2 \end{pmatrix} + \mu_2 \begin{pmatrix} 0 \\ v' \\ 0 \end{pmatrix} + \mu_3 \begin{pmatrix} 0 \\ 0 \\ v' \end{pmatrix}.$$

We get $\lambda_1 = 0$, $\lambda_2 = b_1/v'$, $\lambda_3 = b_2/v'$ and $\mu_1 = 1/ju^2$, $\mu_2 = -a_1/ju^2v'$, $\mu_3 = -a_2/ju^2v'$.

Let $q_1 = (0, b_1, b_2)^t$, $q_2 = (1/j, 0, 0)^t$, $p_1 = (0, v', 0)^t$, and $p_2 = (0, 0, v')^t$. Now, by going modulo $[u^2, a_1, a_2]$ we have

$$\begin{pmatrix} \overline{q_1} \\ \overline{q_2} \end{pmatrix} = \begin{pmatrix} \lambda_2 & \lambda_3 \\ \mu_2 & \mu_3 \end{pmatrix} \begin{pmatrix} \overline{p_1} \\ \overline{p_2} \end{pmatrix} \text{ in } \frac{A^3_{j(1-j)}}{A_{j(1-j)}[u^2, a_1, a_2]},$$

or

$$\begin{pmatrix} \overline{q_1} \\ \overline{q_2} \end{pmatrix} = \begin{pmatrix} b_1/v' & b_2/v' \\ -a_1/ju^2v' & -a_2/ju^2v' \end{pmatrix} \begin{pmatrix} \overline{p_1} \\ \overline{p_2} \end{pmatrix},$$

where $\sigma = \begin{pmatrix} b_1/v' & b_2/v' \\ -a_1/ju^2v' & -a_2/ju^2v' \end{pmatrix} \in \mathrm{SL}_2(A_{j(1-j)})$ is the cocycle associated to the projective module $\frac{A^3}{A[u^2,a_1,a_2]}$. We see that σ splits into a product of two matrices as follows:

$$\begin{pmatrix} b_1/v' & b_2/v' \\ -a_1/ju^2v' & -a_2/ju^2v' \end{pmatrix} = \begin{pmatrix} 1/v' & 0 \\ 0 & 1/u^2v' \end{pmatrix} \begin{pmatrix} b_1 & b_2 \\ -a_1/j & -a_2/j \end{pmatrix} = \delta_1\delta_2,$$

where $\delta_1 \in E_2(A_{1-j})$ and $\delta_2 \in \mathrm{SL}_2(A_j)$. Here σ splits but not in the desired order. Invoking Lemma 14.4, we get $\delta_1\delta_2 = \tau_1\tau_2$ where $\tau_1 \in \mathrm{SL}_2(A_j)$ and $\tau_2 \in \mathrm{SL}_2(A_{1-j})$. So by Theorem 14.4, we see that $[u^2, a_1, a_2]$ is completable. Having settled the $n = 2$ case in detail, the general case proceeds along similar lines but owing to the importance of the proposition we provide details for completeness. We establish the case where n is even. Similarly, one can do the case where n is odd.

$$\text{Let } \alpha = \begin{pmatrix} u^n & 0 & \cdots & 0 \\ a_1 & v' & \cdots & 0 \\ \vdots & & \ddots & \vdots \\ a_n & 0 & \cdots & v' \end{pmatrix} \in \mathrm{SL}_{n+1}(A_{1-j}) \text{ and } \beta = \begin{pmatrix} u^n & 0 & 0 & \cdots & 0 & 1/j \\ a_1 & \lambda_{11} & \lambda_{21} & \cdots & \lambda_{(n-1)1} & 0 \\ a_1 & \lambda_{12} & \lambda_{22} & \cdots & \lambda_{(n-1)2} & 0 \\ \vdots & \vdots & \vdots & & \vdots & \vdots \\ a_1 & \lambda_{1n} & \lambda_{2n} & \cdots & \lambda_{(n-1)n} & 0 \end{pmatrix} \in \mathrm{SL}_{n+1}(A_j),$$

$$\text{where } j = \det \begin{pmatrix} a_1 & \lambda_{11} & \lambda_{21} & \cdots & \lambda_{(n-1)1} \\ a_1 & \lambda_{12} & \lambda_{22} & \cdots & \lambda_{(n-1)2} \\ \vdots & \vdots & \vdots & & \vdots \\ a_1 & \lambda_{1n} & \lambda_{2n} & \cdots & \lambda_{(n-1)n} \end{pmatrix}.$$

$$\text{Let } \alpha^{-1}\beta = \begin{pmatrix} 1 & \nu_1 & \nu_2 & \cdots & \nu_{n-1} & \nu_n \\ 0 & \nu_{11} & \nu_{21} & \cdots & \nu_{(n-1)1} & \nu_{n1} \\ 0 & \nu_{12} & \nu_{22} & \cdots & \nu_{(n-1)2} & \nu_{n2} \\ \vdots & \vdots & \vdots & & \vdots & \vdots \\ 0 & \nu_{1n} & \nu_{2n} & \cdots & \nu_{(n-1)n} & \nu_{nn} \end{pmatrix}.$$

Thus, we have the following equations:

$$\begin{pmatrix} 0 \\ \lambda_{11} \\ \lambda_{12} \\ \vdots \\ \lambda_{1n} \end{pmatrix} = \nu_1 \begin{pmatrix} u^n \\ a_1 \\ a_2 \\ \vdots \\ a_n \end{pmatrix} + \nu_{11} \begin{pmatrix} 0 \\ v' \\ 0 \\ \vdots \\ 0 \end{pmatrix} + \cdots + \nu_{1n} \begin{pmatrix} 0 \\ 0 \\ 0 \\ \vdots \\ v' \end{pmatrix}, \tag{14.24}$$

$$\begin{pmatrix} 0 \\ \lambda_{21} \\ \lambda_{22} \\ \vdots \\ \lambda_{2n} \end{pmatrix} = \nu_2 \begin{pmatrix} u^n \\ a_1 \\ a_2 \\ \vdots \\ a_n \end{pmatrix} + \nu_{21} \begin{pmatrix} 0 \\ v' \\ 0 \\ \vdots \\ 0 \end{pmatrix} + \cdots + \nu_{2n} \begin{pmatrix} 0 \\ 0 \\ 0 \\ \vdots \\ v' \end{pmatrix}, \ldots, \tag{14.25}$$

$$\begin{pmatrix} 0 \\ \lambda_{(n-1)1} \\ \lambda_{(n-1)2} \\ \vdots \\ \lambda_{(n-1)n} \end{pmatrix} = \nu_{n-1} \begin{pmatrix} u^n \\ a_1 \\ a_2 \\ \vdots \\ a_n \end{pmatrix} + \nu_{(n-1)1} \begin{pmatrix} 0 \\ v' \\ 0 \\ \vdots \\ 0 \end{pmatrix} + \cdots + \nu_{(n-1)n} \begin{pmatrix} 0 \\ 0 \\ 0 \\ \vdots \\ v' \end{pmatrix}, \tag{14.26}$$

$$\begin{pmatrix} 1/j \\ 0 \\ 0 \\ \vdots \\ 0 \end{pmatrix} = \nu_n \begin{pmatrix} u^n \\ a_1 \\ a_2 \\ \vdots \\ a_n \end{pmatrix} + \nu_{n1} \begin{pmatrix} 0 \\ v' \\ 0 \\ \vdots \\ 0 \end{pmatrix} + \cdots + \nu_{nn} \begin{pmatrix} 0 \\ 0 \\ v' \\ \vdots \\ 0 \end{pmatrix}. \tag{14.27}$$

□

We get $\nu_1 = 0, \nu_{11} = \lambda_{11}/v', \nu_{12} = \lambda_{12}/v', \ldots, \nu_{1n} = \lambda_{1n}/v'$, $\nu_2 = 0$, $\nu_{21} = \lambda_{21}/v'$, $\nu_{22} = \lambda_{22}/v', \ldots, \nu_{2n} = \lambda_{2n}/v', \ldots, \nu_{n-1} = 0, \nu_{(n-1)1} = \lambda_{(n-1)1}/v', \nu_{(n-1)2} = \lambda_{(n-1)2}/v', \ldots, \nu_{(n-1)n} = \lambda_{(n-1)n}/v'$ and $\nu_n = 1/ju^n$, $\nu_{(n+1)1} = -a_1/ju^n v'$, $\nu_{(n+1)2} = -a_2/ju^n v', \ldots, \nu_{nn} = -a_n/ju^n v'$. We get the following cocycle:

$$\sigma = \begin{pmatrix} \lambda_{11}/v' & \lambda_{12}/v' & \cdots & \lambda_{1n}/v' \\ \lambda_{21}/v' & \lambda_{22}/v' & \cdots & \lambda_{2n}/v' \\ \vdots & \vdots & & \vdots \\ \lambda_{(n-1)1}/v' & \lambda_{(n-1)2}/v' & \cdots & \lambda_{(n-1)n}/v' \\ -a_1/ju^n v' & -a_2/ju^n v' & \cdots & -a_n/ju^n v' \end{pmatrix}.$$

We see that σ splits into a product of two matrices as follows:

$$\begin{pmatrix} \lambda_{11}/v' & \lambda_{12}/v' & \cdots & \lambda_{1n}/v' \\ \lambda_{21}/v' & \lambda_{22}/v' & \cdots & \lambda_{2n}/v' \\ \vdots & \vdots & & \vdots \\ \lambda_{(n-1)1}/v' & \lambda_{(n-1)2}/v' & \cdots & \lambda_{(n-1)n}/v' \\ -a_1/ju^nv' & -a_2/ju^nv' & \cdots & -a_n/ju^nv' \end{pmatrix} =$$

$$\begin{pmatrix} 1/v' & 0 & \cdots & 0 \\ 0 & 1/v' & \cdots & 0 \\ \vdots & \vdots & & \vdots \\ 0 & 0 & \cdots & 1/u^nv' \end{pmatrix} \begin{pmatrix} \lambda_{11} & \lambda_{12} & \cdots & \lambda_{1n} \\ \lambda_{21} & \lambda_{22} & \cdots & \lambda_{2n} \\ \vdots & \vdots & & \vdots \\ \lambda_{(n-1)1} & \lambda_{(n-1)2} & \cdots & \lambda_{(n-1)n} \\ -a_1/j & -a_2/j & \cdots & -a_n/j \end{pmatrix} = \delta_1\delta_2,$$

where $\delta_1 \in E_n(A_{1-j})$ and $\delta_2 \in \mathrm{SL}_n(A_j)$. Here σ splits but not in desired order. Invoking Lemma 14.4, we get $\delta_1\delta_2 = \tau_1\tau_2$ where $\tau_1 \in \mathrm{SL}_n(A_j)$ and $\tau_2 \in \mathrm{SL}_n(A_{1-j})$. So by Theorem 14.4, we see that $[u^n, a_1, \ldots, a_n]$ is completable. □

Theorem 14.5 (Suslin) *Let A be a ring and $[u, a_1, a_2, \ldots, a_n]$ be a unimodular row in A. Then $[u^{n!}, a_1, a_2, \ldots, a_n]$ can be completed to a matrix in $\mathrm{SL}_{n+1}(A)$.*

We refer the reader to the books of Murthy–Gupta [12], or Mandal [21] or Lam [18] or the paper of Suslin [27] for the proof of the above theorem using Proposition 14.1.

We now use the method of cocycles to prove the following theorem of Murthy–Swan [22].

Theorem 14.6 *Let A be a ring and $[x, y, z]$, $[u^2x, y, z]$ be two unimodular rows in A. Then these unimodular rows are in the same $\mathrm{SL}_3(A)$ orbit.*

Proof We are given that $[x, y, z]$ and $[u^2x, y, z]$ are two unimodular rows in A. Let $J = (y, z)$. First we find the cocycle corresponding to $P_1 = \frac{A^3}{A[x,y,z]}$.

Since $[u^2x, y, z]$ is a unimodular row, there exist $x', b_1, b_2 \in A$ such that $(u^2x)x' + yb_1 + zb_2 = 1$. By rewriting the above we have $(x)u^2x' + yb_1 + zb_2 = 1$. Then we have $(x)u^2x' = 1 - j$, where $j \in J$. Or $(x)\frac{u^2x'}{1-j} = 1$, that is, $(x)v' = 1$, where $v' = \frac{u^2x'}{1-j}$. Now, we get completions of the unimodular row $[x, y, z]$ as follows:

$$\begin{pmatrix} x & 0 & 0 \\ y & v' & 0 \\ z & 0 & 1 \end{pmatrix} \in \mathrm{SL}_3(A_{1-j}) \text{ and } \begin{pmatrix} x & 0 & 1/j \\ y & b_1 & 0 \\ z & b_2 & 0 \end{pmatrix} \in \mathrm{SL}_3(A_j).$$

Therefore, we can write

$$\begin{pmatrix} 0 \\ b_1 \\ b_2 \end{pmatrix} = \lambda_{11} \begin{pmatrix} x \\ y \\ z \end{pmatrix} + \lambda_{12} \begin{pmatrix} 0 \\ v' \\ 0 \end{pmatrix} + \lambda_{13} \begin{pmatrix} 0 \\ 0 \\ 1 \end{pmatrix}$$

and

$$\begin{pmatrix}1/j\\0\\0\end{pmatrix} = \lambda_{21}\begin{pmatrix}x\\y\\z\end{pmatrix} + \lambda_{22}\begin{pmatrix}0\\v'\\0\end{pmatrix} + \lambda_{23}\begin{pmatrix}0\\0\\1\end{pmatrix}.$$

We get $\lambda_{11} = 0$, $\lambda_{12} = b_1/v'$, $\lambda_{13} = b_2$ and $\lambda_{21} = 1/jx$, $\lambda_{22} = -y/jxv'$, $\lambda_{23} = -z/jx$. Thus the cocycle of P_1 is

$$\tau_1 = \begin{pmatrix} b_1/v' & b_2 \\ -\frac{y}{jxv'} & -\frac{z}{jx}\end{pmatrix}.$$

Now, we compute the cocycle of $P_2 = \frac{A^3}{A[u^2x,y,z]}$.

As before since $[u^2x, y, z]$ is a unimodular row, there exist $x', b_1, b_2 \in A$ such that $(u^2x)x' + yb_1 + zb_2 = 1$. Then we have $(u^2x)x' = 1 - j$, or $(u^2x)\frac{x'}{1-j} = 1$, that is $(u^2x)v'' = 1$, where $v'' = \frac{x'}{1-j}$. Let $v' = u^2v''$. Now, we get completions of the unimodular row $[u^2x, y, z]$ as follows:

$$\begin{pmatrix}u^2x & 0 & 0\\ y & v'' & 0\\ z & 0 & 1\end{pmatrix} \in \mathrm{SL}_3(A_{1-j}) \text{ and } \begin{pmatrix}u^2x & 0 & 1/j\\ y & b_1 & 0\\ z & b_2 & 0\end{pmatrix} \in \mathrm{SL}_3(A_j).$$

As before we can write

$$\begin{pmatrix}0\\b_1\\b_2\end{pmatrix} = \mu_{11}\begin{pmatrix}u^2x\\y\\z\end{pmatrix} + \mu_{12}\begin{pmatrix}0\\v''\\0\end{pmatrix} + \mu_{13}\begin{pmatrix}0\\0\\1\end{pmatrix}$$

and

$$\begin{pmatrix}1/j\\0\\0\end{pmatrix} = \mu_{21}\begin{pmatrix}u^2x\\y\\z\end{pmatrix} + \mu_{22}\begin{pmatrix}0\\v''\\0\end{pmatrix} + \mu_{23}\begin{pmatrix}0\\0\\1\end{pmatrix}.$$

We get $\mu_{11} = 0$, $\mu_{12} = b_1/v''$, $\mu_{13} = b_2$ and $\mu_{21} = 1/ju^2x$, $\mu_{22} = -y/ju^2xv''$, $\mu_{23} = -z/ju^2x$. Thus the cocycle of P_2 is

$$\tau_2 = \begin{pmatrix} b_1/v'' & b_2 \\ -\frac{y}{ju^2xv''} & -\frac{z}{ju^2x}\end{pmatrix}.$$

As u is a unit in A_{1-j}, there exists $v \in A_{1-j}$ such that $uv = 1$. Note that $v' = u^2v''$. We have $\begin{pmatrix}u & 0\\0 & v\end{pmatrix}\begin{pmatrix} b_1/v' & b_2 \\ -\frac{y}{jxv'} & -\frac{z}{jx}\end{pmatrix}\begin{pmatrix}u & 0\\0 & v\end{pmatrix} = \begin{pmatrix} u^2b_1/v' & b_2 \\ -\frac{y}{jxv'} & -v^2\frac{z}{jx}\end{pmatrix} = \begin{pmatrix} b_1/v'' & b_2 \\ -\frac{y}{jxu^2v''} & -\frac{z}{ju^2x}\end{pmatrix}$ $= \tau_2$. Since $\begin{pmatrix}u & 0\\0 & v\end{pmatrix} \in E_2(A_{1-j'})$, therefore by the result of Lemma 14.5, the cocycles of P_1 and P_2 are equivalent and so by Remark 14.2, the rows $[x, y, z]$ and $[u^2x, y, z]$ are in the same $\mathrm{SL}_3(A)$ orbit. □

Acknowledgements The authors would like to thank Professor Ravi A. Rao for his valuable support during this work. The authors would like to thank Professor Gopala Krishna Srinivasan for giving his time most generously and helping us make this paper more readable. The second named author would like to thank Professor Gopala Krishna Srinivasan for his support and advice during difficult times. The authors would also like to thank the referee for going through the paper carefully and pointing out some mistakes. The second named author also acknowledges the financial support from CSIR, which enabled him to pursue his doctoral studies.

References

1. H. Bass, K-theory and stable algebra. Inst. Hautes Études Sci. Publ. Math. **22**, 5–60 (1964)
2. R. Basu, *Topics in classical algebraic K-theory*. Ph.D. thesis, School of Mathematics, Tata Institute of Fundamental Research, Mumbai (2006)
3. R. Basu, R. Sridharan, On forster's conjecture and related results. Punjab Univ. Res. J. (Sci.) **57**, 13–66 (2007)
4. S.M. Bhatwadekar, M.K. Das, S. Mandal, Projective modules over smooth real affine varieties. Invent. Math. **166**(1), 151–184 (2006)
5. S.M. Bhatwadekar, M.K. Keshari, A question of Nori: projective generation of ideals. K-Theory **28**(4), 329–351 (2003)
6. S.M. Bhatwadekar, H. Lindel, R.A. Rao, The Bass-Murthy question: Serre dimension of Laurent polynomial extensions. Invent. Math. **81**(1), 189–203 (1985)
7. S.M. Bhatwadekar, R. Sridharan, Projective generation of curves in polynomial extensions of an affine domain and a question of Nori. Invent. Math. **133**(1), 161–192 (1998)
8. S.M. Bhatwadekar, R. Sridharan, The Euler class group of a Noetherian ring. Compositio Math. **122**(2), 183–222 (2000)
9. S.M. Bhatwadekar, R. Sridharan, On Euler classes and stably free projective modules, in *Algebra, Arithmetic and Geometry, Part I, II (Mumbai, 2000)*. Tata Inst. Fund. Res. Stud. Math., vol. 16 (Tata Institute of Fundamental Research, Bombay, 2002), pp. 139–158
10. M.K. Das, R. Sridharan, Good invariants for bad ideals. J. Algebr. **323**(12), 3216–3229 (2010)
11. N.S. Gopalakrishnan, *Commutative Algebra* (1984)
12. S.K. Gupta, M.P. Murthy, *Suslin's Work on Linear Groups over Polynomial Rings and Serre Problem*. ISI Lecture Notes, vol. 8 (Macmillan Co. of India Ltd, New Delhi, 1980)
13. M.K. Keshari, *Euler Class Group of a Noetherian Ring*. Ph.D. thesis, School of Mathematics, Tata Institute of Fundamental Research, Mumbai (2001)
14. M. Krusemeyer, Skewly completable rows and a theorem of Swan and Towber. Commun. Algebr. **4**(7), 657–663 (1976)
15. N.M. Kumar, Complete intersections. J. Math. Kyoto Univ. **17**(3), 533–538 (1977)
16. N.M. Kumar, A note on the cancellation of reflexive modules. J. Ramanujan Math. Soc. **17**(2), 93–100 (2002)
17. N.M. Kumar, On a theorem of Seshadri, in *Connected at Infinity*. Texts Read. Math., vol. 25 (Hindustan Book Agency, New Delhi, 2003), pp. 91–104
18. T.Y. Lam, *Serre's Problem on Projective Modules*, Springer Monographs in Mathematics (Springer, Berlin, 2006)
19. S. MacLane, *Homology*, 1st edn. (Springer, Berlin, 1967). Die Grundlehren der mathematischen Wissenschaften, Band 114
20. S. Mandal, Homotopy of sections of projective modules. J. Algebr. Geom. **1**(4), 639–646 (1992). With an appendix by Madhav V. Nori
21. S. Mandal, *Projective Modules and Complete Intersections*. Lecture Notes in Mathematics, vol. 1672 (Springer, Berlin, 1997)
22. M.P. Murthy, R.G. Swan, Vector bundles over affine surfaces. Invent. Math. **36**, 125–165 (1976)
23. D. Quillen, Projective modules over polynomial rings. Invent. Math. **36**, 167–171 (1976)

24. J.P. Serre, Sur les modules projectifs. *Séminaire Dubreil. Algèbre et théorie des nombres* **14**, 1–16, 1960–1961
25. C.S. Seshadri, Triviality of vector bundles over the affine space K^2. Proc. Nat. Acad. Sci. U.S.A. **44**, 456–458 (1958)
26. C.S. Seshadri, Algebraic vector bundles over the product of an affine curve and the affine line. Proc. Am. Math. Soc. **10**, 670–673 (1959)
27. A.A. Suslin, Stably free modules. Mat. Sb. (N.S.) **102**(144)(4), 537–550, 632 (1977)
28. R.G. Swan, Algebraic vector bundles on the 2-sphere. Rocky Mountain J. Math. **23**(4), 1443–1469 (1993)
29. R.G. Swan, J. Towber, A class of projective modules which are nearly free. J. Algebr. **36**(3), 427–434 (1975)
30. W. van der Kallen, A group structure on certain orbit sets of unimodular rows. J. Algebr. **82**(2), 363–397 (1983)
31. W. van der Kallen, A module structure on certain orbit sets of unimodular rows. J. Pure Appl. Algebr. **57**(3), 281–316 (1989)
32. L.N. Vaserstein, Stabilization of unitary and orthogonal groups over a ring with involution. Mat. Sb. (N.S.) **81**(123), 328–351 (1970)
33. L.N. Vaserstein, A.A. Suslin, Serre's problem on projective modules over polynomial rings, and algebraic K-theory. Izv. Akad. Nauk SSSR Ser. Mat. **40**(5), 993–1054, 1199 (1976)
34. C.T.C. Wall, *A Geometric Introduction to Topology* (Addison-Wesley Publishing Co., Reading, 1972)

Chapter 15
On an Algebraic Analogue of the Mayer–Vietoris Sequence

Raja Sridharan, Sumit Kumar Upadhyay and Sunil K. Yadav

15.1 Introduction

Let X be a topological space, $H^0(X, \mathbb{Z})$ be the set of continuous maps from X to $\mathbb{Z}$ and $H^1(X, \mathbb{Z})$ be the set of all homotopy classes of continuous maps from X to S^1. Since $\mathbb{Z}$ and S^1 are abelian groups, $H^0(X, \mathbb{Z})$ and $H^1(X, \mathbb{Z})$ are also abelian groups. In the literature, the group $H^1(X, \mathbb{Z})$ is known as Bruschlinsky group (for details one can see [5]).

Theorem 15.1 *Let U_1 and U_2 be two open sets of a topological space X. Then we have an exact sequence*

$$H^0(U_1 \cup U_2, \mathbb{Z}) \to H^0(U_1, \mathbb{Z}) \oplus H^0(U_2, \mathbb{Z}) \to H^0(U_1 \cap U_2, \mathbb{Z}) \to H^1(U_1 \cup U_2, \mathbb{Z})$$
$$\to H^1(U_1, \mathbb{Z}) \oplus H^1(U_2, \mathbb{Z}) \to H^1(U_1 \cap U_2, \mathbb{Z}).$$

This sequence is known as Mayer–Vietoris sequence.

We refer the reader to see the book of Wall [12] for the definitions and the construction of the Mayer–Vietoris sequence. It is natural to ask that 'Does there exist an algebraic

R. Sridharan
Tata Institute of Fundamental Research, 1, Dr. Homi Bhabha Road,
Mumbai 400005, India
e-mail: sraja@math.tifr.res.in

S. K. Upadhyay
Department of Applied Science, Indian Institute of Information Technology,
Allahabad 211015, Uttar Pradesh, India
e-mail: upadhyaysumit365@gmail.com

S. K. Yadav (✉)
Department of Mathematics, Indian Institute of Technology Bombay,
Powai, Mumbai 400076, India
e-mail: skyadav@math.iitb.ac.in

A. A. Ambily et al. (eds.), *Leavitt Path Algebras and Classical K-Theory*,
Indian Statistical Institute Series,
https://doi.org/10.1007/978-981-15-1611-5_15

analogue of Mayer–Vietoris sequence?' The main aim of this paper is to define two algebraic groups $\Gamma(A)$ and $\pi_1(\mathrm{SL}_2(A))$, where A is an integral domain and also to prove an algebraic analogue of Mayer–Vietoris sequence with the help these groups for an integral domain of dimension 1. The group $\Gamma(A)$ is also discussed by Krusemeyer ([7]) in different context.

By using the theory of symplectic modules, we also give an algebraic analogue of the connecting homomorphism

$$H^1(U_1 \cap U_2, \mathbb{Z}) \to H^2(U_1 \cup U_2, \mathbb{Z}). \tag{15.1}$$

This paper is organized as follows. After recalling some preliminary results in Sect. 15.2, we give an analogue of Theorem 15.1 in Sects. 15.3 and 15.4. In Sect. 15.5, we give an analogue of the map (15.1) and finally in Sect. 15.6, we deduce some corollaries of our results.

15.2 Some Preliminaries

In this section, we give some definitions and preliminary results. Throughout the paper, ring A means commutative ring with identity.

Definition 15.1 1. Let A be a ring. A row $(a_1, a_2, \dots, a_n) \in A^n$ is said to be **unimodular** (of length n) if the ideal $(a_1, a_2, \dots, a_n) = A$. The set of unimodular rows of length n is denoted by $\mathrm{Um}_n(A)$.
2. A unimodular row $(a_1, a_2, \dots, a_n)$ is said to be **completable** if there is a matrix in $\mathrm{SL}_n(A)$ (or in $\mathrm{GL}_n(A)$) whose first row (or first column) is $(a_1, a_2, \dots, a_n)$.
3. We define $E_n(A)$ to be the subgroup of $\mathrm{GL}_n(A)$ generated by all matrices of the form $E_{ij}(\lambda) = I_n + \lambda e_{ij}$, $\lambda \in A$, $i \neq j$, where e_{ij} is a matrix whose (i, j)th entry is 1 and all other entries are 0. The matrices $E_{ij}(\lambda)$ will be referred to as elementary matrices.

We now define the symplectic and elementary symplectic group of a ring. Let e_{ij} be the matrix with 1 in the (i, j) place and zeros elsewhere, e_i the ith row of I_n, and

$$\chi_r = \sum_{i=1}^{r} e_{2i-1,2i} - \sum_{i=1}^{r} e_{2i,2i-1}.$$

We display the case $r = 2$ explicitly below.

$$\begin{aligned}\chi_2 &= \sum_{i=1}^{2} e_{2i-1,2i} - \sum_{i=1}^{2} e_{2i,2i-1}\\ &= e_{12} + e_{34} - e_{21} - e_{43}\end{aligned}$$

$$= \begin{pmatrix} 0 & 1 & 0 & 0 \\ -1 & 0 & 0 & 0 \\ 0 & 0 & 0 & 1 \\ 0 & 0 & -1 & 0 \end{pmatrix}.$$

Definition 15.2 The group of symplectic matrices $\mathrm{Sp}_{2r}(A)$ is given by

$$\mathrm{Sp}_{2r}(A) = \{\alpha \in \mathrm{GL}_{2r}(A) : \alpha^t \chi_r \alpha = \chi_r\},$$

which is clearly a subgroup of of $\mathrm{GL}_{2r}(A)$.

In order to define the elementary symplectic matrices, we use the permutation σ on $2r$-letters given by

$$\sigma(2i) = 2i - 1 \text{ and } \sigma(2i-1) = 2i, \text{ for } 1 \le i \ne j \le 2r.$$

Definition 15.3 1. For each pair $i \ne j$ $(1 \le i \ne j \le 2r)$ the elementary symplectic matrix $se_{ij}(z)$ is given by

$$se_{ij}(z) = \begin{cases} I_{2r} + z \cdot e_{ij} & \text{if } i = \sigma j \\ I_{2r} + z \cdot e_{ij} - (-1)^{i+j} \cdot z \cdot e_{\sigma j, \sigma i} & \text{if } i \ne \sigma j \text{ and } i < j. \end{cases}$$

We shall call these matrices elementary symplectic.
2. The group $\mathrm{ESp}_{2r}(A)$ is then the subgroup of $\mathrm{Sp}_{2r}(A)$ generated by the elementary symplectic matrices over A.

For the case $r = 2$, there are eight such matrices, the matrix $se_{13}(z)$ $(i \ne \sigma(j))$ is displayed below.

$$\begin{pmatrix} 1 & 0 & z & 0 \\ 0 & 1 & 0 & 0 \\ 0 & 0 & 1 & 0 \\ 0 & -z & 0 & 1 \end{pmatrix}.$$

For the other three cases, the positions of $\pm z$ change accordingly. Likewise for $r = 2$ the matrix $se_{43}(z)$ $(i = \sigma(j))$ is displayed below.

$$\begin{pmatrix} 1 & 0 & 0 & 0 \\ 0 & 1 & 0 & 0 \\ 0 & 0 & 1 & 0 \\ 0 & 0 & z & 1 \end{pmatrix}.$$

For the other three cases, the positions of z change accordingly.

Let us recall Quillen's Splitting Lemma [8] with the proof following the exposition of [3]. In what follows, $(\psi_1(X))_t$ denotes the image of $\psi_1(X)$ in $\mathrm{SL}_n(A_{st}[X])$ and $(\psi_2(X))_s$ denotes the image of $\psi_2(X)$ in $\mathrm{SL}_n(A_{st}[X])$.

Lemma 15.1 (see [8]) *Let A be a domain and $s, t \in A$ be such that $sA + tA = A$. Suppose there exists $\sigma(X) \in \mathrm{SL}_n(A_{st}[X])$ with the property that $\sigma(0) = I_n$. Then there exist $\psi_1(X) \in \mathrm{SL}_n(A_s[X])$ with $\psi_1(0) = I_n$ and $\psi_2(X) \in \mathrm{SL}_n(A_t[X])$ with $\psi_2(0) = I_n$ such that $\sigma(X) = (\psi_1(X))_t(\psi_2(X))_s$.*

Proof Since $\sigma(0) = I_n$, $\sigma(X) = I_n + X\tau(X)$, where $\tau(X) \in M_n(A_{st}[X])$, we choose a large integer N_1 such that $\sigma(\lambda s^k X) \in \mathrm{SL}_n(A_t[X])$ for all $\lambda \in A$ and for all $k \geq N_1$. Define $\beta(X, Y, Z) \in \mathrm{SL}_n(A_{st}[X, Y, Z])$ as follows:

$$\beta(X, Y, Z) = \sigma((Y + Z)X)\sigma(YX)^{-1}. \tag{15.2}$$

Then $\beta(X, Y, 0) = I_n$, and hence there exists a large integer N_2 such that for all $k \geq N_2$ and for all $\mu \in A$ we have $\beta(X, Y, \mu t^k Z) \in \mathrm{SL}_n(A_s[X, Y, Z])$. This means

$$\beta(X, Y, \mu t^k Z) = (\sigma_1(X, Y, Z))_t, \tag{15.3}$$

where $\sigma_1(X, Y, Z) \in \mathrm{SL}_n(A_s[X, Y, Z])$ with $\sigma_1(X, Y, 0) = I_n$.

Taking $N = \max(N_1, N_2)$, it follows by the comaximality of sA and tA that $s^N A + t^N A = A$. Pick $\lambda, \mu \in A$ such that $\lambda s^N + \mu t^N = 1$. Setting $Y = \lambda s^N$, $Z = \mu t^N$ in (15.2) and $Z = 1$, $Y = \lambda s^N$ in (15.3) we get

$$\beta(X, \lambda s^N, \mu t^N) = \sigma(X)\sigma(\lambda s^N X)^{-1}$$

and

$$\beta(X, \lambda s^N, \mu t^N) = (\sigma_1(X, \lambda s^N, \mu t^N))_t = (\psi_1(X))_t, \text{ where } \psi_1(X) \in \mathrm{SL}_n(A_s[X]).$$

Hence, we conclude $\sigma(X)\sigma(\lambda s^N X)^{-1} = (\psi_1(X))_t$. Let $\sigma(\lambda s^N X) = (\psi_2(X))_s$, where $(\psi_2(X))_s \in \mathrm{SL}_n(A_t[X])$. Since $\sigma(0) = I_n$, $\psi_1(0) = \psi_2(0) = I_n$, the result follows by using the identity $\sigma(X) = \sigma(X)\sigma(\lambda s^N X)^{-1}\sigma(\lambda s^N X)$.

Lemma 15.2 ([4]) *Let A be a domain and $s, t \in A$ be such that $sA + tA = A$. For each $\sigma \in \mathrm{SL}_n(A_{st})$ and $\varepsilon \in E_n(A_{st})$ there exist $\tau_1 \in \mathrm{SL}_n(A_s)$ and $\tau_2 \in \mathrm{SL}_n(A_t)$ such that $\sigma\varepsilon = \tau_1\sigma\tau_2$.*

Proof Let $\varepsilon = \varepsilon_1\varepsilon_2$, where $\varepsilon_1 \in \mathrm{SL}_n(A_s)$ is chosen such that $\varepsilon_1 = I_n \bmod (t^N)$ for sufficiently large N and $\varepsilon_2 \in \mathrm{SL}_n(A_t)$. So, we have $\sigma\varepsilon = \sigma\varepsilon_1\varepsilon_2 = \sigma\varepsilon_1\sigma^{-1}\sigma\varepsilon_2$. Now, since $\varepsilon_1 = I_n \bmod (t^N)$ for sufficiently large N, $\sigma\varepsilon_1\sigma^{-1} \in \mathrm{SL}_n(A_s)$. Now by taking $\tau_1 = \sigma\varepsilon_1\sigma^{-1}$ and $\tau_2 = \varepsilon_2$, we have $\sigma\varepsilon = \tau_1\sigma\tau_2$.

15.3 The Group $\Gamma(A)$

In this Section, we define the group $\Gamma(A)$ which is an algebraic analogue of the group $H^1(X, \mathbb{Z})$.

Definition 15.4 Let A be a ring. We say a matrix $\alpha \in \mathrm{SL}_2(A)$ can be connected to the identity matrix I_2 if there exists a matrix $\beta(T) \in \mathrm{SL}_2(A[T])$ such that $\beta(0) = I_2$ and $\beta(1) = \alpha$.

Definition 15.5 We say that two unimodular rows (a, b), (c, d) over A are equivalent, written as $(a, b) \sim (c, d)$, if one (and hence both) of the following equivalent conditions hold.

1. There exists $(f_{11}(T), f_{12}(T)) \in \mathrm{Um}_2(A[T])$ such that $(f_{11}(0), f_{12}(0)) = (a, b)$ and $(f_{11}(1), f_{12}(1)) = (c, d)$.
2. There exists a matrix $\alpha \in \mathrm{SL}_2(A)$ which is connected to the identity matrix (that is, there exists a matrix $\beta(T) \in \mathrm{SL}_2(A[T])$ such that $\beta(0) = I_2$ and $\beta(1) = \alpha$) such that $\alpha \begin{pmatrix} a \\ b \end{pmatrix} = \begin{pmatrix} c \\ d \end{pmatrix}$.

The fact that $\sim$ is an equivalence relation will be established later. We first show that these two conditions are equivalent.
(2) $\Longrightarrow$ (1).
Suppose $\beta(T) = \begin{pmatrix} g_{11}(T) & g_{12}(T) \\ g_{21}(T) & g_{22}(T) \end{pmatrix}$ such that $\beta(0) = I_2$ and $\beta(1) = \alpha$, which means

$$ag_{11}(1) + bg_{12}(1) = c \text{ and } ag_{21}(1) + bg_{22}(1) = d.$$

Let

$$\begin{pmatrix} f_{11}(T) \\ f_{12}(T) \end{pmatrix} = \beta(T) \begin{pmatrix} a \\ b \end{pmatrix} = \begin{pmatrix} ag_{11}(T) + bg_{12}(T) \\ ag_{21}(T) + bg_{22}(T) \end{pmatrix}. \tag{15.4}$$

Thus, it is clear that

$$(f_{11}(0), f_{12}(0)) = (a, b) \text{ and } (f_{11}(1), f_{12}(1)) = (c, d).$$

Since (a, b) is unimodular, we have $(a', b') \in A^2$ such that $ab' - ba' = 1$. Then

$$f_{11}(T) f_{22}(T) - f_{12}(T) f_{21}(T) = 1,$$

where $(f_{21}(T), f_{22}(T)) = (a'g_{11}(T) + b'g_{12}(T), a'g_{21}(T) + b'g_{22}(T))$. Thus

$$(f_{11}(T), f_{12}(T)) \in \mathrm{Um}_2(A[T]).$$

Therefore, definition (2) implies definition (1).
(1) $\Longrightarrow$ (2).

Since $(f_{11}(T), f_{12}(T)) \in \mathrm{Um}_2(A[T])$, there exists $(f_{21}(T), f_{22}(T)) \in (A[T])^2$ such that

$$f_{11}(T)f_{22}(T) - f_{12}(T)f_{21}(T) = 1.$$

Thus $af_{22}(0) - bf_{21}(0) = 1$.

Let $\beta(T) = \begin{pmatrix} f_{11}(T) & f_{21}(T) \\ f_{12}(T) & f_{22}(T) \end{pmatrix} \begin{pmatrix} f_{22}(0) & -f_{21}(0) \\ -b & a \end{pmatrix}$. Then $\beta(0) = I_2$ and $\beta(1)\begin{pmatrix} a \\ b \end{pmatrix} = \begin{pmatrix} c \\ d \end{pmatrix}$. For $\alpha = \beta(1)$, the definition (2) follows.

We now turn to proof that $\sim$ is an equivalence relation.

Reflexivity: To show $(a, b) \sim (a, b)$, we use (1) of Definition 15.5 and simply take $(f_{11}(T), f_{12}(T)) = (a, b)$.

Symmetry: Suppose $(a, b) \sim (c, d)$. By (2) of Definition 15.5, there exists a matrix $\alpha \in \mathrm{SL}_2(A)$ which is connected to the identity matrix such that $\alpha\begin{pmatrix} a \\ b \end{pmatrix} = \begin{pmatrix} c \\ d \end{pmatrix}$. Since α^{-1} is also connected to I_2 and $\alpha^{-1}\begin{pmatrix} c \\ d \end{pmatrix} = \begin{pmatrix} a \\ b \end{pmatrix}$, we get $(c, d) \sim (a, b)$.

Transitivity: Suppose $(a, b) \sim (c, d)$ and $(c, d) \sim (e, f)$. Then we have matrices $\alpha, \beta \in \mathrm{SL}_2(A)$ which are connected to the identity matrix such that $\alpha\begin{pmatrix} a \\ b \end{pmatrix} = \begin{pmatrix} c \\ d \end{pmatrix}$ and $\beta\begin{pmatrix} c \\ d \end{pmatrix} = \begin{pmatrix} e \\ f \end{pmatrix}$. Therefore $\beta\alpha\begin{pmatrix} a \\ b \end{pmatrix} = \begin{pmatrix} e \\ f \end{pmatrix}$.

Since α and β are connected to the identity matrix, there exist matrices $\gamma(T), \delta(T) \in \mathrm{SL}_2(A[T])$ such that $\gamma(0) = I_2 = \delta(0)$ and $\gamma(1) = \alpha$, $\delta(1) = \beta$. Take $\theta(T) = \delta(T)\gamma(T)$. Thus $\theta(0) = I_2$ and $\theta(1) = \beta\alpha$, that is, $\beta\alpha$ is connected to the identity matrix. Hence $(a, b) \sim (e, f)$.

Note that a unimodular row will always be denoted by parenthesis and its equivalence class by [,]. Thus the equivalence class of (a, b) is $[a, b]$.

Definition 15.6 Let $\Gamma(A)$ be the set of all equivalence classes of unimodular rows given by the equivalence relation $\sim$ as above. Define a product $*$ in $\Gamma(A)$ as follows.

Let $(a, b), (c, d) \in \mathrm{Um}_2(A)$. Complete these to $\mathrm{SL}_2(A)$ matrices $\sigma = \begin{pmatrix} a & e \\ b & f \end{pmatrix}$ and $\tau = \begin{pmatrix} c & g \\ d & h \end{pmatrix}$. We define product of two elements $[a, b], [c, d] \in \Gamma(A)$ as follows:

$$[a, b] * [c, d] = [\text{first column of } \sigma\tau] = [ac + de, bc + df].$$

Claim. $*$ does not depend on the choice of completions.

Let $\sigma' = \begin{pmatrix} a & e' \\ b & f' \end{pmatrix}$ and $\tau' = \begin{pmatrix} c & g' \\ d & h' \end{pmatrix} \in \mathrm{SL}_2(A)$ be another completion of (a, b) and (c, d), respectively. Since columns of σ and σ' form bases of A^2, columns of σ' can be written as linear combination of columns of σ. Since σ and σ' in $\mathrm{SL}_2(A)$, $\sigma' = \sigma\begin{pmatrix} 1 & \lambda \\ 0 & 1 \end{pmatrix}$ for some $\lambda \in A$. Similarly $\tau' = \tau\begin{pmatrix} 1 & \mu \\ 0 & 1 \end{pmatrix}$ for some $\mu \in A$. Therefore

$$\sigma'\tau' = \sigma \begin{pmatrix} 1 & \lambda \\ 0 & 1 \end{pmatrix} \tau \begin{pmatrix} 1 & \mu \\ 0 & 1 \end{pmatrix}.$$

Consider the matrix

$$\beta(T) = \sigma \begin{pmatrix} 1 & \lambda T \\ 0 & 1 \end{pmatrix} \tau \begin{pmatrix} 1 & \mu T \\ 0 & 1 \end{pmatrix} \tau^{-1}\sigma^{-1} \in \mathrm{SL}_2(A[T]).$$

Thus

$$\beta(0) = \sigma \begin{pmatrix} 1 & 0 \\ 0 & 1 \end{pmatrix} \tau \begin{pmatrix} 1 & 0 \\ 0 & 1 \end{pmatrix} \tau^{-1}\sigma^{-1} = I_2, \text{ and}$$

$$\beta(1)\sigma\tau = \sigma \begin{pmatrix} 1 & \lambda \\ 0 & 1 \end{pmatrix} \tau \begin{pmatrix} 1 & \mu \\ 0 & 1 \end{pmatrix} \tau^{-1}\sigma^{-1}\sigma\tau = \sigma'\tau'.$$

Therefore

$$\beta(1)\sigma\tau \begin{pmatrix} 1 \\ 0 \end{pmatrix} = \sigma'\tau' \begin{pmatrix} 1 \\ 0 \end{pmatrix}.$$

Hence $[(ac + de, bc + df)] = [(ac + de', bc + df')]$. So $*$ does not depend on the choice of completions.

Claim. $*$ is a well-defined operation on $\Gamma(A)$, that is, we have to show that if $(a, b) \sim (a', b')$ and $(c, d) \sim (c', d')$, then

$$[a, b] * [c, d] = [a', b'] * [c', d']. \tag{15.5}$$

Since $(a, b) \sim (a', b')$ and $(c, d) \sim (c', d')$, there exist $(f_{11}(T), f_{12}(T))$ and $(g_{11}(T), g_{12}(T))$ in $\mathrm{Um}_2(A[T])$ such that

$$(f_{11}(0), f_{12}(0)) = (a, b),\ (f_{11}(1), f_{12}(1)) = (a', b'),$$

$$(g_{11}(0), g_{12}(0)) = (c, d),\ (g_{11}(1), g_{12}(1)) = (c', d').$$

Again there exist $f_{21}(T), f_{22}(T), g_{21}(T), g_{22}(T)$ in $A[T]$ such that

$$f_{11}(T)f_{21}(T) - f_{12}(T)f_{22}(T) = 1 \text{ and } g_{11}(T)g_{21}(T) - g_{12}(T)g_{22}(T) = 1.$$

Consider $\sigma(T) = \begin{pmatrix} f_{11}(T) & f_{22}(T) \\ f_{12}(T) & f_{21}(T) \end{pmatrix}$ and $\tau(T) = \begin{pmatrix} g_{11}(T) & g_{22}(T) \\ g_{12}(T) & g_{21}(T) \end{pmatrix}$ in SL_2 $(A[T])$. Thus the first column of the product $\sigma(T)\tau(T)$ is unimodular, that is,

$$(f_{11}(T)g_{11}(T) + f_{22}(T)g_{12}(T), f_{12}(T)g_{11}(T) + f_{21}(T)g_{12}(T)) \in \mathrm{Um}_2(A[T]). \tag{15.6}$$

Setting $T = 0$ and $T = 1$ in (15.6), we get (15.5). Hence the product '$*$' is well defined.

Since matrix multiplication is associative, the product $*$ is associative. Since $[a, b] * [1, 0] = [a, b]$ for every $(a, b) \in \mathrm{Um}_2(A)$, we see that $[1, 0]$ is the identity element. Let $(a, b) \in \mathrm{Um}_2(A)$ and $\sigma = \begin{pmatrix} a & e \\ b & f \end{pmatrix} \in \mathrm{SL}_2(A)$. Then $\sigma^{-1} = \begin{pmatrix} f & -e \\ -b & a \end{pmatrix}$ and $[a, b] * [f, -b] = [1, 0]$. So $[f, -b]$ is the inverse of $[a, b]$ in $(\Gamma(A), *)$. Hence $(\Gamma(A), *)$ forms a group.

Now, let A be an integral domain and $a, b \in A$ be such that $aA + bA = A$. Define the maps

$$\varphi : \Gamma(A) \longrightarrow \Gamma(A_a) \oplus \Gamma(A_b)$$

given by $\varphi(\lambda) = (\lambda, \lambda)$ and

$$\psi : \Gamma(A_a) \oplus \Gamma(A_b) \longrightarrow \Gamma(A_{ab})$$

given by $\psi(\lambda, \mu) = \lambda - \mu$. We would like these maps to be homomorphisms but since $\Gamma(A)$ is not known to be abelian, ψ may not be a homomorphism.

Claim. $\Gamma(A) \xrightarrow{\varphi} \Gamma(A_a) \oplus \Gamma(A_b) \xrightarrow{\psi} \Gamma(A_{ab})$ is an exact sequence of groups.

To prove the claim, suppose we have elements $\lambda \in \Gamma(A_a)$ and $\mu \in \Gamma(A_b)$ which are equal in $\Gamma(A_{ab})$, that is, there is an element $\alpha(T) \in \mathrm{SL}_2(A_{ab}[T])$ such that $\alpha(0) = I_2$, and $\lambda = \alpha(1)\mu$. We split $\alpha(T)$ (by Lemma 15.1) as $\alpha_1(T)\alpha_2(T)$, where $\alpha_1(T) \in \mathrm{SL}_2(A_a[T])$ with $\alpha_1(0) = I_2$ and $\alpha_2(T) \in \mathrm{SL}_2(A_b[T])$ with $\alpha_2(0) = I_2$. Therefore $\alpha_1(1)^{-1}\lambda = \alpha_2(1)\mu$ and these elements patch to yield an element of $\alpha \in \Gamma(A)$. So $\varphi(\alpha) = (\alpha, \alpha) = (\lambda, \mu)$. Hence $\ker(\psi) \subseteq \mathrm{Im}(\phi)$.

By the definition of φ and ψ, it is clear that $\mathrm{Im}(\phi) \subseteq \ker(\psi)$. Hence the claim.

Another way of formulating this is to say that

$$\begin{array}{ccc} \Gamma(A) & \longrightarrow & \Gamma(A_a) \\ \downarrow & & \downarrow \\ \Gamma(A_b) & \longrightarrow & \Gamma(A_{ab}) \end{array}$$

is a fiber product diagram.

Remark 15.1 Let N be the set of $\alpha \in \mathrm{SL}_2(A)$ such that there exists $\beta(T) \in \mathrm{SL}_2(A[T])$ with $\beta(0) = I_2$ and $\beta(1) = \alpha$. Then N is the connected component of I_2 in $\mathrm{SL}_2(A)$ and $N \supset E_2(A)$. The group $\Gamma(A)$ can also be defined to be the quotient group $\mathrm{SL}_2(A)/N$. The reason we cannot take N to be $E_2(A)$ is that $E_2(A)$ is not in general normal in $\mathrm{SL}_2(A)$ and therefore it is necessary to consider a larger group N containing $E_2(A)$.

15.4 On the Group $\pi_1(\mathrm{SL}_2(A))$

In this section, we define the group $\pi_1(\mathrm{SL}_2(A))$ and give a connecting homomorphism between $\pi_1(\mathrm{SL}_2(A))$ and $\Gamma(A)$. Throughout this section, we assume A as an integral domain.

Let L be the set of loops in $\mathrm{SL}_2(A)$ starting and ending at the identity matrix I_2, that is, $L = \{\alpha(T) \in \mathrm{SL}_2(A[T]) \mid \alpha(0) = \alpha(1) = I_2\}$. We say that two loops $\alpha(T), \beta(T) \in L$ are equivalent (that is, written as $\alpha(T) \sim_1 \beta(T)$) if they are homotopic, that is, there exists $\gamma(T, S) \in \mathrm{SL}_2(A[T, S])$ such that $\gamma(T, 0) = \alpha(T), \gamma(T, 1) = \beta(T)$ and $\gamma(0, S) = \gamma(1, S) = I_2$. We call $\gamma(T, S)$ to be a homotopy between $\alpha(T)$ and $\beta(T)$.

We now show that $\sim_1$ is an equivalence relation.

Reflexivity: To show $\alpha(T) \sim_1 \alpha(T)$, we simply take $\gamma(T, S) = \alpha(T) \in \mathrm{SL}_2(A[T, S])$. This is obviously the desired homotopy.

Symmetry: Suppose $\gamma(T, S) \in \mathrm{SL}_2(A[T, S])$ is the homotopy between $\alpha(T)$ and $\beta(T)$. Then $\gamma(T, 1 - S)$ is a homotopy between $\beta(T)$ and $\alpha(T)$.

Transitivity: Let $\alpha(T) \sim_1 \beta(T)$ and $\beta(T) \sim_1 \delta(T)$. Then there exist matrices $\gamma_1(T, S)$, $\gamma_2(T, S)$ in $\mathrm{SL}_2(A[T, S])$ such that $\gamma_1(T, 0) = \alpha(T)$, $\gamma_1(T, 1) = \beta(T)$, $\gamma_1(0, S) = \gamma_1(1, S) = I_2$, $\gamma_2(T, 0) = \beta(T)$, $\gamma_2(T, 1) = \delta(T)$ and $\gamma_2(0, S) = \gamma_2(1, S) = I_2$. Take $\gamma_3(T, S) = \gamma_1(T, S)\beta(T)^{-1}\gamma_2(T, S)$. Hence

$$\gamma_3(T, 0) = \gamma_1(T, 0)\beta(T)^{-1}\gamma_2(T, 0) = \alpha(T),$$
$$\gamma_3(T, 1) = \gamma_1(T, 1)\beta(T)^{-1}\gamma_2(T, 1) = \delta(T), \text{ and}$$
$$\gamma_3(0, S) = \gamma_3(1, S) = I_2 \text{ (since } \beta(0)^{-1} = \beta(1)^{-1} = I_2).$$

Thus $\alpha(T) \sim_1 \delta(T)$.

Definition 15.7 For a domain A, $\pi_1(\mathrm{SL}_2(A))$ is the set of all equivalence classes of loops based on I_2. For $\alpha(T) \in \mathrm{SL}_2(A[T])$ with $\alpha(0) = \alpha(1) = I_2$, we denote its equivalence class in $\pi_1(\mathrm{SL}_2(A))$ by $[\alpha(T)]$.

Theorem 15.2 *The set $\pi_1(\mathrm{SL}_2(A))$ forms an abelian group under the binary operation '$*$' defined as $[\alpha(T)] * [\beta(T)] = [\alpha(T)\beta(T)]$.*

Proof First we show that the operation '$*$' is well defined. Let $\alpha(T) \sim_1 \beta(T)$ and $\gamma(T) \sim_1 \delta(T)$. Then there exist $\gamma_1(T, S), \gamma_2(T, S) \in \mathrm{SL}_2(A[T, S])$ such that $\gamma_1(T, 0) = \alpha(T)$, $\gamma_1(T, 1) = \beta(T)$, $\gamma_1(0, S) = \gamma_1(1, S) = I_2$; $\gamma_2(T, 0) = \gamma(T)$, $\gamma_2(T, 1) = \delta(T)$ and $\gamma_2(0, S) = \gamma_2(1, S) = I_2$. Take $\gamma_3(T, S) = \gamma_1(T, S)\gamma_2(T, S)$, we have $\gamma_3(T, 0) = \alpha(T)\gamma(T)$, $\gamma_3(T, 1) = \beta(T)\delta(T)$ and $\gamma_3(0, S) = \gamma_3(1, S) = I_2$. Hence $\alpha(T)\gamma(T) \sim_1 \beta(T)\delta(T)$.

Since matrix multiplication is associative, '$*$' is also associative. Therefore $\pi_1(\mathrm{SL}_2(A))$ is a group with $[I_2]$ as the identity element and $[\alpha(T)^{-1}]$ is the inverse of the element $[\alpha(T)] \in \pi_1(\mathrm{SL}_2(A))$.

Let $\alpha(T), \beta(T) \in L$. Then we will show that $\alpha(T) \sim_1 \beta(T)\alpha(T)\beta(T)^{-1}$. Consider $\gamma(T, S) = \beta(TS)\alpha(T)\beta(TS)^{-1} \in \mathrm{SL}_2(A[T, S])$. Then,

1. $\gamma(T, 0) = \alpha(T), \ \gamma(T, 1) = \beta(T)\alpha(T)\beta(T)^{-1}$,
2. $\gamma(0, S) = \gamma(1, S) = I_2$.

Therefore $\alpha(T) \sim_1 \beta(T)\alpha(T)\beta(T)^{-1}$ which means $\alpha(T)\beta(T) \sim_1 \beta(T)\alpha(T)$. This implies that $[\alpha(T)] * [\beta(T)] = [\alpha(T)\beta(T)] = [\beta(T)\alpha(T)] = [\beta(T)] * [\alpha(T)]$. Hence $(\pi_1(\mathrm{SL}_2(A)), *)$ is an abelian group. □

Let $a, b \in A$ be such that $aA + bA = A$. Define the maps

$$\varphi_1 : \pi_1(\mathrm{SL}_2(A)) \longrightarrow \pi_1(\mathrm{SL}_2(A_a)) \oplus \pi_1(\mathrm{SL}_2(A_b)), \text{ and}$$

$$\psi_1 : \pi_1(\mathrm{SL}_2(A_a)) \oplus \pi_1(\mathrm{SL}_2(A_b)) \longrightarrow \pi_1(\mathrm{SL}_2(A_{ab}))$$

by $\varphi_1(\lambda) = (\lambda, \lambda)$ and $\psi_1(\lambda, \mu) = \lambda\mu^{-1}$, respectively. As in the case of $\Gamma(A)$, it is easy to show using Quillen's splitting that we have an exact sequence of groups

$$\pi_1(\mathrm{SL}_2(A)) \xrightarrow{\varphi_1} \pi_1(\mathrm{SL}_2(A_a)) \oplus \pi_1(\mathrm{SL}_2(A_b)) \xrightarrow{\psi_1} \pi_1(\mathrm{SL}_2(A_{ab})).$$

Definition 15.8 (*The connecting map* $\Gamma : \pi_1(\mathrm{SL}_2(A_{ab})) \to \Gamma(A)$) Let $\alpha(T) \in \pi_1(\mathrm{SL}_2(A_{ab}))$, that is, $\alpha(T) \in \mathrm{SL}_2(A_{ab}[T])$ such that $\alpha(0) = \alpha(1) = I_2$. Let $\alpha(T) = \alpha_1(T)^{-1}\alpha_2(T)$ be a Quillen splitting, where $\alpha_1(T) \in \mathrm{SL}_2(A_a[T])$ with $\alpha_1(0) = I_2$ and $\alpha_2(T) \in \mathrm{SL}_2(A_b[T])$ with $\alpha_2(0) = I_2$. Then $\alpha(1) = I_2 = \alpha_1(1)^{-1}\alpha_2(1)$. Hence $\alpha_1(1) = \alpha_2(1)$ and $\alpha_1(1)$ and $\alpha_2(1)$ patch up to yield an element $\gamma \in \mathrm{SL}_2(A)$. We define $\Gamma([\alpha(T)]) =$ [first column of γ] in $\Gamma(A)$. We will also write it as $\Gamma([\alpha(T)]) = \alpha_2(1)\begin{pmatrix}1\\0\end{pmatrix}$.

Theorem 15.3 1. *The above association does not depend on the Quillen splitting of* α.
2. Γ *is a well-defined map.*
3. Γ *is a group homomorphism.*
4. *The sequence of groups*

$$\pi_1(\mathrm{SL}_2(A_{ab})) \xrightarrow{\Gamma} \Gamma(A) \xrightarrow{\phi} \Gamma(A_a) \oplus \Gamma(A_b) \tag{15.7}$$

is exact.

Proof (1) Suppose we are given two Quillen splittings of $\alpha(T)$ as follows:

$$\alpha(T) = \alpha_1(T)^{-1}\alpha_2(T); \ \alpha(T) = \beta_1(T)^{-1}\beta_2(T), \tag{15.8}$$

where $\alpha_1(T), \beta_1(T) \in \mathrm{SL}_2(A_a[T])$ with $\alpha_1(0) = \beta_1(0) = I_2$ and $\alpha_2(T), \beta_2(T) \in \mathrm{SL}_2(A_b[T])$ with $\alpha_2(0) = \beta_2(0) = I_2$. Then $\alpha_1(T)^{-1}\alpha_2(T) = \beta_1(T)^{-1}\beta_2(T)$ or we have

$$\beta_1(T)\alpha_1(T)^{-1} = \beta_2(T)\alpha_2(T)^{-1} \tag{15.9}$$

and these patch up to yield $\delta(T) \in \mathrm{SL}_2(A[T])$ such that $\delta(0) = I_2$.

An easy computation using (15.9) yields that multiplication by $\delta(1)$ sends the unimodular row associated to the first Quillen splitting to the unimodular row given by the second Quillen splitting. It now follows by definition that the element $[\Gamma(\alpha)]$ in $\Gamma(A)$ does not depend upon the choice of Quillen splitting.

(2) Now we have to show that Γ is well defined, that is, the homotopic loops in $\mathrm{SL}_2(A_{ab})$ go to the same element of $\Gamma(A)$.

Let $\alpha(T), \beta(T)$ be loops in $\mathrm{SL}_2(A_{ab}[T])$ with $\alpha(0) = \beta(0) = \alpha(1) = \beta(1) = I_2$, which are homotopic as loops. That is, there exists $\gamma(T, S) \in \mathrm{SL}_2(A_{ab}[T, S])$ such that $\gamma(T, 0) = \alpha(T), \gamma(T, 1) = \beta(T)$ and $\gamma(0, S) = I_2 = \gamma(1, S)$. Since $\gamma(0, S) = I_2$, we can write $\gamma(T, S) = \gamma_1(T, S)^{-1}\gamma_2(T, S)$, where $\gamma_1(T, S) \in \mathrm{SL}_2(A_a[T, S])$ with $\gamma_1(0, S) = I_2$ and $\gamma_2(T, S) \in \mathrm{SL}_2(A_b[T, S])$ with $\gamma_2(0, S) = I_2$.

Further,

$$\alpha(T) = \gamma(T, 0) = \gamma_1(T, 0)^{-1}\gamma_2(T, 0), \text{ and}$$

$$\beta(T) = \gamma(T, 1) = \gamma_1(T, 1)^{-1}\gamma_2(T, 1)$$

are Quillen splittings.

Consider the matrix $\gamma' \in \mathrm{SL}_2(A)$ obtained by patching $\gamma_1(1, 0)$ and $\gamma_2(1, 0)$, the matrix $\gamma'' \in \mathrm{SL}_2(A)$ obtained by patching $\gamma_1(1, 1)$ and $\gamma_2(1, 1)$ and $\widetilde{\gamma}(S)$ obtained by patching $\gamma_1(1, S)$ and $\gamma_2(1, S)$. Then the first column of $\widetilde{\gamma}(S)$ is a unimodular row in $A[S]$ which at $S = 0$ is the first column of γ' and at $S = 1$ is the first column of γ''. Thus Γ is well defined.

(3) Let $\alpha(T), \beta(T) \in \mathrm{SL}_2(A_{ab}[T])$ with $\alpha(0) = \beta(0) = I_2$ and $\alpha(1) = \beta(1) = I_2$. Suppose $\alpha(T) = \alpha_1(T)^{-1}\alpha_2(T)$ and $\beta(T) = \beta_1(T)^{-1}\beta_2(T)$ be Quillen splittings of $\alpha(T)$ and $\beta(T)$, respectively. Then $\Gamma([\alpha(T)]) = \alpha_2(1)\begin{pmatrix}1\\0\end{pmatrix}$ and $\Gamma([\beta(T)]) = \beta_2(1)\begin{pmatrix}1\\0\end{pmatrix}$. Thus

$$\Gamma([\alpha(T)]) * \Gamma([\beta(T)]) = (\alpha_2(1)\begin{pmatrix}1\\0\end{pmatrix}) * (\beta_2(1)\begin{pmatrix}1\\0\end{pmatrix}) = \alpha_2(1)\beta_2(1)\begin{pmatrix}1\\0\end{pmatrix},$$

by the definition of $*$ in $\Gamma(A)$. On the other hand, we have

$$\begin{aligned}\alpha(T)\beta(T) &= \alpha_1(T)^{-1}\alpha_2(T)\beta_1(T)^{-1}\beta_2(T) \\ &= \alpha_1(T)^{-1}\alpha_2(T)\beta_1(T)^{-1}\alpha_2(T)^{-1}\alpha_2(T)\beta_2(T).\end{aligned} \tag{15.10}$$

Since $\beta_1(T)$ and hence $\beta_1(T)^{-1}$ can be chosen (see Lemma 15.1) such that $\beta_1(T) \equiv I_2 \pmod{b^N}$ for sufficiently large N, as in Lemma 15.2, we may assume that

$$\alpha_2(T)\beta_1(T)^{-1}\alpha_2(T)^{-1} \in \mathrm{SL}_2(A_a[T]).$$

Therefore the Quillen splitting of $\alpha(T)\beta(T)$ is $\mu(T)\alpha_2(T)\beta_2(T)$, where $\mu(T)$ is a matrix in $\mathrm{SL}_2(A_a[T])$, $\mu(0) = I_2$ and $\alpha_2(T)\beta_2(T) \in \mathrm{SL}_2(A_b[T])$ with $\alpha_2(0)\beta_2(0) = I_2$. Therefore,

$$\Gamma([\alpha(T)\beta(T)]) = \alpha_2(1)\beta_2(1)\begin{pmatrix}1\\0\end{pmatrix} = \Gamma([\alpha(T)]) * \Gamma([\beta(T)]).$$

Hence Γ is a group homomorphism.

(4) By the definition of Γ, it is clear that $\mathrm{Im}(\Gamma) \subseteq \ker(\phi)$. Conversely, let $[(e, f)] \in \ker(\phi)$ that is, $[(e, f)] = [(1, 0)]$ in $\Gamma(A_a)$ and $\Gamma(A_b)$. This implies that we can get matrices $\alpha_1(T) \in \mathrm{SL}_2(A_a[T])$ and $\alpha_2(T) \in \mathrm{SL}_2(A_b[T])$ with $\alpha_1(0) = I_2 = \alpha_2(0)$, $\alpha_1(1)\begin{pmatrix}1\\0\end{pmatrix} = \begin{pmatrix}e\\f\end{pmatrix}$ and $\alpha_2(1)\begin{pmatrix}1\\0\end{pmatrix} = \begin{pmatrix}e\\f\end{pmatrix}$.

We have $\alpha_2(1)^{-1}\alpha_1(1)\begin{pmatrix}1\\0\end{pmatrix} = \begin{pmatrix}1\\0\end{pmatrix}$. This implies that $\alpha_2(1)^{-1}\alpha_1(1) = \begin{pmatrix}1 & \mu\\0 & 1\end{pmatrix}$, where $\mu \in A_{ab}$. Further, we have $\begin{pmatrix}1 & \mu\\0 & 1\end{pmatrix} = \begin{pmatrix}1 & \mu_2\\0 & 1\end{pmatrix}\begin{pmatrix}1 & -\mu_1\\0 & 1\end{pmatrix}$, where $\mu_1 \in A_a$ and $\mu_2 \in A_b$. Thus

$$\alpha_2(1)\begin{pmatrix}1 & \mu_2\\0 & 1\end{pmatrix} = \alpha_1(1)\begin{pmatrix}1 & \mu_1\\0 & 1\end{pmatrix}.$$

Let $\beta_1(T) = \alpha_1(T)\begin{pmatrix}1 & \mu_1 T\\0 & 1\end{pmatrix}$ and $\beta_2(T) = \alpha_2(T)\begin{pmatrix}1 & \mu_2 T\\0 & 1\end{pmatrix}$. Then

$$\Gamma([\beta_1(T)^{-1}\beta_2(T)]) = \begin{pmatrix}e\\f\end{pmatrix}.$$

Hence $\mathrm{Im}(\Gamma) \supseteq \ker(\phi)$. Therefore we have an exact sequence $\pi_1(\mathrm{SL}_2(A)) \xrightarrow{\varphi_1} \pi_1(\mathrm{SL}_2(A_a)) \oplus \pi_1(\mathrm{SL}_2(A_b)) \xrightarrow{\psi_1} \pi_1(\mathrm{SL}_2(A_{ab})) \xrightarrow{\Gamma} \Gamma(A) \xrightarrow{\phi} \Gamma(A_a) \oplus \Gamma(A_b) \xrightarrow{\psi} \Gamma(A_{ab})$. (**) $\square$

15.5 On Cocycles Associated to Alternating Matrices

In this section, we associate cocycles to alternating forms on projective modules.

Let A be a domain and P be a projective A-module of rank 2. Suppose there exist $f_1, f_2 \in A$ such that $f_1A + f_2A = A$ and $P_{f_1} \simeq A^2_{f_1}$, $P_{f_2} \simeq A^2_{f_2}$.

Since P_{f_1} and P_{f_2} are free, there exist bases $\{p_1, p_2\}$ of P_{f_1} and $\{p'_1, p'_2\}$ of P_{f_2}. Therefore we have two bases $\{p_1, p_2\}$ and $\{p'_1, p'_2\}$ of $P_{f_1f_2}$. So we can get a matrix $\sigma \in \mathrm{GL}_2(A_{f_1f_2})$ such that $\sigma\begin{pmatrix}p'_1\\p'_2\end{pmatrix} = \begin{pmatrix}p_1\\p_2\end{pmatrix}$.

Definition 15.9 1. The matrix σ is called cocycle is associated to the projective module P.

2. Two cocycles σ_1 and σ_2 are said to be equivalent if there exist $\mu_1 \in \mathrm{GL}_2(A_{f_1})$ and $\mu_2 \in \mathrm{GL}_2(A_{f_2})$ such that $\sigma_2 = \mu_1\sigma_1\mu_2$. In particular, we say that a cocycle σ splits if σ is equivalent to identity. It is known that a rank 2 projective module P is free if the cocycle associated to P splits.

Now, instead of considering rank 2 projective A-modules one can consider 4×4 invertible alternating matrices over a ring A, where free modules are replaced by

$$\psi_1 \perp \psi_1 = \begin{pmatrix} 0 & 1 & 0 & 0 \\ -1 & 0 & 0 & 0 \\ 0 & 0 & 0 & 1 \\ 0 & 0 & -1 & 0 \end{pmatrix}.$$

Definition 15.10 1. Let α and β be two invertible 4×4 alternating matrices over a domain A. We say that α and β are isometric if there exists $\gamma \in \mathrm{GL}_4(A)$ such that $\gamma\alpha\gamma^t = \beta$.

2. Let $\alpha \in \mathrm{GL}_4(A)$ be an alternating matrix. Suppose there exist $\alpha_1 \in \mathrm{GL}_4(A_{f_1})$ and $\alpha_2 \in \mathrm{GL}_4(A_{f_2})$ such that

$$\alpha_1\alpha\alpha_1^t = \psi_1 \perp \psi_1; \ \alpha_2\alpha\alpha_2^t = \psi_1 \perp \psi_1.$$

Then $\beta = \alpha_1\alpha_2^{-1}$ satisfies $\beta(\psi_1 \perp \psi_1)\beta^t = \psi_1 \perp \psi_1$ and we say β is the cocycle associated to α. Clearly $\beta \in \mathrm{Sp}_4(A_{f_1 f_2})$.

Lemma 15.3 *Let β be the cocycle associated to an invertible alternating matrix α as above. If β splits in $\mathrm{Sp}_4(A_{f_1 f_2})$, then α and $\psi_1 \perp \psi_1$ are isometric.*

Proof Since β splits, there exist $\delta_1 \in \mathrm{Sp}_4(A_{f_1})$ and $\delta_2 \in \mathrm{Sp}_4(A_{f_2})$ such that $\beta = \alpha_1\alpha_2^{-1} = \delta_1^{-1}\delta_2 \Rightarrow \delta_1\alpha_1 = \delta_2\alpha_2$. Suppose $\alpha_1' = \delta_1\alpha_1$ and $\alpha_2' = \delta_2\alpha_2$. Then $\alpha_1'\alpha(\alpha_1')^t = \psi_1 \perp \psi_1$; $\alpha_2'\alpha(\alpha_2')^t = \psi_1 \perp \psi_1$, where $\alpha_1' \in \mathrm{GL}_4(A_{f_1})$ and $\alpha_2' \in \mathrm{GL}_4(A_{f_2})$. Also since $\alpha_1' = \alpha_2'$, we obtain $\widetilde{\alpha} \in \mathrm{GL}_4(A)$ such that $\widetilde{\alpha}\alpha\widetilde{\alpha}^t = \psi_1 \perp \psi_1$. Therefore α and $\psi_1 \perp \psi_1$ are isometric. Thus α is trivial if the cocycle associated to α splits. □

Suppose $\alpha, \beta \in \mathrm{GL}_4(A)$ are alternating and

$$\alpha_1\alpha\alpha_1^t = \psi_1 \perp \psi_1; \ \alpha_2\alpha\alpha_2^t = \psi_1 \perp \psi_1,$$

where $\alpha_1 \in \mathrm{GL}_4(A_{f_1})$ and $\alpha_2 \in \mathrm{GL}_4(A_{f_2})$ and

$$\beta_1\beta\beta_1^t = \psi_1 \perp \psi_1; \ \beta_2\beta\beta_2^t = \psi_1 \perp \psi_1,$$

where $\beta_1 \in \mathrm{GL}_4(A_{f_1})$ and $\beta_2 \in \mathrm{GL}_4(A_{f_2})$.

Let $\gamma_1 = \alpha_1\alpha_2^{-1} \in \mathrm{Sp}_4(A_{f_1 f_2})$ and $\gamma_2 = \beta_1\beta_2^{-1} \in \mathrm{Sp}_4(A_{f_1 f_2})$ be the cocycles associated to α and β. Suppose there exist $\nu_1 \in \mathrm{Sp}_4(A_{f_1})$ and $\nu_2 \in \mathrm{Sp}_4(A_{f_2})$ such that $\nu_1\gamma_1\nu_2 = \gamma_2$, then one can check that α and β are isometric, that is, there exists $\nu \in \mathrm{GL}_4(A)$ such that $\nu\alpha\nu^t = \beta$ (by using same argument as in the proof of Lemma 15.3). This shows that if the cocyles associated to α and β are equivalent, then α and β are isometric.

Remark 15.2 There is a one-to-one correspondence between alternating forms on a free module of rank n over a ring A and alternating matrices of order n with entries in A.

Proposition 15.1 *Let A be a domain of dimension 2. Suppose $f_1A + f_2A = A$ and P, Q are stably free A-modules of rank 2 such that P_{f_1} and P_{f_2} are free and the associated cocycle is $\sigma \in \mathrm{SL}_2(A_{f_1 f_2})$ and Q_{f_1}, Q_{f_2} are free and the associated cocycle is $\tau \in \mathrm{SL}_2(A_{f_1 f_2})$. Let Q' be the projective A-module associated to the cocycle $\sigma\tau$ and s, t, t' be the corresponding alternating forms on P, Q and Q'. Then we have an isometry of alternating forms*

$$(P, s) \perp (Q, t) \simeq (A^2, \psi_1) \perp (Q', t').$$

Proof Since P_{f_1} and P_{f_2} are free, we have isomorphisms

$$P_{f_1} \xrightarrow{i_1} A^2_{f_1};\ P_{f_2} \xrightarrow{i_2} A^2_{f_2}$$

such that the cocycle associated to P is $\sigma \in \mathrm{SL}_2(A_{f_1 f_2})$. Since $\sigma \in \mathrm{SL}_2(A_{f_1 f_2})$, the alternating form $s : P \times P \to A$ is (using the form ψ_1 on $A^2_{f_2}$) given by

$$s(p_1, p_2) = \det(i_1(p_1), i_1(p_2)) = \det(i_2(p_1), i_2(p_2)).$$

Similarly we have isomorphisms

$$Q_{f_1} \xrightarrow{j_1} A^2_{f_1};\ Q_{f_2} \xrightarrow{j_2} A^2_{f_2}$$

such that the cocycle associated to Q is $\tau \in \mathrm{SL}_2(A_{f_1 f_2})$ and alternating form $t : Q \times Q \to A$ is given by

$$t(q_1, q_2) = \det(j_1(q_1), j_1(q_2)) = \det(j_2(q_1), j_2(q_2)).$$

Therefore we get an alternating form $s \perp t$ on $P \oplus Q$. Since $P \oplus Q \simeq A^4$ ([1], Bass Cancellation Theorem), $s \perp t$ yields a matrix $\alpha \in \mathrm{GL}_4(A)$ which is alternating.

Further, the isomorphisms i_1 and j_1 show that $(\alpha)_{f_1} \simeq \psi_1 \perp \psi_2$ and isomorphisms i_2 and j_2 show that $(\alpha)_{f_2} \simeq \psi_1 \perp \psi_1$. It is easy to check that the cocycle associated to α is $\begin{pmatrix} \sigma & 0 \\ 0 & \tau \end{pmatrix} \in \mathrm{Sp}_4(A_{f_1 f_2})$.

Further, there are isomorphisms $Q'_{f_1} \xrightarrow{\theta_1} A^2_{f_1}$ and $Q'_{f_2} \xrightarrow{\theta_2} A^2_{f_2}$ such that the associated cocycle is $\sigma\tau$. The isomorphisms θ_1 and θ_2 induce an alternating form $t' : Q' \times Q' \to A$. Now, since $A^2 \oplus Q' \simeq A^4$, we get an alternating form $\beta = (A^2, \psi_1) \perp (Q', t')$ on A^4, which in view of the isomorphisms θ_1, θ_2 satisfies the property that β_{f_1} and β_{f_2} are both isometric to $\psi_1 \perp \psi_1$ and the cocycle associated to β is $\begin{pmatrix} I_2 & 0 \\ 0 & \sigma\tau \end{pmatrix}$.

Now,

$$\begin{pmatrix} \sigma^{-1} & 0 \\ 0 & \sigma \end{pmatrix} \begin{pmatrix} \sigma & 0 \\ 0 & \tau \end{pmatrix} = \begin{pmatrix} I_2 & 0 \\ 0 & \sigma\tau \end{pmatrix}.$$

Since $\sigma \in \mathrm{SL}_2(A_{f_1 f_2})$ and $\begin{pmatrix} \sigma^{-1} & 0 \\ 0 & \sigma \end{pmatrix} \in \mathrm{ESp}_4(A_{f_1 f_2})$, (by a lemma of Vaserstein [11], see [2, Lemma 1.2.9 c]), so by a Symplectic version of the Bhatwadekar–Lindel–Rao lemma, whose proof follows exactly the linear case Lemma 15.2, the cocycles $\begin{pmatrix} \sigma & 0 \\ 0 & \tau \end{pmatrix}$ and $\begin{pmatrix} I_2 & 0 \\ 0 & \sigma\tau \end{pmatrix}$ are equivalent and therefore the alternating forms $(P, s) \perp (Q, t)$ and $(A^2, \psi_1) \perp (Q', t')$ are equivalent. Therefore, we have proved. □

15.6 On Some Consequences of the Above Results

We saw in the previous section that if A is a ring and $a, b \in A$ are such that $aA + bA = A$, then we can associate $\sigma \in \mathrm{SL}_2(A_{ab})$ to a projective A-module P of trivial determinant together with a non-singular alternating form $\delta : P \times P \to A$.

Now, let A be a domain with $\dim A = 2$ and S be the set of pairs (P, s), where P is a rank 2 projective module and $s : P \times P \to A$ is a non-singular alternating form. Then by theorem of Bass [10, Appendix A.7], the set S is an abelian group with the group structure $+$ given by $(P, s) + (Q, t) = (Q', t')$, where $(P, s) \perp (Q, t) \simeq (A^2, \psi_1) \perp (Q', t')$, where $\perp$ denotes the direct sum of alternating forms.

By Proposition 15.1, we have a homomorphism $H \to S$, where H is the subgroup of $\Gamma(A_{ab})$ corresponding to cocycles corresponding to stably free modules. Since S is abelian group, in particular we have the following:

Corollary 15.1 *Let A be a domain with $\dim A = 2$. Let $a, b \in A$ be such that $aA + bA = A$. Let $\sigma \in \mathrm{SL}_2(A_{ab})$ and $\tau \in \mathrm{SL}_2(A_{ab})$ be cocycles corresponding to stably free modules. Then $\sigma\tau\sigma^{-1}\tau^{-1} = \alpha_1\alpha_2$, where $\alpha_1 \in \mathrm{SL}_2(A_a)$ and $\alpha_2 \in \mathrm{SL}_2(A_b)$.*

Proof Since S is an abelian group, the image of the element of H corresponding to the cocycle $\sigma\tau\sigma^{-1}\tau^{-1}$ in S is the identity element of S that is, the cocycle $\sigma\tau\sigma^{-1}\tau^{-1}$ corresponds to a free module of rank 2 over A. Therefore the cocycle $\sigma\tau\sigma^{-1}\tau^{-1}$ splits, that is, $\sigma\tau\sigma^{-1}\tau^{-1} = \alpha_1\alpha_2$, where $\alpha_1 \in \mathrm{SL}_2(A_a)$ and $\alpha_2 \in \mathrm{SL}_2(A_b)$ ([9], Theorem 14.4). □

It would be interesting to see if the restriction that dim $A = 2$ can be removed in Corollary 15.1.

Next we would like to give conditions under which $\Gamma(A)$ is an abelian group. To obtain such condition observe that if $\Gamma(A)$ is an abelian group and $\sigma, \tau \in \mathrm{SL}_2(A)$, then the columns $v = \sigma\tau\begin{pmatrix}1\\0\end{pmatrix}$ and $w = \tau\sigma\begin{pmatrix}1\\0\end{pmatrix}$ are equal in $\Gamma(A)$, whereby there exists $\alpha(T) \in \mathrm{SL}_2(A[T])$ such that $\alpha(0) = I_2$ and $\alpha(1)v = w$.

Now, since $\Gamma : \pi_1(\mathrm{SL}_2(A_{ab})) \to \Gamma(A)$ is a homomorphism and $\pi_1(\mathrm{SL}_2(A_{ab}))$ is an abelian group, its image in $\Gamma(A)$ under Γ is likewise abelian and so any pair v, w in the image commute. An element of $\Gamma(A)$ lies in this image if it maps to 0 in $\Gamma(A_a)$ and $\Gamma(A_b)$. This will be the case if we can find elementary completions of the corresponding unimodular row in A_a and A_b.

We use these observations to prove the following corollary:

Corollary 15.2 *Let A be a Noetherian domain of dimension one. Then $\Gamma(A)$ is an abelian group.*

Proof Let $[v] = (c, d)$, $[w] = (c', d')$. We want to show that $[v]$ and $[w]$ commute. Since elementary matrices can be connected to the identity matrix, we can perform elementary transformations on v and w without changing the class of v and w in $\Gamma(A)$.

We may, therefore, assume that $d' \neq 0$. Let $\mathfrak{m}_1, \mathfrak{m}_2, \ldots, \mathfrak{m}_r$ be the maximal ideals of A containing d'. By replacing d by $d + \lambda c$, we may assume that $d \notin \mathfrak{m}_i$ for any $1 \leq i \leq r$, which implies that $(d) + (d') = A$.

By the Chinese reminder theorem, we may choose $\widetilde{c} \in A$ such that $\widetilde{c} = c \bmod (d)$ and $\widetilde{c} = c' \bmod (d')$. Then $\widetilde{c} = c + \mu d$ and $\widetilde{c} = c' + \mu' d'$. Therefore, $(c, d) \overset{E_2(A)}{\sim} (\widetilde{c}, d)$ and $(c', d') \overset{E_2(A)}{\sim} (\widetilde{c}, d')$ (This idea is well known but we have given an argument for the convenience of the reader). Since $(\widetilde{c}, d)$ is unimodular, there exist $g, h \in A$ such that $g\widetilde{c} + hd = 1$ and $g', h' \in A$ such that $g'\widetilde{c} + h'd' = 1$.

Let $a = \widetilde{c}$ and $b = (1 - g\widetilde{c})(1 - g'\widetilde{c})$. Then $\widetilde{c}$ is a unit in A_a, d and d' are units in A_b. Thus, $(\widetilde{c}, d)$ and $(\widetilde{c}, d')$ can be completed to elementary matrices in A_a and A_b. Hence $[v] = 0$ in $\Gamma(A_a)$ and $\Gamma(A_b)$ and $[w] = 0$ in $\Gamma(A_a)$ and $\Gamma(A_b)$. Therefore $[v]$ and $[w]$ which are in $\Gamma(A)$ commute proving the corollary. □

Corollary 15.2 leads to the following interesting question:

? Question 1

Does Corollary 15.2 hold for rings of dimension bigger than one?

By using Corollary 15.2, we can say that the exact sequence (∗∗) in Sect. 15.4 for a Noetherian domain of dimension one is an algebraic analogue of the Theorem 15.1.

Remark 15.3 Let A be the coordinate ring of a real affine variety $X = \text{Spec } A$. Then any element $a \in A$ gives a continuous function $a : X(\mathbb{R}) \to \mathbb{R}$. Therefore a unimodular row $(a_1, a_2) \in A^2$ gives a continuous map $(a_1, a_2) : X(\mathbb{R}) \to \mathbb{R}^2 - \{(0, 0)\}$.

Two unimodular rows give the same element of $\Gamma(A)$ if the corresponding maps $(a_1, a_2) : X(\mathbb{R}) \to \mathbb{R}^2 - \{(0, 0)\}$ are homotopic. Thus the group $\Gamma(A)$ can be considered in a certain sense as the algebraic analogue of the set of homotopy classes of continuous maps from $X \to \mathbb{R}^2 - \{(0, 0)\}$ or the homotopy classes of continuous maps X to S^1 or the group $H^1(X, \mathbb{Z})$.

Further, if A is the coordinate ring of a real affine variety $X = \text{Spec } A$ (as above), then an element of $\pi_1(\text{SL}_2(A))$ gives a continuous function from $X(\mathbb{R}) \to \pi_1(\text{SL}_2(\mathbb{R}))$ and $\pi_1(\text{SL}_2(\mathbb{R})) = \mathbb{Z}$. Thus $\pi_1(\text{SL}_2(A)))$ can be considered $H^0(\text{Spec}$ $(A), \pi_1(\text{SL}_2(A)))$ which is the analogue of the group $H^0(X, \mathbb{Z})$ (the set of continuous maps from X to $\mathbb{Z}$ or the free abelian group on the set of connected component of X).

Now the group homomorphism $\Gamma : \pi_1(\text{SL}_2(A_{ab})) \longrightarrow \Gamma(A)$ shows that the $H^1(\text{Spec}(A), \pi_1(\text{SL}_2(A)))$ is connected to the group $H^1(X, \mathbb{Z})$. So one can ask 'is the group $H^2(\text{Spec}(A), \pi_1(\text{SL}_2(A)))$ connected to the group $H^2(X, \mathbb{Z})$?' This was the suggestion of Nori. We elaborate this in the next remark. The cohomology groups are considered in this remark with respect to Zariski topology on $\text{Spec}(A)$.

Remark 15.4 Let A be a domain and $\widetilde{\Gamma}(A) = \{\alpha(T) \in \text{SL}_2(A[T]) : \alpha(1) = I_2\}$, We have a homomorphism $\widetilde{\Gamma}(A) \to \text{SL}_2(A)$ sending $\alpha(T)$ to $\alpha(0)$. A projective A-module P of rank 2 and trivial determinant gives a cocycle $H^1(X, \text{SL}_2)$, where $X = \text{Spec } A$. By Quillen's localization theorem [8], a projective A-module P of rank 2 is free if the 1-cocycle associated to P belonging to $H^1(X, \text{SL}_2)$ can be lifted to $H^1(X, \widetilde{\Gamma})$. Let $N(A)$ be the kernel of the map $\widetilde{\Gamma}(A)$ to $\text{SL}_2(A)$ given above, that is,

$$1 \to N(A) \to \widetilde{\Gamma}(A) \to \text{SL}_2(A) \to 1$$

is exact.

Nori suggested to the first author that one should use the above exact sequence to define a connecting map $H^1(X, \text{SL}_2(A)) \to H^2(X, N/N_0)$, where $N_0(A)$ is the connected component of identity of $N(A)$ and associate to P an obstruction in $H^2(X, N/N_0)$, and show that if dimension of A is 2 and this obstruction vanishes then P is free (Nori also showed that $N(A)/N_0(A) \simeq \pi_1(\text{SL}_2(A)))$. Therefore $H^2(X, N(A)/N_0(A))$ is same as $H^2(\text{Spec}(A), \pi_1(\text{SL}_2(A)))$. This was Nori's original approach to defining a group to evaluate Euler Classes.

We will try to show how Nori's suggestion motivated our work. We consider the following problem:

? Question 2

Can one associate an obstruction to a matrix in $\text{SL}_2(A)$ whose vanishing implies the matrix is trivial in $\Gamma(A)$?

We know that over a local ring B any matrix belonging to $\mathrm{SL}_2(B)$ is elementary, and therefore can be connected to the identity matrix.

Let

$$\Gamma'(A) = \{\beta(T) \in \mathrm{SL}_2(A[T]) : \beta(0) = I_2\}.$$

We have a map $\Gamma'(A) \to \mathrm{SL}_2(A)$ given by $\beta \to \beta(1)$.

A matrix $\alpha \in \mathrm{SL}_2(A)$ can be connected to the identity matrix if α can be lifted to $\Gamma'(A)$ under the above map. Suppose there exist $a, b \in A$ such that $aA + bA = A$, and $\alpha \in \mathrm{SL}_2(A)$ is such that both $(\alpha)_a$ and $(\alpha)_b$ can be connected to the identity matrix, that is, there exist $\beta_1(T) \in \Gamma'(A_a)$ which is a lift of $(\alpha)_a$ and $\beta_2(T) \in \Gamma'(A_b)$ which is a lift of $(\alpha)_b$. Then $\beta_1\beta_2^{-1} \in \pi_1(\mathrm{SL}_2(A_{ab}))$. This leads us to consider the map $\pi_1(\mathrm{SL}_2(A_{ab}))$ to $\Gamma(A)$ discussed in this paper and naturally to the other results of this paper.

Remark 15.5 It would be interesting to know other places where the group $\Gamma(A)$ is used and where it first occurs. We have been able to trace its occurrence to a paper of Krusemeyer [7, Lemma 3.3] who refers to a paper of Karoubi–Villamayor (see [6]).

The exact sequence

$$1 \to \pi_1(\mathrm{SL}_2(A)) \to \Gamma(A) \to \mathrm{SL}_2(A) \to 1$$

occurs in [7, Lemma 3.6]. The main idea of this paper is to write down a Mayer–Vietoris sequence associated to the above exact sequence.

Acknowledgements The authors would like to thank Professor Ravi A. Rao for his valuable support during this work and for bringing to our attention the crucial lemma of Vaserstein used in Sect. 15.5. The authors would like to thank Professor Gopala Krishna Srinivasan for giving his time most generously and helping us make this paper more readable. The third named author would like to thank Professor Gopala Krishna Srinivasan for his support and advice during difficult times. The third named author also acknowledges the financial support from CSIR, which enabled him to pursue his doctoral studies.

References

1. H. Bass, K-theory and stable algebra. Publications Math-matiques de lInstitut des Hautes tudes Scientifiques **22**, 5–60 (1964)
2. R. Basu, Topics in classical algebraic K-theory. PhD thesis School of Mathematics, Tata Institute of Fundamental Research, Mumbai, 2006
3. R. Basu, R. Sridharan, On Forster's conjecture and related results. Punjab Univ. Res. J. (Sci.) **57**, 13–66 (2007)
4. S.M. Bhatwadekar, H. Lindel, R.A. Rao, The Bass-Murthy question: Serre dimension of Laurent polynomial extensions. Invent. Math. **81**(1), 189–203 (1985)
5. S.T. Hu, *Homology Theory: A First Course in Algebraic Topology* (Holden-Day Inc, San Francisco, 1966)
6. M. Karoubi, O. Villamayor, K-theorie algebrique et K-theorie topologique I. Math. Scand. **28**, 265–307 (1972)

7. M.I. Krusemeyer, Fundamental groups, algebraic K-theory, and a problem of Abhyankar. Invent. Math. **19**, 15–47 (1973)
8. D. Quillen, Projective Modules over polynomial rings. Invent. Math. **36**, 167–171 (1976)
9. R. Sridharan, S.K. Yadav, On a theorem of Suslin, to be appear in *Leavitt path Algebras and classical K-theory*, Indian Statistical Institute book series
10. R.G. Swan, Algebraic vector bundles on the 2-sphere. Rocky Mt. J. Math. **23**(4), 1443–1469 (1993)
11. L.N. Vaserstein, Stabilization of unitary and orthogonal groups over a ring with involution, Mat. Sb. (N.S.), **81**(123), 328–351 (1970)
12. C.T.C. Wall, *A Geometric Introduction to Topology* (Addison-Wesley Publishing Co, Reading, 1972)

Chapter 16
On the Completability of Unimodular Rows of Length Three

Raja Sridharan and Sunil K. Yadav

16.1 Introduction

In 1958 (see [25]), Seshadri proved that *if k is a field then finitely generated projective modules over $k[X, Y]$ are free.* This result of Seshadri has resulted among many other things, in the growth of a School of Projective modules at the Tata Institute (see [15] for more details), which led to the construction of a theory of Euler classes of projective modules.

In this paper, we try to complete the circle, that is, we see the relevance of the present to the past by giving a proof of Seshadri's theorem using the theory of Euler classes. We focus on unimodular rows of length 3, to which the theory applies. We use a result of Suslin [33, Lemma A.10] (which roughly says that a unimodular row of length 3 is completable if its Euler class vanishes) to give a proof of Seshadri's theorem.

We give another proof of a result of Bhatwadekar–Keshari [7, Lemma 3.3] and use this to deduce using Suslin's lemma, that any row of the form (a^2, b, c) is completable (The Swan–Towber [29], Krusemeyer [14], Suslin [27] theorem). We then use an argument of Abhyankar to give a proof of Seshadri's theorem using Euler class groups.

This paper is arranged as follows.

In Sect. 16.2, we recall some preliminary results. In Sect. 16.3, we recall some of the basic definitions in the theory of Euler classes. In Sect. 16.4, we recall the

R. Sridharan
Tata Institute of Fundamental Research, 1, Dr. Homi Bhabha Road,
Mumbai 400005, India
e-mail: sraja@math.tifr.res.in

S. K. Yadav (✉)
Department of Mathematics, Indian Institute of Technology Bombay,
Powai, Mumbai 400076, India
e-mail: skyadav@math.iitb.ac.in; skymath.bhu@gmail.com

A. A. Ambily et al. (eds.), *Leavitt Path Algebras and Classical K-Theory*,
Indian Statistical Institute Series,
https://doi.org/10.1007/978-981-15-1611-5_16

proof of Suslin's lemma. In Sect. 16.5, we give various proofs of Seshadri's Theorem including some known ones for completeness and comparison. In Sect. 16.6, we give a proof of a lemma of Bhatwadekar–Keshari. In Sect. 16.7, we recall an argument of Abhyankar and give a proof of Seshadri's theorem using Euler class groups. Finally, we make a Remark 16.6, which puts everything together.

16.2 Some Preliminaries

Definition 16.1 (i) Let A be a ring. A row $(a_1, a_2, \dots, a_n) \in A^n$ is said to be **unimodular** of length n if the ideal $(a_1, a_2, \dots, a_n) = A$. The set of unimodular rows of length n is denoted by $\mathrm{Um_n(A)}$.

(ii) A unimodular row $(a_1, a_2, \dots, a_n)$ is said to be **completable** if there is a matrix in $SL_n(A)$ whose first row is $(a_1, a_2, \dots, a_n)$.

(iii) We define $E_n(A)$ to be the subgroup of $GL_n(A)$ generated by all matrices of the form $e_{ij}(\lambda) = I_n + \lambda E_{ij}, \lambda \in A, i \neq j$, where E_{ij} is a matrix whose (i, j)th entry is 1 and all other entries are zero. The matrices $e_{ij}(\lambda)$ will be referred to as elementary matrices.

Lemma 16.1 (Prime Avoidance Lemma, see [3]) *Let A be a ring $I \subset A$ an ideal. Suppose $I \subset \bigcup_{i=1}^{n} p_i$, where $p_i \in Spec(A)$. Then $I \subset p_i$ for some i, $1 \leq i \leq n$.*

Lemma 16.2 (see [3]) *Let A be a ring, $p_1, p_2, \dots, p_r \in Spec(A)$ and $I = (a_1, a_2, \dots, a_n)$ be an ideal of A such that $I \subsetneq p_i$, $1 \leq i \leq r$. Then there exist $b_2, b_3, \dots, b_n \in A$ such that the element*

$$c = a_1 + a_2b_2 + a_3b_3 + \cdots + a_nb_n \notin \bigcup_{i=1}^{r} p_i.$$

Since $M_n(A)$ acts on A^n via matrix multiplication, the group $E_n(A)$ which is a subset of $M_n(A)$ also acts on A^n. This induces an action of $E_n(A)$ on $\mathrm{Um}_n(A)$. The equivalence relation on $\mathrm{Um}_n(A)$ given by this action is denoted by $\overset{E_n(A)}{\sim}$. Similarly one can define $\overset{GL_n(A)}{\sim}$ and $\overset{SL_n(A)}{\sim}$.

Theorem 16.1 (see [3]) *Let A be a ring and $(a_1, a_2, \dots, a_n) \in A^n$ be a unimodular row of length n which contains a unimodular row of shorter length. Then the row $(a_1, a_2, \dots, a_n)$ is completable. In fact,*

$$(a_1, a_2, \dots, a_n) \overset{E_n(A)}{\sim} (1, 0, \dots, 0).$$

(i) *Let A be a semilocal ring. Then any unimodular row* $(a_1, a_2, \ldots, a_n)$ *of length* $n \geq 2$ *is completable. In fact,*

$$(a_1, a_2, \ldots, a_n) \overset{E_n(A)}{\sim} (1, 0, \ldots, 0).$$

Definition 16.2 Two matrices α and β in $SL_n(A)$ are said to be **connected** if there exists $\sigma(X) \in SL_n(A[X])$ such that $\sigma(0) = \alpha$ and $\sigma(1) = \beta$. By considering the matrix $\sigma(1 - X)$, it follows that if α is connected to β then β is connected to α.

Lemma 16.3 *Any matrix in* $E_n(A)$ *can be connected to the identity matrix.*

Proof Every matrix $\alpha \in E_n(A)$ can be written as a product of elementary matrices of the form $e_{ij}(\lambda) = I_n + \lambda E_{ij}$ for $i \neq j$, that is, $\alpha = \prod_{i=1}^{r} e_{ij}(\lambda)$. We define $\sigma(X) = \prod_{i=1}^{r} e_{ij}(\lambda X)$. Then $\sigma(X) \in SL_n(A[X])$, $\sigma(0) = I_n$ and $\sigma(1) = \alpha$. This proves the lemma. □

Lemma 16.4 *Let A be a ring and I be an ideal of A. Then the map* $E_n(A) \to E_n(A/I)$ *is surjective.*

Proof The proof follows from the fact that the generators $e_{ij}(\bar{\lambda})$ of $E_n(A/I)$ for $\lambda \in A$, can be lifted to generators $e_{ij}(\lambda)$ of $E_n(A)$. □

Let us recall Quillen's Splitting Lemma [23] with the proof following the exposition of [3]. In what follows, $(\psi_1(X))_t$ denotes the image of $\psi_1(X)$ in $GL_n(A_{st}[X])$ and $(\psi_2(X))_s$ denotes the image of $\psi_2(X)$ in $GL_n(A_{st}[X])$.)

Lemma 16.5 (see [23]) *Let A be a domain and* $s, t \in A$ *be such that* $sA + tA = A$. *Suppose there exists* $\sigma(X) \in GL_n(A_{st}[X])$ *with the property that* $\sigma(0) = I_n$. *Then there exist* $\psi_1(X) \in GL_n(A_s[X])$ *with* $\psi_1(0) = I_n$ *and* $\psi_2(X) \in GL_n(A_t[X])$ *with* $\psi_2(0) = I_n$ *such that* $\sigma(X) = (\psi_1(X))_t(\psi_2(X))_s$.

Proof Since $\sigma(0) = I_n$, $\sigma(X) = I_n + X\tau(X)$, where $\tau(X) \in M_n(A_{st}[X])$, we choose a large integer N_1 such that $\sigma(\lambda s^k X) \in GL_n(A_t[X])$ for all $\lambda \in A$ and for all $k \geq N_1$. Define $\beta(X, Y, Z) \in GL_n(A_{st}[X, Y, Z])$ as follows:

$$\beta(X, Y, Z) = \sigma((Y + Z)X)\sigma(YX)^{-1}. \tag{16.1}$$

Then $\beta(X, Y, 0) = I_n$, and hence there exists a large integer N_2 such that for all $k \geq N_2$ and for all $\mu \in A$, we have $\beta(X, Y, \mu t^k Z) \in GL_n(A_s[X, Y, Z])$. This means

$$\beta(X, Y, \mu t^k Z) = (\sigma_1(X, Y, Z))_t, \tag{16.2}$$

where $\sigma_1(X, Y, Z) \in GL_n(A_s[X, Y, Z])$ with $\sigma_1(X, Y, 0) = I_n$.

Taking $N = \max(N_1, N_2)$, it follows by the comaximality of sA and tA that $s^N A + t^N A = A$. Pick $\lambda, \mu \in A$ such that $\lambda s^N + \mu t^N = 1$. Setting $Y = \lambda s^N$, $Z = \mu t^N$ in (16.1) and $Z = 1$, $Y = \lambda s^N$ in (16.2) we get

$$\beta(X, \lambda s^N, \mu t^N) = \sigma(X)\sigma(\lambda s^N X)^{-1}$$

and

$$\beta(X, \lambda s^N, \mu t^N) = (\sigma_1(X, \lambda s^N, \mu t^N))_t = (\psi_1(X))_t, \text{ where } \psi_1(X) \in GL_n(A_s[X]).$$

Hence, we conclude $\sigma(X)\sigma(\lambda s^N X)^{-1} = (\psi_1(X))_t$. Let $\sigma(\lambda s^N X) = (\psi_2(X))_s$, where $(\psi_2(X))_s \in GL_n(A_t[X])$. Since $\sigma(0) = I_n$, $\psi_1(0) = \psi_2(0) = I_n$, the result follows by using the identity $\sigma(X) = \sigma(X)\sigma(\lambda s^N X)^{-1}\sigma(\lambda s^N X)$. □

Lemma 16.6 (see [16]) *Let A be a commutative ring with* 1. *Let e be an idempotent of A and b be any element of A. Then the ideal* (e, b) *of A is principal.*

Proof We claim that $(e, b) = (e + (1 - e)b)$. For multiplying $(e + (1 - e)b)$ by e and using $e(1 - e) = 0$, we see that e is in the ideal $(e + (1 - e)b)$, and therefore $(e + (1 - e)b) = (e, e + (1 - e)b) = (e, b)$. □

Lemma 16.7 (see [16]) *Let A be a Noetherian ring and I be an ideal of A. Suppose there exist* $a_1, \ldots, a_n \in I$ *such that* $(a_1, \ldots, a_n) + I^2 = I$. *Then for any* $b \in A$ *the ideal* (I, b) *is generated by* $n + 1$ *elements.*

Proof *(Sketch)* One can show that $I = (a_1, \ldots, a_n, a)$, where $a(1 - a)$ belongs to $(a_1, \ldots, a_n) = J$, that is, a is an idempotent modulo $(a_1, \ldots, a_n)$. Now after going modulo $(a_1, \ldots, a_n)$, we see by Lemma 16.6 that the image of the ideal (a, b) in A/J is principal. This implies that $(I, b) = (a_1, \ldots, a_n, a, b)$ is generated by $n + 1$ elements. □

Lemma 16.8 *Let A be a domain and* (a_1, a_2, a_3) *be a unimodular row in A. Then* $\frac{A^3}{(a_1,a_2,a_3)}$ *is a torsion free A-module. That is, if* $\lambda(\overline{c}_1, \overline{c}_2, \overline{c}_3) = (\overline{0}, \overline{0}, \overline{0})$ *in* $\frac{A^3}{(a_1,a_2,a_3)}$, *where* $\lambda \in A - \{0\}$, *then* $(\overline{c}_1, \overline{c}_2, \overline{c}_3) = (\overline{0}, \overline{0}, \overline{0})$ *in* $\frac{A^3}{(a_1,a_2,a_3)}$.

Proof We are given $\lambda(\overline{c}_1, \overline{c}_2, \overline{c}_3) = (\overline{0}, \overline{0}, \overline{0})$ in $\frac{A^3}{(a_1,a_2,a_3)}$, where $\lambda \in A$. Therefore, there exists $\mu \in A$ such that $\lambda(c_1, c_2, c_3) = \mu(a_1, a_2, a_3)$. This implies that $\lambda c_1 = \mu a_1$, $\lambda c_2 = \mu a_2$ and $\lambda c_3 = \mu a_3$. As (a_1, a_2, a_3) is unimodular, there exist $b_1, b_2, b_3 \in A$ such that $a_1 b_1 + a_2 b_2 + a_3 b_3 = 1$. This implies that $\mu a_1 b_1 + \mu a_2 b_2 + \mu a_3 b_3 = \mu$. Since $\lambda c_1 = \mu a_1$, $\lambda c_2 = \mu a_2$ and $\lambda c_3 = \mu a_3$, we have $\lambda(c_1 b_1 + c_2 b_2 + c_3 b_3) = \mu$. Now by putting the value of μ in the equation $\lambda(c_1, c_2, c_3) = \mu(a_1, a_2, a_3)$, we get $\lambda(c_1, c_2, c_3) = \lambda(c_1 b_1 + c_2 b_2 + c_3 b_3)(a_1, a_2, a_3)$. Since A is a domain, therefore $(c_1, c_2, c_3) = (c_1 b_1 + c_2 b_2 + c_3 b_3)(a_1, a_2, a_3)$. Hence $(\overline{c}_1, \overline{c}_2, \overline{c}_3) = (\overline{0}, \overline{0}, \overline{0})$ in $\frac{A^3}{(a_1,a_2,a_3)}$. □

16.3 On the Euler Class Group

In this section we give the definition of the Euler class group of a Noetherian ring due to Bhatwadekar–Raja Sridharan and prove Lemma 16.9 [6, Lemma 5.3]. We follow the exposition of Manoj Keshari [13].

Let A be a Noetherian ring with dim $A = n \geq 2$. We define the Euler class group of A, denoted by $E(A)$, as follows.

Let $J \subset A$ be an ideal of height n such that J/J^2 is generated by n elements. Let α and β be two surjections from $(A/J)^n$ to J/J^2. We say that α and β are related if there exists an automorphism σ of $(A/J)^n$ of determinant 1 such that $\alpha\sigma = \beta$. It is easy to see that this is an equivalence relation. If $\alpha : (A/J)^n \twoheadrightarrow J/J^2$ is a surjection, then by $[\alpha]$, we denote the equivalence class of α. We call such an equivalence class $[\alpha]$ a local orientation of J.

Since dim $A/J = 0$ and $n \geq 2$, we have $SL_n(A/J) = E_n(A/J)$ and therefore, the canonical map from $SL_n(A)$ to $SL_n(A/J)$ is surjective. Hence, if a surjection $\alpha : (A/J)^n \twoheadrightarrow J/J^2$ can be lifted to a surjection $\theta : A^n \twoheadrightarrow J$, and α is equivalent to $\beta : (A/J)^n \twoheadrightarrow J/J^2$, then β can also be lifted to a surjection from A^n to J. For, let $\alpha\sigma = \beta$ for some $\sigma \in SL_n(A/J)$. Since dim $A/J = 0$, there exists $\widetilde{\sigma} \in SL_n(A)$ which is a lift of σ. Then $\theta\widetilde{\sigma} : A^n \twoheadrightarrow J$ is a lift of β.

A local orientation $[\alpha]$ of J is called a global orientation of J if the surjection $\alpha : (A/J)^n \twoheadrightarrow J/J^2$ can be lifted to a surjection $\theta : A^n \twoheadrightarrow J$.

We shall also, from now on, identify a surjection α with the equivalence class $[\alpha]$ to which α belongs.

Let $\mathfrak{M} \subset A$ be a maximal ideal of height n and $\mathfrak{N}$ be a $\mathfrak{M}$-primary ideal such that $\mathfrak{N}/\mathfrak{N}^2$ is generated by n elements. Let $w_{\mathfrak{N}}$ be a local orientation of $\mathfrak{N}$. Let G be the free abelian group on the set of pairs $(\mathfrak{N}, w_{\mathfrak{N}})$, where $\mathfrak{N}$ is a $\mathfrak{M}$-primary ideal and $w_{\mathfrak{N}}$ is a local orientation of $\mathfrak{N}$.

Let $J = \cap\mathfrak{N}_i$ be the intersection of finitely many ideals $\mathfrak{N}_i$, where $\mathfrak{N}_i$ is $\mathfrak{M}_i$-primary, $\mathfrak{M}_i \subset A$ being distinct maximal ideals of height n. Assume that J/J^2 is generated by n elements. Let w_J be a local orientation of J. Then w_J gives rise, in a natural way, to a local orientation $w_{\mathfrak{N}_i}$ of $\mathfrak{N}_i$. We associate to the pair (J, w_J), the element $\sum(\mathfrak{N}_i, w_{\mathfrak{N}_i})$ of G. By abuse of notation, we denote the element $\sum(\mathfrak{N}_i, w_{\mathfrak{N}_i})$ by (J, w_J).

Let H be the subgroup of G generated by the set of pairs (J, w_J), where J is an ideal of height n which is generated by n elements and w_J is a global orientation of J. We define the Euler class group of A denoted by $E(A)$, to be G/H. Thus $E(A)$ can be thought of as the quotient of the group of local orientations by the subgroup generated by global orientations.

Let J as above be an ideal of height n and $\alpha : (A/J)^n \twoheadrightarrow J/J^2$ be a surjection giving a local orientation w_J of J. Composing α with an automorphism $\lambda : (A/J)^n \to (A/J)^n$ such that $\det(\lambda) = \overline{a} \in (A/J)^*$, we obtain a local orientation $\alpha\lambda : (A/J)^n \to J/J^2$ which we denote by $(J, \overline{a}w_J)$.

Lemma 16.9 (see [13]) *Let A be a Noetherian ring of dimension $n \geq 2$, $J \subset A$ an ideal of height n and w_J a local orientation of J. Let $\overline{a} \in A/J$ be a unit. Then $(J, w_J) = (J, \overline{a^2}w_J)$ in $E(A)$.*

16.4 On a Lemma of Suslin

In this section, we give a proof of a lemma of Suslin. We begin with some preliminaries on extensions. Let E be an extension of an R-module A by an R-module C, that is, $E \in \mathrm{Ext}(C, A)$.

Lemma 16.10 (see [19]) *For $E \in Ext(C, A)$ and $\alpha : A \to A'$ there is an extension E' of A' by C (called the pushout) and a morphism $\Gamma = (\alpha, \beta, 1_C) : E \to E'$. The pair (Γ, E') is unique up to congruence.*

Proof We are required to fill in the diagram

$$\begin{array}{ccccccccc} E: 0 & \longrightarrow & A & \xrightarrow{\gamma} & B & \xrightarrow{\sigma} & C & \longrightarrow & 0 \\ & & \downarrow{\scriptstyle\alpha} & & \vdots{\scriptstyle\beta} & & \| & & \\ E': 0 & \longrightarrow & A' & \overset{\beta'}{\cdots\cdots>} & ? & \overset{\sigma'}{\cdots\cdots>} & C & \longrightarrow & 0 \end{array}$$

at the question mark and the dotted arrows so as to make the diagram commutative and the bottom row exact. To do so, take in $A' \oplus B$ the submodule N of all elements $(-\alpha(a), \gamma(a))$ for $a \in A$. At the question mark in the diagram put the quotient module $(A' \oplus B)/N$, and write elements of this quotient module as cosets $(a', b) + N$. Then the equations $\beta'(a') = (a', 0) + N$, $\sigma'[(a', b') + N] = \sigma b$ and $\beta b = (0, b) + N$ define the maps which satisfy the required conditions. The E' so constructed is unique up to congruence (natural isomorphism). □

Lemma 16.11 (Schanuel's lemma, see [11]) *Let $0 \to N \to P \to M \to 0$ and $0 \to N' \to P' \to M \to 0$ be two exact sequences of R-modules, where P and P' are projective. Then $P \oplus N' \simeq P' \oplus N$.*

Proof Consider the diagram

$$\begin{array}{ccccccccc} 0 & \longrightarrow & N & \xrightarrow{f} & P & \xrightarrow{g} & M & \longrightarrow & 0 \\ & & \vdots{\scriptstyle\theta'} & & \vdots{\scriptstyle\theta} & & \|{\scriptstyle I_M} & & \\ 0 & \longrightarrow & N' & \xrightarrow{f'} & P' & \xrightarrow{g'} & M & \longrightarrow & 0 \end{array}$$

Since P is projective, there exist a R-linear map $\theta : P \to P'$ such that $g'\theta = g$. Now θ induces an R-linear map $\theta' : N \to N'$.

Define $\psi : P \oplus N' \to P'$ by $\psi(x, y) = \theta(x) - f'(y)$, $x \in P$, $y \in N'$ and $\phi : N \to P \oplus N'$, by $\phi(y) = (f(y), \theta'(y))$. Then the sequence

$$0 \to N \xrightarrow{\phi} P \oplus N' \xrightarrow{\psi} P' \to 0$$

is exact. Since P' is projective, the sequence splits. Hence $P \oplus N' \simeq P' \oplus N$. □

Lemma 16.12 *Let A be a ring. Suppose $a, b \in A$ are such that a is a non zero divisor in A and b is a non-zero divisor in A/aA. Then $xa + yb = 0$ if and only if $x = \lambda b$, $y = -\lambda a$.*

Proof Suppose $xa + yb = 0$, by going modulo $\langle a \rangle$, we have $\overline{y}\overline{b} = 0$. As $\overline{b}$ is non-zero divisor in A/aA, this implies $\overline{y} = 0$, that is $y \in \langle a \rangle$. Hence $y = \lambda a$ for some $\lambda \in A$.

Now, since $xa + yb = 0$, that is $xa + \lambda ab = 0$, that is $(x + \lambda b)a = 0$. Since a is non-zero divisor, we have $x + \lambda b = 0$, which implies $x = -\lambda b$.

Conversely, if $x = \lambda b$ and $y = -\lambda a$ then $xa + yb = 0$. □

Definition 16.3 Let A be a ring. Then $a, b \in A$ satisfying the property of Lemma 16.12 is called a **regular sequence**.

Lemma 16.13 *Let A be a ring and $a, b \in A$ be a regular sequence and $J = (a, b)$. Let $s : A^2 \to J$ be a map given by $s(e_1) = a$ and $s(e_2) = b$. Then*

1. *$ker(s) = (b, -a)A = (be_1 - ae_2)$.*
2. *The sequence*

$$0 \to A \overset{\alpha}{\to} A^2 \overset{s}{\to} J \to 0$$

given by

$$\alpha(1) = (b, -a), s(e_1) = a, s(e_2) = b$$

is exact.

Proof The proof is easy to check. □

Example 16.1 An example on regular sequence Let $a, b \in A$ be a regular sequence and $J = (a, b)$. Let us consider the above exact sequence

$$0 \to A \overset{\alpha}{\to} A^2 \overset{s}{\longrightarrow} J \to 0,$$

where $s(e_1) = a$ and $s(e_2) = b$. We want to classify the projective A-modules P such that the sequence

$$0 \to A \overset{i}{\longrightarrow} P \overset{f}{\longrightarrow} J \to 0$$

is exact. The reason we want to classify such projective modules P is that by Lemma 16.11 such P satisfy $P \oplus A \simeq A^3$ and therefore P is free if a certain unimodular row is completable. We would like to identify the unimodular row corresponding to P.

Let $f(p_1) = a$ and $f(p_2) = b$, where $p_1, p_2 \in P$. We have a commutative diagram

$$\begin{array}{ccccccccc} 0 & \longrightarrow & A & \overset{\alpha}{\longrightarrow} & A^2 & \overset{s}{\longrightarrow} & J & \longrightarrow & 0 \\ & & & & \downarrow{\scriptstyle g} & & \downarrow{\scriptstyle Id} & & \\ 0 & \longrightarrow & A & \overset{i}{\longrightarrow} & P & \overset{f}{\longrightarrow} & J & \longrightarrow & 0. \end{array}$$

Let $g(e_1) = p_1$ and $g(e_2) = p_2$. Then we have a surjection $h = (i, g) : A \oplus A^2 \longrightarrow P$. Now, we want to compute ker(i, g). Let $(-\nu, \lambda, \mu) \in$ ker(i, g) then

$$\begin{aligned}
(i, g)(-\nu, \lambda, \mu) &= i(-\nu) + g(\lambda, \mu) \\
&= -i(\nu) + g(\lambda e_1) + g(\mu e_2) \\
&= -i(\nu) + \lambda p_1 + \mu p_2 \\
\lambda p_1 + \mu p_2 &= i(\nu) \\
f(\lambda p_1 + \mu p_2) &= f(i(\nu)) \\
\lambda f(p_1) + \mu f(p_2) &= 0 \\
\lambda a + \mu b &= 0.
\end{aligned}$$

So, $(\lambda, \mu) = d(b, -a)$ for some $d \in A$. Hence $(-\nu, \lambda, \mu) = (-\nu, d(b, -a)) = (-\nu, db, -da)$. Now, we write $\lambda p_1 + \mu p_2 = dbp_1 - dap_2 = d(bp_1 - ap_2)$. Again we have $\lambda p_1 + \mu p_2 = i(\nu)$, this implies $d(bp_1 - ap_2) = i(\nu)$. Then $\nu = di^{-1}(bp_1 - ap_2)$. Suppose $\nu' = i^{-1}(bp_1 - ap_2)$ then we have $\nu = d\nu'$. This implies that

$$(-\nu, \lambda, \mu) = (-d\nu', \lambda, \mu) = d(-\nu', b, -a).$$

Therefore, ker(i, g) is generated by $(-\nu', b, -a)$, where $\nu' = i^{-1}(bp_1 - ap_2)$.

So, we have an exact sequence

$$0 \to A \xrightarrow{j} A \oplus A^2 \xrightarrow{(i,g)} P \xrightarrow{f} J \to 0,$$

where $j(1) = (-\nu', b, -a)$. Therefore, we have $P \cong \frac{A \oplus A^2}{(-\nu', b, -a)}$. Let $-\nu' = u$ then $P \cong \frac{A \oplus A^2}{(u, b, -a)}$. Therefore, we get the following pushout (see Lemmas 16.10, 16.11) diagram:

$$\begin{array}{ccccccccc}
0 & \longrightarrow & A & \xrightarrow{\alpha} & A^2 & \xrightarrow{s} & J & \longrightarrow & 0 \\
 & & \downarrow{\scriptstyle u} & & \downarrow{\scriptstyle g} & & \| & & \\
0 & \longrightarrow & A & \xrightarrow{i} & P = \frac{A \oplus A^2}{(u, b, -a)} & \longrightarrow & J & \longrightarrow & 0.
\end{array}$$

Lemma 16.14 (Suslin, see [33, Lemma A.10]) *Let A be a ring and J be an ideal generated by two elements a, b. Suppose $(u, b, -a)$ is a unimodular row. Then the unimodular row $(u, b, -a)$ is completable if and only if there exists a set of generators c, d of the ideal J, where $c = \lambda_{12}a + \lambda_{13}b$ and $d = \lambda_{22}a + \lambda_{23}b$, with*

$$det \begin{pmatrix} \lambda_{12} & \lambda_{13} \\ \lambda_{22} & \lambda_{23} \end{pmatrix} = u^{-1} \; mod \; J.$$

Proof (1) We give our first proof by assuming the condition that $a, b \in A$ is a regular sequence. From Example 16.1, we have the following pushout diagram:

$$\begin{array}{ccccccccc} 0 & \longrightarrow & A & \xrightarrow{\alpha} & A^2 & \xrightarrow{s} & J & \longrightarrow & 0 \\ & & \Big\downarrow u & & \Big\downarrow g & & \Big\| & & \\ 0 & \longrightarrow & A & \xrightarrow{i} & P = \frac{A \oplus A^2}{(u,b,-a)} & \longrightarrow & J & \longrightarrow & 0. \end{array}$$

Suppose the unimodular row $(u, b, -a)$ is completable, then the projective module $P = \frac{A^3}{(u,b,-a)}$ is free. Then we have

$$\begin{array}{ccccccccc} 0 & \longrightarrow & A & \xrightarrow{\alpha} & A^2 & \xrightarrow{s} & J & \longrightarrow & 0 \\ & & \Big\downarrow u & & \Big\downarrow g & & \Big\| & & \\ 0 & \longrightarrow & A & \xrightarrow{i} & \frac{A \oplus A^2}{(u,b,-a)} & \xrightarrow{f} & J & \longrightarrow & 0, \end{array}$$

where $f|_{A^2} = s$ and $f(1, 0, 0) = 0$.

Since, the unimodular row $(u, b, -a)$ is completable, there exists a matrix $M \in SL_3(A)$ such that

$$M = \begin{pmatrix} u & b & -a \\ \lambda_{11} & \lambda_{12} & \lambda_{13} \\ \lambda_{21} & \lambda_{22} & \lambda_{23} \end{pmatrix}.$$

Since the rows $(u, b, -a)$, $(\lambda_{11}, \lambda_{12}, \lambda_{13})$, and $(\lambda_{21}, \lambda_{22}, \lambda_{23})$ generate A^3, $f(\lambda_{11}, \lambda_{12}, \lambda_{13})$ and $f(\lambda_{21}, \lambda_{22}, \lambda_{23})$ generate the ideal J. Also, we have $f(1, 0, 0) = 0$. Therefore, $s(\lambda_{12}, \lambda_{13})$ and $s(\lambda_{22}, \lambda_{23})$ generate J. Let $c = s(\lambda_{12}, \lambda_{13})$ and $d = s(\lambda_{22}, \lambda_{23})$. Then $\{c, d\}$ generates the ideal J. We have $c = \lambda_{12}a + \lambda_{13}b$ and $d = \lambda_{22}a + \lambda_{23}b$. Going modulo $J = (a, b)$ and computing the determinant of M we have

$$\overline{u} \det \begin{pmatrix} \lambda_{12} & \lambda_{13} \\ \lambda_{22} & \lambda_{23} \end{pmatrix} = \overline{1} \bmod J \text{ (since } M \in SL_3(A) \text{ and } J = (a, b)).$$

Conversely, suppose there exists a set of generators $\{c, d\}$ of the ideal J, where $c = \lambda_{12}a + \lambda_{13}b$, $d = \lambda_{22}a + \lambda_{23}b$, such that

$$\det \begin{pmatrix} \lambda_{12} & \lambda_{13} \\ \lambda_{22} & \lambda_{23} \end{pmatrix} = u^{-1} \bmod J,$$

or

$$u \det \begin{pmatrix} \lambda_{12} & \lambda_{13} \\ \lambda_{22} & \lambda_{23} \end{pmatrix} = 1 \bmod J.$$

We have to show that $(u, b, -a)$ is completable.

$$\text{Let } M = \begin{pmatrix} u & b & -a \\ \lambda_{11} & \lambda_{12} & \lambda_{13} \\ \lambda_{21} & \lambda_{22} & \lambda_{23} \end{pmatrix}.$$

Thus

$$\det(M) = u\det\begin{pmatrix}\lambda_{12} & \lambda_{13}\\ \lambda_{22} & \lambda_{23}\end{pmatrix} - \lambda_{11}\det\begin{pmatrix} b & -a\\ \lambda_{22} & \lambda_{23}\end{pmatrix} + \lambda_{21}\det\begin{pmatrix} b & -a\\ \lambda_{12} & \lambda_{13}\end{pmatrix}$$
$$= u(\lambda_{12}\lambda_{23} - \lambda_{13}\lambda_{21}) - \lambda_{11}(a\lambda_{22} + b\lambda_{23}) + \lambda_{21}(a\lambda_{12} + b\lambda_{13}).$$

Since $c = a\lambda_{12} + b\lambda_{13}$ and $d = a\lambda_{22} + b\lambda_{23}$, then

$$\det(M) = u(\lambda_{12}\lambda_{23} - \lambda_{13}\lambda_{21}) - \lambda_{11}d + \lambda_{21}c.$$

So, we have $\det(M) = u(\lambda_{12}\lambda_{23} - \lambda_{13}\lambda_{21}) = 1 \bmod J$. Now, since $J = (c, d)$, we can choose λ_{11} and λ_{21} such that $\det(M) = 1$. This implies that $(u, b, -a)$ is completable. □

Proof (2) Now we assume that (a, b) is not a regular sequence. Suppose $(u, b, -a)$ is completable and $J = (a, b)$. Assume that the projective A-module $P = \frac{A^3}{(u,b,-a)}$ is free. We have a surjection $f : \frac{A^3}{(u,b,-a)} \twoheadrightarrow J$ given by $f(e_1) = 0$, $f(e_2) = a$ and $f(e_3) = b$. Now, suppose that the matrix M, where

$$M = \begin{pmatrix} u & b & -a\\ \lambda_{11} & \lambda_{12} & \lambda_{13}\\ \lambda_{21} & \lambda_{22} & \lambda_{23}\end{pmatrix},$$

is a completion of the unimodular row $(u, b, -a)$ in $SL_3(A)$. Then rows $(u, b, -a)$, $(\lambda_{11}, \lambda_{12}, \lambda_{13})$, and $(\lambda_{21}, \lambda_{22}, \lambda_{23})$ generate A^3. Thus $\overline{(\lambda_{11}, \lambda_{12}, \lambda_{13})}$ and $\overline{(\lambda_{21}, \lambda_{22}, \lambda_{23})}$ generate $\frac{A^3}{(u,b,-a)}$. Since the map $f : \frac{A^3}{(u,b,-a)} \twoheadrightarrow J$ is surjective, $f(\overline{(\lambda_{11}, \lambda_{12}, \lambda_{13})})$ and $f(\overline{(\lambda_{21}, \lambda_{22}, \lambda_{23})})$ that generate J. As $f(\overline{e_1}) = f(e_1) = 0$, we have $f(\overline{(0, \lambda_{12}, \lambda_{13})})$ and $f(\overline{(0, \lambda_{22}, \lambda_{23})})$ generate J. That is, $(\lambda_{12} f(e_2) + \lambda_{13} f(e_3))$ and $(\lambda_{22} f(e_2) + \lambda_{23} f(e_3))$ generate J, that is, $\lambda_{12}a + \lambda_{13}b$ and $\lambda_{22}a + \lambda_{23}b$ generate J. Let $c = \lambda_{12}a + \lambda_{13}b$ and $d = \lambda_{22}a + \lambda_{23}b$. Then $J = (c, d)$ and

$$\begin{pmatrix} c\\ d\end{pmatrix} = \begin{pmatrix}\lambda_{12} & \lambda_{13}\\ \lambda_{22} & \lambda_{23}\end{pmatrix}\begin{pmatrix} a\\ b\end{pmatrix}.$$

Since $\det(M) = 1$, we have

$$u\det\begin{pmatrix}\lambda_{12} & \lambda_{13}\\ \lambda_{22} & \lambda_{23}\end{pmatrix} - b\det\begin{pmatrix}\lambda_{11} & \lambda_{13}\\ \lambda_{22} & \lambda_{23}\end{pmatrix} - a\det\begin{pmatrix}\lambda_{11} & \lambda_{12}\\ \lambda_{21} & \lambda_{22}\end{pmatrix} = 1.$$

This implies

$$\det\begin{pmatrix}\lambda_{12} & \lambda_{13}\\ \lambda_{22} & \lambda_{23}\end{pmatrix} = u^{-1} \bmod J.$$

The converse is proved as above. □

Corollary 16.1 (Suslin, (see [33, Lemma A.11])) *Let A be a ring and (u, a, b) be a unimodular row in A, where $u = u_1u_2$. Let J be the ideal (a, b) of the ring A. Suppose the following properties hold;*

(1) (u_1, a, b) is completable to a matrix in $SL_3(A)$.

(2) for any $c, d \in A$ such that (u_2, c, d) is unimodular, (u_2, c, d) is completable to a matrix in $SL_3(A)$.

Then (u, a, b) is completable to a matrix in $SL_3(A)$.

Proof Since $J = (a, b)$ and (u_1, a, b) is completable to a matrix in $SL_3(A)$. So, by Lemma 16.14, there exists a new set of generators $\{c, d\}$ of the ideal $J = (a, b)$ such that

$$\begin{pmatrix} c \\ d \end{pmatrix} = \begin{pmatrix} \lambda_{11} & \lambda_{12} \\ \lambda_{21} & \lambda_{22} \end{pmatrix} \begin{pmatrix} a \\ b \end{pmatrix} \tag{16.3}$$

and

$$\det \begin{pmatrix} \lambda_{11} & \lambda_{12} \\ \lambda_{21} & \lambda_{22} \end{pmatrix} = u_1^{-1} \bmod (a, b).$$

Now, look at the unimodular row (u_2, c, d), which is also completable (by (2)). Therefore, again by using Lemma 16.14, there exists a new set of generators $\{c', d'\}$ of the ideal $J = (c, d)$ such that

$$\begin{pmatrix} c' \\ d' \end{pmatrix} = \begin{pmatrix} \mu_{11} & \mu_{12} \\ \mu_{21} & \mu_{22} \end{pmatrix} \begin{pmatrix} c \\ d \end{pmatrix} \tag{16.4}$$

and

$$\det \begin{pmatrix} \mu_{11} & \mu_{12} \\ \mu_{21} & \mu_{22} \end{pmatrix} = u_2^{-1} \bmod (c, d).$$

Suppose

$$\begin{pmatrix} \lambda_{11} & \lambda_{12} \\ \lambda_{21} & \lambda_{22} \end{pmatrix} \begin{pmatrix} \mu_{11} & \mu_{12} \\ \mu_{21} & \mu_{22} \end{pmatrix} = \begin{pmatrix} \nu_{11} & \nu_{12} \\ \nu_{21} & \nu_{22} \end{pmatrix}.$$

Thus from (16.3) and (16.4), we get

$$\begin{pmatrix} c' \\ d' \end{pmatrix} = \begin{pmatrix} \mu_{11} & \mu_{12} \\ \mu_{21} & \mu_{22} \end{pmatrix} \begin{pmatrix} \lambda_{11} & \lambda_{12} \\ \lambda_{21} & \lambda_{22} \end{pmatrix} \begin{pmatrix} a \\ b \end{pmatrix} = \begin{pmatrix} \nu_{11} & \nu_{12} \\ \nu_{21} & \nu_{22} \end{pmatrix} \begin{pmatrix} a \\ b \end{pmatrix}$$

and

$$\det \begin{pmatrix} \mu_{11} & \mu_{12} \\ \mu_{21} & \mu_{22} \end{pmatrix} \det \begin{pmatrix} \lambda_{11} & \lambda_{12} \\ \lambda_{21} & \lambda_{22} \end{pmatrix} = \det \begin{pmatrix} \nu_{11} & \nu_{12} \\ \nu_{21} & \nu_{22} \end{pmatrix} = u_2^{-1} u_1^{-1} \bmod (a, b).$$

That is, we get a set of generators $\{c', d'\}$ of the ideal $J = (a, b)$ such that

$$\begin{pmatrix} c' \\ d' \end{pmatrix} = \begin{pmatrix} \nu_{11} & \nu_{12} \\ \nu_{21} & \nu_{22} \end{pmatrix} \begin{pmatrix} a \\ b \end{pmatrix}$$

and

$$\det\begin{pmatrix} v_{11} & v_{12} \\ v_{21} & v_{22} \end{pmatrix} = (u_1u_2)^{-1} \bmod (a, b).$$

Therefore, by the converse of Lemma 16.14, $(u_1u_2, a, b) = (u, a, b)$ is completable to a matrix in $SL_3(A)$. □

16.5 On Various Proofs of Seshadri's Theorem

In this section, we record various proofs of Seshadri's theorem [25].

Theorem 16.2 (Seshadri's Theorem Version 1) *Let A be a principal ideal domain and $\mathfrak{M} \subset A[T]$ be a maximal ideal of height 2. Suppose*

$$0 \to A[T] \to P \to \mathfrak{M} \to 0$$

is an exact sequence. Then

1. *$\mathfrak{M} \cap A = (s)$, where (s) is a maximal ideal of height 1 in A.*
2. *The image $\overline{\mathfrak{M}}$ of $\mathfrak{M}$ in $\frac{A}{(s)}[T]$ is principal.*
3. *$\overline{\mathfrak{M}} = (\overline{g(T)})$ and therefore $\mathfrak{M} = (s, g(T))$.*
4. *$P = \frac{A[T]^3}{(u(T),s,g(T))}$.*
5. *We have $(u(T), s, g(T))$ is completable and P is free.*

Proof 1. Since A is principal ideal domain, $\mathfrak{M} \cap A$ is principal ideal. Suppose $\mathfrak{M} \cap A$ is generated by an element s, that is $\mathfrak{M} \cap A = (s)$. Then (s) is a nonzero prime ideal. Therefore (s) is a maximal ideal of A.

2. Since $\frac{A}{(s)}$ is a field, so $\frac{A}{(s)}[T]$ is a principal ideal domain. This gives that the image $\overline{\mathfrak{M}}$ of $\mathfrak{M}$ in $\frac{A}{(s)}[T]$ is principal.
3. Since $\overline{\mathfrak{M}} = (\overline{g(T)})$ for some element $\overline{g(T)} \in \frac{A}{(s)}[T]$ taking a lift in $A[T]$ we have $\mathfrak{M} = (s, g(T))$.
4. This follows by Example 16.1.
5. Since $\frac{A}{(s)}$ is field and $(\overline{u(T)}, \overline{g(T)}) \overset{E_2(\frac{A}{(s)}[T])}{\sim} (\overline{1}, \overline{0})$. Therefore, there exists a matrix $\alpha \in E_2(\frac{A}{(s)}[T])$ such that

$$\alpha\begin{pmatrix} \overline{u(T)} \\ \overline{g(T)} \end{pmatrix} = \begin{pmatrix} \overline{1} \\ \overline{0} \end{pmatrix},$$

where bar denotes reduction modulo (s). Since the canonical map $A[T] \to \frac{A}{(s)}[T]$ is surjective, we can lift α to σ in $E_2(A[T])$. We, therefore, get

$$\sigma \begin{pmatrix} u(T) \\ g(T) \end{pmatrix} = \begin{pmatrix} 1 + sh_1(T) \\ sh_2(T) \end{pmatrix},$$

where $h_1(T), h_2(T) \in A[T]$. Therefore

$$\begin{pmatrix} 1 & 0 \\ 0 & \sigma \end{pmatrix} \begin{pmatrix} s \\ u(T) \\ g(T) \end{pmatrix} = \begin{pmatrix} s \\ 1 + sh_1(T) \\ sh_2(T) \end{pmatrix} \overset{E_3(A)}{\sim} \begin{pmatrix} 1 \\ 0 \\ 0 \end{pmatrix}.$$

$$\text{Hence } (u(T), s, g(T)) \overset{E_3(A)}{\sim} (1, 0, 0)$$

and P is free. □

Theorem 16.3 (see [25]) *Let k be a field and $(f_1(X, Y), f_2(X, Y), \dots, f_n(X, Y))$ be a unimodular row in $k[X, Y]$. Then $(f_1(X, Y), f_2(X, Y), \dots, f_n(X, Y))$ is completable to a matrix in $SL_n(k[X, Y])$.*

We prove this in the case where $n = 3$.

Theorem 16.4 (Seshadri's Theorem Version 2) *Let k be a field and $v(X, Y) = (u(X, Y), f(X, Y), g(X, Y))$ be a unimodular row in $k[X, Y]$. Then $v(X, Y)$ is completable.*

Proof Let $A = k[X, Y]$. We assume for simplicity that k is algebraically closed. If the ideal $(f(X, Y), g(X, Y)) = A$, then by Theorem 16.1, the row $(u(X, Y), f(X, Y), g(X, Y))$ is completable. If not, by adding suitable multiples of $u(X, Y)$ to $f(X, Y)$ and $g(X, Y)$ we can assume that $\text{ht}(f(X, Y), g(X, Y)) = 2$. Also note that the ideal $(f(X, Y), g(X, Y))$ of A is contained in finitely many maximal ideals. Assume without loss of generality that $(f(X, Y), g(X, Y)) \subset (X, Y)$, that is, $f(0, 0) = g(0, 0) = 0$ and that $l(\frac{A}{(f,g)})$ is finite. We have to show that $(u(X, Y), f(X, Y), g(X, Y))$ is completable. Put $Y = 0$ and since $f(0, 0) = 0 = g(0, 0)$, one can check that

$$(f(X, 0), g(X, 0)) \subset (X).$$

This implies that we can transform the row $(f(X, 0), g(X, 0))$ to $(0, X^t h(X))$ via elementary transformations, where $h(X) \in k[X]$. Performing the same transformations on $(f(X, Y), g(X, Y))$, we may assume $f(X, 0) = 0$. This implies $f(X, Y) = Yf'(X, Y)$, where $f'(X, Y) \in k[X, Y]$. Now, adding multiples of $f(X, Y)$ to $u(X, Y)$ assume $\text{ht}(u(X, Y), g(X, Y)) = 2$.

We want to show the row $(f(X, Y), u(X, Y), g(X, Y))$ is completable, that is, the projective module

$$\frac{A^3}{(f(X, Y), u(X, Y), g(X, Y))} = \frac{A^3}{(Yf'(X, Y), u(X, Y), g(X, Y))}$$

is free. We have a pushout diagram

$$\begin{array}{ccccccccc} 0 & \longrightarrow & A & \longrightarrow & A^2 & \longrightarrow & (u, g) & \longrightarrow & 0 \\ & & \downarrow{\scriptstyle Yf'} & & \downarrow & & \| & & \\ 0 & \longrightarrow & A & \longrightarrow & P & \longrightarrow & (u, g) & \longrightarrow & 0. \end{array}$$

This can be broken up as

$$\begin{array}{ccccccccc} 0 & \longrightarrow & A & \longrightarrow & A^2 & \longrightarrow & (u, g) & \longrightarrow & 0 \\ & & \downarrow{\scriptstyle f'} & & \downarrow & & \| & & \\ 0 & \longrightarrow & A & \longrightarrow & P' & \longrightarrow & (u, g) & \longrightarrow & 0 \\ & & \downarrow{\scriptstyle Y} & & \downarrow & & \| & & \\ 0 & \longrightarrow & A & \longrightarrow & P & \longrightarrow & (u, g) & \longrightarrow & 0, \end{array}$$

where $P' = A^3/(f', u, g) = A^3/(u, f', g)$ (by using Lemma 16.11). Since (f', g) has fewer solutions than (f, g), that is, $l(\frac{A}{(f',g)}) < l(\frac{A}{(f,g)})$, therefore by induction, $A^3/(u, f', g)$ is free. This implies that

$$0 \to A \to P' \to (u, g) \to 0$$

is equivalent to

$$0 \to A \to A^2 \xrightarrow{s'} (\widetilde{u}, \widetilde{g}) \to 0,$$

for some surjection s'. Now

$$0 \to A \to P \to (\widetilde{u}, \widetilde{g}) \to 0$$

is obtained as a pushout

$$\begin{array}{ccccccccc} 0 & \longrightarrow & A & \longrightarrow & A^2 & \xrightarrow{s'} & (\widetilde{u}, \widetilde{g}) & \longrightarrow & 0 \\ & & \downarrow{\scriptstyle Y} & & \downarrow & & \| & & \\ 0 & \longrightarrow & A & \longrightarrow & P & \longrightarrow & (\widetilde{u}, \widetilde{g}) & \longrightarrow & 0, \end{array}$$

where $P = A^3/(Y, \widetilde{u}, \widetilde{g})$ (again by using Lemma 16.11). Now, from Seshadri's Theorem Version 1, part (5), as $(Y, \widetilde{u}, \widetilde{g})$ is completable, so $P = A^3/(Y, \widetilde{u}, \widetilde{g})$ is free. This implies $\frac{A^3}{(f(X,Y), u(X,Y), g(X,Y))}$ is free. Hence $v(X, Y)$ is completable. □

Remark 16.1 Instead of using pushout diagrams one can use Corollary 16.1 instead where $u_1 = Y$, $u_2 = f'$ and go through the above proof.

We give some arguments similar to the one given by Seshadri to prove that projective modules over $k[X, Y]$ are free. We begin by sketching Roitman's proof of Seshadri's theorem (see [18]).

Theorem 16.5 (Seshadri's theorem) *Let k be an algebraically closed field and $v(X, Y) = (f_1(X, Y), f_2(X, Y), f_3(X, Y))$ be a unimodular row in $k[X, Y]$. Then $v(X, Y)$ is completable to a matrix in $SL_3(k[X, Y])$.*

Proof We remark that if we can show that, a row which is obtained from $v(X, Y)$ by performing elementary transformations is completable, then we can also show that $v(X, Y)$ is completable.

We have to show that $v(X, Y)$ is completable to a matrix in $SL_3(k[X, Y])$. We denote by $k(X)$ the quotient field of the ring $k[X]$. Then $k(X)[Y]$ is a Euclidean domain. Therefore $v(X, Y)$ is completable to a matrix $\alpha'(X, Y) \in SL_3(k(X)[Y])$. Now, multiplying the second and third rows of the matrix $\alpha'(X, Y)$ by suitable multiples of polynomials of $k[X]$, we get a matrix $\alpha(X, Y) \in M_3(k[X, Y])$, whose first row is $v(X, Y)$ and $\det(\alpha(X, Y)) \in k[X]$.

If $\det(\alpha(X, Y)) \in k^*$ then there exists a matrix in $GL_3(k[X, Y])$ whose first row is $v(X, Y)$. Let $\det(\alpha(X, Y)) = c$, then by multiplying the second or third row of the matrix $\alpha(X, Y)$ by c^{-1}, we can get a matrix in $SL_3(k[X, Y])$, whose first row is $v(X, Y)$. So, we are done in this case.

Otherwise, suppose $\det(\alpha(X, Y)) = g(X)$, for some polynomial $g(X) \in k[X]$. Since k is an algebraically closed field, $g(X)$ can be written as a product of linear polynomials in $k[X]$. So, we can write $g(X) = \lambda(X - \lambda_1)(X - \lambda_2) \cdots (X - \lambda_n)$. For simplicity, we assume that $\lambda_1 = 0$. Thus we have $g(X) = \lambda X(X - \lambda_2) \cdots (X - \lambda_n)$. Now, going modulo X namely, by putting $X = 0$, we have $v(0, Y) = (f_1(0, Y), f_2(0, Y), f_3(0, Y))$. Since $v(X, Y)$ is a unimodular row in $k[X, Y]$, therefore $v(0, Y)$ is a unimodular row in $k[Y]$. Since $k[Y]$ is a Euclidean domain, therefore $v(0, Y)$ can be transformed to $(1, 0, 0)$ using a matrix in $E_3(k[Y])$. That is, there exists a matrix $\varepsilon_1(Y) \in E_3(k[Y])$ such that $\varepsilon_1(Y)v(0, Y)^t = (1, 0, 0)^t$. Therefore, $\varepsilon_1(Y)\alpha^t(X, Y)$ is of the form

$$\alpha_1^t(X, Y) = \begin{pmatrix} 1 + Xh_{11}(X, Y) & * & * \\ Xh_{21}(X, Y) & * & * \\ Xh_{31}(X, Y) & * & * \end{pmatrix}.$$

By the preliminary remark, we may assume that

$$\alpha_1(0, Y) = \begin{pmatrix} 1 & 0 & 0 \\ a_{21}(Y) & a_{22}(Y) & a_{23}(Y) \\ a_{31}(Y) & a_{32}(Y) & a_{33}(Y) \end{pmatrix}.$$

Further, multiplying the first row by $-a_{21}(Y)$ and adding it to the second row and multiplying the first row by $-a_{31}(Y)$ and adding it to the third row, and performing the same transformations on $\alpha_1(X, Y)$ we may assume

$$\alpha_1(0, Y) = \begin{pmatrix} 1 & 0 & 0 \\ 0 & a_{22}(Y) & a_{23}(Y) \\ 0 & a_{32}(Y) & a_{33}(Y) \end{pmatrix}.$$

Since $k[Y]$ is a Euclidean domain, the greatest common divisor of $a_{22}(Y)$ and $a_{32}(Y)$ exists. Suppose g.c.d $(a_{22}(Y), a_{32}(Y)) = f'(Y)$, then there exists a matrix $\varepsilon_2(Y) \in E_2(k[Y])$ such that $\varepsilon_2(Y)\begin{pmatrix} a_{22}(Y) \\ a_{32}(Y) \end{pmatrix} = \begin{pmatrix} f'(Y) \\ 0 \end{pmatrix}$. Therefore, we have

$$\begin{pmatrix} 1 & 0 \\ 0 & \varepsilon_2(Y) \end{pmatrix} \begin{pmatrix} 1 & 0 & 0 \\ 0 & a_{22}(Y) & a_{23}(Y) \\ 0 & a_{32}(Y) & a_{33}(Y) \end{pmatrix} = \begin{pmatrix} 1 & 0 & 0 \\ 0 & f'(Y) & a'_{23}(Y) \\ 0 & 0 & a'_{33}(Y) \end{pmatrix}.$$

Now, since the $\det(\alpha_1(0, Y)) = 0$, we have $\det \begin{pmatrix} f'(Y) & a'_{23}(Y) \\ 0 & a'_{33}(Y) \end{pmatrix} = 0$. This implies that $a'_{33}(Y) = 0$. So, we have

$$\alpha_2(0, Y) = \begin{pmatrix} 1 & 0 & 0 \\ 0 & f'(Y) & a'_{23}(Y) \\ 0 & 0 & 0 \end{pmatrix},$$

where

$$\alpha_2(X, Y) = \begin{pmatrix} 1 & 0 \\ 0 & \varepsilon(Y) \end{pmatrix} \alpha_1(X, Y).$$

Therefore, by this procedure we obtain a matrix $\alpha_2(X, Y)$ whose first row is elementarily equivalent to $v(X, Y)$ and last row is divisible by X. By cancelling this X, we obtain a matrix $\gamma(X, Y)$ whose first row is elementarily equivalent to $v(X, Y)$ and $\det(\gamma(X, Y)) = \lambda(X - \lambda_2)(X - \lambda_3) \cdots (X - \lambda_n)$. Therefore by the induction on n the proof is complete. □

Remark 16.2 Let A be a principal ideal domain and $(f_1(X), f_2(X), \ldots, f_n(X))$ be a unimodular row in $A[X]$. Then a modification of the above proof shows that $(f_1(X), f_2(X), \ldots, f_n(X))$ is completable to a matrix in $SL_n(A[X])$.

Remark 16.3 Seshadri's theorem can also be proved as follows. This proof was inspired by [15] and is perhaps close to the original proof given by Seshadri for unimodular rows, which does not appear in Seshadri's paper. The first author learned about this proof of Seshadri from a lecture of Ramanan. Let us consider the following special case, which will illustrate the general proof.

Let k be a field and $v(X, Y) = (f_1(X, Y), f_2(X, Y), f_3(X, Y))$ be a unimodular row in $k[X, Y]$. Then we have to show that $v(X, Y)$ is completable to a matrix in $SL_3(k[X, Y])$. Suppose there is a matrix $\alpha(X, Y) \in M_3(k[X, Y])$ such that

$$\alpha(X,Y)=\begin{pmatrix} f_1(X,Y) & & \\ f_2(X,Y) & p_1(X,Y) & p_2(X,Y) \\ f_3(X,Y) & & \end{pmatrix}$$

and $\det(\alpha(X,Y))=X$. We will show how to construct a matrix $\beta(X,Y)\in SL_3(k[X,Y])$, whose first column is $v(X,Y)^t$. Going modulo X or putting $X=0$, we have the unimodular row $v(0,Y)$ and

$$\alpha(0,Y)=\begin{pmatrix} f_1(0,Y) & & \\ f_2(0,Y) & p_1(0,Y) & p_2(0,Y) \\ f_3(0,Y) & & \end{pmatrix}$$

such that $\det(\alpha(0,Y))=0$. Since $\det(\alpha(0,Y))=0$, there exist $\lambda_1(Y)$, $\lambda_2(Y)$ and $\lambda_3(Y)$ in $k[Y]$ such that $(\lambda_1(Y),\lambda_2(Y),\lambda_3(Y))\neq(0,0,0)$ and

$$\alpha(0,Y)\begin{pmatrix}\lambda_1(Y)\\ \lambda_2(Y)\\ \lambda_3(Y)\end{pmatrix}=\begin{pmatrix}0\\0\\0\end{pmatrix}.$$

This implies

$$\lambda_1(Y)v(0,Y)+\lambda_2(Y)p_1(0,Y)+\lambda_3(Y)p_2(0,Y)=0. \tag{16.5}$$

Therefore, we can write

$$\lambda_1(Y)v(X,Y)+\lambda_2(Y)p_1(X,Y)+\lambda_3(Y)p_2(X,Y)=Xw(X,Y), \tag{16.6}$$

where $w(X,Y)\in k[X,Y]$.

Now, suppose that the greatest common divisor of $\lambda_2(Y)$ and $\lambda_3(Y)$ is $\lambda(Y)$. So, we can write $\lambda_2(Y)=\lambda(Y)g_1(Y)$ and $\lambda_3(Y)=\lambda(Y)g_2(Y)$ for some $g_1(Y), g_2(Y)\in k[Y]$. Now, by going modulo $v(0,Y)$ to (16.5), we get $\lambda_2(Y)\overline{p_1(0,Y)}+\lambda_3(Y)\overline{p_2(0,Y)}=\overline{0}$ in $\frac{k[Y]^3}{v(0,Y)}$, where bar denotes reduction modulo $(v(0,Y))$. Substituting the value of $\lambda_2(Y)$ and $\lambda_3(Y)$, we get

$$\lambda(Y)(g_1(Y)\overline{p_1(0,Y)}+g_2(Y)\overline{p_2(0,Y)})=(\overline{0},\overline{0},\overline{0}) \text{ in } \frac{k[Y]^3}{v(0,Y)}.$$

Now, $\frac{k[Y]^3}{v(0,Y)}$ being projective is torsion free (by Lemma 16.8). So, we have $(g_1(Y)\overline{p_1(0,Y)}+g_2(Y)\overline{p_2(0,Y)})=(\overline{0},\overline{0},\overline{0})$ in $\frac{k[Y]^3}{v(0,Y)}$. So, without loss of generality, we may assume that $\lambda_2(Y)$ and $\lambda_3(Y)$ are relatively prime in $k[Y]$. Therefore there exists $\lambda_2'(Y),\lambda_3'(Y)\in k[Y]$ such that $\det\begin{pmatrix}\lambda_2(Y) & \lambda_2'(Y)\\ \lambda_3(Y) & \lambda_3'(Y)\end{pmatrix}=1$. Now, consider the matrix $\begin{pmatrix} 1 & \lambda_1(Y) & 0 \\ 0 & \lambda_2(Y) & \lambda_2'(Y) \\ 0 & \lambda_3(Y) & \lambda_3'(Y) \end{pmatrix}$ whose determinant is 1. Therefore the first column of the product

$$\begin{pmatrix} f_1(X, Y) & & \\ f_2(X, Y) & p_1(X, Y) & p_2(X, Y) \\ f_3(X, Y) & & \end{pmatrix} \begin{pmatrix} 1 & \lambda_1(Y) & 0 \\ 0 & \lambda_2(Y) & \lambda_2'(Y) \\ 0 & \lambda_3(Y) & \lambda_3'(Y) \end{pmatrix} \tag{16.7}$$

is $v(X, Y)^t$. Using (16.6), the second column of the above product is a multiple of X. Now, dividing the second column of the product (16.7) by X, we get a matrix in $SL_3(k[X, Y])$, whose first column is $v(X, Y)^t$. Hence $v(X, Y)$ is completable to a matrix in $SL_3(k[X, Y])$.

Remark 16.4 Let A be a Dedekind domain and $\mathfrak{M} \subset A[X]$ be a maximal ideal of height 2. Then $\mathfrak{M} \cap A$ is a prime ideal $\mathfrak{p}$ of A of height 1. Since A is a Dedekind domain, $\mathfrak{p}/\mathfrak{p}^2$ is generated by a single element $s \in \mathfrak{p}$. Since $A/\mathfrak{p}$ is a field, the image of $\mathfrak{M}$ in $A/\mathfrak{p}[T]$ is generated by $\overline{g(T)}$. Therefore $\mathfrak{M} = (\mathfrak{p}[T], g(T))$.

Since $\bar{s}$ generates $\mathfrak{p}/\mathfrak{p}^2$ by Lemma 16.7, $\mathfrak{M} = (s, h(T))$. Therefore by Example 16.1, any projective $A[T]$-module P of rank 2 and of trivial determinant mapping onto $\mathfrak{M}$ is given by $P \simeq \frac{A[T]^3}{(s, h(T), h_1(T))}$.

Since $\frac{A}{s}$ modulo it's nilradical is a product of fields, the unimodular row

$$(\overline{h(T)}, \overline{h_1(T)}) \overset{E_2(\frac{A}{s}[T])}{\sim} (1, 0).$$

This implies as we have seen before that P is free.

Theorem 16.6 (Seshadri [26], Bass [1, Section 22], Serre [24, Section 1]) *Let A be a Dedekind domain. Let $w = [v(T), f(T), g(T)]$ be a unimodular row in $A[T]$. Then w is completable.*

Proof Without loss of generality, we may assume that $\text{ht}(f(T), g(T)) = 2$. Let $\sqrt{(f(T), g(T))} = \mathfrak{M}_1 \cap \mathfrak{M}_2 \cap \cdots \cap \mathfrak{M}_t$, where $\mathfrak{M}_i \subset A[T]$ are maximal ideals. Reorder $\mathfrak{M}_1, \mathfrak{M}_2, \ldots, \mathfrak{M}_t$ such that $\mathfrak{M}_1 \cap A = \mathfrak{M}_2 \cap A = \cdots = \mathfrak{M}_r \cap A = \mathfrak{p}$ and $\mathfrak{M}_i \cap A = \mathfrak{p}_i$ is different from $\mathfrak{p}$ for $i > r$.

By using Prime Avoidance Lemma, we can choose $s \in \mathfrak{p}$, $s \notin \mathfrak{p}^2$ such that $(s) + \mathfrak{p}_i = A$ for all $i > r$. Let bar denote reduction modulo (s). Then since $(s) + \mathfrak{p}_i = A$ for $i > r$, we have $(s) + \mathfrak{M}_i = A[T]$ for $i > r$. Therefore $\sqrt{(\overline{f(T)}, \overline{g(T)})} = \overline{\mathfrak{M}_1} \cap \overline{\mathfrak{M}_2} \cap \cdots \cap \overline{\mathfrak{M}_r}$. First note that since $A/\mathfrak{p}$ is a field, the image of $\mathfrak{M}_i$ in $A/\mathfrak{p}[T]$ is principal. Now since $\mathfrak{p}/\mathfrak{p}^2$ is generated by s, $\mathfrak{M}_i = (s, h_i(T))$ for suitable $h_i(T) \in A[T]$ for $i = 1, 2, \ldots, r$ (see Lemma 16.7). Therefore, if bar denotes reduction modulo (s), then $\overline{\mathfrak{M}_i} = (\overline{h_i(T)})$ is principal. It follows from primary decomposition that $(\overline{f(T)}, \overline{g(T)})$ is a principal ideal.

Assume that $(\overline{f(T)}, \overline{g(T)}) = (\overline{h(T)})$. Then $\overline{f(T)} = \overline{h(T)} \cdot \overline{\lambda_1(T)}$ and $\overline{g(T)} = \overline{h(T)} \cdot \overline{\lambda_2(T)}$ and since $(\overline{f(T)}, \overline{g(T)}) = (\overline{h(T)})$. We have

$$\overline{h(T)} \cdot \overline{\lambda_1(T)} \cdot \overline{\lambda_1(T)'} + \overline{h(T)} \cdot \overline{\lambda_2(T)} \cdot \overline{\lambda_2(T)'} = \overline{h(T)}$$
$$\overline{h(T)}(\overline{1} - \overline{\lambda_1(T)} \cdot \overline{\lambda_1(T)}' - \overline{\lambda_2(T)} \cdot \overline{\lambda_2(T)}') = \overline{0} \tag{16.8}$$

Now, since $\mathrm{ht}(\mathfrak{M}_i) = 2$, for $i = 1, 2, \ldots, r$ and $(s, h(T)) = (s, f(T), g(T))$ has height 2, we have $\overline{h(T)}$ that does not belong to any minimal prime ideal of $\frac{A}{(s)}[T]$. Then by (16.8), we have $\overline{1} - \overline{\lambda_1(T)} \cdot \overline{\lambda_1(T)}' - \overline{\lambda_2(T)} \cdot \overline{\lambda_2(T)}'$ that belong to every minimal prime ideal of $\frac{A}{(s)}[T]$ and therefore is nilpotent. This implies that $\overline{\lambda_1(T)} \cdot \overline{\lambda_1(T)}' - \overline{\lambda_2(T)} \cdot \overline{\lambda_2(T)}' = 1 + a$, where $a \in \frac{A}{(s)}[T]$, is a nilpotent element. This means that $1 + a$ is a unit of $\frac{A}{(s)}[T]$.

Hence $(\overline{\lambda_1(T)}, \overline{\lambda_2(T)}) \in \mathrm{Um}_2(\frac{A}{(s)}[T])$. Since $\dim \frac{A}{(s)} = 0$, we have

$$(\overline{\lambda_1(T)}, \overline{\lambda_2(T)}) \overset{E_2(\frac{A}{(s)}[T])}{\sim} (\overline{1}, \overline{0}).$$

This implies

$$(\overline{f(T)}, \overline{g(T)}) = (\overline{h(T)} \cdot \overline{\lambda_1(T)}, \overline{h(T)} \cdot \overline{\lambda_2(T)}) \overset{E_2(\frac{A}{(s)}[T])}{\sim} (\overline{h(T)}, \overline{0}).$$

Therefore

$$(f(T), g(T)) \overset{E_2(A[T])}{\sim} (h'(T), s\widetilde{\lambda}(T)).$$

This implies

$$w = (v(T), f(T), g(T)) \overset{E_2(A[T])}{\sim} (v(T), h'(T), s\widetilde{\lambda}(T)).$$

To verify the hypotheses, we will use Corollary 16.1, with $u_1 = \widetilde{\lambda}(T)$ and $u_2 = s$ and the previous induction method of proof of Seshadri's theorem. The proof is complete provided we verify that the hypotheses of Corollary 16.1 are satisfied.

The hypotheses of the Corollary 16.1 are satisfied since A/s modulo nilpotents is a product of fields. Therefore, any unimodular row of length two in $\frac{A}{s}[T]$ is elementary equivalent to $(\overline{1}, \overline{0})$ and hence any unimodular row in $A[T]$ of length 3 with first entry s is completable. Further since

$$l\left(\frac{A[T]}{(h'(T), \widetilde{\lambda(T)})}\right) < l\left(\frac{A[T]}{(h'(T), s\widetilde{\lambda(T)})}\right) = l\left(\frac{A[T]}{(f(T), g(T))}\right),$$

we see by induction that $(v(T), h'(T), \widetilde{\lambda(T)})$ is completable. This proves the theorem. □

16.6 On a Result of Bhatwadekar–Keshari

In this section, we give a different proof of a lemma of Bhatwadekar–Keshari and use this to deduce via Suslin's lemma [33] the Swan–Towber [29], Krusemeyer [14], Suslin [27] theorem.

Lemma 16.15 (Bhatwadekar–Keshari [7, Lemma 3.3]) *Let A be a domain $f, g \in A$ and suppose that $(u, f, g) \in A^3$ is unimodular. Then there exist $f', g' \in J$ such that $J = (f', g')$ and $f' - uf \in J^2$, $g' - ug \in J^2$.*

Proof Since (u, f, g) is a unimodular row. Then there exist $v, \lambda, \mu \in A$ such that $uv + f\lambda + g\mu = 1$. Let $j = 1 - uv$, then $uv = 1 - j$ and in the ring A_{1-j}, u is a unit and $J_{1-j} = (uf, ug)$, where $J = (f, g)$.

Note since $j \in J = (f, g)$, then $1 \in (f, g)A_j$ and $J_j = A_j$. Therefore (f, g) is a unimodular in A_j, whereby there exists a matrix $\sigma \in SL_2(A_j)$ such that

$$\sigma \begin{pmatrix} f \\ g \end{pmatrix} = \begin{pmatrix} 1 \\ 0 \end{pmatrix}.$$

Next, $(uf, ug)A_{1-j} = J_{1-j}$ and hence

$$(uf, ug)A_{j(1-j)} = J_{j(1-j)} = A_{j(1-j)}.$$

Let $v' = \frac{v}{1-j}$ then $uv' = 1$. Since $(f, g) \in A_j^2$ is a unimodular row, there exists

$$\alpha = \begin{pmatrix} f & f' \\ g & g' \end{pmatrix} \in SL_2(A_j).$$

Then

$$\beta = \begin{pmatrix} uf & v'f' \\ ug & v'g' \end{pmatrix} \in SL_2(Aj(1-j)).$$

We can write β as

$$\beta = \begin{pmatrix} f & f' \\ g & g' \end{pmatrix} \begin{pmatrix} u & 0 \\ 0 & v' \end{pmatrix} = \alpha\varepsilon.$$

Also we have

$$\beta \begin{pmatrix} 1 \\ 0 \end{pmatrix} = \begin{pmatrix} uf \\ ug \end{pmatrix}.$$

Thus

$$\varepsilon^{-1}\alpha^{-1} \begin{pmatrix} uf \\ ug \end{pmatrix} = \begin{pmatrix} 1 \\ 0 \end{pmatrix}.$$

Let $\delta = \alpha^{-1}$ and $\varepsilon' = \varepsilon^{-1} \in E_2(A_{j(1-j)})$. Thus

$$\delta^{-1}\varepsilon'\delta\begin{pmatrix}uf\\ ug\end{pmatrix}=\delta^{-1}\begin{pmatrix}1\\ 0\end{pmatrix}.$$

Claim: $\delta^{-1}\varepsilon'\delta=\varepsilon_1\varepsilon_2$, where $\varepsilon_1\in SL_2(A_j)$, $\varepsilon_2\in SL_2(A_{1-j})$ and

$$\varepsilon_2=\begin{pmatrix}1+\mu_{11} & \mu_{12}\\ \mu_{21} & 1+\mu_{22}\end{pmatrix}, \tag{16.9}$$

where $\mu_{ij}\in J_{1-j}$.

Assuming the claim, we see that

$$\varepsilon_1\varepsilon_2\begin{pmatrix}uf\\ ug\end{pmatrix}=\delta^{-1}\begin{pmatrix}1\\ 0\end{pmatrix}.$$

Therefore

$$\varepsilon_2\begin{pmatrix}uf\\ ug\end{pmatrix}=\varepsilon_1^{-1}\delta^{-1}\begin{pmatrix}1\\ 0\end{pmatrix}. \tag{16.10}$$

Note that $\varepsilon_2\begin{pmatrix}uf\\ ug\end{pmatrix}\in A_{1-j}^2$ and since $\varepsilon_1^{-1}\delta^{-1}\in SL_2(A_j)$, we have $\varepsilon_1^{-1}\delta^{-1}\begin{pmatrix}1\\ 0\end{pmatrix}\in A_j^2$. Now patching together $\varepsilon_2\begin{pmatrix}uf\\ ug\end{pmatrix}$ and $\varepsilon_1^{-1}\delta^{-1}\begin{pmatrix}1\\ 0\end{pmatrix}$ in $A_{j(1-j)}$, we get (f',g') which generate J.

Now we claim that $f'-uf\in J^2$ and $g'-ug\in J^2$. To show this we take ε_2 as in (16.9). Therefore from (16.9) and (16.10), we get

$$(1+\mu_{11})(uf)+\mu_{12}(ug)=f';\ (\mu_{21})(uf)+(1+\mu_{22})(ug)=g'.$$

Multiplying by suitable power of $1-j$, to clear denominators, we get

$$(1+\nu_{11})(uf)+\nu_{12}(ug)=(1-j')f';\ (\nu_{21})(uf)+(1+\nu_{22})(ug)=(1-j')g'$$

for some $j'\in J$, where $\nu_{ij}\in J$. This implies that $f'-uf\in J^2$ and $g'-ug\in J^2$. This proves the lemma assuming the claim. □

Now we prove (16.9) as follows.

Lemma 16.16 (Mandal, see [20]) *Suppose $\nu(T)\in SL_2(A_{j(1-j)}[T])$ and $\nu(0)=I_2$. Then $\nu(1)=\varepsilon_1\varepsilon_2$, where $\varepsilon_1\in SL_2(A_j)$ and $\varepsilon_2\in SL_2(A_{1-j})$ is of the form $\begin{pmatrix}1+\mu_{11} & \mu_{12}\\ \mu_{21} & 1+\mu_{22}\end{pmatrix}$, where $\mu_{ij}\in J_{1-j}$.*

Proof Let $s=j$, $t=1-j$ and $\lambda,\mu\in A$ be chosen so that $\lambda s^k+\mu t^k=1$. Then by Lemma 16.5, $\nu(T)=\nu_1(T)\nu_2(T)$, where $\nu_1(T)\in SL_2(A_j[T])$ and $\nu_2(T)=\nu(\lambda s^kT)$. Let

$$\nu(T)=\begin{pmatrix}1+T\lambda_{11}(T) & T\lambda_{12}(T)\\ T\lambda_{21}(T) & 1+T\lambda_{22}(T)\end{pmatrix}$$

then by construction, $\nu_2(T) = \nu(\lambda s^k T)$, that is,

$$\nu_2(T) = \begin{pmatrix} 1 + \lambda s^k T \lambda_{11}(\lambda s^k T) & \lambda s^k T \lambda_{12}(\lambda s^k T) \\ \lambda s^k T \lambda_{21}(\lambda s^k T) & 1 + \lambda s^k T \lambda_{22}(\lambda s^k T) \end{pmatrix}.$$

Therefore $\nu(1) = \varepsilon_1 \varepsilon_2$, where $\varepsilon_1 \in SL_2(A_j)$ and $\varepsilon_2 = \nu_2(1)$ is of the required form (since $s = j$). □

Corollary 16.2 (Swan–Towber [29], Krusemeyer [14], Suslin [27]) *Let A be a ring $f, g \in A$. Let $v \in A$ be such that $(v^2, f, g) \in A^3$ is a unimodular row. Then (v^2, f, g) is completable.*

Proof Let $J = (f, g)$ and $u = v^{-1} \bmod (f, g)$. Then $\{\overline{uf}, \overline{ug}\}$ is a set of generators of J/J^2 which can be lifted to a set of generators of J, (Lemma 16.15). Suppose $\{f', g'\}$ is the lift of the set of generators $\{\overline{uf}, \overline{ug}\}$ of J/J^2 to a set of generators of J. Thus we have $f' = uf \bmod J^2$ and $g' = ug \bmod J^2$, that is, $f' - uf \in J^2$, $g' - ug \in J^2$ and $J = (f', g')$. Therefore, we can write $f' - uf = \lambda_{11} f + \lambda_{12} g$ and $g' - ug = \lambda_{21} f + \lambda_{22} g$, or $f' = (u + \lambda_{11}) f + \lambda_{12} g$ and $g' = \lambda_{21} f + (u + \lambda_{22}) g$, where $\lambda_{ij} \in J$. We can write this in matrix form as follows:

$$\begin{pmatrix} f' \\ g' \end{pmatrix} = \begin{pmatrix} u + \lambda_{11} & \lambda_{12} \\ \lambda_{21} & u + \lambda_{22} \end{pmatrix} \begin{pmatrix} f \\ g \end{pmatrix},$$

where

$$\det \begin{pmatrix} u + \lambda_{11} & \lambda_{12} \\ \lambda_{21} & u + \lambda_{22} \end{pmatrix} = u^2 + u\lambda_{11} + u\lambda_{22} + \lambda_{11}\lambda_{22} - \lambda_{21}\lambda_{12}.$$

Now, by going modulo J, we get $\det \begin{pmatrix} u + \lambda_{11} & \lambda_{12} \\ \lambda_{21} & u + \lambda_{22} \end{pmatrix} = u^2 \bmod J$. Therefore, we get a new set of generators $\{f', g'\}$ of the ideal J, where $f' = (u + \lambda_{11}) f + \lambda_{12} g$ and $g' = \lambda_{21} f + (u + \lambda_{22}) g$, such that

$$\det \begin{pmatrix} u + \lambda_{11} & \lambda_{12} \\ \lambda_{21} & u + \lambda_{22} \end{pmatrix} = u^2 \bmod J.$$

Hence by Lemma 16.14, (v^2, f, g) is completable, where $v = u^{-1} \bmod (f, g)$. □

More generally, we have

Corollary 16.3 *Let A be a ring $f, g \in A$ and $J = (f, g)$. Let $(u, f, g) \in A^3$ be a unimodular row. Suppose the set of generators $\{\overline{uf}, \overline{g}\}$ of J/J^2, can be lifted to a set of generators of J. Suppose $v = u^{-1}$ mod (f, g). Then (v, f, g) is completable.*

Proof We are given that $\{\overline{uf}, \overline{g}\}$ is a set of generators of J/J^2 which can be lifted to a set of generators of J. Suppose $\{f', g'\}$ is a lift of the set of generators $\{\overline{uf}, \overline{g}\}$ of J/J^2 such that $J = (f', g')$. Thus we have $f' = uf \bmod J^2$ and $g' = g \bmod J^2$, that is,

$f' - uf \in J^2$, $g' - g \in J^2$ and $J = (f', g')$. Therefore, we can write $f' - uf = \lambda_{11} f + \lambda_{12} g$ and $g' - g = \lambda_{21} f + \lambda_{22} g$, where $\lambda_{11}, \lambda_{12}, \lambda_{21}, \lambda_{22} \in J$. We can write this in matrix form as follows:

$$\begin{pmatrix} f' \\ g' \end{pmatrix} = \begin{pmatrix} u + \lambda_{11} & \lambda_{12} \\ \lambda_{21} & 1 + \lambda_{22} \end{pmatrix} \begin{pmatrix} f \\ g \end{pmatrix},$$

where

$$\det \begin{pmatrix} u + \lambda_{11} & \lambda_{12} \\ \lambda_{21} & 1 + \lambda_{22} \end{pmatrix} = u + \lambda_{11} + u\lambda_{22} + \lambda_{11}\lambda_{22} - \lambda_{21}\lambda_{12}.$$

Now, by going modulo J, we get: $\det \begin{pmatrix} u + \lambda_{11} & \lambda_{12} \\ \lambda_{21} & 1 + \lambda_{22} \end{pmatrix} = u \bmod J$. So, we get new set of generators $\{f', g'\}$ of the ideal J, where $f' = (u + \lambda_{11}) f + \lambda_{12} g$ and $g' = \lambda_{21} f + (1 + \lambda_{22}) g$, such that

$$\det \begin{pmatrix} u + \lambda_{11} & \lambda_{12} \\ \lambda_{21} & 1 + \lambda_{22} \end{pmatrix} = u \bmod J.$$

Hence by Lemma 16.14, (v, f, g) is completable, where $v = u^{-1} \bmod (f, g)$. □

16.7 Seshadri's Theorem and Euler Class Groups

In this section, we give a proof of Seshadri's theorem using the theory of Euler classes. The proof we give is inspired by an argument of Abhyankar.

Proposition 16.1 (Abhyankar, see [10, Proposition 6.2]) *Let k be an algebraically closed field and $I \subset k[X, Y] = A$ be an ideal such that $\dim A/I = 0$. Suppose that I/I^2 is generated by two elements. Then I is generated by two elements.*

Proof *(Sketch of the proof following* [4]*)* Let $f_1, f_2 \in I$ generate I/I^2. We may assume that $(f_1, f_2) = I \cap I'$, where $I + I' = A$ and $I' = \mathfrak{M}_1 \cap \ldots \cap \mathfrak{M}_r$, $r < \infty$, $\mathfrak{M}_i$ are maximal ideals of A. We have $\mathfrak{M}_i = (X - a_i, Y - b_i)$ and by a change of variables we may assume that $\mathfrak{M}_1 = (X, Y)$. The ideal $(f_1(X, 0), f_2(X, 0))$ of $k[X]$ is contained in (X). Therefore, by using the Euclidean algorithm we can transform $(f_1(X, 0), f_2(X, 0))$ to $(X^l h(X), 0)$, where $h(X) \in k[X]$.

In order to prove that I is generated by two elements we need to reduce the number of maximal ideals $\mathfrak{M}_i$ one by one. We do this as follows. Considering the elements of $E_2(k[X])$ as elements of $E_2(k[X, Y])$, we can transform the row $(f_1(X, Y), f_2(X, Y))$ to another row $(h_1(X, Y), h_2'(X, Y))$ such that

(1) The ideal $(f_1(X, Y), f_2(X, Y)) = (h_1(X, Y), h_2'(X, Y))$;

(2) $h_1(X, 0) = X^l h(X)$, $h_2'(X, 0) = 0$, i.e. $h_2'(X, Y) = Y h_2(X, Y)$.

This implies $(h_1(X, Y), Y h_2(X, Y)) = I \cap \mathfrak{M}_1 \cap \ldots \cap \mathfrak{M}_r$. Since $\mathfrak{M}_1 = (X, Y)$, by a linear change of variables, for example, replacing Y by $Y + cX$, we may assume in the beginning that the line $Y = 0$ does not pass through the finitely many points belonging to $V(I)$ and the points (a_i, b_i), $2 \le i \le r$ (notice that $Y \in \mathfrak{M}_1$). This implies in particular that the element $Y \in A$ is a unit modulo I.

Now, since elements $h_1(X, Y), Yh_2(X, Y)$ generate I/I^2 and $\mathfrak{M}_i/\mathfrak{M}_i^2, 2 \le i \le r$. We have $h_1(X, Y), h_2(X, Y)$ generate I/I^2 and $\mathfrak{M}_i/\mathfrak{M}_i^2$, $2 \le i \le r$. Since

$$(h_1(X, Y), Yh_2(X, Y)) + \mathfrak{M}_1^2 = \mathfrak{M}_1$$

and $\mathfrak{M}_1 = (X, Y)$, it follows that $Yh_2(X, Y) \notin \mathfrak{M}_1^2$ and hence $h_2(0, 0) \neq 0$. Hence $h_2(X, Y) \notin \mathfrak{M}_1$. It follows that

$$(h_1(X, Y), h_2(X, Y)) = I \cap \mathfrak{M}_2 \cap \ldots \cap \mathfrak{M}_r.$$

By continuing this process, we see that I is generated by two elements. □

Theorem 16.7 (Seshadri's Theorem for Euler class groups) *Let A be a Cohen–Macaulay domain of dimension 2. Let $f \in A$ be a non-zero prime element (that is, (f) is a prime ideal). Suppose that $A/(f)$ is a principal ideal domain with $SL_2(A/(f)) = E_2(A/(f))$. Let $J \subset A$ be an ideal of height 2 such that J/J^2 is generated by 2 elements and w_J be a local orientation of J. Suppose $(J, w_J) = 0$ in $E(A_f)$. Then $(J, w_J) = 0$ in $E(A)$.*

Proof *(Sketch)* We may assume as in [6, Lemma 5.6], that $J + (f) = A$, and there exist $b_1, b_2 \in J$ such that $(b_1, b_2) = J \cap J'$, where J' contains a power of f that is $f^n \in J$, and the local orientation of J given by b_1, b_2 is w_J.

If $n = 0$, that is, $f^0 = 1 \in J'$, then $J' = A$ and $J = (b_1, b_2)$ with $(J, w_J) = 0$ in $E(A)$. Now suppose that $n > 0$. Let bar denote reduction modulo (f). Then we may operate $(\overline{b_1}, \overline{b_2})$ by an element of $E_2(A/(f))$ to transform $(\overline{b_1}, \overline{b_2})$ by $(\overline{c_1}, \overline{0})$. Therefore we transform (b_1, b_2) to (c_1', c_2) via elementary transformations and assume that c_2 is multiple of f, that is $c_2 = fc_2'$. Then $(c_1', c_2') = J \cap J''$, where $l(A/J'') < l(A/J)$. The local orientation of J given by c_1', c_2' is $f^{-1}w_J$. Continuing this process we obtain a set of generators $\{d_1, d_2\}$ of J which give the local orientation $(J, f^{-k}w_J)$ that is $(J, f^{-k}w_J) = 0$ in $E(A)$.

If k is even, $(J, w_J) = 0$ in $E(A)$ (by Lemma 16.9). If k is odd, we have $(J, f^{-1}w_J) = 0$ in $E(A)$. Now by multiplying f^2 we get by Lemma 16.9, $(J, fw_J) = 0$ in $E(A)$ and fw_J is given by the set of generators $\{t_1, t_2\}$ of J. Then (J, w_J) is given by the set of generators $(f^{-1}t_1, t_2)$ of J/J^2. Using the fact that $SL_2(A/(f)) = E_2(A/(f))$. It follows that $(\overline{t_1}, \overline{t_2}) \overset{E_2(A/(f))}{\sim} (\overline{1}, \overline{0})$. Hence $(f, t_1, t_2) \overset{E_3(A)}{\sim} (1, 0, 0)$. Therefore (f, t_1, t_2) is completable and hence by Lemma 16.14, the set of generators $f^{-1}t_1, t_2$ of J/J^2 can be lifted to a set of generators of J showing that $(J, w_J) = 0$ in $E(A)$. □

Remark 16.5 If k is an algebraically closed field of characteristic $\neq 2$, then it is known that one can give a proof of Seshadri's theorem as follows:

Let $v(X, Y) = (u_1(X, Y), u_2(X, Y), u_3(X, Y))$ be a unimodular row in $A = k[X, Y]$. We can assume that $\text{ht}(u_2(X, Y), u_3(X, Y)) = 2$. Let $J = (u_2(X, Y),$

$u_3(X, Y))$. Then A/J modulo its nilradical is a product $k \times k \times k \times \cdots \times k$ of algebraically closed fields. Thus the element $\overline{u_1(X, Y)}$ of A/J is a square. Thus, it follows using the Swan–Towber, Krusemeyer, Suslin theorem that $v(X, Y)$ is completable.

Remark 16.6 Let A be a Noetherian ring with $\dim A = 2$. Let $(u, f, g) \in A^3$ be a unimodular row with $\mathrm{ht}(f, g) = 2$. Let $v = u^{-1} \bmod (f, g)$. Then the Euler class of (u, f, g) is given by the set of generators $\{vf, g\}$ of J/J^2. If the set of generators $\{vf, g\}$ of J/J^2 can be lifted to the set of generators of J then the Euler class of (u, f, g) is trivial and the unimodular row (u, f, g) is completable. This relates the lemma of Suslin to the theory of Euler class groups.

Remark 16.7 We describe the results schematically as follows:
Euler class groups $\longrightarrow$ Suslin's lemma $\longrightarrow$ Seshadri's theorem.
Suslin's lemma $\xrightarrow{\text{(via Bhatwadekar–Keshari)}}$ Swan–Towber, Krusemeyer, Suslin theorem $\longrightarrow$ Seshadri's theorem.
Swan–Towber, Krusemeyer, Suslin theorem $\xrightarrow{\text{(via Euler class groups)}}$ Seshadri's theorem.

Acknowledgements The authors would like to thank Professor Ravi A. Rao for his valuable support during this work. The authors would like to thank Professor Gopala Krishna Srinivasan for giving his time most generously and helping us make this paper more readable. The second named author would like to thank Professor Gopala Krishna Srinivasan for his support and advice during difficult times. The authors would also like to thank the referee for going through the paper carefully and pointing out some mistakes. The second named author also acknowledges the financial support from CSIR which enabled him to pursue his doctoral studies.

References

1. H. Bass, K-theory and stable algebra. Inst. Hautes Études Sci. Publ. Math. **22**, 5–60 (1964)
2. R. Basu, Topics in classical algebraic K-theory. PhD thesis, School of Mathematics, Tata Institute of Fundamental Research, Mumbai, 2006
3. R. Basu, R. Sridharan, On Forster's conjecture and related results. Punjab Univ. Res. J. (Sci.) **57**, 13–66 (2007)
4. S.M. Bhatwadekar, R. Sridharan, On Euler classes and stably free projective modules, *Algebra, Arithmetic and Geometry, Part I, II (Mumbai, 2000)*. Tata Institute of Fundamental Research Studies in Mathematics, vol. 16 (Tata Institute of Fundamental Research, Bombay, 2002), pp. 139–158
5. S.M. Bhatwadekar, R. Sridharan, Projective generation of curves in polynomial extensions of an affine domain and a question of Nori. Invent. Math. **133**(1), 161–192 (1998)
6. S.M. Bhatwadekar, R. Sridharan, The Euler class group of a Noetherian ring. Compos. Math. **122**(2), 183–222 (2000)
7. S.M. Bhatwadekar, M.K. Keshari, A question of Nori: projective generation of ideals. K-Theory, **28**(4), 329–351 (2003)
8. S.M. Bhatwadekar, H. Lindel, R.A. Rao, The Bass-Murthy question: Serre dimension of Laurent polynomial extensions. Invent. Math. **81**(1), 189–203 (1985)
9. S.M. Bhatwadekar, M.K. Das, S. Mandal, Projective modules over smooth real affine varieties. Invent. Math. **166**(1), 151–184 (2006)

10. M.K. Das, R. Sridharan, Good invariants for bad ideals. J. Algebra **323**(12), 3216–3229 (2010)
11. N.S. Gopalakrishnan, *Commutative Algebra* (1984)
12. S.K. Gupta, M.P. Murthy, *Suslin's Work on Linear Groups over Polynomial Rings and Serre Problem*. ISI Lecture Notes, vol. 8 (Macmillan Co. of India Ltd, New Delhi, 1980)
13. M.K. Keshari, Euler class group of a Noetherian ring. PhD thesis, School of Mathematics, Tata Institute of Fundamental Research, Mumbai, 2001
14. M. Krusemeyer, Skewly completable rows and a theorem of Swan and Towber. Commun. Algebra **4**(7), 657–663 (1976)
15. N.M. Kumar, On a theorem of Seshadri, *Connected at Infinity*. Texts and Readings in Mathematics, vol. 25 (Hindustan Book Agency, New Delhi, 2003), pp. 91–104
16. N.M. Kumar, Complete intersections. J. Math. Kyoto Univ. **17**(3), 533–538 (1977)
17. N.M. Kumar, A note on the cancellation of reflexive modules. J. Ramanujan Math. Soc. **17**(2), 93–100 (2002)
18. T.Y. Lam, *Serre's Problem on Projective Modules*, Springer Monographs in Mathematics (Springer, Berlin, 2006)
19. S. MacLane, *Homology*. Die Grundlehren der mathematischen Wissenschaften, vol. 114, 1st edn. (Springer, Berlin-New York, 1967)
20. S. Mandal, Homotopy of sections of projective modules. J. Algebraic Geom. **1**(4), 639–646 (1992). With an appendix by Madhav V. Nori
21. S. Mandal, *Projective modules and complete intersections*. Lecture Notes in Mathematics, vol. 1672 (Springer, Berlin, 1997)
22. M.P. Murthy, R.G. Swan, Vector bundles over affine surfaces. Invent. Math. **36**, 125–165 (1976)
23. D. Quillen, Projective modules over polynomial rings. Invent. Math. **36**, 167–171 (1976)
24. J.P. Serre, Sur les modules projectifs. SÃl'minaire Dubreil. AlgÃíbre et thÃl'orie des nombres **14**, 1–16 (1960–1961)
25. C.S. Seshadri, Triviality of vector bundles over the affine space K^2. Proc. Natl. Acad. Sci. USA **44**, 456–458 (1958)
26. C.S. Seshadri, Algebraic vector bundles over the product of an affine curve and the affine line. Proc. Am. Math. Soc. **10**, 670–673 (1959)
27. A.A. Suslin, Stably free modules. Mat. Sb. (N.S.) **102**(144)(4), 537–550, 632 (1977)
28. R.G. Swan, Algebraic vector bundles on the 2-sphere. Rocky Mt. J. Math. **23**(4), 1443–1469 (1993)
29. R.G. Swan, J. Towber, A class of projective modules which are nearly free. J. Algebra **36**(3), 427–434 (1975)
30. W. van der Kallen, A group structure on certain orbit sets of unimodular rows. J. Algebra **82**(2), 363–397 (1983)
31. W. van der Kallen, A module structure on certain orbit sets of unimodular rows. J. Pure Appl. Algebra **57**(3), 281–316 (1989)
32. L.N. Vaserstein, Stabilization of unitary and orthogonal groups over a ring with involution. Mat. Sb. (N.S.) **81**(123), 328–351 (1970)
33. L.N. Vaserstein, A.A. Suslin, Serre's problem on projective modules over polynomial rings, and algebraic K-theory. Izv. Akad. Nauk SSSR Ser. Mat. **40**(5), 993–1054, 1199 (1976)
34. C.T.C. Wall, *A Geometric Introduction to Topology* (Addison-Wesley Publishing Co, Reading, 1972)

Chapter 17
On a Group Structure on Unimodular Rows of Length Three over a Two-Dimensional Ring

Anjan Gupta, Raja Sridharan and Sunil K. Yadav

17.1 Introduction

Let Y be a set, G be a group and $Q : Y \to G$ be a one–one onto map of sets. Then, we can use the group structure on G to define a group structure on Y. This paper is based on the above idea.

Let A be a Noetherian domain of dimension 2 and Y be the orbit space $\mathrm{Um}_3(\mathrm{A})/\mathrm{SL}_3(\mathrm{A})$, where $\mathrm{Um}_3(\mathrm{A})$ is the set of unimodular rows of length 3 over A on which the group $\mathrm{SL}_3(\mathrm{A})$ acts. Then, it is proved in [6, 7], that $\varphi : \mathrm{Um}_3(\mathrm{A})/\mathrm{SL}_3(\mathrm{A}) \to \mathrm{E(A)}$ is a homomorphism of groups, where $\mathrm{E(A)}$ is the Euler class group of A (see [6, 7]) and the group structure on $\mathrm{Um}_3(\mathrm{A})/\mathrm{SL}_3(\mathrm{A})$ is defined either by using Bass' theory of Symplectic modules [28, Appendix], or using the Vaserstein symbol [33, Sect. 4]. Both these methods yield the same group structure on $\mathrm{Um}_3(\mathrm{A})/\mathrm{SL}_3(\mathrm{A})$.

In this paper, we give yet another method of giving a group structure on $\mathrm{Um}_3(\mathrm{A})/\mathrm{SL}_3(\mathrm{A})$, namely we use the set theoretic map φ to pull back the group structure on $\mathrm{E(A)}$ to obtain a group structure on $\mathrm{Um}_3(\mathrm{A})/\mathrm{SL}_3(\mathrm{A})$.

A. Gupta
Department of Mathematics, Institute of Science Education and Research Bhopal, Bhopal, India
e-mail: agmath@gmail.com

R. Sridharan
School of Mathematics, Tata Institute of Fundamental Research, 1, Dr. Homi Bhabha Road, Mumbai 400005, India
e-mail: sraja@math.tifr.res.in

S. K. Yadav (✉)
Department of Mathematics, Indian Institute of Technology Bombay, Powai, Mumbai 400076, India
e-mail: skyadav@math.iitb.ac.in; skymath.bhu@gmail.com

A. A. Ambily et al. (eds.), *Leavitt Path Algebras and Classical* K-*Theory*,
Indian Statistical Institute Series,
https://doi.org/10.1007/978-981-15-1611-5_17

It would be interesting to see if this method applies to unimodular rows of other sizes, thereby giving another way of understanding some of the results of van der Kallen (See [30, 31]).

In Sect. 17.2, we record some preliminaries. In Sects. 17.3 and 17.4, we recall some results of Bhatwadekar–Raja Sridharan on the Euler Class group of a Noetherian ring. In Sects. 17.5 and 17.6, we give the definition of a group structure on $\mathrm{Um}_3(A)/\mathrm{SL}_3(A)$ and prove that this structure defines a group.

17.2 Some Preliminaries

Definition 17.1

(i) Let A be a ring. A row $(a_1, a_2, \ldots, a_n) \in A^n$ is said to be **unimodular** of length n if the ideal generated by $a_1, a_2, \ldots, a_n$ is A. The set of unimodular rows of length n is denoted by $\mathrm{Um_n}(A)$.

(ii) A unimodular row $(a_1, a_2, \ldots, a_n)$ is said to be **completable** if there is a matrix in $\mathrm{SL_n}(A)$ whose first row is $(a_1, a_2, \ldots, a_n)$.

(iii) We define $\mathrm{E_n}(A)$ to be the subgroup of $\mathrm{GL_n}(A)$ generated by all matrices of the form $e_{ij}(\lambda) = I_n + \lambda E_{ij}, \lambda \in A, i \neq j$, where E_{ij} is a matrix whose (i, j)th entry is 1 and all other entries are zero. The matrices $e_{ij}(\lambda)$ will be referred to as elementary matrices.

Lemma 17.1 (Prime Avoidance Lemma, see [3]) *Let A be a ring $I \subset A$ an ideal. Suppose $I \subset \bigcup_{i=1}^{n} p_i$, where $p_i \in \mathrm{Spec}(A)$. Then $I \subset p_i$ for some i, $1 \leq i \leq n$.*

Lemma 17.2 (see [3]) *Let A be a ring, $p_1, p_2, \ldots, p_r \in \mathrm{Spec}(A)$ and $I = (a_1, a_2, \ldots, a_n)$ be an ideal of A such that $I \subsetneq p_i$, $1 \leq i \leq r$. Then there exist $b_2, b_3, \ldots, b_n \in A$ such that the element $c = a_1 + a_2b_2 + a_3b_3 + \cdots + a_nb_n \notin \bigcup_{i=1}^{r} p_i$.*

Since $\mathrm{M}_n(A)$ acts on A^n via matrix multiplication, the group $\mathrm{E}_n(A)$ which is a subset of $\mathrm{M}_n(A)$ also acts on A^n. This induces an action of $\mathrm{E}_n(A)$ on $\mathrm{Um}_n(A)$. The equivalence relation on $\mathrm{Um}_n(A)$ given by this action is denoted by $\overset{\mathrm{E_n}(A)}{\sim}$. Similarly one can define $\overset{\mathrm{SL}_n(A)}{\sim}$ and $\overset{\mathrm{GL}_n(A)}{\sim}$.

Theorem 17.1 (see [3])

(i) *Let A be a ring and $(a_1, a_2, \ldots, a_n) \in A^n$ be a unimodular row of length n which contains a unimodular row of shorter length. Then the row $(a_1, a_2, \ldots, a_n)$ is completable. In fact, $(a_1, a_2, \ldots, a_n) \overset{\mathrm{E_n}(A)}{\sim} (1, 0, \ldots, 0)$.*

(ii) *Let A be a semilocal ring. Then any unimodular row $(a_1, a_2, \ldots, a_n)$ of length $n \geq 2$ is completable. In fact, $(a_1, a_2, \ldots, a_n) \overset{\mathrm{E_n}(A)}{\sim} (1, 0, \ldots, 0)$.*

Definition 17.2 Two matrices α and β in $\mathrm{SL_n}(A)$ are said to be **connected** if there exists $\sigma(X) \in \mathrm{SL_n}(A[X])$ such that $\sigma(0) = \alpha$ and $\sigma(1) = \beta$. By considering the matrix $\sigma(1 - X)$, it follows that if α is connected to β then β is connected to α.

Lemma 17.3 *Any matrix in* $\mathrm{E_n(A)}$ *can be connected to the identity matrix.*

Proof Every matrix $\alpha \in \mathrm{E_n(A)}$ can be written as a product of elementary matrices of the form $e_{ij}(\lambda) = I_n + \lambda E_{ij}$ for $i \neq j$, that is, $\alpha = \prod_{i=1}^{r} e_{ij}(\lambda)$. We define $\sigma(X) = \prod_{i=1}^{r} e_{ij}(\lambda X)$. Then $\sigma(X) \in \mathrm{SL_n(A[X])}$, $\sigma(0) = I_n$ and $\sigma(1) = \alpha$. This proves the lemma. □

Lemma 17.4 *Let A be a ring and I be an ideal of A. Then the map* $\mathrm{E_n(A)} \rightarrow \mathrm{E_n(A/I)}$ *is surjective.*

Proof The proof follows from the fact that the generators $e_{ij}(\bar{\lambda})$ of $\mathrm{E_n(A/I)}$ for $\lambda \in A$, can be lifted to generators $e_{ij}(\lambda)$ of $\mathrm{E_n(A)}$. □

Let us recall Quillen's Splitting Lemma [23] with the proof following the exposition of [3]. In what follows, $(\psi_1(X))_t$ denotes the image of $\psi_1(X)$ in $\mathrm{GL_n(A_{st}[X])}$ and $(\psi_2(X))_s$ denotes the image of $\psi_2(X)$ in $\mathrm{GL_n(A_{st}[X])}$.)

Lemma 17.5 (see [23]) *Let A be a domain and* $s, t \in A$ *be such that* $sA + tA = A$. *Suppose there exists* $\sigma(X) \in \mathrm{GL_n(A_{st}[X])}$ *with the property that* $\sigma(0) = I_n$. *Then there exist* $\psi_1(X) \in \mathrm{GL_n(A_s[X])}$ *with* $\psi_1(0) = I_n$ *and* $\psi_2(X) \in \mathrm{GL_n(A_t[X])}$ *with* $\psi_2(0) = I_n$ *such that* $\sigma(X) = (\psi_1(X))_t(\psi_2(X))_s$.

Proof Since $\sigma(0) = I_n$, $\sigma(X) = I_n + X\tau(X)$, where $\tau(X) \in \mathrm{M_n(A_{st}[X])}$, we choose a large integer N_1 such that $\sigma(\lambda s^k X) \in \mathrm{GL_n(A_t[X])}$ for all $\lambda \in A$ and for all $k \geq N_1$. Define $\beta(X, Y, Z) \in \mathrm{GL_n(A_{st}[X, Y, Z])}$ as follows.

$$\beta(X, Y, Z) = \sigma((Y + Z)X)\sigma(YX)^{-1}. \tag{17.1}$$

Then $\beta(X, Y, 0) = I_n$, and hence there exists a large integer N_2 such that for all $k \geq N_2$ and for all $\mu \in A$ we have $\beta(X, Y, \mu t^k Z) \in \mathrm{GL_n(A_s[X, Y, Z])}$. This means

$$\beta(X, Y, \mu t^k Z) = (\sigma_1(X, Y, Z))_t, \tag{17.2}$$

where $\sigma_1(X, Y, Z) \in \mathrm{GL_n(A_s[X, Y, Z])}$ with $\sigma_1(X, Y, 0) = I_n$. Taking $N = \max(N_1, N_2)$, it follows by the comaximality of sA and tA that $s^N A + t^N A = A$. Pick $\lambda, \mu \in A$ such that $\lambda s^N + \mu t^N = 1$. Setting $Y = \lambda s^N$, $Z = \mu t^N$ in (17.1) and $Z = 1$, $Y = \lambda s^N$ in (17.2), we get

$$\beta(X, \lambda s^N, \mu t^N) = \sigma(X)\sigma(\lambda s^N X)^{-1}, \text{ and}$$

$$\beta(X, \lambda s^N, \mu t^N) = (\sigma_1(X, \lambda s^N, \mu t^N))_t = (\psi_1(X))_t,$$

where $\psi_1(X) \in \mathrm{GL_n(A_s[X])}$. Hence, we conclude that $\sigma(X)\sigma(\lambda s^N X)^{-1} = (\psi_1(X))_t$. Let $\sigma(\lambda s^N X) = (\psi_2(X))_s$, where $(\psi_2(X))_s \in \mathrm{GL_n(A_t[X])}$. Since $\sigma(0) = I_n$, $\psi_1(0) = \psi_2(0) = I_n$, the result follows by using the identity $\sigma(X) = \sigma(X)$ $\sigma(\lambda s^N X)^{-1}\sigma(\lambda s^N X)$. □

Remark 17.1 In above lemma by interchanging the roles of s and t we can write $\sigma(X) = (\tau_1(X))_s(\tau_2(X))_t$, where $\tau_1(X) \in GL_n(A_t[X])$ with $\tau_1(0) = I_n$ and $\tau_2(X) \in \mathrm{GL_n(A_s[X])}$ with $\tau_2(0) = I_n$.

Lemma 17.6 *Let A be a domain and $s, t \in A$ be such that $sA + tA = A$. If $\sigma_1 \in \mathrm{SL_n(A_s)}$, $\sigma_2 \in \mathrm{E_n(A_t)}$, then $\sigma_1\sigma_2 = \beta_1\beta_2$, where $\beta_1 \in \mathrm{SL_n(A_t)}$ and $\beta_2 \in \mathrm{SL_n(A_s)}$.*

Proof We can write $\sigma_1\sigma_2 = \sigma_1\sigma_2\sigma_1^{-1}\sigma_1$. Therefore, it suffices to show that $\sigma_1\sigma_2\sigma_1^{-1} = \gamma_1\gamma_2$, where $\gamma_1 \in \mathrm{SL_n(A_t)}$ and $\gamma_2 \in \mathrm{SL_n(A_s)}$. Then the result follows by setting $\beta_1 = \gamma_1$ and $\beta_2 = \gamma_2\sigma_1$. Since any elementary matrix can be connected to the identity matrix, we can find $\alpha(X) \in \mathrm{SL_n(A_t[X])}$ such that $\alpha(0) = I_n$ and $\alpha(1) = \sigma_2$. Let $\delta(X) = \sigma_1\alpha(X)\sigma_1^{-1}$. Then $\delta(1) = \sigma_1\sigma_2\sigma_1^{-1}$. Since $\delta(X) \in \mathrm{SL_n(A_{st}[X])}$ and $\delta(0) = I_n$, by Remark 17.1, $\delta(X) = \delta_1(X)\delta_2(X)$, where $\delta_1(X) \in \mathrm{SL_n(A_t[X])}$ and $\delta_2(X) \in \mathrm{SL_n(A_s[X])}$. Let $\gamma_1 = \delta_1(1)$ and $\gamma_2 = \delta_2(1)$. Hence the lemma follows. □

Lemma 17.7 (see [9]) *Let A be a domain and $s, t \in A$ be such that $sA + tA = A$. Let $\sigma \in \mathrm{SL_n(A_{st})}$ and $\varepsilon \in \mathrm{E_n(A_{st})}$. Then $\sigma\varepsilon = \tau_1\sigma\tau_2$, where $\tau_1 \in \mathrm{SL_n(A_s)}$ and $\tau_2 \in \mathrm{SL_n(A_t)}$.*

Proof Let $\varepsilon = \varepsilon_1\varepsilon_2$, where $\varepsilon_1 \in \mathrm{SL_n(A_s)}$ is chosen such that $\varepsilon_1 = I_n \bmod (t^N)$ for sufficiently large N and $\varepsilon_2 \in \mathrm{SL_n(A_t)}$. So, we have

$$\sigma\varepsilon = \sigma\varepsilon_1\varepsilon_2 = \sigma\varepsilon_1\sigma^{-1}\sigma\varepsilon_2.$$

Now, since $\varepsilon_1 = I_n \bmod (t^N)$ for sufficiently large N, therefore $\sigma\varepsilon_1\sigma^{-1} \in \mathrm{SL_n(A_s)}$. Now by taking $\tau_1 = \sigma\varepsilon_1\sigma^{-1}$ and $\tau_2 = \varepsilon_2$, we have $\sigma\varepsilon = \tau_1\sigma\tau_2$. □

Lemma 17.8 *Let A be a domain and I be an ideal of A. Let $a, c \in A$ be such that $aA + cA = A$. Then*

$$\begin{array}{ccc} I & \longrightarrow & I_a \\ \downarrow & & \downarrow \\ I_c & \longrightarrow & I_{ac} \end{array}$$

is a pullback diagram. This means that if two elements $x \in I_a$, $y \in I_c$ are equal in I_{ac}, then there exists a unique $z \in I$ such that $\frac{z}{1} = x$ in I_a and $\frac{z}{1} = y$ in I_b.

Proof Let $x = \frac{b}{a^r}$ and $y = \frac{d}{c^s}$ be such that $\frac{b}{a^r} = \frac{d}{c^s}$ in I_{ac}, where $b, c \in A$. Hence $bc^s = da^r$ in A. Since $aA + cA = A$, $a^rA + c^sA = A$. we choose $\lambda, \mu \in A$ such that $\lambda a^r + \mu c^s = 1$. Let $z = \lambda b + \mu d$. Then

$$a^r z = a^r\lambda b + a^r\mu d = a^r\lambda b + c^s\mu b = b(a^r\lambda + c^s\mu) = b \text{ and}$$

$$c^s z = c^s\lambda b + c^s\mu d = a^r\lambda d + c^s\mu d = d(a^r\lambda + c^s\mu) = d.$$

Hence we have $\frac{z}{1} = \frac{b}{a^r}$ in I_a and $\frac{z}{1} = \frac{d}{c^s}$ in I_c. The uniqueness is proved in a similar manner. □

17.3 On the Euler Class Group

In this section, we give the definition of the Euler class group of a Noetherian ring due to Bhatwadekar–Raja Sridharan and prove Lemma 17.9, [6, Lemma 5.3]. We follow the exposition of Manoj Keshari [13].

Let A be a Noetherian ring with dim $A = n \geq 2$. We define the Euler class group of A, denoted by E(A), as follows:

Let $J \subset A$ be an ideal of height n such that J/J^2 is generated by n elements. Let α and β be two surjections from $(A/J)^n$ to J/J^2. We say that α and β are related if there exists an automorphism σ of $(A/J)^n$ of determinant 1 such that $\alpha\sigma = \beta$. It is easy to see that this is an equivalence relation. If $\alpha : (A/J)^n \twoheadrightarrow J/J^2$ is a surjection, then by $[\alpha]$, we denote the equivalence class of α. We call such an equivalence class $[\alpha]$ a local orientation of J.

Since dimA/J = 0 and $n \geq 2$, we have $\mathrm{SL_n(A/J)} = \mathrm{E_n(A/J)}$ and therefore, the canonical map from $\mathrm{SL_n(A)}$ to $\mathrm{SL_n(A/J)}$ is surjective. Hence, if a surjection $\alpha : (A/J)^n \twoheadrightarrow J/J^2$ can be lifted to a surjection $\theta : A^n \twoheadrightarrow J$, and α is equivalent to $\beta : (A/J)^n \twoheadrightarrow J/J^2$, then β can also be lifted to a surjection from A^n to J. Let $\alpha\sigma = \beta$ for some $\sigma \in \mathrm{SL_n(A/J)}$. Since dim $A/J = 0$, there exists $\widetilde{\sigma} \in \mathrm{SL_n(A)}$ which is a lift of σ. Then $\theta\widetilde{\sigma} : A^n \twoheadrightarrow J$ is a lift of β.

A local orientation $[\alpha]$ of J is called a global orientation of J if the surjection $\alpha : (A/J)^n \twoheadrightarrow J/J^2$ can be lifted to a surjection $\theta : A^n \twoheadrightarrow J$.

We shall also, from now on, identify a surjection α with the equivalence class $[\alpha]$ to which α belongs.

Let $\mathfrak{M} \subset A$ be a maximal ideal of height n and $\mathfrak{N}$ be a $\mathfrak{M}$-primary ideal such that $\mathfrak{N}/\mathfrak{N}^2$ is generated by n elements. Let $w_{\mathfrak{N}}$ be a local orientation of $\mathfrak{N}$. Let G be the free abelian group on the set of pairs $(\mathfrak{N}, w_{\mathfrak{N}})$, where $\mathfrak{N}$ is a $\mathfrak{M}$-primary ideal and $w_{\mathfrak{N}}$ is a local orientation of $\mathfrak{N}$.

Let $J = \cap\mathfrak{N}_i$ be the intersection of finitely many ideals $\mathfrak{N}_i$, where $\mathfrak{N}_i$ is $\mathfrak{M}_i$-primary, $\mathfrak{M}_i \subset A$ being distinct maximal ideals of height n. Assume that J/J^2 is generated by n elements. Let w_J be a local orientation of J. Then w_J gives rise, in a natural way, to a local orientation $w_{\mathfrak{N}_i}$ of $\mathfrak{N}_i$. We associate to the pair (J, w_J), the element $\sum(\mathfrak{N}_i, w_{\mathfrak{N}_i})$ of G. By abuse of notation, we denote the element $\sum(\mathfrak{N}_i, w_{\mathfrak{N}_i})$ by (J, w_J).

Let H be the subgroup of G generated by the set of pairs (J, w_J), where J is an ideal of height n which is generated by n elements and w_J is a global orientation of J. We define the Euler class group of A denoted by E(A), to be G/H. Thus E(A) can be thought of as the quotient of the group of local orientations by the subgroup generated by global orientations.

Let J as above be an ideal of height n and $\alpha : (A/J)^n \twoheadrightarrow J/J^2$ be a surjection giving a local orientation w_J of J. Composing α with an automorphism $\lambda : (A/J)^n \to (A/J)^n$ such that $\det(\lambda) = \overline{a} \in (A/J)^*$, we obtain a local orientation $\alpha\lambda : (A/J)^n \to J/J^2$ which we denote by $(J, \overline{a}w_J)$.

Lemma 17.9 (see [13]) *Let A be a Noetherian ring of dimension $n \geq 2$, $J \subset A$ an ideal of height n and w_J a local orientation of J. Let $\overline{a} \in A/J$ be a unit. Then $(J, w_J) = (J, \overline{a^2} w_J)$ in* E(A).

17.4 On Some Results of Bhatwadekar–Raja Sridharan

In this section, we recall some results of Bhatwadekar–Raja Sridharan on the Euler class group. We follow [3, 5–7, 13].

Lemma 17.10 (The SL_2 Lemma) *Let A be a ring and let J be a proper ideal of A. Let $J = (a, b) = (c, d)$. Suppose $[a, b] = [c, d]$ mod J^2. Then there exists an automorphism Δ of A^2 such that*

(i) $\Delta \begin{pmatrix} a \\ b \end{pmatrix} = \begin{pmatrix} c \\ d \end{pmatrix}$,
(ii) $det(\Delta) = 1$.

Proof We have $a - c, b - d \in J^2$. So we can write $a - c = aa_1 + ba_2$ and $b - d = aa_3 + ba_4$, where $a_i \in J$ for $1 \leq i \leq 4$. Let $u = 1 - a_1, v = -a_2, w = -a_3$, and $x = 1 - a_4$. Then we have the following equation:

$$\begin{pmatrix} u & v \\ w & x \end{pmatrix} \cdot \begin{pmatrix} a \\ b \end{pmatrix} = \begin{pmatrix} c \\ d \end{pmatrix}.$$

Now, we see that $ux - vw = 1 - f$, for some $f \in J$. There exist $t_1, t_2 \in A$ such that $f = dt_2 - ct_1$. The endomorphism Δ of A^2 given by

$$\begin{pmatrix} u + bt_2 & v - at_2 \\ w + bt_1 & x - at_1 \end{pmatrix}$$

is an automorphism of determinant 1 with $\Delta \begin{pmatrix} a \\ b \end{pmatrix} = \begin{pmatrix} c \\ d \end{pmatrix}$. □

Corollary 17.1 *Let A be a ring and let J be a proper ideal of A. Let $J = (a, b) = (c, d)$. Suppose there exists a matrix $\beta \in$* SL₂(A/J) *such that $\beta \begin{pmatrix} \overline{a} \\ \overline{b} \end{pmatrix} = \begin{pmatrix} \overline{c} \\ \overline{d} \end{pmatrix}$ in J/J^2, then there exists a matrix $\alpha \in \mathrm{SL}_2(\mathrm{A})$ such that $\alpha \begin{pmatrix} a \\ b \end{pmatrix} = \begin{pmatrix} c \\ d \end{pmatrix}$.*

Proof We are given that there exists a matrix $\beta \in \mathrm{SL}_2(\mathrm{A/J})$ such that $\beta \begin{pmatrix} \overline{a} \\ \overline{b} \end{pmatrix} = \begin{pmatrix} \overline{c} \\ \overline{d} \end{pmatrix}$.

Therefore we can take $\beta = \begin{pmatrix} \overline{\lambda_1} & \overline{\lambda_2} \\ \overline{\lambda_3} & \overline{\lambda_4} \end{pmatrix} \in \mathrm{SL}_2(\mathrm{A/J})$ such that

$$\begin{pmatrix} \overline{\lambda_1} & \overline{\lambda_2} \\ \overline{\lambda_3} & \overline{\lambda_4} \end{pmatrix} \begin{pmatrix} \overline{a} \\ \overline{b} \end{pmatrix} = \begin{pmatrix} \overline{c} \\ \overline{d} \end{pmatrix}.$$

This implies $\overline{c} = \overline{\lambda}_1\overline{a} + \overline{\lambda}_2\overline{b}$ and $\overline{d} = \overline{\lambda}_3\overline{a} + \overline{\lambda}_4\overline{b}$. Now, taking lifts in A, we get $c = (\lambda_1 + \mu_1)a + (\lambda_2 + \mu_2)b$ and $d = (\lambda_3 + \mu_3)a + (\lambda_4 + \mu_4)b$, where $\mu_1, \mu_2, \mu_3, \mu_4 \in J$. We can write this in matrix form as

$$\begin{pmatrix} \lambda_1 + \mu_1 & \lambda_2 + \mu_2 \\ \lambda_3 + \mu_3 & \lambda_4 + \mu_4 \end{pmatrix} \begin{pmatrix} a \\ b \end{pmatrix} = \begin{pmatrix} c \\ d \end{pmatrix},$$

and $\det \begin{pmatrix} \lambda_1 + \mu_1 & \lambda_2 + \mu_2 \\ \lambda_3 + \mu_3 & \lambda_4 + \mu_4 \end{pmatrix} = 1 \pmod{J}$. Therefore, by the proof of the SL_2 lemma, there exists a matrix $\alpha \in \mathrm{SL}_2(\mathrm{A})$ such that $\alpha \begin{pmatrix} a \\ b \end{pmatrix} = \begin{pmatrix} c \\ d \end{pmatrix}$. □

Theorem 17.2 (Addition Principle) *Let A be a Noetherian ring of dimension* 2. *Let $J_1 = (a_1, a_2)$ and $J_2 = (b_1, b_2)$ be ideals of A with* $\mathrm{ht}(\mathrm{a}_1, \mathrm{a}_2) = 2 = \mathrm{ht}(\mathrm{b}_1, \mathrm{b}_2)$ *and $J_1 + J_2 = A$. Then $J_1 \cap J_2 = (c_1, c_2)$ such that $c_1 = a_1 \bmod J_1^2$, $c_2 = a_2 \bmod J_1^2$, $c_1 = b_1 \bmod J_2^2$ and $c_2 = b_2 \bmod J_2^2$. In particular if J_1, J_2 are generated by two elements then $J_1 \cap J_2$ is also generated by two elements.*

Proof Suppose $J_1 = (a_1, a_2)$ and $J_2 = (b_1, b_2)$. Since $\mathrm{ht}(\mathrm{J}_2) = 2$ and $\dim \mathrm{A} = 2$, it follows that J_2 is contained in finitely many maximal ideals $\mathfrak{m}_1, \mathfrak{m}_2, \ldots, \mathfrak{m}_r$ of A. Since $J_1 + J_2 = A$, J_1 not contained in $\mathfrak{m}_i$ for every i, that is $J_1 = (a_1, a_2)$ is not a subset of $\cup_{i=1}^r \mathfrak{m}_i$. Therefore by Lemma 17.2, we may choose $\lambda \in A$ such that $a_1 + \lambda a_2 \notin \cup_{i=1}^r \mathfrak{m}_i$. Then $J_1 = (a_1', a_2)$, where $a_1' = a_1 + \lambda a_2$ satisfies $(a_1') + \mathfrak{m}_i = A$ for $i = 1, 2, \ldots, r$ or $(a_1') + J_2 = A$ that is $(a_1', b_1, b_2) \in \mathrm{Um}_3(\mathrm{A})$.

As $(a_1', b_1, b_2) \in \mathrm{Um}_3(\mathrm{A})$, there exist $\lambda_1, \lambda_2, \lambda_3 \in A$ such that $\lambda_1 a_1' + \lambda_2 b_1 + \lambda_3 b_2 = 1$. This implies that $\lambda_2 b_1 + \lambda_3 b_2 = 1 - \lambda_1 a_1'$. This implies that $1 - \lambda_1 a_1' \in (b_1, b_2) = J_2$. Now let $c = \lambda_1 a_1'$ and $d = \lambda_2 b_1 + \lambda_3 b_2$. Then we have $c + d = 1$, where $c \in J_1$ and $d \in J_2$.

Now, $J_1 \cap J_2 = (a_1', a_2) \cap (b_1, b_2)$ and $c \in J_1$. This implies that $(J_1 \cap J_2)_c = (b_1, b_2)_c$. As $d \in (b_1, b_2)$, it follows that $(b_1, b_2) \in \mathrm{Um}_2(\mathrm{A}_\mathrm{d})$.

Since any unimodular row of length 2 is completable, we have $\tau \in \mathrm{SL}_2(\mathrm{A}_\mathrm{d})$ such that

$$\tau \begin{pmatrix} b_1 \\ b_2 \end{pmatrix} = \begin{pmatrix} 1 \\ 0 \end{pmatrix}. \tag{17.3}$$

Since, $J_1 \cap J_2 = (a_1', a_2) \cap (b_1, b_2)$ and $d \in J_2$, this implies that $(J_1 \cap J_2)_d = (a_1', a_2)_d$. Now, as c is a unit in A_c and $c = \lambda_1 a_1'$, so a_1' is also a unit in A_c. So there exists $\varepsilon_1 \in \mathrm{E}_2(\mathrm{A}_\mathrm{c})$ such that

$$\varepsilon_1 \begin{pmatrix} a_1' \\ a_2 \end{pmatrix} = \begin{pmatrix} 1 \\ 0 \end{pmatrix}. \tag{17.4}$$

Combining (17.3) and (17.4), we get

$$\tau^{-1}\varepsilon_1 \begin{pmatrix} a_1' \\ a_2 \end{pmatrix} = \begin{pmatrix} b_1 \\ b_2 \end{pmatrix}.$$

Let $\sigma = \tau^{-1}\varepsilon_1$, where $\tau \in SL_2(A_d)$ and $\varepsilon_1 \in E_2(A_c)$. Then by Lemma 17.6, there exist $\varepsilon_2 \in SL_2(A_c)$ and $\varepsilon_3 \in SL_2(A_d)$ such that $\tau^{-1}\varepsilon_1 = \varepsilon_2\varepsilon_3$. Then we have

$$\varepsilon_2\varepsilon_3 \begin{pmatrix} a_1' \\ a_2 \end{pmatrix} = \begin{pmatrix} b_1 \\ b_2 \end{pmatrix},$$

or

$$\varepsilon_3 \begin{pmatrix} a_1' \\ a_2 \end{pmatrix} = \varepsilon_2^{-1} \begin{pmatrix} b_1 \\ b_2 \end{pmatrix},$$

where $\varepsilon_3 \begin{pmatrix} a_1' \\ a_2 \end{pmatrix} \in (J_1 \cap J_2)_d$ and $\varepsilon_2^{-1} \begin{pmatrix} b_1 \\ b_2 \end{pmatrix} \in (J_1 \cap J_2)_c$. Now patching $\varepsilon_3 \begin{pmatrix} a_1' \\ a_2 \end{pmatrix}$ and $\varepsilon_2^{-1} \begin{pmatrix} b_1 \\ b_2 \end{pmatrix}$ and using Lemma 17.8, we have $J_1 \cap J_2$ is generated by two elements c_1, c_2. For the proof that we can choose generators c_1, c_2 such that $c_1 = a_1 \bmod J_1^2$, $c_2 = a_2 \bmod J_1^2$, $c_1 = b_1 \bmod J_2^2$ and $c_2 = b_2 \bmod J_2^2$, we refer to [6].

To understand the above proof better consider the following diagram:

$$\begin{array}{ccc} J_1 \cap J_2 & \longrightarrow & (J_1 \cap J_2)_c \\ \downarrow & & \downarrow \\ (J_1 \cap J_2)_d & \longrightarrow & (J_1 \cap J_2)_{cd} \end{array}.$$

We are patching generators of $(J_1 \cap J_2)_c$ and $(J_1 \cap J_2)_d$ via a split automorphism to obtain generators of $J_1 \cap J_2$. □

Theorem 17.3 (Subtraction Principle, see [13]) *Let A be a Noetherian domain with* $\dim A = 2$. *Let J_1 and J_2 be ideals of height 2 in A such that $J_1 + J_2 = A$. Suppose $J_1 = (a_1, a_2)$ and $J_1 \cap J_2 = (c_1, c_2)$ with $c_1 = a_1 \bmod J_1^2$ and $c_2 = a_2 \bmod J_1^2$. Then $J_2 = (b_1, b_2)$, where b_1, b_2 satisfy the property that $b_1 = c_1 \bmod J_2^2$ and $b_2 = c_2 \bmod J_2^2$.*

Proof We may assume by replacing a_1 with $a_1 + \lambda' a_2$ that $(a_1) + J_2 = A$. Choose $d \in J_2$ such that $\lambda a_1 + d = 1$, where $\lambda \in A$. Putting $\lambda a_1 = c$, we have $c + d = 1$. Then $(J_2)_c = (J_1 \cap J_2)_c$ (since $c \in J_1$), and this implies $(J_2)_c = (c_1, c_2)_c$. Since $d \in J_2$, $(J_2)_d = A_d$ and this implies $(J_2)_d$ is generated by $(1, 0)$. We have

$$(J_2)_{cd} = A_{cd} = (c_1, c_2)_{cd} = (1, 0)_{cd}.$$

The theorem will be proved (as in the case of the Addition principle) if we show that there is a matrix $\sigma \in \mathrm{SL}_2(\mathrm{A_{cd}})$ which splits such that $\sigma(c_1, c_2)^t = (1, 0)^t$.

We have $(c_1, c_2)_d = (a_1, a_2)_d$. Therefore by Lemma 17.10, there exists $\varepsilon_3 \in SL_2(A_d)$ such that

$$\varepsilon_3 \begin{pmatrix} a_1 \\ a_2 \end{pmatrix} = \begin{pmatrix} c_1 \\ c_2 \end{pmatrix}. \tag{17.5}$$

Further since $c = \lambda a_1$, there exists $\varepsilon_1 \in E_2(A_c)$ such that

$$\varepsilon_1 \begin{pmatrix} a_1 \\ a_2 \end{pmatrix} = \begin{pmatrix} 1 \\ 0 \end{pmatrix}. \tag{17.6}$$

Combining (17.5) and (17.6), we get

$$\varepsilon_1 \varepsilon_3^{-1} \begin{pmatrix} c_1 \\ c_2 \end{pmatrix} = \begin{pmatrix} 1 \\ 0 \end{pmatrix}.$$

Let $\sigma = \varepsilon_1 \varepsilon_3^{-1}$, where $\varepsilon_1 \in E_2(A_c)$ and $\varepsilon_3 \in SL_2(A_d)$. Then by Lemma 17.6, there exist $\tau \in SL_2(A_d)$ and $\varepsilon_2 \in SL_2(A_c)$ such that $\varepsilon_1 \varepsilon_3^{-1} = \tau \varepsilon_2$. Therefore, we have

$$\tau \varepsilon_2 \begin{pmatrix} c_1 \\ c_2 \end{pmatrix} = \begin{pmatrix} 1 \\ 0 \end{pmatrix},$$

or

$$\varepsilon_2 \begin{pmatrix} c_1 \\ c_2 \end{pmatrix} = \tau^{-1} \begin{pmatrix} 1 \\ 0 \end{pmatrix},$$

where $\varepsilon_2 \begin{pmatrix} c_1 \\ c_2 \end{pmatrix} \in (J_2)_c$ and $\tau^{-1} \begin{pmatrix} 1 \\ 0 \end{pmatrix} \in (J_2)_d$. Now patching $\varepsilon_2 \begin{pmatrix} c_1 \\ c_2 \end{pmatrix}$ and $\tau^{-1} \begin{pmatrix} 1 \\ 0 \end{pmatrix}$ and using Lemma 17.8, we get that J_2 is generated by two elements b_1, b_2. For the proof that we can choose generators such that $b_1 = c_1 \bmod J_2^2$ and $b_2 = c_2 \bmod J_2^2$, we refer to [6].

To understand the proof better consider the following diagram:

$$\begin{array}{ccc} J_2 & \longrightarrow & (J_2)_c \\ \downarrow & & \downarrow \\ (J_2)_d & \longrightarrow & (J_2)_{cd} \end{array}$$

Remark 17.2 Now we see how the proof of the Subtraction principle comes from that of the Addition principle. We look at the Addition principle. Let $J_1 = (a_1, a_2)$ and $J_2 = (b_1, b_2)$. By replacing a_1 to $a_1 + \lambda' a_2$ we may assume $(a_1) + J_2 = A$. Choose $d \in J_2$ such that $\lambda a_1 + d = 1$, where $\lambda \in A$. Put $\lambda a_1 = c$, we have $c + d = 1$.

Now, from the proof of the Addition principle we get $\varepsilon_1 \in E_2(A_c)$ such that $\varepsilon_1 \begin{pmatrix} a_1 \\ a_2 \end{pmatrix} = \begin{pmatrix} 1 \\ 0 \end{pmatrix}$ and $\tau \in SL_2(A_d)$ such that $\tau \begin{pmatrix} b_1 \\ b_2 \end{pmatrix} = \begin{pmatrix} 1 \\ 0 \end{pmatrix}$. Then we have $\tau^{-1}\varepsilon_1 \begin{pmatrix} a_1 \\ a_2 \end{pmatrix} = \begin{pmatrix} b_1 \\ b_2 \end{pmatrix}$. As $\varepsilon_1 \in E_2(A_c)$ and $\tau \in SL_2(A_d)$, by Lemma 17.6, there exist

$\varepsilon_2 \in SL_2(A_c)$ and $\varepsilon_3 \in SL_2(A_d)$ such that $\tau^{-1}\varepsilon_1 = \varepsilon_2\varepsilon_3$. Then we have $\varepsilon_2\varepsilon_3\begin{pmatrix}a_1\\a_2\end{pmatrix} = \begin{pmatrix}b_1\\b_2\end{pmatrix}$ or $\varepsilon_3\begin{pmatrix}a_1\\a_2\end{pmatrix} = \varepsilon_2^{-1}\begin{pmatrix}b_1\\b_2\end{pmatrix} = \begin{pmatrix}c_1\\c_2\end{pmatrix}$.

In the proof of the subtraction principle we are given (a_1, a_2) and (c_1, c_2) and we want (b_1, b_2). That is, we are given ε_3 and ε_1 and we want ε_2. To do this we rewrite the equation $\tau^{-1}\varepsilon_1 = \varepsilon_2\varepsilon_3$ as $\tau\varepsilon_2 = \varepsilon_1\varepsilon_3^{-1}$ and split $\varepsilon_1\varepsilon_3^{-1}$ to obtain ε_2 (and also τ).

Lemma 17.11 *Let A be a Noetherian ring of dimension* 2. *Let* (a_1, a_2, a_3) *and* (d_1, d_2, d_3) *be two unimodular rows of length* 3. *Then there exist matrices* $\sigma, \tau \in \mathrm{E}_3(\mathrm{A})$ *such that* $\sigma\begin{pmatrix}a_1\\a_2\\a_3\end{pmatrix} = \begin{pmatrix}a_1'\\a_2'\\a_3'\end{pmatrix}$ *and* $\tau\begin{pmatrix}d_1\\d_2\\d_3\end{pmatrix} = \begin{pmatrix}d_1'\\d_2'\\d_3'\end{pmatrix}$, *where* $\mathrm{ht}(\mathrm{a}_2', \mathrm{a}_3') = \mathrm{ht}(\mathrm{d}_2', \mathrm{d}_3') = 2$ *and* $(a_2', a_3') + (d_2', d_3') = A$.

Proof Adding suitable multiples of a_1 to a_2, a_3, we may transform (a_1, a_2, a_3) to (a_1, a_2', a_3') with $\mathrm{ht}(\mathrm{a}_2', \mathrm{a}_3') = 2$. Let $J = (a_2', a_3')$ and bar denote reduction modulo J. Then $(\overline{d_1}, \overline{d_2}, \overline{d_3}) \in \mathrm{Um}_3(\mathrm{A/J})$, and since $\mathrm{ht(J)} = 2$, $\dim \mathrm{A/J} = 0$, we can find $\tau' \in \mathrm{E}_3(\mathrm{A/J})$ such that $\tau'\begin{pmatrix}\overline{d_1}\\\overline{d_2}\\\overline{d_3}\end{pmatrix} = \begin{pmatrix}\overline{0}\\\overline{0}\\\overline{1}\end{pmatrix}$. Lifting τ' to an element of $\mathrm{E}_3(\mathrm{A})$ we find $\tau'' \in \mathrm{E}_3(\mathrm{A})$ such that $\tau''\begin{pmatrix}d_1\\d_2\\d_3\end{pmatrix} = \begin{pmatrix}\widetilde{d_1}\\\widetilde{d_2}\\\widetilde{d_3}\end{pmatrix}$. Thus $\widetilde{d_1} = 0 \bmod J$, $\widetilde{d_2} = 0 \bmod J$, $\widetilde{d_3} = 1 \bmod J$. Therefore, if $\lambda_2, \lambda_3 \in A$ then $\widetilde{d_2} + \lambda_2\widetilde{d_1} = 0 \bmod J$ and $\widetilde{d_3} + \lambda_3\widetilde{d_1} = 1 \bmod J$. Choosing λ_2, λ_3 suitably, we may assume that $\mathrm{ht}(\widetilde{d_2} + \lambda_2\widetilde{d_1}, \widetilde{d_3} + \lambda_3\widetilde{d_1}) = 2$.

Let $a_1' = a_1$, a_2', and a_3' be as above, $d_1' = \widetilde{d_1}$, $d_2' = \widetilde{d_2} + \lambda_2\widetilde{d_1}$, and $d_3' = \widetilde{d_3} + \lambda_3\widetilde{d_1}$. Then $\mathrm{ht}(\mathrm{a}_2', \mathrm{a}_3') = 2 = \mathrm{ht}(\mathrm{d}_2', \mathrm{d}_3')$. Further, since $d_3' = 1 \bmod (a_2', a_3')$, we have $(a_2', a_3') + (d_2', d_3') = A$. Thus the lemma follows. □

17.5 The Definition of the Group Structure on $\mathrm{Um}_3(\mathrm{A})/\mathrm{SL}_3(\mathrm{A})$

In this section, we define a certain operation "$*$" on the set $\mathrm{Um}_3(\mathrm{A})/\mathrm{SL}_3(\mathrm{A})$. We will show in the next two sections that "$*$" defines a group structure on $\mathrm{Um}_3(\mathrm{A})/\mathrm{SL}_3(\mathrm{A})$.

The method we adopt to define a group structure is described below. In [6, 7], it is proved that there is a well-defined group homomorphism $\varphi : \mathrm{Um}_3(\mathrm{A})/\mathrm{SL}_3(\mathrm{A}) \to \mathrm{E(A)}$. We use the fact that φ is well defined to define a group structure on $\mathrm{Um}_3(\mathrm{A})/\mathrm{SL}_3(\mathrm{A})$ as follows:

(Note that a unimodular row will always be denoted by parenthesis and its equivalence class by [,]. Thus the equivalence class of (a, b) is $[a, b]$.) Let $[v]$ and $[w]$

be two elements of $\mathrm{Um}_3(\mathrm{A})/\mathrm{SL}_3(\mathrm{A})$, we define $[v] * [w] = \varphi^{-1}[\varphi([v]) + \varphi([w])]$, where $+$ is the structure on $\mathrm{E}(\mathrm{A})$.

We begin by recalling the results from [7], where the map φ is well defined.

Let A be a Noetherian ring of dimension 2. Let $(a_1, a_2, a_3) \in \mathrm{Um}_3(\mathrm{A})$ and $P = \frac{A^3}{(a_1, a_2, a_3)}$. Let $s : P \twoheadrightarrow J$ be a surjective map with $\mathrm{ht(J)} = 2$. We have an induced surjection $\overline{s} : P/JP \twoheadrightarrow J/J^2$. Let bar denote reduction modulo J and choose $(\lambda_{11}, \lambda_{12}, \lambda_{13}), (\lambda_{21}, \lambda_{22}, \lambda_{23}) \in A^3$ such that $\begin{pmatrix} \overline{a_1} & \overline{a_2} & \overline{a_3} \\ \overline{\lambda_{11}} & \overline{\lambda_{12}} & \overline{\lambda_{13}} \\ \overline{\lambda_{21}} & \overline{\lambda_{22}} & \overline{\lambda_{23}} \end{pmatrix} \in SL_3(A/J) = E_3(A/J)$.

Let $\alpha : (A/J)^2 \twoheadrightarrow J/J^2$ be given as $\alpha(\overline{e_1}) = \overline{s}(\overline{\lambda_{11}}, \overline{\lambda_{12}}, \overline{\lambda_{13}})$ and $\alpha(\overline{e_2}) = \overline{s}(\overline{\lambda_{21}}, \overline{\lambda_{22}}, \overline{\lambda_{23}})$. Let $\varphi([a_1, a_2, a_3])$ be the set of generators of J/J^2 given by $\alpha(\overline{e_1})$, $\alpha(\overline{e_2})$. Then $\varphi([a_1, a_2, a_3])$ is an element of $E(A)$. It is proved in [6] that φ yields a well-defined homomorphism $\mathrm{Um}_3(\mathrm{A})/\mathrm{SL}_3(\mathrm{A})$ to $E(A)$.

We compute φ explicitly as follows:

Let $(a_1, a_2, a_3) \in \mathrm{Um}_3(\mathrm{A})$. By adding a suitable multiples of a_1 to a_2 and a_3 we may assume that height of (a_2, a_3) is 2. We define $s : \frac{A^3}{(a_1, a_2, a_3)} \twoheadrightarrow J = (a_2, a_3)$ given by $s(e_1) = 0$, $s(e_2) = a_3$, and $s(e_3) = -a_2$. Let $a_1 b_1 = 1 \bmod (a_2, a_3)$, then $\begin{pmatrix} \overline{a_1} & \overline{a_2} & \overline{a_3} \\ \overline{0} & \overline{b_1} & \overline{0} \\ \overline{0} & \overline{0} & \overline{1} \end{pmatrix} \in SL_3(A/(a_2, a_3))$. Using this we compute

$$\varphi([a_1, a_2, a_3]) = (\overline{s}(\overline{0}, \overline{b_1}, \overline{0}), \overline{s}(\overline{0}, \overline{0}, \overline{1})) = (\overline{b_1 a_3}, -\overline{a_2}) \in \mathrm{E(A)},$$

where $\overline{b_1 a_3}, -\overline{a_2}$ generate J/J^2, $J = (a_2, a_3)$ (we are identifying a surjection $(A/J)^2 \twoheadrightarrow J/J^2$ with the images of $\overline{e_1}$ and $\overline{e_2}$). Now since $\begin{pmatrix} \overline{b_1^{-1}} & \overline{0} \\ \overline{0} & \overline{b_1} \end{pmatrix} \in \mathrm{E_2(A/J)}$, $(\overline{b_1 a_3}, -\overline{a_2}) = (\overline{a_3}, -\overline{b_1 a_2})$ in $\mathrm{E(A)}$. Now, since $\begin{pmatrix} 0 & 1 \\ -1 & 0 \end{pmatrix} \in \mathrm{E_2(A)}$, $(\overline{a_3}, -\overline{b_1 a_2}) = (\overline{b_1 a_2}, \overline{a_3})$ in $\mathrm{E(A)}$ and since $(J, w_J) = (J, a_1^2 w_J)$ in $\mathrm{E(A)}$ (by Lemma 17.9), we have $(\overline{b_1 a_2}, \overline{a_3}) = (\overline{a_1 a_2}, \overline{a_3})$ in $\mathrm{E(A)}$.

Now suppose, we have two unimodular rows $v = (a_1, a_2, a_3)$ and $w = (b_1, b_2, b_3)$. Operating on v and w by elementary matrices, we may assume (by Lemma 17.11) that $\mathrm{ht}(a_2, a_3) = 2 = \mathrm{ht}(b_2, b_3)$ and $(a_2, a_3) + (b_2, b_3) = A$. Let u be chosen so that $u = a_1 \bmod (a_2, a_3)$ and $u = b_1 \bmod (b_2, b_3)$. Such a choice of u is possible by the Chinese remainder theorem. Then, since $u - a_1 \in (a_2, a_3)$, $(a_1, a_2, a_3) \overset{\mathrm{E_3A}}{\sim} (u, a_2, a_3)$, similarly $(b_1, b_2, b_3) \overset{\mathrm{E_3}(A)}{\sim} (u, b_2, b_3)$. Let $J_1 = (a_2, a_3)$, $J_2 = (b_2, b_3)$ and $(a_2, a_3) \cap (b_2, b_3) = (c_2, c_3) = J$ with $c_2 = a_2 \bmod J_1^2$, $c_3 = a_3 \bmod J_1^2$, $c_2 = b_2 \bmod J_2^2$, and $c_3 = b_2 \bmod J_2^2$. We have (u, c_2, c_3) is a unimodular row and

$$\begin{aligned}\varphi([u, c_2, c_3]) &= (J, w_J)\\ &= (\overline{uc_2}, \overline{c_3})\\ &= (u\overline{a_2}, a_3) + (u\overline{b_2}, \overline{b_3})\\ &= (J_1, w_{J_1}) + (J_2, w_{J_2})\\ &= \varphi([u, a_2, a_3]) + \varphi([u, b_2, b_3])\\ &= \varphi([a_1, a_2, a_3]) + \varphi([b_1, b_2, b_3]).\end{aligned}$$

Further by the injectivity of φ (Lemma 17.14), there exists a unique element $[\widetilde{v}]$ of $\mathrm{Um}_3(\mathrm{A})/\mathrm{SL}_3(\mathrm{A})$ such that $\varphi[\widetilde{v}] = \varphi([v]) + \varphi([w])$. It follows (see next paragraph) that there is a well defined group structure on $\mathrm{Um}_3(\mathrm{A})/\mathrm{SL}_3(\mathrm{A})$, by defining $[v] * [w] = \varphi^{-1}(\varphi([v]) + \varphi([w]))$, where $\varphi : \mathrm{Um}_3(\mathrm{A})/\mathrm{SL}_3(\mathrm{A}) \to \mathrm{E(A)}$ is the above map.

If $[v] \overset{\mathrm{SL}_3(\mathrm{A})}{\sim} [v']$ and $[w] \overset{\mathrm{SL}_3(\mathrm{A})}{\sim} [w']$, then we have

$$\varphi([v] * [w]) = \varphi([v]) + \varphi([w]) = \varphi([v']) + \varphi([w']) = \varphi([v'] * [w']).$$

Therefore by the injectivity of φ, $[v] * [w] = [v'] * [w']$ and hence φ is well defined.

The proof that "$*$" defines a group structure on E(A) will occupy the rest of this section and the next. In Sect. 17.6, we will show that the definition "$*$" is independent of the various choices that we here made.

(i) To identify the **identity element** with the above definition of the addition of two unimodular rows, let (u_1, a_1, a_2) and $(1, b_1, b_2)$ be two unimodular rows such that the ideals (a_1, a_2) and (b_1, b_2) are of height 2 and comaximal, that is $(a_1, a_2) + (b_1, b_2) = A$. Now, let $(a_1, a_2) \cap (b_1, b_2) = (c_1, c_2)$ with $a_1 = c_1 \bmod J_1^2$, $a_2 = c_2 \bmod J_1^2$ and $b_1 = c_1 \bmod J_2^2$, $b_2 = c_2 \bmod J_2^2$, where $J_1 = (a_1, a_2)$ and $J_2 = (b_1, b_2)$. By using the Chinese remainder theorem, we get an element $u \in A$ such that $u = u_1 \bmod J_1$ and $u = 1 \bmod J_2$. Therefore (u, c_1, c_2) is unimodular row. By definition $(u_1, a_1, a_2) * (1, b_1, b_2) = (u, c_1, c_2)$. By Theorem 17.6, (u, c_1, c_2) and (u_1, a_1, a_2) are in same $\mathrm{SL}_3(\mathrm{A})$ orbit. This proves that $(1, 0, 0) \overset{\mathrm{SL}_3(\mathrm{A})}{\sim} (1, b_1, b_2)$ acts as the identity element.

(ii) In order to determine the **inverse** of a given unimodular row (u, a_1, a_2), we prove the following theorem whose proof is motivated by [9, Lemma 3.6].

Theorem 17.4 *Let A be a domain and $(x, y, z) \in A^3$ be a unimodular row with* $\mathrm{ht}(\mathrm{y}, \mathrm{z}) = 2$. *Then the inverse of (x, y, z) under above operation is $(x, -y, z)$.*

Proof We have $(x, -y) + (y, z) = (x, y, z) = A$ and $(x, -y) + (z) = A$. Therefore we can choose $\lambda \in A$ such that $-y + \lambda^2 x^2 = y'$ does not belong to the maximal ideals of A containing (y, z) and does not belong to the height 1 prime ideals of A containing (z). Therefore, we have

(1) $(x, y', z) = (x, -y + \lambda^2 x^2, z)$ is unimodular.
(2) $(y') + (y, z) = A$, hence $(y', z) + (y, z) = A$.
(3) $\mathrm{ht}(\mathrm{y}', \mathrm{z}) = 2$.

First note that (x, yy', z) is unimodular and

$$(x, yy', z) = (x, -y^2 + \lambda^2 x^2 y, z) \overset{E_3(A)}{\sim} (x, -y^2, z) \overset{SL_3(A)}{\sim} (1, 0, 0).$$

Now, we prove that the element of E(A) associated to (x, yy', z) is the sum of elements in E(A) associated to (x, y, z) and (x, y', z). Since $(x, yy', z) \overset{SL_3(A)}{\sim} (1, 0, 0)$, this will show that the inverse of (x, y, z) is $(x, -y, z)$.

Let $J_1 = (y, z)$ and $J_2 = (y', z)$. The element of E(A) is associated to $(x, y, z) = (J_1, w_{J_1})$, where w_{J_1} is the set of generators of J_1/J_1^2 given by $x^{-1}y, z$ or xy, z (by Lemma 17.9). The element of E(A) is associated to $(x, y', z) = (J_2, w_{J_2})$, where w_{J_2} is the set of generators of J_2/J_2^2 given by $x^{-1}y', z$ or xy', z.

Let us compute the element of $E(A)$ associated to (x, yy', z). Since $(y, z) + (y', z) = A$, going modulo (z), we see that $(\overline{y}) \cap (\overline{y'}) = (\overline{yy'})$, and hence $(y, z) \cap (y', z) = (yy', z)$. We have $yy' = -y^2 + \lambda^2 x^2 y$ and $yy' = \lambda^2 x^2 y \bmod J_1^2$. Also

$$yy' = y'(-y' + \lambda^2 x^2) = (-y'^2 + \lambda^2 x^2 y') = \lambda^2 x^2 y' \bmod J_2^2.$$

Hence the element of E(A) associated to (x, yy', z) is $(J_1, \widetilde{w_{J_1}}) + (J_2, \widetilde{w_{J_2}})$, where $\widetilde{w_{J_1}}$ is the set of generators of J_1/J_1^2 given by $(x\lambda^2 x^2 y, z)$ and $\widetilde{w_{J_2}}$ is the set of generators of J_2/J_2^2 given by $(x\lambda^2 x^2 y', z)$. Further

$$(\lambda^2 x^2, y, z) = (-y + \lambda^2 x^2, y, z) = (y', y, z) = A.$$

and

$$(\lambda^2 x^2, y', z) = (\lambda^2 x^2, -y + \lambda^2 x^2, z) = (\lambda^2 x^2, y, z) = A.$$

Hence $\lambda^2 x^2$ which is a square is a unit modulo (y, z) and (y', z). Hence by Lemma 17.9,

$$(J_1, \widetilde{w_{J_1}}) + (J_2, \widetilde{w_{J_2}}) = (J_1, w_{J_1}) + (J_2, w_{J_2})$$

in $E(A)$ and the element of E(A) associated to the completable row (x, yy', z) is the sum of the elements in E(A) associated to (x, y, z) and (x, y', z). □

17.6 The Conclusion of the Proof of the Existence of a Group Structure

In this section, we complete the proof of the existence of a group structure on $Um_3(A)/SL_3(A)$. We prove Theorem 17.6 which implies the existence of an identity element and Lemma 17.14 which implies that the map $\varphi : Um_3(A)/SL_3(A) \to E(A)$ is injective. We begin with

Theorem 17.5 *Let A be a Noetherian domain with* $\dim A = 2$. *Let $J \subset A$ be an ideal of height 2 such that $J = (a_1, a_2) = (b_1, b_2)$. Let $u_1, u_1' \in A$ be units modulo J and $u_1 v_1 = u_1' v_1' = 1$ mod J. Suppose the generators $\{\overline{v_1 a_1}, \overline{a_2}\}$ and $\{\overline{v_1' b_1}, \overline{b_2}\}$ of J/J^2 are connected by a matrix $\sigma \in \mathrm{SL}_2(\mathrm{A/J})$, that is,*

$$\sigma \begin{pmatrix} \overline{v_1 a_1} \\ \overline{a_2} \end{pmatrix} = \begin{pmatrix} \overline{v_1' b_1} \\ \overline{b_2} \end{pmatrix}.$$

Then we have an isomorphism of projective A-modules

$$\frac{A^3}{(u_1, a_2, -a_1)} \simeq \frac{A^3}{(u_1', b_2, -b_1)}.$$

In fact, there exists a matrix $\alpha \in \mathrm{SL}_3(\mathrm{A})$ such that

$$\alpha \begin{pmatrix} u_1 \\ a_2 \\ -a_1 \end{pmatrix} = \begin{pmatrix} u_1' \\ b_2 \\ -b_1 \end{pmatrix}.$$

Proof We have $u_1 v_1 = 1 - j_1$ and $u_1' v_1' = 1 - j_2$ with $j_1, j_2 \in J$. Let

$$(1 - j_1)(1 - j_2) = 1 - j,\ P = \frac{A^3}{(u_1, a_2, -a_1)} \text{ and } Q = \frac{A^3}{(u_1', b_2, -b_1)}.$$

Now, we compute the cocycle associated to the projective module $P = \frac{A^3}{(u, a_2, -a_1)}$. We have a surjection $f : \frac{A^3}{(u, a_2, -a_1)} \twoheadrightarrow (a_1, a_2) = J$ given by $f(e_1) = 0,\ f(e_2) = a_1,\ f(e_3) = a_2$. This gives a surjection $f_{1-j} : P_{1-j} \twoheadrightarrow J_{1-j}$.

Since $u_1 v_1 = 1 - j_1$ and $(1 - j_1)(1 - j_2) = 1 - j$, we have $u_1 v' = 1$, where $v' = \frac{v_1(1-j_2)}{1-j}$. So we can take a completion of $(u_1, a_2, -a_1)$ in $SL_3(A_{1-j})$ as $\begin{pmatrix} u_1 & a_2 & -a_1 \\ 0 & v' & 0 \\ 0 & 0 & 1 \end{pmatrix}$. Let $q_1 = (0, v', 0)$ and $q_2 = (0, 0, 1)$, then $\{q_1, q_2\}$ is a basis of the free module P_{1-j}.

Since $j \in (a_1, a_2)$, so $(J)_j = A_j$. This implies that the unimodular row $(u, a_2, -a_1)$ in A_j consists a unimodular row of shorter length. Hence $(u, a_2, -a_1)$ is completable in A_j and hence P_j is free. Choose a basis $\{p_1, p_2\}$ of P_j such that $f_j(p_1) = 1,\ f_j(p_2) = 0$. For this purpose we choose any basis of P_j. The image of this basis is a unimodular row in A_j. We transform the unimodular row to $(1, 0)$ via a matrix in $\mathrm{SL_2(A_j)}$ and correspondingly transform the basis of P_j to obtain p_1, p_2 such that $f_j(p_1) = 1,\ f_j(p_2) = 0$. In particular, we have $\begin{pmatrix} u & a_2 & -a_1 \\ & p_1 & \\ & p_2 & \end{pmatrix} \in \mathrm{GL_3(A_j)}$.

We can modify p_2 by a unit in A_j and assume that the condition $f_j(p_2) = 0$ still holds whereby the above matrix belongs to $\mathrm{SL_3(A_j)}$. Thus $\{p_1, p_2\}$ is a basis of P_j.

Since we have bases for P_j and P_{1-j}, we compute the cocycle associated to P. Let $q_1 = \lambda_{11} p_1 + \lambda_{12} p_2$ and $q_2 = \lambda_{21} p_1 + \lambda_{22} p_2$, then $\begin{pmatrix} \lambda_{11} & \lambda_{12} \\ \lambda_{21} & \lambda_{22} \end{pmatrix} \in \mathrm{SL_2(A_{j(1-j)})}$ is the cocycle associated to P.

We have

$$f_{j(1-j)}(q_1) = \lambda_{11} f_j(p_1) = \lambda_{11},\ f_{j(1-j)}(0, v', 0) = v'a_1 \text{ and}$$

$$f_{j(1-j)}(q_2) = \lambda_{21} f_j(p_1) = \lambda_{21},\ f_{j(1-j)}(0, 0, 1) = a_2.$$

Thus $\lambda_{11} = v'a_1$ and $\lambda_{21} = a_2$. Hence the cocycle associated to P is

$$\sigma_1 = \begin{pmatrix} v'a_1 & \lambda_{12} \\ a_2 & \lambda_{22} \end{pmatrix} \in \mathrm{SL_2(A_{j(1-j)})}.$$

Similarly, the cocycle associated to the projective module $Q = \frac{A^3}{(u'_1, b_2, -b_1)}$ is

$$\sigma_2 = \begin{pmatrix} v''b_1 & \mu_{12} \\ b_2 & \mu_{22} \end{pmatrix} \in \mathrm{SL_2(A_{j(1-j)})},\ \text{where } \mathrm{v''} = \frac{\mathrm{v'_1(1-j_1)}}{\mathrm{(1-j)}}.$$

By Corollary 17.1, there exists $\sigma \in SL_2(A_{1-j})$ such that

$$\sigma \begin{pmatrix} v'a_1 & \lambda_{12} \\ a_2 & \lambda_{22} \end{pmatrix} = \begin{pmatrix} v''b_1 & \widetilde{\mu_{12}} \\ b_2 & \widetilde{\mu_{22}} \end{pmatrix}.$$

So, (by Step(6) of the proof of the Theorem 17.6), we can find an element $\varepsilon \in \mathrm{E_2(A_{j(1-j)})}$ such that $\varepsilon\sigma\sigma_1 = \sigma_2$. This implies by Lemma 17.7, that the cocycle σ_1 is equivalent to σ_2. Hence $P \simeq Q$ and the two rows are in the same $\mathrm{SL_3(A)}$ orbit. □

Theorem 17.6 (Addition Principle) *Let A be a domain of dimension 2. Let $J_1 = (a_1, a_2)$ and $J_2 = (b_1, b_2)$ be two comaximal ideals of height 2 in A such that $(b_1) + (a_1, a_2) = A$. Let $(a_1, a_2) \cap (b_1, b_2) = (c_1, c_2) = J_3$ with $c_1 = a_1 \bmod J_1^2$, $c_2 = a_2 \bmod J_1^2$, $c_1 = b_1 \bmod J_2^2$ and $c_2 = b_2 \bmod J_2^2$. Let u be a unit modulo (a_1, a_2) such that $u = 1 \bmod J_2$. Then the unimodular rows (u, a_1, a_2) and (u, c_1, c_2) are in the same* $\mathrm{SL_3(A)}$ *orbit.*

Proof We prove this in several steps the method being to show that the cocycles associated to the two unimodular rows are equivalent.

Step(1): Since $(u, a_1, a_2) = A$, we can choose $j_1 \in (a_1, a_2)$ such that $1 - j_1 = \lambda' u$. Since $(b_1) + (a_1, a_2) = A$, $(b_1, a_1, a_2) = A$ and we can choose $j_2 \in (a_1, a_2)$ such that $1 - j_2 = \mu' b_1$. Let $1 - j = (1 - j_1)(1 - j_2)$. Then $j \in (a_1, a_2)$ and $1 - j = vu$ and $1 - j = \mu b_1$.

Step(2): In this step, we compute the cocycle associated to the projective module $P = \frac{A^3}{(u,a_1,a_2)}$. Let us consider the surjective map $f : \frac{A^3}{(u,a_1,a_2)} \twoheadrightarrow (a_1, a_2) = J_1$ given by $f(e_1) = 0$, $f(e_2) = a_2$, $f(e_3) = -a_1$. Since by Step(1) we have $uv = 1 - j$, we can take a completion of (u, a_1, a_2) as $\begin{pmatrix} u & a_1 & a_2 \\ 0 & \frac{v}{1-j} & 0 \\ 0 & 0 & 1 \end{pmatrix}$ in $\mathrm{SL_3(A_{1-j})}$. Let $q_1 = (0, \frac{v}{1-j}, 0)$ and $q_2 = (0, 0, 1)$. Then $\{q_1, q_2\}$ is a basis for the free module P_{1-j}.

Now, since $j \in (a_1, a_2)$, so $(J_1)_j = A_j$. This implies that the unimodular row (u, a_1, a_2) contains a unimodular row of shorter length in A_j. So, (u, a_1, a_2) is completable in A_j and hence P_j is free. Choose a basis $\{p_1, p_2\}$ of P_j such that $f_j(p_1) = 1$, $f_j(p_2) = 0$. For this purpose we choose any basis of P_j. The image of this basis is a unimodular row in A_j. We transform the unimodular row to $(1, 0)$ via a matrix in $\mathrm{SL_2(A_j)}$ and correspondingly transform the basis of P_j to obtain p_1, p_2 such that $f_j(p_1) = 1$, $f_j(p_2) = 0$. In particular, we have $\begin{pmatrix} u \; a_1 \; a_2 \\ p_1 \\ p_2 \end{pmatrix} \in GL_3(A_j)$.

So, we can modify p_2 by a unit in A_j assume that the condition $f_j(p_2) = 0$ still holds and the above matrix belongs to $\mathrm{SL_3(A_j)}$. Thus $\{p_1, p_2\}$ is a basis of P_j.

Since we have basis of P_j and P_{1-j}, we can use these two bases to compute the cocycle of P. Let $q_1 = \lambda_{11}p_1 + \lambda_{12}p_2$ and $q_2 = \lambda_{21}p_1 + \lambda_{22}p_2$, then

$$\begin{pmatrix} \lambda_{11} & \lambda_{12} \\ \lambda_{21} & \lambda_{22} \end{pmatrix} \in \mathrm{SL_2(A_{j(1-j)})}$$

is the cocycle of P.

We have $f_{j(1-j)}(q_1) = \lambda_{11} f_j(p_1) = \lambda_{11}$, $f_{j(1-j)}(0, v', 0) = \frac{v}{1-j}a_2$ and $f_{j(1-j)}(q_2) = \lambda_{21} f_j(p_1) = \lambda_{21}$, $f_{j(1-j)}(0, 0, 1) = -a_1$. Thus $\lambda_{11} = \frac{v}{1-j}a_2$ and $\lambda_{21} = -a_1$. Hence the cocycle associated to P is

$$\begin{pmatrix} \frac{v}{1-j}a_2 & \lambda_{12} \\ -a_1 & \lambda_{22} \end{pmatrix} \in \mathrm{SL_2(A_{j(1-j)})}.$$

Step(3): We recall the statement of the $\mathrm{SL_2}$ lemma. Let $J = (f, g) = (f', g')$. Suppose there exists a matrix $\alpha \in M_2(A)$ with $\det(\alpha) = 1 \bmod J$ and $\alpha \begin{pmatrix} f \\ g \end{pmatrix} = \begin{pmatrix} f' \\ g' \end{pmatrix}$. Then there exists $\beta \in \mathrm{SL_2(A)}$ such that $\beta \begin{pmatrix} f \\ g \end{pmatrix} = \begin{pmatrix} f' \\ g' \end{pmatrix}$.

In particular, if $J = (f, g) = (f', g')$, $f' = f \bmod J^2$ and $g' = g \bmod J^2$, then there exists $\beta \in SL_2(A)$ such that $\beta \begin{pmatrix} f \\ g \end{pmatrix} = \begin{pmatrix} f' \\ g' \end{pmatrix}$.

Step(4): Since $(a_1, a_2) \cap (b_1, b_2) = (c_1, c_2)$ and $j \in (a_1, a_2)$, we have $(b_1, b_2)_j = (c_1, c_2)_j$. Now, since $b_i = c_i \bmod J_2^2$, then by the $\mathrm{SL_2}$ lemma there exists $\sigma \in \mathrm{SL_2(A_j)}$ such that $\sigma \begin{pmatrix} b_1 \\ b_2 \end{pmatrix} = \begin{pmatrix} c_1 \\ c_1 \end{pmatrix}$. Hence the unimodular rows $(u, b_1, b_2)_j$ and

$(u, c_1, c_2)_j$ are in the same $\mathrm{SL}_3(\mathrm{A_j})$ orbit. Since $u = 1 \bmod (b_1, b_2)$, then (u, b_1, b_2) is completable to a matrix in $\mathrm{SL}_3(\mathrm{A_j})$. Hence (u, c_1, c_2) is completable to a matrix in $\mathrm{SL}_3(\mathrm{A_j})$.

Let $\alpha = \begin{pmatrix} u & c_1 & c_2 \\ \lambda'_{11} & \lambda'_{12} & \lambda'_{13} \\ \lambda'_{21} & \lambda'_{22} & \lambda'_{23} \end{pmatrix} \in SL_3(A_j)$ and let $p'_1 = (\lambda'_{11}, \lambda'_{12}, \lambda'_{13})$ and $p'_2 = (\lambda'_{21}, \lambda'_{22}, \lambda'_{23})$. Thus $\{p'_1, p'_2\}$ forms a basis for Q_j, where Q is the projective module $\frac{A^3}{(u,c_1,c_2)}$.

Step(5): Let $g : \frac{A^3}{(u,c_1,c_2)} \twoheadrightarrow J_3 = (c_1, c_2)$ be the surjection given by $g(e_1) = 0$, $g(e_2) = c_2$, and $g(e_3) = -c_1$. Since $\alpha \in \mathrm{SL}_3(\mathrm{A_j})$ (Step(4)) and $g(u, c_1, c_2) = 0$, the elements $g(\lambda'_{11}, \lambda'_{12}, \lambda'_{13})$ and $g(\lambda'_{21}, \lambda'_{22}, \lambda'_{23})$ generate $(J_3)_j$. We have

$$g(\lambda'_{11}, \lambda'_{12}, \lambda'_{13}) = \lambda'_{12}c_2 - \lambda'_{13}c_1 \text{ and } g(\lambda'_{21}, \lambda'_{22}, \lambda'_{23}) = \lambda'_{22}c_2 - \lambda'_{23}c_1.$$

Let $g_1 = \lambda'_{12}c_2 - \lambda'_{13}c_1$ and $g_2 = \lambda'_{22}c_2 - \lambda'_{23}c_1$. We can write this in the matrix form as

$$\begin{pmatrix} -\lambda'_{13} & \lambda'_{12} \\ -\lambda'_{23} & \lambda'_{22} \end{pmatrix} \begin{pmatrix} c_1 \\ c_2 \end{pmatrix} = \begin{pmatrix} g_1 \\ g_2 \end{pmatrix} \tag{17.7}$$

and since $\det \alpha = 1$,

$$\det \begin{pmatrix} -\lambda'_{13} & \lambda'_{12} \\ -\lambda'_{23} & \lambda'_{22} \end{pmatrix} = \det \begin{pmatrix} \lambda'_{12} & \lambda'_{13} \\ \lambda'_{22} & \lambda'_{23} \end{pmatrix} = u^{-1} \bmod (c_1, c_2)_j.$$

Now, since $(b_1, b_2)_j = (c_1, c_2)_j$ and $u = 1 \bmod (b_1, b_2)_j$, therefore $u = 1 \bmod (c_1, c_2)_j$. Hence $u^{-1} = 1 \bmod (c_1, c_2)_j$. Then by Corollary 17.1 and the SL_2 lemma (Step(3)), $\begin{pmatrix} c_1 \\ c_2 \end{pmatrix}_j$ and $\begin{pmatrix} g_1 \\ g_2 \end{pmatrix}_j$ are in the same $\mathrm{SL}_2(\mathrm{A_j})$ orbit. By Step(4), $\begin{pmatrix} c_1 \\ c_2 \end{pmatrix}_j$ and $\begin{pmatrix} b_1 \\ b_2 \end{pmatrix}_j$ are in the same $\mathrm{SL}_2(\mathrm{A_j})$ orbit. Hence $\begin{pmatrix} c_1 \\ c_2 \end{pmatrix}_j$, $\begin{pmatrix} g_1 \\ g_2 \end{pmatrix}_j$ and $\begin{pmatrix} b_1 \\ b_2 \end{pmatrix}_j$ are in the same $\mathrm{SL}_2(\mathrm{A_j})$ orbit.

Step(6): Let A be a ring, σ and τ are two matrices in $\mathrm{SL}_2(\mathrm{A})$ having the same first column. Then $\sigma^{-1}\tau$ is elementary. This holds since $\sigma^{-1}\tau \begin{pmatrix} 1 \\ 0 \end{pmatrix} = \begin{pmatrix} 1 \\ 0 \end{pmatrix}$, we have $\sigma^{-1}\tau = \begin{pmatrix} 1 & \mu^* \\ 0 & 1 \end{pmatrix}$ which is elementary.

Step(7): In this step, we compute the cocycle associated to the projective module $Q = \frac{A^3}{(u,c_1,c_2)}$. Let us consider the surjective map $g : \frac{A^3}{(u,c_1,c_2)} \twoheadrightarrow (c_1, c_2) = J_3$ given by $g(e_1) = 0$, $g(e_2) = c_2$, and $g(e_3) = -c_1$. Since by Step(1) we have $uv = 1 - j$, we can take a completion of (u, c_1, c_2) in $\mathrm{SL}_3(\mathrm{A}_{1-\mathrm{j}})$ as $\begin{pmatrix} u & c_1 & c_2 \\ 0 & \frac{v}{1-j} & 0 \\ 0 & 0 & 1 \end{pmatrix}$. Let $q_1' = (0, \frac{v}{1-j}, 0)$ and $q_2' = (0, 0, 1)$, then $\{q_1', q_2'\}$ is a basis for the free module Q_{1-j}.

Now from Step(4), we have a completion $\alpha \in \mathrm{SL}_3(\mathrm{A}_\mathrm{j})$ of (u, c_1, c_2) whose second and third rows are p_1', p_2'. Thus $\{p_1', p_2'\}$ is a basis of Q_j. Let $q_1' = \delta_{11}p_1' + \delta_{12}p_2'$ and $q_2' = \delta_{21}p_1' + \delta_{22}p_2'$. Then $\begin{pmatrix} \delta_{11} & \delta_{12} \\ \delta_{21} & \delta_{22} \end{pmatrix} \in \mathrm{SL}_2(\mathrm{A}_{\mathrm{j}(1-\mathrm{j})})$ is the cocycle associated to the projective module Q. Also we have $g(q_1') = \frac{v}{1-j}c_2$ and $g(q_2') = -c_1$. On the other hand

$$\begin{pmatrix} g(q_1') \\ g(q_2') \end{pmatrix} = \begin{pmatrix} \delta_{11} & \delta_{12} \\ \delta_{21} & \delta_{22} \end{pmatrix} \begin{pmatrix} g(p_1') \\ g(p_2') \end{pmatrix}.$$

By Step(5), $g(p_1') = g_1$ and $g(p_2') = g_2$. Hence

$$\begin{pmatrix} \frac{v}{1-j}c_2 \\ -c_1 \end{pmatrix} = \begin{pmatrix} \delta_{11} & \delta_{12} \\ \delta_{21} & \delta_{22} \end{pmatrix} \begin{pmatrix} g_1 \\ g_2 \end{pmatrix},$$

where $\begin{pmatrix} \delta_{11} & \delta_{12} \\ \delta_{21} & \delta_{22} \end{pmatrix} \in SL_2(A_{j(1-j)})$ is the cocycle associated to $Q = \frac{A^3}{(u,c_1,c_2)}$.

Step(8): We have $\begin{pmatrix} \frac{v}{1-j}c_2 \\ -c_1 \end{pmatrix}$ and $\begin{pmatrix} \frac{v}{1-j}a_2 \\ -a_1 \end{pmatrix}$ are in the same $\mathrm{SL}_2(\mathrm{A}_{1-\mathrm{j}})$ orbit. This follows since $(c_1, c_2)_{1-j} = (a_1, a_2)_{1-j}$ (because $1 - j \in (b_1, b_2)$), and $c_i = a_i \bmod J_1^2$, where $J_1 = (a_1, a_2)$.

Step(9): In this step, we show that the cocycles associated to P and Q are equivalent. Recall that the cocycle associated to $P = \frac{A^3}{(u,a_1,a_2)}$ is

$$\begin{pmatrix} \lambda_{11} & \lambda_{12} \\ \lambda_{21} & \lambda_{22} \end{pmatrix} = \begin{pmatrix} \frac{v}{1-j}a_2 & \lambda_{12} \\ -a_1 & \lambda_{22} \end{pmatrix}$$

and the cocycle associated to $Q = \frac{A^3}{(u,c_1,c_2)}$ is

$$\begin{pmatrix} \delta_{11} & \delta_{12} \\ \delta_{21} & \delta_{22} \end{pmatrix}, \text{ where } \begin{pmatrix} \delta_{11} & \delta_{12} \\ \delta_{21} & \delta_{22} \end{pmatrix} \begin{pmatrix} g_1 \\ g_2 \end{pmatrix} = \begin{pmatrix} \frac{v}{1-j}c_2 \\ -c_1 \end{pmatrix}.$$

Now, we want to show that $\begin{pmatrix} \lambda_{11} & \lambda_{12} \\ \lambda_{21} & \lambda_{22} \end{pmatrix}$ and $\begin{pmatrix} \delta_{11} & \delta_{12} \\ \delta_{21} & \delta_{22} \end{pmatrix}$ are equivalent. Since from Step(1) $1 - j = \mu b_1$, there exists an elementary matrix $\varepsilon \in \mathrm{E}_2(\mathrm{A}_{1-\mathrm{j}})$ such that $\varepsilon \begin{pmatrix} b_1 \\ b_2 \end{pmatrix} = \begin{pmatrix} 1 \\ 0 \end{pmatrix}$. Therefore $\varepsilon^{-1} \begin{pmatrix} 1 \\ 0 \end{pmatrix} = \begin{pmatrix} b_1 \\ b_2 \end{pmatrix}$. From Step(5), there exists $\nu \in \mathrm{SL}_2(\mathrm{A}_\mathrm{j})$ such that $\nu \begin{pmatrix} b_1 \\ b_2 \end{pmatrix} = \begin{pmatrix} g_1 \\ g_2 \end{pmatrix}$ and from Step(7), $\begin{pmatrix} \delta_{11} & \delta_{12} \\ \delta_{21} & \delta_{22} \end{pmatrix} \begin{pmatrix} g_1 \\ g_2 \end{pmatrix} = \begin{pmatrix} \frac{v}{1-j}c_2 \\ -c_1 \end{pmatrix}$. From Step(8), there exists $\beta \in \mathrm{SL}_2(\mathrm{A}_{1-\mathrm{j}})$ such that

$$\beta \begin{pmatrix} \frac{v}{1-j}c_2 \\ -c_1 \end{pmatrix} = \begin{pmatrix} \frac{v}{1-j}a_2 \\ -a_1 \end{pmatrix} = \begin{pmatrix} \lambda_{11} & \lambda_{12} \\ \lambda_{21} & \lambda_{22} \end{pmatrix} \begin{pmatrix} 1 \\ 0 \end{pmatrix}.$$

Hence we have

$$\beta \begin{pmatrix} \delta_{11} & \delta_{12} \\ \delta_{21} & \delta_{22} \end{pmatrix} \nu\varepsilon^{-1} \begin{pmatrix} 1 \\ 0 \end{pmatrix} = \begin{pmatrix} \lambda_{11} & \lambda_{12} \\ \lambda_{21} & \lambda_{22} \end{pmatrix} \begin{pmatrix} 1 \\ 0 \end{pmatrix}.$$

Hence by Lemma 17.7, the cocycles $\beta \begin{pmatrix} \delta_{11} & \delta_{12} \\ \delta_{21} & \delta_{22} \end{pmatrix} \nu\varepsilon^{-1}$ and $\begin{pmatrix} \lambda_{11} & \lambda_{12} \\ \lambda_{21} & \lambda_{22} \end{pmatrix}$ are equivalent (since by Step(6), these matrices differ by an elementary matrix). Again by Lemma 17.7, the cocycle $\beta \begin{pmatrix} \delta_{11} & \delta_{12} \\ \delta_{21} & \delta_{22} \end{pmatrix} \nu\varepsilon^{-1}$ is equivalent to the cocycle $\beta \begin{pmatrix} \delta_{11} & \delta_{12} \\ \delta_{21} & \delta_{22} \end{pmatrix} \nu$. Since $\beta \in SL_2(A_{1-j})$ and $\nu \in SL_2(A_j)$, the cocycles $\beta \begin{pmatrix} \delta_{11} & \delta_{12} \\ \delta_{21} & \delta_{22} \end{pmatrix} \nu$ and $\begin{pmatrix} \delta_{11} & \delta_{12} \\ \delta_{21} & \delta_{22} \end{pmatrix}$ are equivalent. Hence the cocycles $\begin{pmatrix} \delta_{11} & \delta_{12} \\ \delta_{21} & \delta_{22} \end{pmatrix}$ and $\begin{pmatrix} \lambda_{11} & \lambda_{12} \\ \lambda_{21} & \lambda_{22} \end{pmatrix}$ are equivalent. This implies $P \simeq Q$ and the unimodular rows (u, a_1, a_2) and (u, c_1, c_2) are in the same $SL_3(A)$ orbit. □

We now wish to prove Theorem 17.14. This is proved using Theorem 17.6 and the ideas of [5–7]. We recall the lemmas of [5–7] that are needed with sketches of proofs.

The following lemma holds for general n. We state it only in the case where $n = 2$ because that is the case we use here.

Lemma 17.12 (Moving Lemma see [6], [5, Proposition 4.10]) *Let A be a Noetherian ring of dimension 2. Suppose $J_1, J_2 \subset A$ are comaximal ideals of height 2. Suppose $J_1 \cap J_2 = (a_1, a_2)$, where a_1, a_2 give the orientation w_{J_1} of J_1 and the orientation w_{J_2} of J_2. Suppose further that there are ideals $J_3, J_4 \subset A$ of height 2 such that*

(1) *J_1, J_2, J_3, J_4 are pairwise comaximal.*
(2) *$J_2 \cap J_3 = (b_1, b_2)$, where b_1, b_2 give the orientation w_{J_2} of J_2 and the orientation w_{J_3} of J_3.*
(3) *$J_3 \cap J_4 = (c_1, c_2)$, where c_1, c_2 gives the orientation w_{J_3} of J_3 and the orientation w_{J_4} of J_4.*

Then $J_1 \cap J_4 = (d_1, d_2)$, where d_1, d_2 and c_1, c_2 give the same orientation of J_4 and d_1, d_2 and a_1, a_2 give the same orientation of J_1.

Remark 17.3 Roughly the lemma says that if

$$[J_1] + [J_2] = 0,\ [J_2] + [J_3] = 0,\ \text{and } [J_3] + [J_4] = 0 \text{ in E(A)}.$$

Then $[J_1] + [J_4] = 0$ in E(A), where symbols are to be interpreted approximately. The idea is to interpret the identity $-(-a) = a$ in a suitable way.

Proof (*Sketch of the proof of Lemma* 17.12) Since $J_1 \cap J_2$ and $J_3 \cap J_4$ are generated by 2 elements, we see by the Addition principle that $J_1 \cap J_2 \cap J_3 \cap J_4$ is also generated by 2 elements. Now since $J_2 \cap J_3$ is generated by 2 elements, using the Subtraction principle one can show that $J_1 \cap J_4$ is also generated by 2 elements. Further, at each stage one can keep track of the orientations to prove the lemma. □

Remark 17.4 Another way of interpreting Lemma 17.12 is the following: Suppose we are given that $J_1 \cap J_2$ and $J_2 \cap J_3$ (which are two ideals having common component J_3) are generated by two elements. Then, we can replace J_2 by J_4 and assume that $J_1 \cap J_4$ and $J_3 \cap J_4$ are generated by two elements.

Lemma 17.13 (see [6], [5, Proposition 4.10]) *Let A be a Noetherian ring of dimension* 2. *Let* $J \subset A$ *be an ideal of height* 2 *such that* J/J^2 *is generated by* 2 *elements and* w_J *be the corresponding orientation of* J/J^2. *Suppose* $(J, w_J) = 0$ *in* E(A). *Then* J *is generated by* 2 *elements* a_1, a_2 *and the orientation of* J *given by* a_1, a_2 *is the same as* w_J.

Proof (*Sketch*) Since $(J, w_J) = 0$ in E(A), we have

$$(J, w_J) + \sum_k (J_k, w_{J_k}) = \sum_k (J'_k, w'_{J'_k})$$

in G, where G is as in the definition of the Euler class group. The ideals J_i have height 2 and are generated by 2 elements, and the w_{J_i} are the orientations of J_i given by those 2 elements. If some (J_k, w_{J_k}) and $(J'_k, w'_{J'_k})$ have a common $\mathfrak{m}$-primary component for some maximal ideal $\mathfrak{m}$ of A and w_{J_i} and $w'_{J'_k}$ give the same orientation of the $\mathfrak{m}$-primary ideal corresponding to that component, then by Lemma 17.12, we replace and change that component in J_i and J'_i to obtain ideals $\widetilde{J_i}$ with an orientation $\widetilde{w_{J_i}}$ and $\widetilde{J'_i}$ with an orientation $\widetilde{w'_{J'_i}}$ such that the equation

$$(J, w_J) + \sum_k (J_k, w_{J_k}) = \sum_k (J'_k, w'_{J'_k})$$

holds when (J_i, w_{J_i}) and $(J'_i, w'_{J'_i})$ are replaced by $(\widetilde{J_i}, \widetilde{w_{J_i}})$ and $(\widetilde{J'_i}, \widetilde{w'_{J'_i}})$, and then assume by doing this that the ideals J and $(J_i)_i$ (indexed by i) are mutually comaximal and the $(J'_k)_k$ (indexed by k) are mutually comaximal with $J \cap (\cap J_k) = \cap J'_k$.

By the Addition principle the ideal $\cap J_i$ is generated by 2 elements with the appropriate orientations and so is $\cap J'_i$. By the Subtraction principle J is generated by 2 elements a_1, a_2, which give the orientation w_J of J. □

Now we come to the main lemma that is used to prove the existence of a group structure on $\mathrm{Um}_3(\mathrm{A})/\mathrm{SL}_3(\mathrm{A})$. The main ingredients of the proof, are Theorems 17.6, 17.5 and the technique of the proof of Lemma 17.13. The reader who reads the proof of Lemma 17.13 which is given in [5] will able to understand the details. We do not give all the details to render the proof readable.

Lemma 17.14 *Let A be a Noetherian domain of dimension* 2. *Let* (a_1, a_2, a_3) *and* (b_1, b_2, b_3) *be two unimodular rows in A such that* $P_1 = \frac{A^3}{(a_1,a_2,a_3)}$ *and* $P_2 = \frac{A^3}{(b_1,b_2,b_3)}$. *Suppose the elements associated to P_1 and P_2 in* E(A) *are the same. Then* (a_1, a_2, a_3) *and* (b_1, b_2, b_3) *are in the same* $\mathrm{SL}_3(\mathrm{A})$ *orbit.*

Proof Let $J_1 = (a_2, a_3)$ and $J_2 = (b_2, b_3)$. Thus by Lemma 17.11, we may assume that $\mathrm{ht}(\mathrm{J}_1) = 2 = \mathrm{ht}(\mathrm{J}_2)$ and that $J_1 + J_2 = A$. The elements of E(A) associated to (a_1, a_2, a_3) and (b_1, b_2, b_3) are (J_1, w_{J_1}) and (J_2, w_{J_2}) with orientations which are given by $a_1^{-1}a_2, a_3$ and $b_1^{-1}b_2, b_3$ respectively.

Since $(J_1, w_{J_1}) = (J_2, w_{J_2})$ in E(A). We have

$$(J_1, w_{J_1}) + \sum_k (J'_k, w'_k) = (J_2, w_{J_2}) + \sum_l (\widetilde{J}_l, w_l), \tag{17.8}$$

where for every k, l, the ideals J'_k and $\widetilde{J}_l$ of A are of height 2 and generated by 2 elements and w'_k, w_l are the trivial orientations of J_k and $\widetilde{J}_l$ given by these elements. Since J_1 and J_2 are comaximal, using Lemma 17.12, we may change the J'_k, $\widetilde{J}_l$ and assume that the ideals J_1 and J'_k are mutually comaximal and the ideals J_2 and $\widetilde{J}_l$ are mutually comaximal and (17.8) still holds.

By the Addition principle $\cap_k J'_k$ is generated by 2 elements and also $J_1 \cap (\cap_k J'_k)$ is generated by 2 elements d_2, d_3 with the generators giving the trivial orientations on each component. Similarly, $J_2 \cap (\cap_l \widetilde{J}_l) = (g_2, g_3)$ with the generators giving the trivial orientations on each component. Let $u_1 = a_1 \bmod J_1, u_1 = 1 \bmod \cap_k J'_k$ and $u_2 = b_1 \bmod J_2, u_2 = 1 \bmod \cap_l \widetilde{J}_l$. Then, since $u_1 = a_1 \bmod J_1$, by Theorem 17.6,

$$(a_1, a_2, a_3) \overset{\mathrm{SL}_3(\mathrm{A})}{\sim} (u_1, d_2, d_3),$$

where (d_2, d_3) are generators of $J_1 \cap (\cap_k J'_k)$ which give the orientation (a_2, a_3) of J_1 and the trivial orientation w'_k of J'_k.

Further, since $u_2 = b_1 \bmod J_2$, by Theorem 17.6,

$$(b_1, b_2, b_3) \overset{\mathrm{SL}_3(\mathrm{A})}{\sim} (u_2, g_2, g_3),$$

where (g_2, g_3) are generators of $J_2 \cap (\cap_l \widetilde{J}_l)$ which give the orientation (b_2, b_3) of J_2 and the trivial orientation w_l of $\widetilde{J}_l$. Since

$$(J_1, w_{J_1}) + \sum_k (J'_k, w'_k) = (J_2, w_{J_2}) + \sum_l (\widetilde{J}_l, w_l)$$

in G, we have

(1) $(g_2, g_3) = (d_2, d_3)$
(2) The orientations $(u_1^{-1}d_2, d_3)$ and $(u_2^{-1}g_2, g_3)$ of (g_2, g_3) and (d_2, d_3) respectively, are the same. Therefore, by Theorem 17.5, $(u_1, d_2, d_3) \overset{SL_3(A)}{\sim} (u_2, g_2, g_3)$ and hence $(a_1, a_2, a_3) \overset{SL_3(A)}{\sim} (b_1, b_2, b_3)$.

Acknowledgements The authors would like to thank Professor Ravi A. Rao for his valuable support during this work. The authors would like to thank Professor Gopala Krishna Srinivasan for giving his time most generously and helping us make this paper more readable. The third named author would like to thank Professor Gopala Krishna Srinivasan for his support and advice during difficult times. The authors would also like to thank the referee for going through the paper carefully and pointing out some mistakes. The third named author also acknowledges the financial support from C.S.I.R. which enabled him to pursue his doctoral studies.

References

1. H. Bass, K-theory and stable algebra. Inst. Hautes Études Sci. Publ. Math. **22**, 5–60 (1964)
2. R. Basu, Topics in classical algebraic K-theory. Ph.D. thesis, School of Mathematics, Tata Institute of Fundamental Research, Mumbai (2006)
3. R. Basu, R. Sridharan, On Forster's conjecture and related results. Punjab Univ. Res. J. (Sci.) **57**, 13–66 (2007)
4. S.M. Bhatwadekar, M.K. Keshari, A question of Nori: projective generation of ideals. K-Theory **28**(4), 329–351 (2003)
5. S.M. Bhatwadekar, R. Sridharan, Projective generation of curves in polynomial extensions of an affine domain and a question of Nori. Invent. Math. **133**(1), 161–192 (1998)
6. S.M. Bhatwadekar, R. Sridharan, The Euler class group of a Noetherian ring. Compos. Math. **122**(2), 183–222 (2000)
7. S.M. Bhatwadekar, R. Sridharan, On Euler classes and stably free projective modules, *Algebra, Arithmetic and Geometry, Part I, II (Mumbai, 2000)*. Tata Institute of Fundamental Research Studies in Mathematics, vol. 16 (Tata Institute of Fundamental Research, Bombay, 2002)
8. S.M. Bhatwadekar, H. Lindel, R.A. Rao, The Bass-Murthy question: Serre dimension of Laurent polynomial extensions. Invent. Math. **81**(1), 189–203 (1985)
9. S.M. Bhatwadekar, M.K. Das, S. Mandal, Projective modules over smooth real affine varieties. Invent. Math. **166**(1), 151–184 (2006)
10. M.K. Das, R. Sridharan, Good invariants for bad ideals. J. Algebra **323**(12), 3216–3229 (2010)
11. N.S. Gopalakrishnan, Commutative algebra (1984)
12. S.K. Gupta, M.P. Murthy, *Suslin's Work on Linear Groups over Polynomial Rings and Serre Problem*. ISI Lecture Notes, vol. 8 (Macmillan Co. of India Ltd, New Delhi, 1980)
13. M.K. Keshari, Euler class group of a Noetherian ring. Ph.D. thesis, School of Mathematics, Tata Institute of Fundamental Research, Mumbai (2001)
14. M. Krusemeyer, Skewly completable rows and a theorem of Swan and Towber. Commun. Algebra **4**(7), 657–663 (1976)
15. N.M. Kumar, Complete intersections. J. Math. Kyoto Univ. **17**(3), 533–538 (1977)
16. N.M. Kumar, A note on the cancellation of reflexive modules. J. Ramanujan Math. Soc. **17**(2), 93–100 (2002)
17. N.M. Kumar, On a theorem of Seshadri, *Connected at Infinity*. Texts and Readings in Mathematics, vol. 25 (Hindustan Book Agency, New Delhi, 2003), pp. 91–104
18. T.Y. Lam, *Serre's Problem on Projective Modules*. Springer Monographs in Mathematics (Springer, Berlin, 2006)

19. S. MacLane, *Homology*. Die Grundlehren der mathematischen Wissenschaften, vol. 114, 1st edn. (Springer, Berlin, 1967)
20. S. Mandal, Homotopy of sections of projective modules. J. Algebr. Geom. **1**(4), 639–646 (1992). With an appendix by Madhav V. Nori
21. S. Mandal, *Projective Modules and Complete Intersections*. Lecture Notes in Mathematics, vol. 1672 (Springer, Berlin, 1997)
22. M.P. Murthy, R.G. Swan, Vector bundles over affine surfaces. Invent. Math. **36**, 125–165 (1976)
23. D. Quillen, Projective modules over polynomial rings. Invent. Math. **36**, 167–171 (1976)
24. J.P. Serre, Sur les modules projectifs. SÃl'minaire Dubreil. AlgÃíbre et thÃl'orie des nombres **14**, 1–16 (1960–1961)
25. C.S. Seshadri, Triviality of vector bundles over the affine space K^2. Proc. Natl. Acad. Sci. USA **44**, 456–458 (1958)
26. C.S. Seshadri, Algebraic vector bundles over the product of an affine curve and the affine line. Proc. Am. Math. Soc. **10**, 670–673 (1959)
27. A.A. Suslin, Stably free modules. Mat. Sb. (N.S.) **102**(144)(4), 537–550, 632 (1977)
28. R.G. Swan, Algebraic vector bundles on the 2-sphere. Rocky Mt. J. Math. **23**(4), 1443–1469 (1993)
29. R.G. Swan, J. Towber, A class of projective modules which are nearly free. J. Algebra **36**(3), 427–434 (1975)
30. W. van der Kallen, A group structure on certain orbit sets of unimodular rows. J. Algebra **82**(2), 363–397 (1983)
31. W. van der Kallen, A module structure on certain orbit sets of unimodular rows. J. Pure Appl. Algebra **57**(3), 281–316 (1989)
32. L.N. Vaserstein, Stabilization of unitary and orthogonal groups over a ring with involution. Mat. Sb. (N.S.) **81** (123), 328–351 (1970)
33. L.N. Vaserstein, A.A. Suslin, Serre's problem on projective modules over polynomial rings, and algebraic K-theory. Izv. Akad. Nauk SSSR Ser. Mat. **40**(5), 993–1054, 1199 (1976)
34. C.T.C. Wall, *A Geometric Introduction to Topology* (Addison-Wesley Publishing Co., Reading, 1972)

Chapter 18
Relating the Principles of Quillen–Suslin Theory

Ravi A. Rao and Sunil K. Yadav

18.1 Introduction

Serre's problem on projective modules ranks among the most celebrated problems in commutative algebra since its inception in his famous 1955 paper ([6], p. 243):

"Signalons que, lorsque $V = K^r$ (auquel cas $A = K[X_I, \ldots, X_n]$), on ignore s'il existe des A-modules projectifs de type fini qui ne soient pas libres, ou, ce qui revient au méme, s'il existe des espaces fibrés algébriques à fibrés vectorielles, de base K^r, et non triviaux."

The book of T.Y. Lam [7] has a comprehensive description of the developments on this problem, which evolved through several intermediate stages beginning with the work of C.S. Seshadri [12] in 1958 when $n = 2$, and finally settled by Quillen and Suslin [10, 14] independently. The book of Ischebeck–Rao [5] gives Serre's motivation (in the subject of set-theoretic complete intersection questions in affine space) for suggesting the problem, and the subsequent developments in that direction till the early 1980s.

The solution to Serre's problem on the freeness of finitely generated projective modules over a polynomial extension of a field rests on two pillars, namely Horrock's Monic Inversion Principle [4] and the Quillen's Local–Global Principle [10]. These principles were used by Suslin [15] to show that the special linear group over a polynomial extension of a field consists of elementary matrices, for matrices of size at least 3. Suslin proved both the Local–Global Principle and the Monic Inver-

R. A. Rao (✉)
School of Mathematics, Tata Institute of Fundamental Research,
1, Dr. Homi Bhabha Road, Mumbai 400 005, India
e-mail: ravi@math.tifr.res.in

S. K. Yadav
Department of Mathematics, Indian Institute of Technology Bombay,
Powai, Mumbai 400 076, India
e-mail: skyadav@math.iitb.ac.in

A. A. Ambily et al. (eds.), *Leavitt Path Algebras and Classical* K-*Theory*,
Indian Statistical Institute Series,
https://doi.org/10.1007/978-981-15-1611-5_18

sion Principle for the pair $(\mathrm{GL}_n(R[X]), \mathrm{E}_n(R[X]))$, $n \geq 3$. (The latter being a highly non-trivial argument based on establishing first a Bruhat type decomposition for $\mathrm{E}_n(R[X, X^{-1}], \mathfrak{m}[X, X^{-1}])$, when $(R, \mathfrak{m})$ is a local ring.)

In [11], we shall provide an alternative approach to proving the Monic Inversion Principle. This work was sketched in two one hour talks by the first author in the International workshop on "Leavitt path algebras and K-theory" on July 1, 2017 at CUSAT, Kerala. The talks have been recorded in this article. In particular, no proofs are given in this article, and only the results are announced.

18.2 The Local–Global Principle and Normality

We shall follow the standard notations for the various classical groups, for instance, see [2].

It was established in [2] that the Local–Global Principle holds for the following pairs of classical groups
• the linear case $(\mathrm{SL}_n(R[X]), \mathrm{E}_n(R[X]))$, $n \geq 3$,
• the symplectic case $(\mathrm{Sp}_{2n}(R[X]), \mathrm{ESp}_{2n}(R[X]))$, $n \geq 3$,
• the orthogonal case $(\mathrm{SO}_{2n}(R[X]), \mathrm{EO}_{2n}(R[X]))$, $n \geq 3$,
follows from any of the set of equivalent conditions listed in Theorem 1 stated below.

Moreover, one could establish that the Local–Global Principle for these linear groups is equivalent to the Normality of the Elementary Linear groups in these cases.

The notation used in the statement of Theorem 18.1 is from [2]. We shall assume $n \geq 3$ in the linear case, and $n \geq 6$ in the symplectic and orthogonal cases below.

We begin by recalling that a ring R with unit is said to be *almost commutative* if it is finite over its center (denoted by $C(R)$).

Theorem 18.1 *The following are equivalent for an almost commutative ring R:*

1. **(Normality):** $\mathrm{E}(n, R)$ *is a normal subgroup of* $S(n, R)$.
2. $I_n + M(v, w) \in \mathrm{E}(n, R)$ *if* $v \in \mathrm{Um}_n(R)$ *and* $\langle v, w\rangle = 0$.
3. **(Local–Global Principle):** *If* $\alpha(X) \in \mathrm{S}(n, R[X])$, $\alpha(0) = I_n$ *and* $\alpha_{\mathfrak{m}}(X) \in \mathrm{E}(n, R_{\mathfrak{m}}[X])$ *for all* $\mathfrak{m} \in \mathrm{Max}\,(C(R))$, *then* $\alpha(X) \in \mathrm{E}(n, R[X])$. (*Note that* $R_{\mathfrak{m}}$ *denotes* $S^{-1}R$, *where* $S = C(R) \setminus \mathfrak{m}$.)
4. **(Dilation Principle):** *Let* $\alpha(X) \in \mathrm{S}(n, R[X])$, *with* $\alpha(0) = I_n$. *Let* $\alpha_s(X) \in \mathrm{E}(n, R_s[X])$ *for some non-nilpotent* $s \in C(R)$. *Then* $\alpha(bX) \in \mathrm{E}(n, R[X])$ *for* $b \in (s^l)C(R)$, $l \gg 0$. (*Actually, we mean there exists some* $\beta(X) \in \mathrm{E}(n, R[X])$ *such that* $\beta(0) = I_n$ *and* $\beta_s(X) = \alpha(bX)$. *But, since there is no ambiguity, for simplicity we are using the notation* $\alpha(bX)$ *instead of* $\beta_s(X)$).
5. *If* $\alpha(X) = I_n + X^d M(v, w)$, *where* $v \in \mathrm{E}(n, R)e_1$ *and* $\langle v, w\rangle = 0$, *then* $\alpha(X) \in \mathrm{E}(n, R[X])$ *and is a product decomposition of the form* $\Pi ge_{ij}(Xh(X))$ *for* $d \gg 0$.
6. $I_n + M(v, w) \in \mathrm{E}(n, R)$ *if* $v \in \mathrm{E}(n, R)e_1$ *and* $\langle v, w\rangle = 0$.
7. $I_n + M(v, w) \in \mathrm{E}(n, R)$ *if* $v \in \mathrm{S}(n, R)e_1$ *and* $\langle v, w\rangle = 0$.

Remark 18.1 Statement (6) is true for any associative ring with 1. In particular, an almost commutative ring with identity satisfies statements (1)–(7).

18.3 The Monic Inversion Principle

We give a number of equivalent statements below, which are related to the Quillen–Suslin Monic Inversion Principle for $\mathrm{K}_1(R[X])$ and prove the equivalence of these statements assuming the Local–Global Principle for the pair

$$(\mathrm{GL}_n(R[X]), \mathrm{E}_n(R[X])), \qquad \text{for } n \text{ with } n \geq 3.$$

Remark 18.2 We believe that after appropriate modifications of the conditions (3), (4) below, these statements would be equivalent even without assuming the Local–Global Principle.

Theorem 18.2 *For a local ring A with the unique maximal ideal* $\mathfrak{m}$, *the following statements are equivalent, for* $n \geq 3$:

1. *Splitting of* $\mathrm{E}_n(A[X, X^{-1}], \mathfrak{m}[X, X^{-1}])$. *That is*

$$\mathrm{E}_n(A[X, X^{-1}], \mathfrak{m}[X, X^{-1}]) = G_+ G_-,$$

   where $G_+ = E_n(A[X], \mathfrak{m}[X])$ *and* $G_- = \mathrm{E}_n(A[X^{-1}], \mathfrak{m}[X^{-1}])$.
2. *Let* $\alpha \in \mathrm{GL}_n(A[X])$, $n \geq 3$. *Suppose there exists* $\beta \in \mathrm{GL}_n(A[X^{-1}])$ *such that* $\alpha\beta^{-1} \in \mathrm{E}_n(A[X, X^{-1}])$. *Then* $\alpha \in \mathrm{GL}_n(A)\,\mathrm{E}_n(A[X])$.
3. *Monic Inversion Principle: Let B be a commutative ring with identity. Let* $f := f(X) \in B[X]$ *be a monic polynomial. Let* $\alpha \in \mathrm{GL}_n(B[X])$, *be such that* $\alpha \in \mathrm{E}_n(B[X, \frac{1}{f(X)}])$. *Then* $\alpha \in \mathrm{E}_n(B[X])$.
4. *Let L be a field. Then for any m,* $\mathrm{SL}_n(L[X_1, \ldots, X_m]) = \mathrm{E}_n(L[X_1, \ldots, X_m])$.
5. *Let* $\alpha \in \mathrm{E}_n(A[X])$.

   a. $\delta_i \alpha \delta_i^{-1} \in \mathrm{GL}_n(A[X]) \Rightarrow \delta_i \alpha \delta_i^{-1} \in \mathrm{E}_n(A[X])$.
   b. $\delta_i^{-1} \alpha \delta_i \in \mathrm{GL}_n(A[X]) \Rightarrow \delta_i^{-1} \alpha \delta_i \in \mathrm{E}_n(A[X])$.

6. $\mathrm{E}_n(A[X, X^{-1}])$ *normalizes* $G_+ G_-$.

The proof we shall provide in [11] is far more advantageous in terms of its applicability to a wide range of groups, in fact all one needs is a Matsumoto-type theorem (see [8]) for the respective K_2 in hand.

We feel that our approach would enable one to establish the Monic Inversion Principle in other groups too such as the case when G is an isotropic reductive algebraic group over a polynomial extension of a local ring ([13], Corollary 5.2).

In Theorem 18.2, we enlist a number of equivalent statements, one of which is the Monic Inversion Principle for the elementary linear group, and another is the Bruhat decomposition for $\mathrm{E}_n(A[X, X^{-1}], \mathfrak{m}[X, X^{-1}])$, when $n \geq 3$. Assuming the Local–Global Principle, we show that these statements are equivalent.

One of the statements in Theorem 18.2 is to show that the linear group over a polynomial extension of a field is the elementary linear group. We recall an argument of M.P. Murthy in [3] to prove this via using Mennicke symbols. Thus, (4) can be established using the Local–Global Principle.

Consequently, all the equivalent statements in Theorem 18.2 are established; in particular, one establishes the Monic Inversion Principle for the linear pair of groups $(\mathrm{SL}_n(A[X]), \mathrm{E}_n(A[X]))$, for $n \geq 3$; as well as (1) the splitting property for $(\mathrm{E}_n(A[X, X^{-1}], \mathfrak{m}[X, X^{-1}]), n \geq 3$.

We believe that this is a fairly easier path toward establishing a Bruhat decomposition for $\mathrm{E}_n(A[X, X^{-1}], \mathfrak{m}[X, X^{-1}])$, when $(A, \mathfrak{m})$ is a local ring. This approach could be used to proving such decomposition for a variety of other groups.

We believe that with suitable modifications of the statements in Theorem 18.2, we should be able to prove that these statements are equivalent *without assuming the Local–Global Principle*.

These equivalent statements are restated for the special linear group in Theorem 18.1. The proof of the equivalence of these statements in [2] did use most of the methods needed to prove each of the statements. However, in the proof of the equivalence of the statements in Theorem 18.2 one uses very little of the methods used earlier to prove each of those statements. In particular, one seems to get the Bruhat-type decomposition result *gratis*.

We believe that the equivalent conditions of Theorem 18.2 imply the equivalent conditions of Theorem 18.1. For instance, it is possible via Theorem 18.2-(5) to show that $\mathrm{SL}_{n-1}(A)$ normalizes $\mathrm{E}_n(A)$.

We thus conclude that philosophically the Monic Inversion Principle, the Local–Global Principle, and the Normality of the elementary linear groups should all be equivalent properties; in the sense that if one holds so do the other two; thereby emphasizing the philosophy that an information given at the point at infinity is equivalent to prescribing it at any finite set of points covering the space.

Acknowledgements The second named author would like to thank Professor Raja Sridharan and Professor Gopala Krishna Srinivasan for their constant encouragement and attention to details during this work. He also acknowledges the financial support from CSIR, which enabled him to pursue his doctoral studies. The first named author thanks Neeraj Kumar who pointed out, during the talk, that the statements (3), (4) of Theorem 2 needed to be revised appropriately.

References

1. H. Bass, *Algebraic K-Theory* (W. A. Benjamin Inc, New York, 1968)
2. R. Basu, R. Khanna, R.A. Rao, On Quillen's local global principle. *Commutative Algebra and Algebraic Geometry*. Contemporary Mathematics, vol. 390, (American Mathematical Society, Providence, 2005), pp. 17–30

3. S.K. Gupta, M.P. Murthy, *Suslin's Work on Linear Groups Over Polynomial Rings and Serre's Problem*, vol. 8, ISI Lecture Notes (Macmillan Co. of India Ltd, New Delhi, 1980)
4. G. Horrocks, Projective modules over an extension of a local ring. Proc. Lond. Math. Soc. (3) **14**, 714–718 (1964)
5. F. Ischebeck, R.A. Rao, *Ideals and Reality, Projective Modules and Number of Generators of Ideals*, Springer Monographs in Mathematics (Springer, Berlin, 2005)
6. J-P. Serre, Faisceaux algébriques cohérents. (French). Ann. Math. **61**(2), 197–278 (1955)
7. T.Y. Lam, *Serre's Conjecture*. Lecture Notes in Mathematics, vol. 635 (Springer, Berlin, 1978); Revised edition: *Serre's Problem on Projective Modules*. Springer Monographs in Mathematics. ISBN 3-540-23317-2
8. H. Matsumoto, *Sur les sous-groupes arithmétiques des groupes semi-simples déployés. (French)*, Ann. Sci. École Norm. Sup. (4) **2**, 1–62 (1969)
9. J. Milnor, *Introduction to Algebraic K-Theory*. Annals of Mathematics Studies, No. 72. (Princeton University Press, Princeton; University of Tokyo Press, Tokyo, 1971)
10. D. Quillen, Projective modules over polynomial rings. Invent. Math. **36**, 167–171 (1976)
11. R.A. Rao, S. Yadav, An alternative approach to the Quillen–Suslin monic inversion principle, in preparation
12. C.S. Seshadri, Triviality of vector bundles over the affine space K^2. Proc. Natl. Acad. Sci. USA **44**, 456–458 (1958)
13. A. Stavrova, Homotopy invariance of non-stable K_1-functors. J. K-Theory **13**(2), 199–248 (2014)
14. A.A. Suslin, Projective modules over polynomial rings are free (Russian). Dokl. Akad. Nauk SSSR **229**(5), 1063–1066 (1976)
15. A.A. Suslin, The structure of the special linear group over rings of polynomials. Izv. Akad. Nauk SSSR Ser. Mat. **41**, 235–252 (1977)

MIX
Papier aus verantwortungsvollen Quellen
Paper from responsible sources
FSC® C105338

If you have any concerns about our products, you can contact us on
ProductSafety@springernature.com

In case Publisher is established outside the EU, the EU authorized representative is:
Springer Nature Customer Service Center GmbH
Europaplatz 3, 69115 Heidelberg, Germany

Printed by Libri Plureos GmbH
in Hamburg, Germany